Protein Analysis using Mass Spectrometry

WILEY SERIES ON PHARMACEUTICAL SCIENCE AND BIOTECHNOLOGY: PRACTICES, APPLICATIONS, AND METHODS

Series Editor:

Mike S. Lee
Milestone Development Services

Mike S. Lee • *Integrated Strategies for Drug Discovery Using Mass Spectrometry*

Birendra Pramanik, Mike S. Lee, and Guodong Chen • *Characterization of Impurities and Degradants Using Mass Spectrometry*

Mike S. Lee and Mingshe Zhu • *Mass Spectrometry in Drug Metabolism and Disposition: Basic Principles and Applications*

Mike S. Lee (editor) • *Mass Spectrometry Handbook*

Wenkui Li and Mike S. Lee • *Dried Blood Spots—Applications and Techniques*

Mike S. Lee and Qin C. Ji . • *Protein Analysis using Mass Spectrometry: Accelerating Protein Biotherapeutics from Lab to Patient*

Ayman F. El-Kattan • *Oral Bioavailability Assessment: Basics and Strategies for Drug Discovery and Development*

Protein Analysis using Mass Spectrometry

Accelerating Protein Biotherapeutics from Lab to Patient

Edited By Mike S. Lee and Qin C. Ji

WILEY

This edition first published 2017

Registered Office
John Wiley & Sons, Inc., 111 River Street, Hoboken, NJ 07030, USA

Editorial Office
111 River Street, Hoboken, NJ 07030, USA

For details of our global editorial offices, customer services, and more information about Wiley products visit us at www.wiley.com.

Wiley also publishes its books in a variety of electronic formats and by print-on-demand. Some content that appears in standard print versions of this book may not be available in other formats.

Library of Congress Cataloging-in-Publication Data

Names: Lee, Mike S., 1960- editor. | Ji, Qin C., editor.
Title: Protein analysis using mass spectrometry : accelerating protein biotherapeutics from lab to patient / edited by Mike S. Lee, Qin C. Ji.
Description: 1st edition. | Hoboken, NJ : John Wiley & Sons, Inc., 2017. | Includes bibliographical references and index.
Identifiers: LCCN 2016058010| ISBN 9781118605196 (cloth) | ISBN 9781119359357 (epub)
Subjects: LCSH: Proteins--Analysis. | Proteins--Therapeutic use. | Mass spectrometry.
Classification: LCC QP551 .P748 2017 | DDC 572/.6--dc23 LC record available at https://lccn.loc.gov/2016058010

Cover image: © me4o/Gettyimages;
(Inset Image) Courtesy of Yongjun Xue
Cover design by Wiley

Set in 10/12pt Warnock by SPi Global, Chennai, India

10 9 8 7 6 5 4 3 2 1

Contents

List of Contributors *xiii*
Foreword *xvii*
Preface *xix*

1 **Contemporary Protein Analysis by Ion Mobility Mass Spectrometry** *1*
Johannes P.C. Vissers and James I. Langridge
1.1 Introduction *1*
1.2 Traveling-Wave Ion Mobility Mass Spectrometry *1*
1.3 IM–MS and LC–IM–MS Analysis of Simple and Complex Mixtures *2*
1.3.1 Cross Section and Structure *2*
1.3.2 Separation *4*
1.3.3 Sensitivity *5*
1.4 Outlook *7*
Acknowledgment *8*
References *8*

2 **High-Resolution Accurate Mass Orbitrap and Its Application in Protein Therapeutics Bioanalysis** *11*
Hongxia Wang and Patrick Bennett
2.1 Introduction *11*
2.2 Triple Quadrupole Mass Spectrometer and Its Challenges *11*
2.3 High-Resolution Mass Spectrometers *12*
2.4 Quantitation Modes on Q Exactive Hybrid Quadrupole Orbitrap *13*
2.5 Protein Quantitation Approaches Using Q Exactive Hybrid Quadrupole Orbitrap *14*
2.6 Data Processing *16*
2.7 Other Factors That Impact LC–MS-based Quantitation *16*
2.7.1 Sample Extraction to Reduce Matrices *16*
2.7.2 Internal Standard *17*
2.8 Conclusion and Perspectives of LC–HRMS in Regulated Bioanalysis *18*
References *18*

3 **Current Methods for the Characterization of Posttranslational Modifications in Therapeutic Proteins Using Orbitrap Mass Spectrometry** *21*
Zhiqi Hao, Qiuting Hong, Fan Zhang, Shiaw-Lin Wu, and Patrick Bennett
3.1 Introduction *21*
3.2 Characterization of PTMs Using Higher-Energy Collision Dissociation *23*
3.2.1 Oxidation *24*
3.2.2 Deamidation *24*
3.3 Application of Electron Transfer Dissociation to the Characterization of Labile PTMs *26*
3.3.1 Performing ETD Experiments in Orbitrap Instruments *27*
3.3.2 Structure Elucidation of Glycopeptides Using Multiple Fragmentation Mechanisms in Orbitrap Instruments *28*

3.4 Conclusion *31*
Acknowledgment *32*
References *32*

4 Macro- to Micromolecular Quantitation of Proteins and Peptides by Mass Spectrometry *35*
Suma Ramagiri, Brigitte Simons, and Laura Baker
4.1 Introduction *35*
4.2 Key Challenges of Peptide Bioanalysis *36*
4.2.1 Key Benefits of the LC/MS/MS Peptide Quantitation Workflow *38*
4.3 Key Features of LC/MS/MS-Based Peptide Quantitation *38*
4.3.1 Sensitivity *39*
4.3.2 Selectivity *39*
4.3.2.1 MRM^3 *39*
4.3.2.2 Differential Mobility Spectrometry (DMS) *39*
4.3.3 High-Resolution Accurate-Mass Spectrometry *39*
4.3.4 Software *40*
4.4 Advantages of the Diversity of Mass Spectrometry Systems *41*
4.5 Perspectives for the Future *41*
References *42*

5 Peptide and Protein Bioanalysis Using Integrated Column-to-Source Technology for High-Flow Nanospray *45*
Shane R. Needham and Gary A. Valaskovic
5.1 Introduction – LC–MS Has Enabled the Field of Protein Biomarker Discovery *45*
5.2 Integration of Miniaturized LC with Nanospray ESI-MS Is a Key for Success *46*
5.3 Micro- and Nano-LC Are Well Suited for Quantitative Bioanalysis *47*
5.4 Demonstrating Packed-Emitter Columns Are Suitable for Bioanalysis *49*
5.5 Future Outlook 51
References *52*

6 Targeting the Right Protein Isoform: Mass Spectrometry-Based Proteomic Characterization of Alternative Splice Variants *55*
Jiang Wu
6.1 Introduction *55*
6.2 Alternative Splicing and Human Diseases *55*
6.3 Identification of Splice Variant Proteins *56*
6.3.1 Global Profiling of Splicing Variant Proteins *56*
6.3.2 Characterization of Relative Expression of Protein Splice Variants *57*
6.3.3 Quantitation of Splice Variants by MRM-MS *62*
6.4 Conclusion *64*
References *64*

7 The Application of Immunoaffinity-Based Mass Spectrometry to Characterize Protein Biomarkers and Biotherapeutics *67*
Bradley L. Ackermann and Michael J. Berna
7.1 Introduction *67*
7.1.1 The Importance of Protein Measurement *67*
7.1.2 Ligand Binding Assays *68*
7.1.3 The Introduction of Hybrid IA-MS Methods *68*
7.2 Overview of IA-MS Methods *69*
7.2.1 Classification of IA-MS Methods *69*
7.2.2 Stable-Isotope-Labeled Internal Standards *71*
7.2.3 IA Capture Formats *71*
7.2.4 Liquid Chromatography *72*
7.2.5 MS Detection *74*

7.3 IA-MS Applications – Biomarkers *74*
7.3.1 Peptide Biomarkers *74*
7.3.2 Protein Biomarkers – Anti-Protein Capture *78*
7.3.3 Protein Biomarkers – Anti-Peptide Capture *80*
7.4 IA-MS Applications – Biotherapeutics *81*
7.4.1 Therapeutic Peptides *81*
7.4.2 Therapeutic Antibodies *83*
7.4.3 Antibody–Drug Conjugates *84*
7.5 Future Direction *84*
References *85*

8 Semiquantification and Isotyping of Antidrug Antibodies by Immunocapture-LC/MS for Immunogenicity Assessment *91*
Jianing Zeng, Hao Jiang, and Linlin Luo
8.1 Introduction *91*
8.2 Multiplexing Direct Measurement of ADAs by Immunocapture-LC/MS for Immunogenicity Screening, Titering, and Isotyping *93*
8.3 Indirect Measurement of ADAs by Quantifying ADA Binding Components *95*
8.4 Use of LC–MS to Assist in Method Development of Cell-Based Neutralizing Antibody Assays *96*
8.5 Conclusion and Future Perspectives *97*
References *97*

9 Mass Spectrometry-Based Assay for High-Throughput and High-Sensitivity Biomarker Verification *99*
Xuejiang Guo and Keqi Tang
9.1 Background *99*
9.2 Sample Processing Strategies *100*
9.3 Advanced Electrospray Ionization Mass Spectrometry Instrumentation *102*
9.4 Conclusion *105*
References *105*

10 Monitoring Quality of Critical Reagents Used in Ligand Binding Assays with Liquid Chromatography Mass Spectrometry (LC–MS) *107*
Brian Geist, Adrienne Clements-Egan, and Tong-Yuan Yang
10.1 Introduction *107*
10.2 Case Study Examples *114*
10.2.1 Case Study #1: Confirmation of Correct Reagent Construct Prior to Use in Development of an LBA Method *114*
10.2.2 Case Study #2: Monitoring the Integrity of the Reagent Cell Line Production System *116*
10.2.3 Case Study #3: Investigation of the Loss of LBA Specificity During Clinical Development *116*
10.2.3.1 Prestudy Investigation *116*
10.2.3.2 In-Study Investigation *119*
10.2.4 Case Study #4: Monitoring the Incorporation Ratio of Conjugated Critical Reagent Used in LBAs *122*
10.3 Discussion *122*
10.3.1 Keys to Reagent Management *122*
10.3.2 Importance of LC–MS Characterization *123*
10.3.3 The Analytical Toolbox and a "Fit-for-Purpose" Approach for Reagent Management *125*
Acknowledgment *126*
References *126*

11 Application of Liquid Chromatography-High Resolution Mass Spectrometry in the Quantification of Intact Proteins in Biological Fluids *129*
Stanley (Weihua) Zhang, Jonathan Crowther, and Wenying Jian
11.1 Introduction *129*
11.2 Workflows for Quantification of Proteins Using Full-Scan LC-HRMS *131*
11.2.1 Sample Preparation *131*

11.2.1.1 Solid-Phase Extraction (SPE) *131*
11.2.1.2 Affinity Enrichment *131*
11.2.1.3 Depletion of High-Abundant Proteins *131*
11.2.1.4 Solution Fractionation *132*
11.2.1.5 Protein Precipitation for PEGylated Proteins *132*
11.2.2 LC-HRMS *132*
11.2.2.1 HPLC *132*
11.2.2.2 Full-Scan HRMS Data Acquisition and Analysis *133*
11.3 Internal Standard Strategy *133*
11.3.1 Stable Isotope Labeled Protein *134*
11.3.2 Protein Analog *135*
11.4 Calibration and Quality Control (QC) Sample Strategy *135*
11.5 Common Issues in Quantification of Proteins Using LC-HRMS *135*
11.5.1 Stability *135*
11.5.2 Adsorption *136*
11.5.3 Specific Protein Binding *136*
11.5.4 Posttranslational Modifications (PTMs) *136*
11.6 Examples of LC-HRMS-Based Intact Protein Quantification *137*
11.7 Conclusion and Future Perspectives *138*
Acknowledgment *140*
References *140*

12 LC–MS/MS Bioanalytical Method Development Strategy for Therapeutic Monoclonal Antibodies in Preclinical Studies *145*
Hongyan Li, Timothy Heath, and Christopher A. James
12.1 Introduction: LC-MS/MS Bioanalysis of Therapeutic Monoclonal Antibodies *145*
12.2 Highlights of Recent Method Development Strategies *146*
12.2.1 Strategy for Surrogate Peptide Selection and Optimization *146*
12.2.2 Sample Preparation *148*
12.2.2.1 Immunoaffinity-Based Sample Preparation *148*
12.2.2.2 Nonimmunoaffinity-Based Sample Preparation *151*
12.2.3 Accelerated Trypsin Digestion *152*
12.2.4 Internal Standard Selection *153*
12.2.4.1 SIL-Peptide IS *154*
12.2.4.2 Cleavable Flanking SIL-Peptide IS *154*
12.2.4.3 SIL-mAb IS *154*
12.3 Case Studies of Preclinical Applications of LC–MS/MS for Monoclonal Antibody Bioanalysis *154*
12.3.1 Case Study #1 *154*
12.3.1.1 Key Analytical Method Features *154*
12.3.2 Case Study #2 *155*
12.3.2.1 Key Analytical Method Features *155*
12.4 Conclusion and Future Perspectives *156*
References *158*

13 Generic Peptide Strategies for LC–MS/MS Bioanalysis of Human Monoclonal Antibody Drugs and Drug Candidates *161*
Michael T. Furlong
13.1 Introduction *161*
13.2 A Universal Peptide LC–MS/MS Assay for Bioanalysis of a Diversity of Human Monoclonal Antibodies and Fc Fusion Proteins in Animal Studies *161*
13.2.1 Identification of a Candidate Universal Surrogate Peptide to Enable Quantification of Human mAb and Fc Fusion Protein Drug Candidates *161*

13.2.2 Application of an Exploratory Universal (Peptide 1) LC–MS/MS Assay to a Monkey Pharmacokinetic Study *162*
13.2.3 Potential Applicability of a Peptide 1 Variant to Bioanalysis of Human IgG2-Based mAbs and Fc Fusion Proteins *163*
13.2.4 Impact of Peptide 1 Asparagine Deamidation on Human mAb Quantification Can Be Mitigated *164*
13.3 An Improved "Dual" Universal Peptide LC–MS/MS Assay for Bioanalysis of Human mAb Drug Candidates in Animal Studies *165*
13.3.1 Identification and Evaluation of "Dual" Universal Peptide LC–MS/MS Assay Candidates *165*
13.3.2 Quantitative Evaluation and Comparison of Light and Heavy Chain Dual Universal Peptide Candidates *167*
13.3.3 Assessing the Level of Quantitative Agreement Between Peptide 1 and Peptide 2 in Assay Performance Evaluation Runs *167*
13.3.4 Deployment of the Exploratory Dual Universal Peptide Assay in Support of a Monkey Pharmacokinetic Study *168*
13.3.5 Considerations for Calibration Curve/QC Replicate Acceptance Criteria When a Dual Peptide Assay Is Employed *168*
13.3.6 Interpreting and Reporting Study Sample Concentration Data Generated with a Dual Peptide Assay *168*
13.3.7 Related Studies: Generic LC–MS/MS Assays for Human mAb Bioanalysis in Animal Studies *169*
13.4 Extending the Universal Peptide Assay Concept to Human mAb Bioanalysis in Human Studies *170*
13.4.1 Potential Expansion of the Universal LC–MS/MS Assay Concept into Human Studies *170*
13.4.2 Development and Evaluation of an Exploratory Universal IgG4 Clinical LC–MS/MS Assay *171*
13.4.3 Evaluation of the Impact of Anti-mAb Antibodies on Exploratory Universal IgG4 LC–MS/MS Assay Performance *173*
13.5 Internal Standard Options for Generic Peptide LC–MS/MS Assays *173*
13.5.1 Stable Isotopically Labeled Peptide Internal Standards *173*
13.5.2 Stable Isotopically Labeled Protein Internal Standards *174*
13.5.3 "Flanked" Stable Isotopically Labeled Peptide Internal Standards *175*
13.6 Sample Preparation Strategies for Generic Peptide LC–MS/MS Assays *175*
13.6.1 Direct Digestion, Pellet Digestion, and Solid-Phase Extraction *175*
13.6.2 Affinity Capture *176*
13.6.3 Additional Sample Preparation Approaches for Generic Peptide LC–MS/MS Assays *176*
13.7 Limitations of Generic Peptide LC–MS/MS Assays *177*
13.8 Conclusion *178*
Acknowledgments *178*
References *178*

14 Mass Spectrometry-Based Methodologies for Pharmacokinetic Characterization of Antibody Drug Conjugate Candidates During Drug Development *183*
Yongjun Xue, Priya Sriraman, Matthew V. Myers, Xiaomin Wang, Jian Chen, Brian Melo, Martha Vallejo, Stephen E. Maxwell, and Sekhar Surapaneni
14.1 Introduction *183*
14.2 Mechanism of Action *183*
14.2.1 Linker Chemistry *185*
14.2.2 Toxins *185*
14.2.3 ADME *185*
14.2.4 Unique Bioanalytical Challenges *185*
14.3 Mass Spectrometry Measurement for DAR Distribution of Circulating ADCs *186*
14.3.1 Immunocapture of ADCs from Plasma or Serum *186*
14.3.2 Deglycosylation for Captured ADCs *187*
14.3.3 Mass Spectrometry Measurement for DAR Distribution of Circulating ADCs *188*
14.4 Total Antibody Quantitation by Ligand Binding or LC–MS/MS *189*
14.4.1 Ligand Binding Assay *189*
14.4.2 LC–MS/MS Assay for Total Antibody Quantitation *190*

14.4.2.1 Predigestion Treatment *190*
14.4.2.2 Enzymatic Digestion *191*
14.4.2.3 Postdigestion Treatment *191*
14.4.2.4 LC–MS/MS Analysis *191*
14.4.3 Ligand Binding versus LC–MS/MS Assays *192*
14.5 Total Conjugated Drug Quantitation by Ligand Binding or LC–MS/MS *193*
14.5.1 Ligand Binding Assays for ADC Quantitation *193*
14.5.1.1 DAR-Sensitive Total Conjugated Drug Assay *193*
14.5.1.2 DAR-Insensitive Total Conjugated Antibody Assay *193*
14.5.2 LC–MS/MS for the Total Conjugated Drug Quantitation *194*
14.5.2.1 Predigestion Treatment *194*
14.5.2.2 Enzymatic or Chemical Digestion *194*
14.5.2.3 Postdigestion Treatment *195*
14.5.2.4 LC–MS/MS Analysis *195*
14.5.3 Ligand Binding versus LC–MS/MS *195*
14.6 Catabolite Quantitation by LC–MS/MS *196*
14.6.1 Sample Preparation *196*
14.6.2 LC–MS/MS Analysis *197*
14.7 Preclinical and Clinical Pharmacokinetic Support *197*
14.8 Conclusion and Future Perspectives *198*
References *198*

15 Sample Preparation Strategies for LC–MS Bioanalysis of Proteins *203*
Long Yuan and Qin C. Ji
15.1 Introduction *203*
15.2 Sample Preparation Strategies to Improve Assay Sensitivity *205*
15.2.1 Protein Precipitation *205*
15.2.2 Solid-Phase Extraction *205*
15.2.3 Derivatization *206*
15.2.4 Depletion of High-Abundance Proteins *207*
15.2.5 Immunoaffinity Purification *208*
15.2.5.1 Immunocapture of a Specific Peptide *208*
15.2.5.2 Immunocapture of a Specific Protein *208*
15.2.5.3 Generic Immunocapture *210*
15.2.6 Online Sample Preparation *211*
15.3 Sample Preparation Strategies to Differentiate Free, Total, and ADA-Bound Proteins *213*
15.4 Sample Preparation Strategies to Overcome Interference from Antidrug Antibodies or Soluble Target *214*
15.5 Protein Digestion Strategies *214*
15.6. Conclusion *215*
Acknowledgment *216*
References *216*

16 Characterization of Protein Therapeutics by Mass Spectrometry *221*
Wei Wu, Hangtian Song, Thomas Slaney, Richard Ludwig, Li Tao, and Tapan Das
16.1 Introduction *221*
16.2 Variants Associated with Cysteine/Disulfide Bonds in Protein Therapeutics *221*
16.2.1 Thiolation Isoforms *222*
16.2.2 Disulfide Isoforms *222*
16.2.3 Free Sulfhydryl *224*
16.2.4 Thioether/Trisulfide Bond *224*
16.2.5 Disulfide Bond in Antibody Drug Conjugates *224*
16.3 N–C-Terminal Variants *225*
16.4 Glycation *226*

16.5 Oxidation *226*
16.5.1 Methionine Oxidation *227*
16.5.2 Metal-Catalyzed Oxidation (MCO) *227*
16.5.3 Photooxidation *227*
16.5.4 Deamidation *228*
16.5.5 Effect of Sequence and Structure on Deamidation *228*
16.6 Discoloration *228*
16.7 Sequence Variants *230*
16.8 Glycosylation *232*
16.8.1 Glycoprotein Structure *232*
16.8.2 Intact Glycoprotein Analysis *235*
16.8.3 Glycopeptide Analysis *237*
16.8.4 Tandem MS of Glycopeptides *237*
16.8.5 Free Glycan Analysis *238*
16.8.6 Release of Glycans from Glycoproteins *238*
16.8.7 Detailed Sequence and Linkage Analysis of Glycans *239*
16.9 Conclusion *240*
References *240*

Index *251*

List of Contributors

Bradley L. Ackermann
Eli Lilly and Company
Indianapolis, IN, USA

Laura Baker
SCIEX
Framingham, MA, USA

Patrick Bennett
PPD,
Richmond, VA, USA

Michael J. Berna
Eli Lilly and Company
Indianapolis, IN, USA

Jian Chen
Celgene
Summit, NJ, USA

Adrienne Clements-Egan
Janssen Research & Development, LLC
Spring House, PA, USA

Jonathan Crowther
Ortho Clinical Diagnostics
Raritan, NJ, USA

Tapan Das
Molecular and Analytical Development
Bristol-Myers Squibb Company, USA

Michael T. Furlong
PPD Bioanalytical Lab, Middleton, WI, USA

Brian Geist
Janssen Research & Development, LLC
Spring House, PA, USA

Xuejiang Guo
Pacific Northwest National Laboratory
Richland, WA, USA

and

Nanjing Medical University
Nanjing, PR China

Zhiqi Hao
Thermo Fisher Scientific
San Jose, CA, USA

Timothy Heath
Amgen Inc.
Thousand Oaks, CA, USA

Qiuting Hong
Eurofins Lancaster Laboratories, Inc.
Lancaster, PA, USA

Christopher A. James
Amgen Inc.
Thousand Oaks, CA, USA

Qin C. Ji
Analytical & Bioanalytical Operations
Bristol-Myers Squibb
Princeton, NJ 08543, USA

Wenying Jian
Janssen Research and Development, Johnson & Johnson
Spring House, PA, USA

Hao Jiang
Analytical and Bioanalytical Operations, Bristol-Myers Squibb Co.
Princeton, NJ, USA

James I. Langridge
Waters Corporation
Wilmslow, UK

Hongyan Li
Amgen Inc.
Thousand Oaks, CA, USA

Richard Ludwig
Molecular and Analytical Development
Bristol-Myers Squibb Company, USA

Linlin Luo
Analytical and Bioanalytical Operations, Bristol-Myers Squibb Co.
Princeton, NJ, USA

Stephen E. Maxwell
Celgene
Summit, NJ, USA

Brian Melo
Celgene
Summit, NJ, USA

Matthew V. Myers
Celgene
Summit, NJ, USA

Shane R. Needham
Alturas Analytics
Moscow, ID, USA

Suma Ramagiri
SCIEX
Framingham, MA, USA

Brigitte Simons
SCIEX,
Framingham, MA, USA

Thomas Slaney
Molecular and Analytical Development
Bristol-Myers Squibb Company, USA

Hangtian Song
Molecular and Analytical Development
Bristol-Myers Squibb Company, USA

Priya Sriraman
Celgene
Summit, NJ, USA

Sekhar Surapaneni
Celgene
Summit, NJ, USA

Keqi Tang
Pacific Northwest National Laboratory
Richland, WA, USA

Li Tao
Molecular and Analytical Development
Bristol-Myers Squibb Company, USA

Gary A. Valaskovic
New Objective Inc.
Woburn, MA, USA

Martha Vallejo
Celgene
Summit, NJ, USA

Johannes P.C. Vissers
Waters Corporation
Wilmslow, UK

Hongxia Wang
Thermo Fisher Scientific
San Jose, CA, USA

Xiaomin Wang
Celgene
Summit, NJ, USA

Jiang Wu
Shire Pharmaceuticals
Lexington, MA, USA

Shiaw-Lin Wu
BioAnalytix Inc.
Cambridge, MA, USA

and

Northeastern University
Boston, MA, USA

Wei Wu
Molecular and Analytical Development
Bristol-Myers Squibb Company, USA

Y.-J. Xue
Celgene
Summit, NJ, USA

Long Yuan
Analytical & Bioanalytical Operations
Bristol-Myers Squibb
Princeton, NJ, USA

Tong-Yuan Yang
Janssen Research & Development, LLC
Spring House, PA, USA

Fan Zhang
Northeastern University
Boston, MA, USA

Stanley (Weihua) Zhang
Ortho Clinical Diagnostics
Raritan, NJ, USA

Jianing Zeng
Analytical and Bioanalytical Operations, Bristol-Myers Squibb Co.
Princeton, NJ, USA

Foreword

This book explores recent advances in mass spectrometry and related technology, and the innovative approaches used in measuring and characterizing peptides and proteins as part of bringing new medicines to patients in need. Qin and Mike have brought together a wide range of leading scientists to provide a clear picture of the variety and depth of technology and techniques.

As you will see in each chapter, fundamental LC–MS knowledge has been used in each innovative advance. Sample preparation techniques for peptides and proteins rely on the core of historic approaches used for small molecule drug analyses but have been expanded to address a host of requirements related to protein structure, including reduction and alkylations, acid dissociation, protein digestion, and the specificity possible with immunocapture. Liquid chromatography techniques from regular to ultrahigh-performance approaches and downward to micro- and nanoflow are covered, as well as utilization of 2-D chromatography. Triple quadrupole and high-resolution mass spectrometers, with their recent advances in sensitivity and selectivity, are prominent in the discussions as their advances are central to making possible many advances in peptide and protein analyses.

I hope that the readers find this book to be an engaging learning experience; one that provides insights and causes a cascade to the discovery of further advances in peptide and protein analysis by liquid chromatography mass spectrometry.

Mark E. Arnold
Bioanalytical Solution Integration LLC
mark.arnoldcs@gmail.com
www.linkedin.com/in/markearnoldphd

Preface

We had a discussion on LCMS analysis of proteins for drug development dating back to the early 2000s. At that time, Qin's group at Abbott Laboratories had just published a manuscript in analytical chemistry for an LCMS bioanalytical method for a small protein (MW > 10 kDa). Through the years, multiple discussions on the topic continued at various conferences, including conversations held at several Annual Land O'Lakes Bioanalytical Conferences where Mike was invited to give lectures. Although mass spectrometry protein analysis has been a popular topic in proteomic research for several decades, it was only in the late 2000s it started to receive increasing attention of scientists in drug development. In this book, we present 16 chapters from industry leaders who have first-hand experience in developing new mass spectrometry technologies, knowing the issues and needs of the analysis in drug discovery and development, forming assay strategies, and interpreting assay results with their respective project teams.

The authors of Chapters 1–4 have experience and expertise with mass spectrometry instrumentation as well as with analytical research and development. Johannes and James from Waters discussed extensively the history and theory of ion mobility mass spectrometry and its application in protein analysis. As they pointed out, "The next few years should see significant improvements in both the technology, and the informatics and workflows to use the information generated from ion mobility mass spectrometry for both qualitative and quantitative analyses." In Chapters 2 and 3, Jessica, Zhiqi, and their colleagues discuss the characteristics and capabilities of high-resolution mass spectrometry, especially, the Thermo Orbitrap mass spectrometry and its application in protein therapeutics bioanalysis and the characterization of posttranslational modifications in therapeutic proteins. In Chapter 4, Suma and her colleagues from SCIEX discuss the workflow of quantitative analysis of proteins using mass spectrometry, especially the triple quadrupole time-of-flight mass spectrometry system. Although the benefit of using low flow liquid chromatography mass spectrometry has been well understood theoretically and widely used in the proteomic research area, the application of this technology in quantitative analysis of proteins in biological matrix is still not widely accepted. In Chapter 5, Shane and Gary describe the success and routine usage of New Objective's integrated nanoflow LC column and nanoelectrospray emitter system for the bioanalysis of proteins in biological matrices with excellent assay ruggedness and high assay throughputs. Jiang at Shire is one of the industry leaders in drug discovery mass spectrometry. Jiang comments that understanding relative expression and structure–function relationship of the splice isoforms are essential for the discovery and development of more specific therapeutics and biomarkers. In Chapter 6, Jiang describes the advanced mass spectrometry characterization of gene splice variants in conjunction with high-throughput transcriptomics as an example of protein mass spectrometry analysis in proteomic research for supporting drug discovery. Bradley and Michael from Lilly are among the pioneers in mass spectrometry biomarker analysis. In Chapter 7, they provide a comprehensive review of the immunoaffinity mass spectrometry technology and its application in protein biomarkers and biotherapeutics characterization. Immunogenicity refers to immune responses of humans or animals to antigens, such as biotherapeutics. The technologies, methodology, and regulatory requirements for the immunogenicity test evolved rapidly in recent years. In Chapter 8, Jianing and her coworkers at BMS describe recent advances in using immunocapture LCMS for immunogenicity assessment from "semiquantitative analysis of antidrug antibody" to "assisting the method development of cell-based neutralizing antibody assays." Keqi is well known in the mass spectrometry field for his design of mass spectrometry ionization sources and ion optics for high ion transfer efficiency. In Chapter 9, Xuejiang Guo and Keqi from PNNL discuss recent advances in methodology and mass spectrometry instrumentation for the sensitive and high-throughput mass spectrometry biomarker analysis. In Chapter 10, Tong-Yuan and his coworkers at JNJ describe the mass spectrometry ligand binding assay reagent characterization, which is one of the fast growing areas in the bioanalytical scientific field and has shown significant impacts on improving ligand binding bioanalytical assays. In Chapter 11, Stanley and his coworkers at JNJ describe the recent advances in using high-resolution mass spectrometry in improving selectivity for the mass spectrometry bioanalysis of proteins in biological matrices. In Chapter 12, Hongyan and his coworkers at Amgen discuss the advantages and their assay development strategy of LCMS quantitative analysis of therapeutic monoclonal antibodies (mAbs) in biological matrices in supporting preclinical studies. In Chapter 13, Michael at PPD discusses generic peptide strategies (he is one of the pioneers who developed this approach) for LC–MS bioanalysis of human monoclonal antibody drugs and drug candidates. The advantages of this strategy include significant cost saving and accelerated progress for drug discovery and early drug development. In Chapter 14, Y-J and his coworkers at Celgene describe comprehensively the strategy and methodology of mass spectrometry support of antidrug conjugate (ADC) drug development, one of the most active areas recently in drug development. In Chapter 15, Long and Qin at BMS provide a survey of the sample preparation strategies for LCMS protein bioanalysis, which range from traditional organic solvent protein precipitation, solid-phase extraction to more advanced chemical derivatization, and immunocapture sample preparation. In Chapter 16, Wei and his coworkers at BMS describe the mass spectrometry characterizations of protein therapeutics in drug manufacturing process to ensure the quality and integrity of dug product ingredients.

We would like to take this opportunity to thank all the authors for their diligent work in describing the advances in the protein mass spectrometry analysis in supporting from early-stage basic researches to delivering the safe, efficacious drug to patient bedside. We also would like to thank Wiley for the opportunity to bring this book to our readers, which will further stimulate the advances of mass spectrometry technology and methodology to benefit patients' lives.

Mike S. Lee and Qin C. Ji
December 2016
Princeton, NJ

1

Contemporary Protein Analysis by Ion Mobility Mass Spectrometry

Johannes P.C. Vissers and James I. Langridge

Waters Corporation, Wilmslow, UK

1.1 Introduction

The use of ion mobility as an analytical technique to detect and separate biomolecules dates back to the break of the century with the application of the method for proteomics (Valentine et al. 2006; McLean et al. 2005; Gabryelski and Froese 2003), glycomics (Taraszka et al. 2001; Jin et al. 2005; Hoaglund et al. 1997), and metabolomics (Dwivedi et al. 2008). It is a technique that separates gas-phase ions based upon their mobility in a buffer gas. This separation is related to ion size, shape, as well as *m/z*, and charge. The basis for separation by traditional drift tube ion mobility at a low electric limit can be derived from the Mason–Schamp equation:

$$K = \frac{3}{16}\left(\frac{2\pi}{\mu k_B T}\right)^{1/2} \frac{ze}{N\Omega}$$

where K = drift velocity v_d/electric field strength E, μ = reduced mass of the ion (neutral given by $(m_{neutral}m_{ion})/(m_{neutral}+m_{ion})$, k_B = Boltzmann constant, T = temperature, z = charge state of the analyte ion, e = charge on an electron, N = number density of the drift gas, and Ω = average collision cross section. The hyphenation of ion mobility spectrometry (IMS) with MS is often referred to as ion mobility–mass spectrometry (IM–MS). The most common mass analyzer coupled with IMS comprises a time-of-flight (TOF) instrument due to the inherent high sampling rate, although other mass detection systems have been described (Kanu et al. 2008). Four different methods of ion mobility separation are currently used in combination with MS, including drift-time ion mobility spectrometry (DTIMS), aspiration ion mobility spectrometry (AIMS), differential mobility spectrometry (DMS), also called field-asymmetric waveform ion mobility spectrometry (FAIMS), and traveling-wave ion mobility spectrometry (TWIMS). A description of these methods is beyond the scope of this chapter, particularly since they have been reviewed in great detail elsewhere (Kanu et al. 2008; Lanucara et al. 2014).

The innovative demonstration of protein conformer separation by means of IMS by Clemmer et al. 1995 has prompted instrumental IM–MS designs and the broader application of IMS as an analytical tool. The designs by Pringle et al. 2007 and Baker et al. 2007, both orthogonal acceleration time-of-flight (oa-TOF) based IM–MS platform, but utilizing different IMS geometries, have been commercialized and applied for numerous applications and include drug metabolism/metabolites (Dear et al. 2010), lipids (Kliman et al. 2011), trace impurities (Eckers et al. 2007), carbohydrates (Vakhrushev et al. 2008, Schenauer et al. 2009), macromolecular protein species and viruses (Ruotolo et al. 2005, Bereszczak et al. 2014), metal-based anticancer drugs (Williams et al. 2009), and PEGylated conjugates (Bagal et al. 2008). In this chapter, the application of IMS for the identification, quantification, and characterization of proteins is illustrated by application examples that demonstrate the benefits of integrating IMS into the analytical schema in terms of increased resolution and sensitivity, as well as those obtained from collision cross section measurements.

1.2 Traveling-Wave Ion Mobility Mass Spectrometry

The principle of TWIMS is briefly discussed as it forms the basis of subsequent sections. A schematic of the device is shown in Figure 1.1. Details can be found in the papers of Pringle et al. 2007 and Giles et al. 2004. Ions are formed by electrospray ionization (ESI) in the source and subsequently pass through a quadrupole for mass selection before injection into the ion mobility cell. Unlike our other instruments, which use a uniform electric field across the cell for ion mobility experiments, so-called

Protein Analysis using Mass Spectrometry: Accelerating Protein Biotherapeutics from Lab to Patient, First Edition.
Edited by Mike S. Lee and Qin C. Ji.

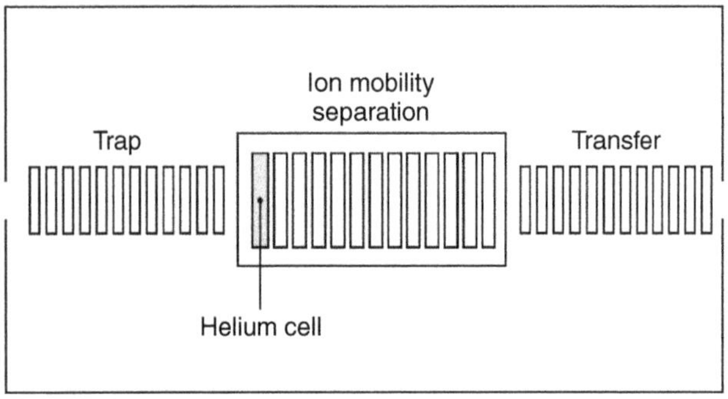

Figure 1.1 Triwave ion mobility optics detail comprising a trap, helium cell, ion mobility separator and transfer. (*Source*: Williams et al. 2012. Reproduced with permission of GIT.)

drift tube IMS, this device uses traveling-wave (T-wave) technology. The T-wave cell consists of a stacked-ring radio frequency (RF) ion guide, which incorporates a repeating sequence of transient voltages applied to the ring electrodes. These voltage pulses result in a traveling electric field that propels ions through the background gas present in the mobility cell. The total time taken for an ion to drift through the cell depends on its mobility, as well as the wave period and height, and the gas pressure. Ions with high mobility are better able to keep up with traveling waves and are pushed more quickly through the cell. Ions with low mobility crest over the waves more often and have to wait for subsequent waves to push them forward, resulting in longer drift times. To measure an arrival time distribution (ATD), ions are gated into the mobility cell using an up-front stacked-ring RF device that traps ions before releasing them into the IMS cell. The oa-TOF pulses in an asynchronous manner, sending ions that have exited the mobility cell into the TOF mass analyzer and finally to the detector. The sum of all detected ions is the ion mobility chromatogram, or mobilogram. Selecting a peak in the ion mobility chromatogram displays the underlying TOF mass spectrum. Resolution enhancements to the device are recently described (Giles et al. 2011).

1.3 IM–MS and LC–IM–MS Analysis of Simple and Complex Mixtures

1.3.1 Cross Section and Structure

By measuring the mobility of an ion, information about the rotationally averaged collision cross section, which is given by shape and size, can be determined. The relationship between the mobility of an ion and its collision cross section has been derived in detail using kinetic theory (Mason and McDaniel 1988). When all experimental IM parameter values are held constant, a dependence of the ion mobility constant results only from the average cross section with $K \sim 1/\Omega$ (Bowers et al.; Henderson et al. 1999; Verbeck et al. 2002), where K = drift velocity v_d/electric field strength E and Ω = average collision cross section. A detailed description of kinetic theory is beyond the scope of this discussion. Ruotolo et al. 2005 were among the first describing the use of IM–MS to decipherer protein complex structure. The analysis of the Trp RNA-binding attenuation protein (TRAP) provided compelling evidence that many features of protein assemblies, including quaternary structure, can be preserved in the absence of solvent molecules. The researchers made use of TWIMS coupled to a modified TOF mass spectrometer to measure the CCS of four charge states of an 11-mer complex, demonstrating that the lowest charge state exhibited the largest CCS, with a value in close agreement to that estimated for a ring structure determined by X-ray crystallography. To investigate the sensitivity of the various conformers to changes in internal energy, they examined collision cross sections of the apo TRAP complex as a function of activation energy by manipulating their acceleration in the atmospheric pressure interface of the instrument, shown in Figure 1.2. The experiment illustrated that when an internal energy is imparted to 22+ ions, an expansion of the collapsed state occurred, while for 19+ ions they could partially drive the structural transitions observed for the ring structure as a function of protein charge state. IM–MS has proved to be extremely useful for the structural analysis of proteins and protein assemblies as illustrated in a number of recent reviews (Lanucara et al. 2014; Zhong et al. 2012; Uetrecht et al. 2010; Snijder and Heck 2014).

Collision cross section measurements and structure IM–MS experiments are not restricted to the analysis of large molecules but have been applied to other molecule classes and applications as well. For example, Valentine et al. 1999 used IMS to measure collision cross sections for 660 peptide ions generated by tryptic digestion proteins. Measured cross sections were compiled into a database that contains peptide molecular weight and sequence information and can be used to generate average intrinsic contributions to cross section for different amino acid residues. This was achieved by relating unknown contributions of individual residues to the sequences and cross sections of database peptides. Size parameters were combined with information about amino acid composition to calculate cross sections for database peptides. Figures 1.3(a) and (b) summarize the work showing cross sections as a function of molecular weight for the singly and doubly charged database peptides, respectively (Valentine et al. 1999). A strong correlation of increasing cross section with increasing molecular weight was observed, suggesting that (predicted) cross section can be

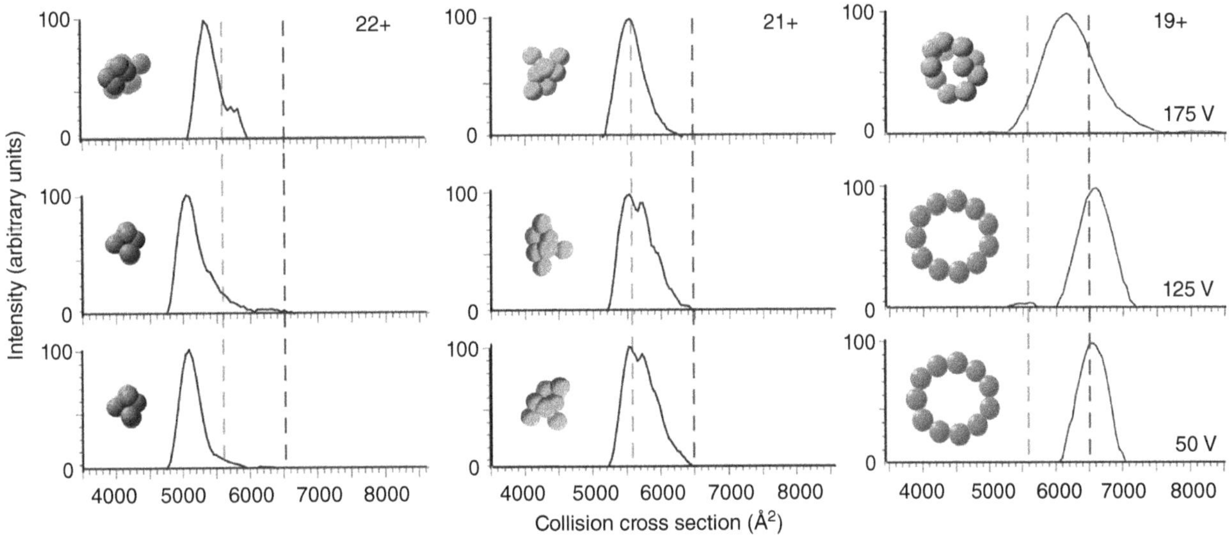

Figure 1.2 Ion mobility data for selected charge states of apo-TRAP (19+, 21+, and 22+) as a function of activation energy (175, 125, and 50V) applied in the high-pressure, sampling cone region of the instrument. The light gray and dark gray dashed lines represent the collision cross sections for collapsed and ring structures. (*Source*: Ruotolo et al. 2005. Reproduced with permission of The American Association for the Advancement of Science.)

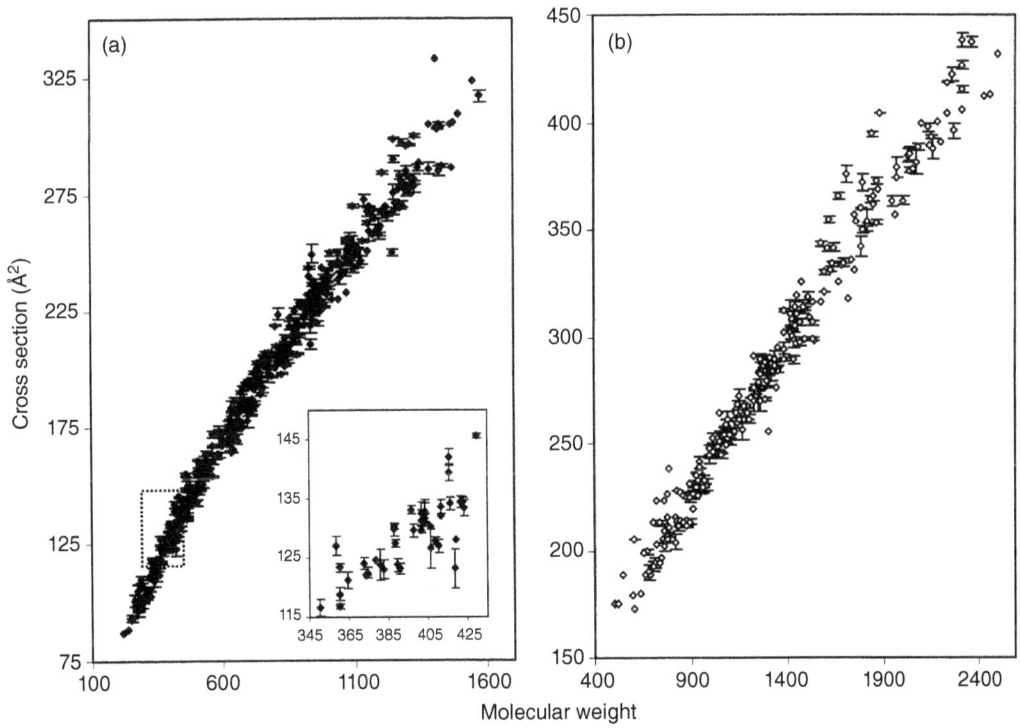

Figure 1.3 (a) Cross sections for 420 $[M+H]^+$ peptides (solid diamonds) as a function of molecular weight. Uncertainties correspond to one standard deviation or to a range. The inset shows variations in cross sections for $[M+H]^+$ peptides over a smaller molecular weight range (defined by the dashed-line box). (b) Cross-sectional measurements for 240 $[M+2H]^{2+}$ peptides (open diamonds) as a function of molecular weight. (*Source*: Valentine et al. 1999. Reproduced with permission of Springer.).

used as an additional search parameter for peptide identification. A follow-up study proposed that the method that employs intrinsic amino acid size parameters to obtain ion mobility predictions can be used to rank candidate peptide ion assignments. Intrinsic amino acid size parameters were determined for doubly charged peptide ions from the complete annotated yeast proteome. The use of the predictive enhancement as a means to aid peptide ion identification was discussed and a simple peptide ion scoring scheme presented.

1.3.2 Separation

The work of Clemmer and coworkers (Liu et al. 2007; Valentine et al. 2001, 2006) demonstrates the use of IMS for the separation and profiling of plasma proteins. The integration of IMS into an LC–MS schema is described to increase the separation power of a platform. The setup comprised off-line strong cation exchange (SCX) and in-line LC–IM–MS separation of trypsin digested plasma proteins. The SCX–LC–IM–MS setup is described in great detail as well as how the additional IMS separation dimension increased the available experimental peak capacity. The experimental two-dimensional LC–IM peak capacity was estimated to be ~6000–9000 obtained from a partial $t_r(t_d)$ base-peak plot derived from a single LC–IM–MS analysis, which greatly exceeds that of a single LC or IMS experiment. Also discussed is the use and creation of a relational table or database that comprises physico-chemical analyte information such as SCX retention time $t_{r,SCX}$, reversed-phase (RP) retention time $t_{r,RP}$, drift time t_d, and m/z. This information can be stored in a multidimensional space as shown in Figure 1.4. Knowledge of the positions of peaks will further corroborate assignments of other data sets. In addition, the accumulation of data provides valuable information for future work that would aim to predict SCX retention times, LC retention times, and mobilities based on sequences and charge states. The contribution of IM for the identification peptides as an additional search and identification parameter has been discussed in detail (Valentine et al. 1999, 2011). These concepts have been applied by Thalassinos et al. 2012 for the identification and quantitation of peptides and proteins across two similar mammalian species and Paglia et al. 2014 for the identification of the key metabolites potentially involved in cancer. The increase in system peak capacity, experimentally derived, for a multidimensional LC IM–MS system has been described and demonstrated by Rodríguez-Suárez et al. 2013.

Ion mobility-assisted data-independent analysis (DIA) LC–MS (Geromanos et al. 2009; Distler et al. 2014a) can be seen as an extension to the work of Clemmer and co-workers. Here, however, IMS is additionally used to align precursor and product ions to increase the specificity of a DIA workflow using TWIMS. In other words, it not only increases system peak capacity but also enhances the selectivity of DIA. In this experiment, to maximize duty cycle, peptide precursor ions are not isolated by the quadrupole mass analyzer positioned in front of the TWIMS cell. The ions undergo separation first in the mobility section and are either not fragmented or collision induced dissociated (CID) in the transfer region. This process is repeated at a fixed frequency, thereby generating so-called low and elevated energy precursor and product ion spectra, respectively. Thus, precursor and product ions share identical t_d, which can be used to entangle the multiplexed product ion spectra. Briefly, precursor and product ion mass extracted chromatograms are created in the t_r and t_d domains. Precursor and product ion that share the same drift and retention time are correlated, which simplifies the multiplexed CID spectra prior to a database search for identification of peptides and proteins. As an example, Yang et al. 2014 applied label-free LC–IM–DIA–MS to demonstrate that RSL3 binds to and inhibits GPX4, which regulates ferroptotic cancer cell death. Figure 1.5 contains a 3D representation of the isotopic clusters of peptide ILAFPCNQFGK from GPX4 analyzed by LC–DIA–IM–MS. Detection and

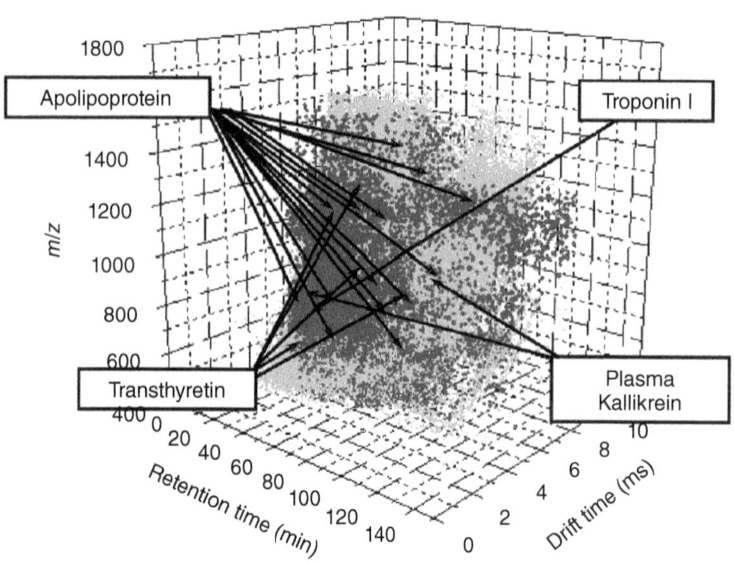

Figure 1.4 3D dot plot representation of the positions of peaks (in the retention time, drift time, and *m/z* dimensions) that are obtained from the 1×10^5 most intense features (light gray) observed during the triplicate LC-IMS-MS analyses of all SCX fractions associated with Sample 1. Superimposed on the plot are the positions for >10,000 features that have been assigned to peptides (dark gray). The arrows indicate some of the precursor ion positions of peptides identified for the four proteins labeled. This representation is intended to provide the reader with the impression that the possible existence of abundant protein in plasma could be tested at many positions in the map and therefore upon comparison there should be little ambiguity regarding its detection, whereas a low-abundance protein may be represented at only a single position, leading to uncertainty about its detection. (*Source*: Liu et al. 2007. Reproduced with permission of Springer.).

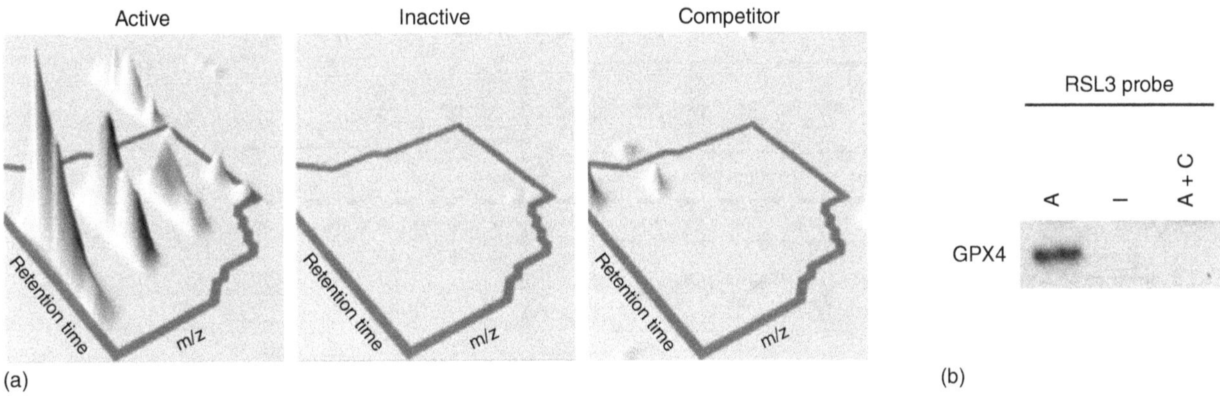

Figure 1.5 Confirmation of GPX4 binding to an active affinity probe. (a) Cell lysates prepared from cells treated with active probe (A), inactive probe (I), or active probe in the presence of competitor (A + C) that were affinity purified by α-fluorescein antibodies and probed for GPX4 by western blot using GPX4-specific antibody. (b) 3D visualization of isotopic clusters of peptide ILAFPCNQFGK from GPX4 as analyzed by LC–DIA–IM–MS. (*Source*: Yang et al. 2014. Reproduced with permission of Elsevier.)

identification was conducted by dedicated software. The results shown in Figure 1.5 illustrate the presence of GPX4 with RSL3 active probe treatment and its absence when the probe was inactive or a competitor was present. It was derived and concluded that RSL3 to inhibit GPX4, a protein essential for cancer cell viability. Numerous applications describe the use of LC–DIA–IM–MS for the label-free quantification, as described in a recent review describing DIA and its application (Distler et al. 2014b).

1.3.3 Sensitivity

A more recent application of IM–MS is described by Helm et al. 2014 who used the technique to increase MS/MS sensitivity in untargeted data-dependent analysis (DDA) and targeted parallel reaction monitoring (PRM) such as proteomic LC–MS experiments on a commercial hybrid quadrupole – ion mobility – time-of-flight mass spectrometer. This technique, as will be demonstrated, enhances the duty cycle of the oa-TOF analyzer and thus sensitivity. Briefly, as shown previously, TWIMS separation is strongly dependent upon ion charge z. Moreover, ions are nested for a given charge state by mass and drift time. This charge state separation and nesting can be used to discriminate against single charge background and to exclusively select multiply charged peptides for tandem MS. Subsequently, precursor ions are sequentially selected by the quadrupole mass analyzer and fragmented by CID in the first stacked-ring ion guide of the triwave device and prior to reaching the ion mobility cell. Product ions are trapped within this first travelling wave region of the triwave device and gated into the high-pressure ion mobility cell where they are separated according to their gas-phase mobility within the cell. As a result, as illustrated in Figure 1.6, fragment ions of the same mobility exit the cell as a series of compact packets. Hence, by synchronizing the pusher pulse that accelerates the fragment ions into the oa-TOF mass analyzer with the arrival of product ions from the TWIMS cell into the pusher region, fragment ions are sequentially injected into the TOF analyzer with greatly enhanced duty cycle (~100%) across the mass scale. This synchronization leads to a concomitant increase in sensitivity, which is reflected by the results shown in Figure 1.7, where the percent identified DDA spectra versus amount protein digest on column is contrasted. On average, a 10-fold increase in peptide MS/MS sensitivity can be observed (Helm et al. 2014). Since the ion mobility time frame is in the order of milliseconds, it nests well between the second time frame of liquid chromatography and that of the oa-TOF mass spectrometer that operates in the microsecond time frame.

An example of an IM-enabled targeted high-resolution multiple reaction monitoring (HR-MRM) experiment is shown in Figure 1.8. In HR-MRM, the last quadrupole of a tandem quadrupole instrument is substituted with a high-resolution mass analyzer to allow parallel detection of all product ions in a single, high-resolution, accurate mass experiment. Here, unlike the previously described experiment, peptide precursor masses, including internal standards, are predefined, along with their retention time and CID collision energy profile. The principle of product ion enrichment to increase duty cycle and MS/MS sensitivity is identical. In this particular example, a number of putative cardiovascular disease plasma proteins were quantified (Domanski et al. 2012). As an example, shown in the top pane of Figure 1.8, are the summed product ion extracted mass chromatograms of ATEHLSTLSEK from Apolipoprotein A-1 and its labeled internal standard analog, as well as product ion spectra of both peptides. The calibration curve of heavy labeled ATEHLSTLSE[K] is shown in the bottom pane of Figure 1.8 from which an Apolipoprotein A-I serum concentration of 1.403 mg/mL can be calculated.

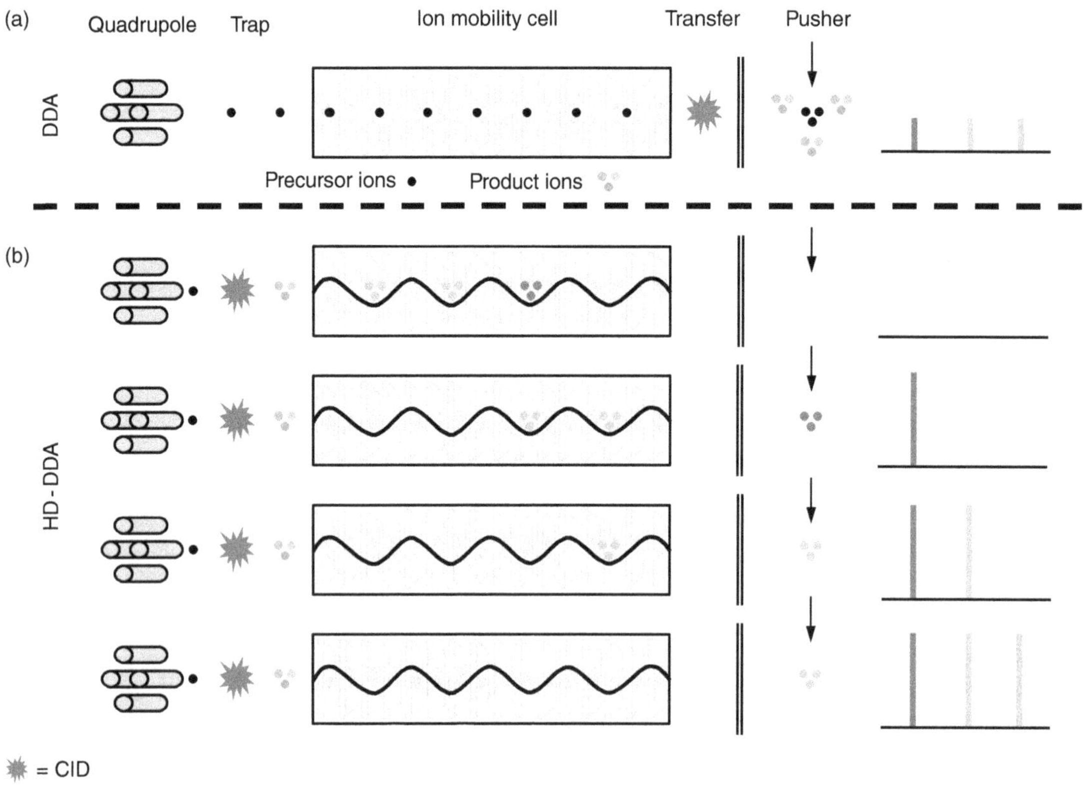

Figure 1.6 Principle IM-enabled DDA with asynchronous pusher operation (a) and synchronization of a pusher pulse with product ion drift time (b).

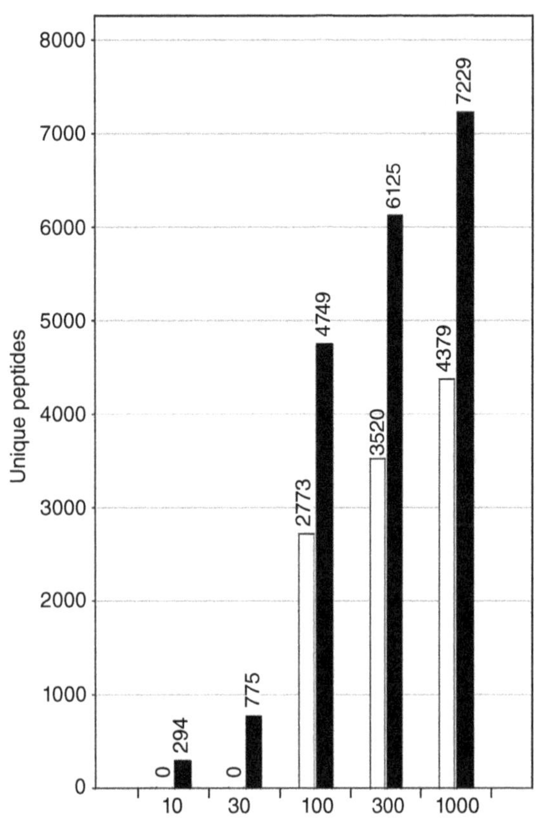

Figure 1.7 Increased MS/MS sensitivity expressed as number of identified peptides for normal DDA (white) and IMS-enabled DDA (black).

This concentration estimation compares well with an average reported literature value of 1.400 mg/mL (Domanski et al. 2012). The linear dynamic range was at least three orders of magnitude; however, the great benefit of high-resolution PRM is its high selectivity in the mass-to-charge domain since high-resolution MS can often separate out interferences from the product ion of interest (Mbasu et al. 2016).

1.4 Outlook

Ion mobility coupled with mass spectrometry has made significant strides since the turn of the twenty-first century. Despite IMS coupled with TOF MS first being described in the 1960s, it is only in the last 20 years that improvements in electronics and the performance of TOF mass analyzers has allowed IMS–MS to become a mainstream platform. As electronic improvements still continue to track Moore's law (Waldrop 2016), this should allow concomitant improvements in the underlying performance of both IMS and oa-TOF technologies, resulting in higher resolving power, as well as faster sampling rates. Improvements in IMS resolution have already been described and the unique hybrid combination of IMS with TOF should allow the comprehensive profiling of complex heterogeneous samples.

The outlook for the future is promising. The use of rotationally averaged collision cross sections as a means

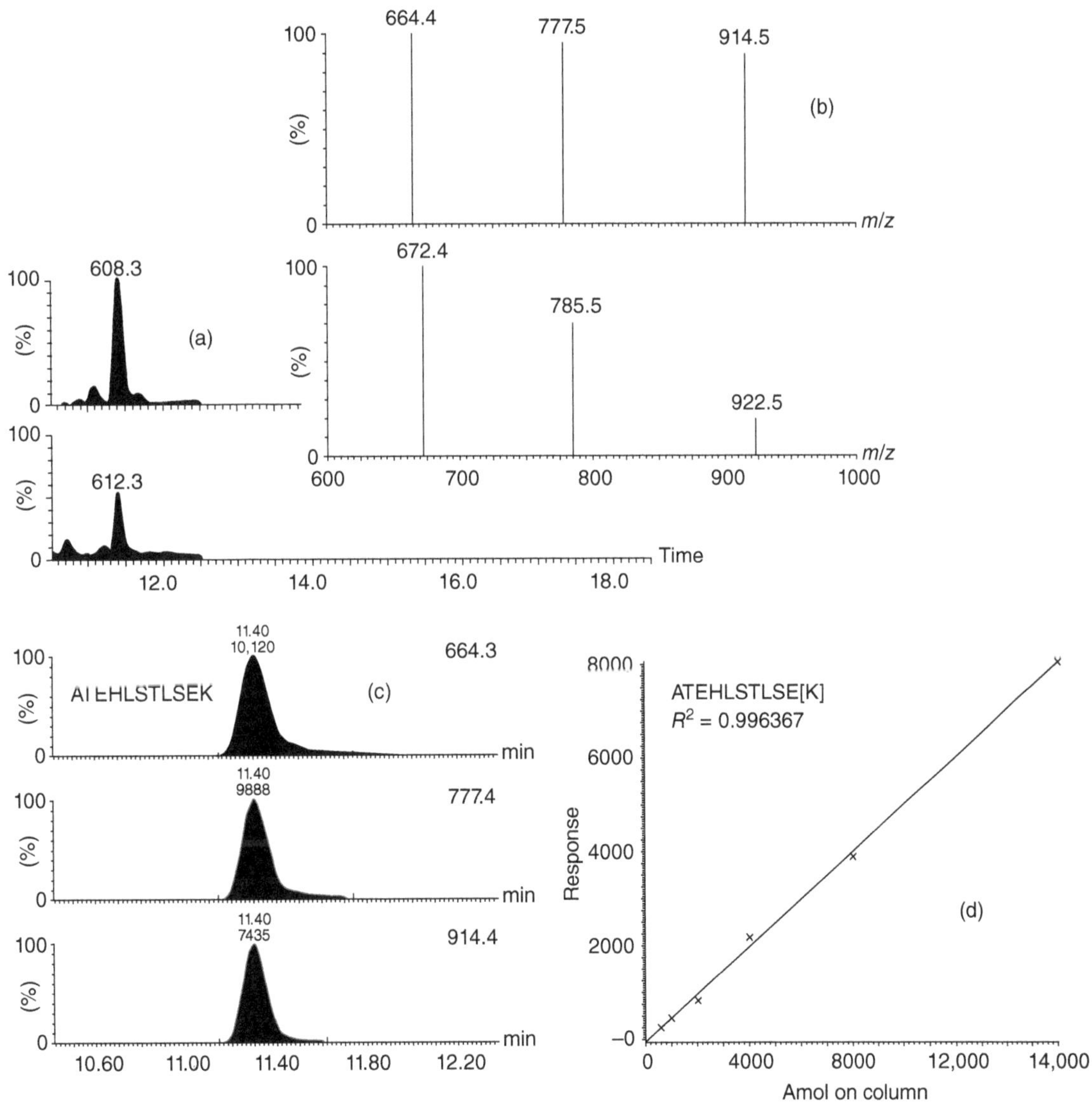

Figure 1.8 Summed product ion extracted mass chromatograms of "light" labeled ATEHLSTLSEK and "heavy" labeled ATEHLSTLSE[K] (a), corresponding part product ion MS/MS spectra, respectively (b), extracted product ion mass chromatograms "light" ATEHLSTLSEK (c), and calibration curve "heavy" ATEHLSTLSE[K] (d).

for identification and confirmation of compound identity is an intriguing prospect, providing a physicochemical supplement to retention time and tandem MS information. The limiting factor is currently the lack of CCS measurements populated into compound libraries and the lack of computational tools to rapidly generate CCS values from compound structure. Developments in these areas will undoubtedly occur and make the routine use of IMS information for identification purposes a powerful technology. In summary, the next few years should see significant improvements in both the technology and the informatics and workflows to use the information generated for both qualitative and quantitative analyses.

Acknowledgment

The Scientific Operations group of Waters Corporation is kindly acknowledged for their support and help in obtaining experimental data.

References

Bagal D, Zhang H and Schnier PD (2008) "Gas-phase proton-transfer chemistry coupled with TOF mass spectrometry and ion mobility-MS for the facile analysis of poly(ethylene glycols) and PEGylated polypeptide conjugates." *Anal Chem* **80**(7):2408–2418.

Baker ES, Clowers BH, Li F, et al. (2007) "Ion mobility spectrometry-mass spectrometry performance using electrodynamic ion funnels and elevated drift gas pressures." *J Am Soc Mass Spectrom* **18**(7):1176–1187.

Bereszczak JZ, Havlik M, Weiss VU, et al. (2014) "Sizing up large protein complexes by electrospray ionisation-based electrophoretic mobility and native mass spectrometry: morphology selective binding of Fabs to hepatitis B virus capsids." *Anal Bioanal Chem* **406**(5):1437–1446.

Bowers MT, Marshall AG and McLafferty FW (1996) "Mass spectrometry: recent advances and future directions". *J Phys Chem*, **100**(31):12897–12910.

Clemmer DE, Hudgins RR and Jarrold MF (1995). "Naked protein conformations: cytochrome c in the gas phase." *J Am Soc Mass Spectrom* **117**: 10141–10142.

Dear GJ, Munoz-Muriedas J, Beaumont C, et al. (2010). "Sites of metabolic substitution: investigating metabolite structures utilising ion mobility and molecular modelling." *Rapid Commun Mass Spectrom* **24**(21):3157–3162.

Distler U, Kuharev J, Navarro P, et al. (2014a) "Drift time-specific collision energies enable deep-coverage data-independent acquisition proteomics." *Nat Methods* **11**(2):167–170.

Distler U, Kuharev J, Tenzer S (2014b) "Biomedical applications of ion mobility-enhanced data-independent acquisition-based label-free quantitative proteomics." *Expert Rev Proteomics* **18**:1–10.

Domanski D, Percy AJ, Yang J, et al. (2012) "MRM-based multiplexed quantitation of 67 putative cardiovascular disease biomarkers in human plasma." *Proteomics* **12**(8):1222–1243.

Dwivedi P, Wu P, Klopsch SJ, et al. (2008). "Metabolic profiling by ion mobility mass spectrometry (IMMS)". *Metabolomics* **4**(1): 63–80.

Eckers C, Laures AM, Giles K, et al. (2007) "Evaluating the utility of ion mobility separation in combination with high-pressure liquid chromatography/mass spectrometry to facilitate detection of trace impurities in formulated drug products." *Rapid Commun Mass Spectrom* **21**(7):1255–1263.

Gabryelski W and Froese KL (2003). Rapid and sensitive differentiation of anomers, linkage, and position isomers of disaccharides using high-field asymmetric waveform ion mobility spectrometry (FAIMS). *J Am Soc Mass Spectrom* **14**(3):265–277.

Geromanos SJ, Vissers JP, Silva JC, et al. (2009) "The detection, correlation, and comparison of peptide precursor and product ions from data independent LC-MS with data dependant LC-MS/MS." *Proteomics* **9**(6):1683–1695.

Giles K, Pringle SD, Worthington KR, et al. (2004) "Applications of a travelling wave-based radio-frequency-only stacked ring ion guide." *Rapid Commun Mass Spectrom* **18**(20):2401–2414.

Giles K, Williams JP and Campuzano I (2011) "Enhancements in travelling wave ion mobility resolution." *Rapid Commun Mass Spectrom* **25**(11):1559–1566.

Helm D, Vissers JP, Hughes CJ, et al. (2014) Ion mobility tandem mass spectrometry enhances performance of bottom-up proteomics. *Mol Cell Proteomics* **13**(12):3709–3715.

Henderson SC, Valentine SJ, Counterman AE, et al. (1999) "ESI/ion trap/ion mobility/time-of-flight mass spectrometry for rapid and sensitive analysis of biomolecular mixtures." *Anal Chem* **71**:291–301.

Hoaglund CS, Valentine SJ and Clemmer DE (1997). "An ion trap interface for ESI-ion mobility experiments." *Anal Chem* **69**(20): 4156–4161.

Jin L, Barran PE, Deakin JA, et al. (2005). "Conformation of glycosaminoglycans by ion mobility mass spectrometry and molecular modelling." *Phys. Chem. Chem. Phys.* **7**(19):3464–3471.

Kanu AB, Dwivedi P, Tam M, et al. (2008). "Ion mobility-mass spectrometry." *J Mass Spectrom* **43**(1):1–22.

Kliman M, May JC and McLean JA (2011) "Lipid analysis and lipidomics by structurally selective ion mobility-mass spectrometry." *Biochim Biophys Acta* **1811**(11):935–945.

Lanucara F, Holman SW, Gray CJ, et al. (2014). "The power of ion mobility-mass spectrometry for structural characterization and the study of conformational dynamics." *Nat Chem* **6**(4):281–294.

Liu X, Valentine SJ, Plasencia MD, et al. (2007) "Mapping the human plasma proteome by SCX-LC-IMS-MS." *J Am Soc Mass Spectrom* 2007 **18**(7):1249–1264.

Mason EA and McDaniel EW (1988). "*Transport Properties of Ions in Gases*." Wiley, New York.

Mbasu RJ, Heaney LM, Molloy BJ, et al. (2016) "Advances in quadrupole and time-of-flight mass spectrometry for peptide MRM based translational research analysis." *Proteomics*, **16**(15–16):2206–2220.

McLean JA, Ruotolo BT, Gillig KJ, et al. (2005). "Ion mobility mass spectrometry: a new paradigm for proteomics." *Int J Mass Spectrom* **240**(3): 301–315.

Paglia G, Williams JP, Menikarachchi L, et al. (2014) "Ion mobility derived collision cross sections to support metabolomics applications." *Anal Chem* **86**(8):3985–3993.

Pringle, SD, Giles, K, Wildgoose, JL, et al. (2007). "An investigation of the mobility separation of some peptide and protein ions using a new hybrid quadrupole/ travelling wave IMS/OA-TOF instrument." *Int J Mass Spectrom* **261**, 1–12.

Rodríguez-Suárez E, Hughes C, Gethings L, et al. (2013). "An ion mobility assisted data independent LC–MS strategy for the analysis of complex biological samples." *Curr Anal Chem* **2**:199–211.

Ruotolo BT, Giles K, Campuzano I, et al. (2005 "Evidence for macromolecular protein rings in the absence of bulk water." *Science* **310**(5754):1658–1661.

Schenauer MR, Meissen JK, Seo Y, et al. (2009) "Heparan sulfate separation, sequencing, and isomeric differentiation: ion mobility spectrometry reveals specific iduronic and glucuronic acid-containing hexasaccharides." *Anal Chem* **81**(24):10179–10185.

Snijder J and Heck AJR (2014) "Analytical approaches for size and mass analysis of large protein assemblies." *Annu Rev Anal Chem* **7**: 43–64.

Taraszka JA, Counterman AE and Clemmer DE (2001). "Gas-phase separations of complex tryptic peptide mixtures." *Fresenius J Anal Chem* **369**(3–4):234–245.

Thalassinos K, Vissers JP, Tenzer S, et al. (2012). "Design and application of a data-independent precursor and product ion repository." *J Am Soc Mass Spectrom* **23**(10):1808–1820.

Uetrecht C, Rose RJ, van Duijn E, et al. (2010) "Ion mobility mass spectrometry of proteins and protein assemblies." *Chem Soc Rev* **39**(5): 1633–1655.

Vakhrushev SY, Langridge J, Campuzano I, et al. (2008) "Ion mobility mass spectrometry analysis of human glycourinome." *Anal Chem* **80**(7):2506–2513.

Valentine SJ, Counterman AE and Clemmer DE (1999) "A database of 660 peptide ion cross sections: use of intrinsic size parameters for bona fide predictions of cross sections." *J Am Soc Mass Spectrom* **10**(11):1188–1211.

Valentine SJ, Ewing MA, Dilger JM, et al. (2011) "Using ion mobility data to improve peptide identification: intrinsic amino acid size parameters." *J Proteome Res* 2011 **10**(5):2318–2329.

Valentine SJ, Kulchania M, Srebalus Barnes CA, et al. (2001) "Multidimensional separations of complex peptide mixtures: a combined high-performance liquid chromatography/ion mobility/time-of-flight mass spectrometry approach." *Int J Mass Spectrom* **212**(1–3):97–109.

Valentine SJ, Plasencia MD, Liu XY, et al. (2006). "Toward plasma proteome profiling with ion mobility-mass spectrometry." *J Proteome Res* **5**(11):2977–2984.

Verbeck GF, Ruotolo BT, Sawyer HA, et al. (2002) "A fundamental introduction to ion mobility mass spectrometry applied to the analysis of biomolecules." *J Biomol Technol* **13**(2):56–61.

Waldrop MM (2016). "The chips are down for Moore's law." *Nature* **530**, 144–147.

Williams JP, Bugarcic T, Habtemariam A, et al. (2009) "Isomer separation and gas-phase configurations of organoruthenium anticancer complexes: ion mobility mass spectrometry and modeling." *J Am Soc Mass Spectrom* **20**(6):1119–1122.

Williams JP, Kipping M and Vissers JP (2012) "Traveling wave ion mobility mass spectrometry – proteomic and biopharmaceutical applications" *GIT Lab J* (http://www.laboratory-journal.com/file/track/8062/1).

Yang WS, SriRamaratnam R, Welsch ME, et al. (2014) "Regulation of ferroptotic cancer cell death by GPX4." *Cell* **156**(1–2):317–331.

Zhong Y, Hyung SJ and Ruotolo BT (2012) "Ion mobility-mass spectrometry for structural proteomics." *Expert Rev Proteomics* **9**(1):47–58.

2

High-Resolution Accurate Mass Orbitrap and Its Application in Protein Therapeutics Bioanalysis

Hongxia Wang[1] *and Patrick Bennett*[2]

[1] *Thermo Fisher Scientific, San Jose, CA, USA*
[2] *PPD, Richmond, VA, USA*

2.1 Introduction

Since the introduction of first recombinant protein therapeutic-human insulin in 1978, protein therapeutics has increased dramatically in both number and frequency to treat many major diseases including cancers, infections, and immune disorders (Beck et al. 2010, Pillay et al. 2011, Scott et al. 2012). At present, over 130 different proteins or peptides are approved for clinical use by the United States Food and Drug Administration (FDA), and many more are in development (Leader et al. 2008). Within this group, antibodies and antibody drug conjugates (ADCs) account for almost half of the sales (Voynov et al. 2012). As more peptide- or protein-based therapeutics move into preclinical and clinical development, the demand to quantify these drugs in biological matrices as well as the identification and relative quantitation of related metabolites is increasing. These applications include the absolute quantitation and pharmacokinetic studies of drug and relevant biomarker assessment. Bioanalysis data provides important information for drug development.

Similar to small molecules, the bioanalysis of protein therapeutics requires assays that have high selectivity and sensitivity, good precision and accuracy, wider linear dynamic range, and high throughput. Two major techniques used for the bioanalysis of protein therapeutics are ligand binding assays (LBAs) and liquid chromatography–mass spectrometry (LC–MS) assays. LBAs measure the concentration of analyte through the specific interaction of capture and detection reagent to the analyte of interest. Historically, LBAs have been adopted to quantify protein in biological matrices due to its high sensitivity, robustness, and throughput (DeSilva et al. 2003, Damen et al. 2009). However, this technique does have drawbacks. The development of highly specific reagents is time-consuming and expensive. Hence, it is challenging to meet time requirements of obtaining bioanalytical results in the fast-paced drug discovery stage (Zheng et al. 2014). A nonspecific LBA might measure the sum of both drug and its metabolites because of insufficient selectivity and may have difficulty discriminating a drug from its metabolites such as truncation of amino acids, cleavage, and modifications. In addition, the accuracy of LBA assay may be compromised by endogenous components and antidrug–antibody (ADA) formation, which can result in underestimation of the true concentration by blocking binding of protein analyte to either the capture or detection reagents (Ezan and Bitsch 2009, Zheng et al. 2014). LC–MS-based strategies, specifically, an LC–high-resolution mass spectrometry (HRMS) strategy, would overcome these drawbacks and are emerging as an important alternative for the quantitation of protein therapeutics.

2.2 Triple Quadrupole Mass Spectrometer and Its Challenges

For protein or peptide bioanalysis, liquid chromatography–tandem mass spectrometry (LC–MS/MS) offers several advantages over traditional LBAs and is the only approach for large molecule quantitation with good selectivity and sensitivity when an appropriate binding reagent is not available. The combination of LC and MS resolving power can differentiate structure-related isoforms and metabolites with modifications or degradation products missing one or two amino acids. It is difficult for LBAs to discriminate these components if binding reagents are not developed against each specific change. Before the advent of modern high-resolution instruments, MS-based quantitation was performed using tandem MS fragmentation on either triple quadrupole or ion trap mass spectrometers.

Protein Analysis using Mass Spectrometry: Accelerating Protein Biotherapeutics from Lab to Patient, First Edition.
Edited by Mike S. Lee and Qin C. Ji.

However, those instruments are unit mass resolution and can only separate ions at a single mass-to-charge ratio (m/z) (e.g., 100 and 101 Da). Because many compounds have identical nominal mass, it is impossible to differentiate them on a single MS dimension without LC separation. In the majority of cases, tandem MS fragmentation generates unique fragment ions in the MS/MS spectrum, which can be used for quantitation. Quantitation of targeted analytes by unique fragment ions in ion trap mass spectrometers has been reported (Wang et al. 2010).

LC-selected reaction monitoring (SRM)-based triple quadrupole MS has been considered as the gold standard for small molecule quantitative analysis for the last 20 years (Jemal 2000, Korfmacher 2005, Zhou et al. 2005). A triple quadrupole mass spectrometer contains three quadrupoles Q1, Q2, and Q3. When operated in the SRM mode, the first quadrupole is used to select the precursor ion of interest. The second quadrupole corresponds to the collision cell where selected precursor ion is fragmented into product ions. The third quadrupole acts as a mass filter and allows only the selected fragment ions to be detected. By monitoring and using the peptide transitions between precursor and fragment ions for quantitation, the assay selectivity is greatly improved, which leads to higher sensitivity for analytes in complex biological matrices.

Unlike small molecules bioanalysis, LC–MS/MS encounters additional challenges for protein therapeutics bioanalysis (Ewles and Goodwin 2011). (i) Typical 30–1500 acquisition mass range is not enough to detect higher mass peptides. (ii) Large proteins such as monoclonal antibody have multiple charges under electrospray ionization and wide charge distributions, which go beyond this low acquisition mass range. Furthermore, peptide fragmentation patterns and efficiency are different for precursor ions at different charge states. A single predominant and high-intensity product ion is the desired fragmentation pattern for the SRM-based triple quadrupole method. (iii) It is not possible to quantify intact protein due to poor fragmentation efficiency of a large protein in the collision cell. (iv) Endogenous protein interferences in biological matrices require higher selectivity. The high background noise from complex matrix decreases selectivity on low-resolution triple quadrupole MS, preventing the detection of low-abundant peptides targets (Lange et al. 2008). (v) In addition, extensive metabolism, modifications, and lack of isotope-labeled internal standard make it hard for method development and instrument optimization. Therefore, quantitative LC–MS/MS methods have been mainly used to quantify peptides from biological samples following enzymatic digestion of the analyte.

2.3 High-Resolution Mass Spectrometers

The main high-resolution mass spectrometers include Fourier transform ion cyclotron resonance (FT-ICR), Orbitrap, and time-of-flight (TOF). The low sensitivity and narrow dynamic range on early versions of HR mass spectrometers limited their application in quantitative analysis. With the development of cutting-edge technologies, high resolution accurate mass (HRAM) Orbitrap mass spectrometers have been applied to both qualitative and quantitative applications with enhanced sensitivity, wider dynamic range, faster scan speed, and higher resolution. Compared to SRM on triple quadrupoles, HRMS offers distinct advantages for both protein and peptide analysis with significantly reduced method development with no requirement of compound-specific tuning. Recent advancement on HRMS greatly expands its applications from purely qualitative to allowing simultaneous quantitative and qualitative analysis (Quan/Qual) (Korfmacher 2011). Current HRMS instruments with increasingly innovative technology have proven capabilities for absolute quantitation of small molecules, peptides, and proteins in complex matrices (Fung et al. 2011, Ruan et al. 2011, Huang et al. 2013, Wang and Bennett 2013, Ciccimaro et al. 2014).

In FT-ICR, ions are trapped by crossed magnetic and electric fields and then in circular trajectories perpendicular to the magnetic field. The excited ions move in circular orbits and the decay of their image currents is detected as time domain signals, which are converted into mass-to-charge ratio by FT (Marshall et al. 1998). Because image current detection does not cause ion loss, ions can be stored in the ICR for an extended time, leading to super high resolution and great mass accuracy. However, FT-ICR MS requires a strong magnetic field, expensive maintenance, a long method optimization time, and large storage space, which greatly limit the standard laboratory usage.

TOF is one of the most widely used HR mass spectrometers because of its fast scan speed and expanded dynamic range. It separates ions solely based on their velocities after acceleration from a high potential. The time for the ion traveling down the flight tube to the detector is measured and correlated to m/z. Small ions travel faster and are detected earlier. High resolving power is achieved by extending the flight path. Reflectron TOF has higher resolution (15,000–30,000) than linear TOF. Due to fundamental principles of ion detection, sensitivity can be lost during longer travel time in the flight tube when trying to achieve higher resolutions. Mass accuracy is within 5 ppm with internal lock mass calibration.

Since the commercial release of the first Thermo Scientific™ LTQ™ Orbitrap Mass Spectrometer in 2005, Orbitrap technology has been significantly improved. Orbitrap is a novel Kingdon trap analyzer. The trapping field is only electrostatic and no RF or magnetic field is required. Ions are trapped and oscillated along the central spindle-like electrode. Similar to FT-ICR, broadband image current detection is used. Mass-to-charge ratio is simply related to the frequency of ion oscillation along the z-axis. Orbitrap MS has ultrahigh resolving power up to 450,000, very high mass accuracy (3 ppm external and 1 ppm internal calibration), and increased space charge capacity at higher masses due to independent trapping potential on m/z (Hu et al. 2005). Compared to FT-ICR, Orbitrap has a larger tapping volume, no magnetic field requirement, small size, easy method development, and robustness, which are the attributes making it increasingly popular. While Orbitrap platforms were originally primarily used for qualitative analysis, it is now demonstrating a powerful capability for quantitative analysis. The next-generation Orbitrap, Thermo Scientific™ benchtop Q Exactive™ series mass spectrometers expanded its use for routine analysis. In addition, the parallel filing and detection, enhanced Fourier transform algorithms allow for increased scan speed to meet the demands of ultrahigh pressure chromatography for bioanalysis.

2.4 Quantitation Modes on Q Exactive Hybrid Quadrupole Orbitrap

The scheme of a hybrid quadruple Orbitrap is shown in Figure 2.1. For full-scan analysis, after electrospray ionization, ions of interest are selected in a quadrupole, accumulated in the C-trap, and detected in the Orbitrap. In fragmentation mode, selected precursor ion is fragmented in the higher-energy collisional dissociation (HCD) cell and fragment ions are stored in the C-trap prior to injection into Orbitrap for detection. Low-abundant ions are detected and quantified with high sensitivity by controlling the automatic gain control (AGC) that allows the appropriate number of charges to be stored in the C-trap, the ion inject time (IT), and the Orbitrap detection time. Even though Orbitrap is a trap-based instrument, there is no low mass cutoff because ions are filtered by a quadrupole instead of an ion trap.

Four instrument methods are available for quantitation. Selected ion monitoring (SIM) and parallel reaction monitoring (PRM) are targeted quantitation methods. The remaining two methods, full-scan and data-independent acquisition (DIA), can perform both quantitative and qualitative analysis from a single run. Full-scan and DIA are used for nontargeted methods. Multiplexed SIM/PRM is also available to increase the number of scans per peak of interest for coeluting compounds. Unlike triple quadruples, no compound-specific tuning is needed on a Q Exactive MS. All acquired data has accurate mass (AM). This greatly reduces the method development time.

In SIM, a very narrow isolation window from the center m/z of the target ion is set on the quadrupole and the mass spectrometer scans at very high resolution, such as 70,000 (at m/z 200) or above to ensure the selectivity for samples in complex matrices.

In PRM, full MS/MS spectrum with all fragment ions is typically acquired at a resolution of 17,500. The presence of all fragment ions eliminates the need for selecting transition, leading to easy and fast method development. The noise-free feature is common for the majority of targets. Near-zero interference in the extracted ion chromatograms is commonly observed on HRAM Orbitrap, which results in more accurate quantification at higher sensitivity. The HRAM full MS/MS spectrum provides extremely high selectivity, ensuring confident confirmation and quantitation from a single injection. Sensitivity can be further improved by extracting and summing more unique fragment ions and using that value for quantitation.

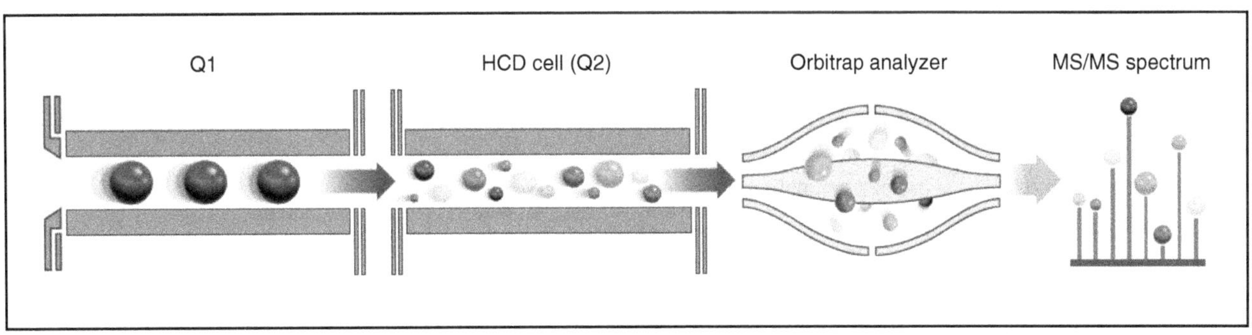

Figure 2.1 Ion path on Q Exactive hybrid quadrupole Orbitrap.

Full scan allows intact protein quantitation with a mass range from 50 to 6000 *m/z*. The charge envelope of a large protein such as a denatured monoclonal antibody (mAb) is typically below 4000 *m/z*. The full-scan method allows simultaneous Quan/Qual analysis because molecular mass information is also obtained during the quantitation of the protein. The combination of full-scan and data-dependent Top N MS/MS is the major approach for simultaneous Quan/Qual analysis. Full mass spectrum provides parent ion information for quantitation and MS/MS spectrum is employed to confirm the identity of the precursor ion. The approach is also useful for the troubleshooting during method development because it allows accurate mass detection of coeluting species. Matrix effects can be eliminated for small molecule quantitation in biological matrices by monitoring lipids elution profile during LC method development (Wang and Bennett 2013). This approach should still be of value for large molecule quantitation. The semiquantitation feature is a useful approach for pharmacokinetic (PK) studies.

DIA is a nontargeted protein quantitation approach. It is currently used for both qualitative and quantitative analysis. A typical mass window of 20–50 *m/z* is isolated in the quadrupole. All ions within the window are fragmented in the collision cell and detected in one spectrum. The advantage of DIA for quantitation is similar to SIM because no method development is necessary. However, sensitivity is compromised for DIA experiments as the intrascan dynamic range will limit the detection of low-abundant species in the presence of highly abundant interference ions. Furthermore, data analysis requires the identification of targeted peptide fragments using retention time and a reference spectrum library before quantitation.

2.5 Protein Quantitation Approaches Using Q Exactive Hybrid Quadrupole Orbitrap

Depending on the required sensitivity, the stage of drug development and the size of the protein, three major approaches, bottom-up, intact, and middle-up, can be employed to quantify protein therapeutics (see Table 2.1). These names are derived from the sample treatment (e.g., digestion, subunit generation) prior to MS analysis.

The bottom-up approach quantifies protein at peptide level. It is the most popular method due to its ability of providing high sensitivity and wider dynamic range for accurate quantitation. In this approach, protein is digested to peptides by proteolytic enzymes. For monoclonal antibodies, the denaturing step is necessitated to unfold the large protein, followed by the reduction of disulfide bonds by dithiothreitol (DTT) and alkylation by iodoacetamide (IAA). Commonly used proteases and their cleavage site on protein sequences should be considered for different application purposes. Serine protease trypsin specifically cleaves protein at the carboxyl side of lysine and arginine residues; chymotrypsin cleavage of hydrophobic regions (tyrosine, phenylalanine, tryptophan, and leucine)

Table 2.1 High-resolution mass spectrometry protein quantitation approaches.

Approach	Targeted analyte	Advantages	Limitations	HR/AM instrument acquisition methods
Bottom-up	Peptide	Accurate quantitation	Introduce unexpected modifications	SIM, PRM, and DIA
		Wider linear dynamic range		
		Highest sensitivity		
Middle-up	Antibody fragments (Light/heavy chain; Fab/Fc)	Simple sample preparation	Lower linear dynamic range	Full scan
		Accurate quantitation for isotopic resolved mass spectra		
		Keep partial original sequence information		
Intact	Intact proteins	Near-zero sample preparation	Accuracy may be affected by unresolved glycoforms, modifications, and truncation of large intact protein	Full scan
		Minimal modifications from enzymatic digestion	Low sensitivity due to wide charge distribution	
		Obtaining mass and spectral information	Lower linear dynamic range	

provides complementary peptides to trypsin. Lys-C and Glu-C cleave the carboxyl side of lysine and glutamic acid residues, respectively, generating larger peptides than those by trypsin alone. Asp-N produces the peptide after the amino side cleavage of aspartic acid. Acid protease pepsin is not a site-specific enzyme. Either one or the combination of two proteases is generally used to digest proteins. Another reason for the popularity of this approach is that triple quadrupole MS is limited to peptide level quantitation because of mass range, specificity, and selectivity.

SIM and PRM are the primary choices for targeted quantitation with HRMS methods. SIM offers the most convenient approach to quantify peptide by only selectively monitoring the peptide parent ion in a narrow *m/z* window. There is no need to optimize collision energy for selected transitions, ion source parameters, for example, sheath gas, auxiliary gas, sweep gas, and spray voltage. Generic settings of the ion source parameters, based on LC flow rate, are recommended for any operation mode on Q Exactive mass spectrometer. In SIM, previous research has shown that 50,000 resolution on Exactive Orbitrap HRMS provided better selectivity than triple MS/MS for the quantitation of seven different veterinary drugs in complex matrices (Kaufmann et al. 2010). A resolution of 70,000 on Q Exactive has enough selectively for samples in most biological matrices after sample cleanup. A higher resolution of 140,000 or 280,000 is the alternative for more complex matrices. When samples contain abundant interference with a mass difference to the ion of interest within 0.001 Da, further selectivity is required. PRM provides higher selectivity than SRM with an additional level of selectivity from accurate mass. Peptide sequences are confidently confirmed simultaneously using the full MS/MS spectrum. This data acquired for quantification may also provide potentially significant characterization for troubleshooting information. Data can be reinterrogated at any time without the need of performing data acquisition again. On triple quadrupoles, additional peptides from specific region of protein would need to be monitored to provide structure-related information and samples would have to be reanalyzed to obtain additional information for other species.

Intact protein quantitation is an appealing approach. The elimination of protein digestion minimizes unexpected modifications and greatly reduces method development cycle and increases sample analysis throughput. This approach has been applied to quantify small protein–insulin with decent sensitivity and four orders of magnitude linear dynamic range from 0.01 to 200 ng on column as shown in Figure 2.2. Benefiting from high

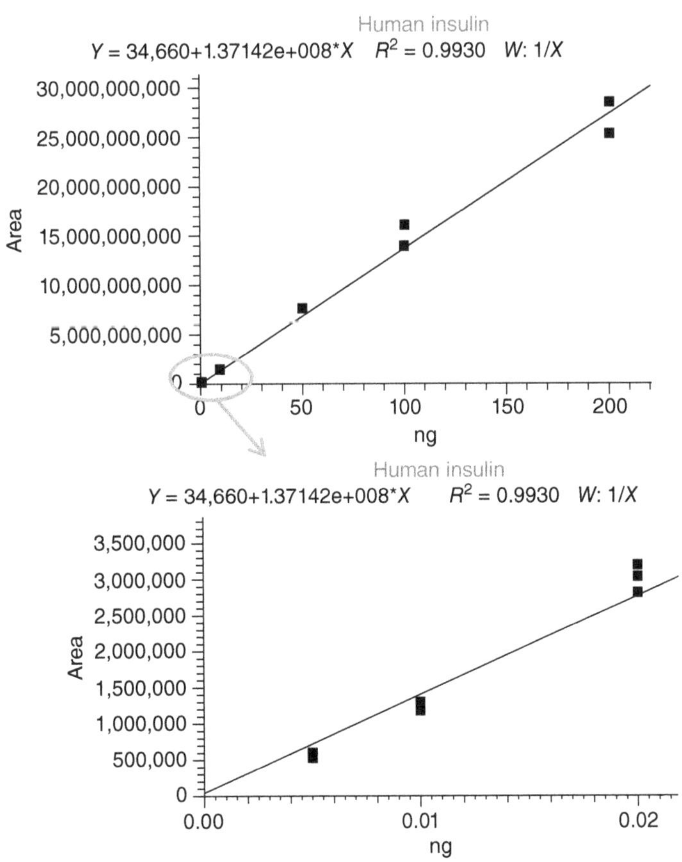

Figure 2.2 Intact quantitation of human insulin by Q Exactive MS.

resolving power, up to 30 kDa protein can be isotopically resolved by 140,000 resolution at m/z 200 on Q Exactive MS. The ultrahigh field Orbitrap further increases the resolution to 240,000 with a faster scan speed, which can resolve up to 50 kDa protein. Accurate quantitation with significantly increased sensitivity is achieved by summing all resolved isotopes at different charge states. Intact mass and certain structural information are provided with full-scan mass spectrum. The disadvantage is that accuracy may be affected by unresolved glycoforms, modifications, and truncation of large intact protein, such as monoclonal antibody (mAb). The linear dynamic range is two to three orders of magnitude depending on the size of protein.

A middle-up strategy fills in the gap between bottom-up and intact approaches. Data is acquired by a full-scan method at very high resolutions (i.e., 140,000 or above). The protein is cleaved into only a few subunits and quantitation is performed using the isotopically resolved high-resolution data of the desired subunit. This method fully utilizes the mass spectrometer's high resolving power capabilities. Absolute quantitation of an mAb can be achieved on a light chain after a simple reduction reaction. Alternatively, an mAb can be quantified by antigen binding fragment (Fab) or crystallizable fragment (Fc/2) generated by combining Ides or papain enzyme cleavage at the hinge region with a reduction reaction. This approach is a powerful way to quantify large proteins and is a very useful technique to quantify bispecific mAb, fusion proteins, and ADCs. Even though it has lower sensitivity compared to bottom-up approach, it is applicable where sensitivity is not critical during drug discovery and development.

2.6 Data Processing

HRAM postacquisition data processing is an area that requires a different knowledge compared to those for triple quadrupole MS users. For discovery research, the identification of peptide targets in proteolytic digested protein samples is achieved by searching HR full-scan and data-dependent MS/MS spectra against SEQUEST or MASCOT databases. The identity of peptide is confirmed by comparing experimental MS/MS spectrum to theoretical one generated from database. The surrogate peptide is usually selected with a unique sequence that has been verified by a basic local alignment search tool (BLAST) search. The length of surrogate peptide is typically 6–20 amino acids, but ideally 8–15, considering ionization and chromatographic retention. The lysine and arginine in the peptide sequence should not be followed by a proline on the carboxyl side to avoid incomplete digestion. Peptides having a stable sequence and the best signal/noise ratio are chosen for targeted quantitation (Zheng et al. 2014).

The extracted ion chromatogram (XIC) of a precursor or fragment ions is typically generated with a mass tolerance window of 5 ppm. The coeluting interferences in biological matrices can be separated from the target ion with the additional selectivity from high-resolution accurate mass data. Even though the charge envelope of large proteins is widely spread in the mass spectrum, resolved isotopes at different charge states can be summed to increase sensitivity. Poor fragmentation efficiency of molecules within the collision cell is not a concern for HR-based intact protein quantitation.

Simpler chromatographic peak integration is possible with HRAM MS, because of the almost nonexistent baseline noise. Conversely, it can require significant time and effort to identify a single set of parameters to properly integrate an entire sample batch for traditional SRM-based LC–MS/MS (Voelker et al. 2013). The unmatched selectivity, better accuracy, precision, and wide linear dynamic range have led to the successful acceptance of HRAM MS for bioanalysis.

2.7 Other Factors That Impact LC–MS-based Quantitation

2.7.1 Sample Extraction to Reduce Matrices

The challenge of increasing LC–MS assay sensitivity can be addressed by improving sample preparation. Insufficient sensitivity of LC–MS results from background noise and signal suppression from endogenous high-abundant proteins, immunoglobulin and albumin. Matrix effects are one of the major issues that affect LC–MS assay sensitivity. It causes signal suppression or enhancement and can change the charge state of peptide analyte in different biological samples. In order to reduce background interference, extraction methods are very important for the method development of proteins/peptides. Depending on the type of protein or peptide samples, protein precipitation, solid-phase extraction (SPE), and affinity purification are the commonly used approaches.

Protein precipitation is usually employed to remove contaminants commonly found in the plasma or serum (Anderson and Anderson 2002). However, it has limited application in the sample preparation of proteins or peptides in biological matrices. Two approaches are possible because of different properties of protein therapeutics. Hydrophilic small proteins and PEGylated proteins are present in the liquid phase following protein precipitation with a mixture of water and organic solvent. The supernatant containing the targeted proteins can be retained for digestion and analysis. Conversely, if the targeted proteins are not water soluble, they are precipitated in the pellet. Further pellet digestion is required. The pellet is suspended in digestion buffer and then digested with enzyme prior to LC–MS analysis. Direct digestion of protein in biological matrices requires minimum method development (Heudi et al. 2008). This approach is simple but results in low sensitivity. It is suitable for assays that do not require high sensitivity.

SPE is widely used for small molecule purification and also well established for peptides (Wang et al. 2012). SPE can be utilized following protein digestion to extract peptides from biological matrices. Depending on the analytes, full method development is needed to select the appropriate sorbent and optimize loading, washing, and eluting conditions. Generally, a dry-down step is required to eliminate excessive volume of elution buffer and subsequent reconstitution of sample into a small volume of solvent for the LC–MS analysis. For protein analysis, peptide loss is noticeable during this step. Microelution plates can reduce this low peptide recovery issue by efficiently eluting the peptide in a small volume that eliminates the need for evaporation and preventing nonspecific binding and adsorption.

Affinity purification provides better assay sensitivity by specifically binding only the target of interest to the reagent antibody. It combines both a sample cleanup and a concentration step. Both protein and peptide affinity purifications have been applied to LC–MS quantification. Protein affinity capture is to enrich targeted proteins and is generally performed before enzymatic digestion. Peptide affinity capture is used to isolate peptide and typically performed after digestion. Affinity purification greatly reduces interferences in plasma and tissue samples after capturing the analyte, which can then be eluted from capture surface or digested directly (Ackermann and Berna 2007). Protein A and G, as generic capture reagents, are widely used in biopharmaceutical laboratories. Protein A is a bacteria-derived protein that binds with high affinity and specificity to the Fc portion of antibodies, particularly those of the IgG class. They bind to Fc-region of antibodies with high affinity, allowing other proteins, lipids, and carbohydrates to be effectively removed. This approach can be utilized to purify proteins containing human Fc-region, such as mAbs, Fc-fusion protein, and ADCs. Based on this concept, Li et al. 2012 developed a generic therapeutic mAb quantitation method with the universal whole antibody being internal standard and applied it to preclinical studies.

The available offline forms of affinity purification are magnetic beads, mass spectrometric immunoassay (MSIA) tips, agarose resin or resin-based spin column and plates. Magnetic beads can be used both manually with a magnetic stand and with automated platforms such as Thermo Scientific KingFisher Instruments for high-throughput workflows. MSIA tips enable effective removal of nonspecific binding and have the flexibility for manual operation or automation with Versette liquid handling platform. Online column-based enrichment methods are typically used to enrich peptides. Excess amounts of capture reagent relative to highest concentration of analyte are needed for both offline and online modes. The online mode involves the regeneration of affinity capture reagent. Therefore, it is necessary to evaluate carryover and recovery (Jenkins et al. 2014).

2.7.2 Internal Standard

Selection of a proper internal standard (IS) is critical for protein quantitation. For LC–MS analysis, it is widely utilized to correct the variation during sample preparation and the ionization efficiency resulting from matrix effects. The type of IS and spiking order influence the assay result. The ideal internal standard is stable isotope-labeled (SIL) protein/peptides. ^{15}N, ^{13}C, or ^{18}O are preferred over deuterium mainly because deuterium-labeled analyte undergoes hydrogen deuterium exchange in aqueous solutions and the chromatographic retention time shift with multiple labeling, resulting in reduced compensation for matrix effect. To ensure enough selectivity between the IS and analyte, the degree of labeling for internal standard is usually taken into consideration for triple quadrupole-based quantitation (Li et al. 2012). The reason is that multiply charged peptide ion in mass spectrum decreases the mass difference between IS and analyte. However, this is not an issue on HRMS, due to the fact that ions can be separated within 0.001 Da.

A white paper (Jenkins et al. 2014) summarized different internal standards for MS protein quantitation. In general, three types of SIL IS have been used for peptide quantitation. SIL protein is generated by a stable isotope labeling with amino acids in cell culture (SILAC) technique during cell culture. SILAC means incorporating heavy isotope-labeled amino acids into protein sequence during recombinant synthesis (Ong et al. 2002). The SIL protein is spiked into sample at the beginning of sample preparation, and it can correct for analyte loss and the deviations that may occur at each step including purification, digestion efficiency, and matrix effect. Multiple peptides are labeled during cell culture. SIL protein is applicable to any type of quantitation approach. The shortcoming of SILAC-introduced SIL proteins is time-consuming biological synthesis as it takes several cell culture cycles to fully incorporate SIL amino acids in cells. In addition, the product has minor structure changes to protein analytes.

Flanking SIL peptide standards and SIL peptides using chemically synthesized IS are commonly used for bottom-up approach because of its commercial availability.

Flanking SIL peptide is the SIL peptide analyte with extra amino acids on either the C- or N-terminal. It is added to the samples before digestion to minimize variation from lot-to-lot digestion efficiency. Following digestion, the additional amino acids are cleaved to form SIL peptide for subsequent LC–MS quantitation. SIL peptide can be spiked either before or after digestion. It may correct any digestion-related instability issues if added before digestion but not digestion efficiency. Sometimes, nonlabeled protein/peptide analogs with very similar structure are selected when a SIL internal standard is not available. The advantage of this approach is to potentially correct for any matrix effects. A universal whole antibody with SIL amino acids is ideal for bottom-up, middle-up, and intact protein quantitation.

2.8 Conclusion and Perspectives of LC–HRMS in Regulated Bioanalysis

To achieve a high sensitivity LC–MS assay, each step from sample preparation, chromatographic separation, MS data acquisition to postdata processing should be evaluated. The advantages of high-resolution accurate MS in protein therapeutics quantitation have been discussed. However, one aspect that must be considered is the data file size when using HR instruments. Data storage is still the challenge for most laboratories and need to be addressed. By minimizing the acquisition window of HRMS for only the compound of interest, the data file size can be significantly decreased.

High-resolution accurate MS is currently primarily used in nonregulated drug discovery of protein therapeutics. It has been demonstrated as a very important tool to assist bioanalytical methods of mAb and ADC. Because both targeted protein and other critical quality attributes, such as posttranslational modifications and truncations, need to be considered during method development. The Quan/Qual feature is particularly important for surrogate peptide screening, catabolites, structural modifications identification, and pharmacokinetics studies. Simultaneous Quan/Qual approach is an attractive option in drug discovery considering its cost-effective feature to screen candidates at early stage.

When protein drugs move into the development stage, more regulated quantitation assays are involved. LC–MS is increasingly accepted as an important and complementary technique to LBAs for protein quantitation. However, unlike small molecule LC–MS quantitation and LBA methods, there is no regulatory guidance available at present because it is a relatively new technique for protein therapeutics bioanalysis. A recently published white paper from industry experts provides recommendations regarding LC–MS/MS for bioanalysis of protein therapeutics at the peptide level in a regulated environment (Jenkins et al. 2014). New FDA validation guidance from Crystal City V meeting, however, only focuses on small molecule bioanalysis.

Recent works demonstrated that HRMS could meet the regulated bioanalysis validation acceptance criteria in terms of sensitivity, precision, accuracy, selectivity, and matrix effect for small molecules (Fung et al. 2011, Voelker et al. 2013). The shift from LC–MS/MS to LC–HRMS for small-molecules bioanalysis has started in DMPK/drug discovery and clinical laboratories (Ramanathan et al. 2011, Josephs 2012, Ramanathan and Korfmacher 2012, Rochat et al. 2012, Huang et al. 2013). LC–HRMS may change the current landscape of bioanalysis and drug metabolism due to the current shift from LC–MS/MS to LC–HRMS in regulated bioanalysis. New-generation HRMS has shown its tremendous capability for quantitating peptide and protein therapeutics (Dillen et al. 2012, Ciccimaro et al. 2014). This demonstrated great potential of HRMS for both method development and validation. Although many technical differences on method development and validation must be considered for protein therapeutic quantitation, we believe that the increased adoption of HRMS in bioanalytical laboratories will help promote the need for a regulatory guidance.

References

Ackermann BL and Berna MJ (2007). "Coupling immunoaffinity techniques with MS for quantitative analysis of low-abundance protein biomarkers." *Expert Rev Proteomics* **4**(2): 175–186.

Anderson NL and Anderson NG (2002). "The human plasma proteome: history, character, and diagnostic prospects." *Molec Cell Proteomics* **1**(11): 845–867.

Beck A, Wurch T, Bailly C, et al. (2010). "Strategies and challenges for the next generation of therapeutic antibodies." *Nat Rev Immunol* **10**(5): 345–352.

Ciccimaro E, Ranasinghe A, D'Arienzo C, et al. (2014). "Strategy to improve the quantitative LC–MS analysis of molecular ions resistant to gas-phase collision induced dissociation: application to disulfide-rich cyclic peptides." *Anal Chem*, **86**(23): 11523–11527.

Damen CWN, de Groot ER, Heij M, et al. (2009). "Development and validation of an enzyme-linked immunosorbent assay for the quantification of trastuzumab in human serum and plasma." *Anal Biochem* **391**(2): 114–120.

DeSilva B, Smith W, Weiner R, et al. (2003). "Recommendations for the bioanalytical method validation of ligand-binding assays to support pharmacokinetic assessments of macromolecules." *Pharm Res* **20**(11): 1885–1900.

Dillen L, Cools W, Vereyken L, et al. (2012). "Comparison of triple quadrupole and high-resolution TOF-MS for quantification of peptides." *Bioanalysis* **4**(5): 565–579.

Ewles M and Goodwin L (2011). "Bioanalytical approaches to analyzing peptides and proteins by LC-MS/MS." *Bioanalysis* **3**(12): 1379–1397.

Ezan E and Bitsch F (2009). "Critical comparison of MS and immunoassays for the bioanalysis of therapeutic antibodies." *Bioanalysis* **1**(8): 1375–1388.

Fung EN, Xia Y-Q, Aubry A-F, et al. (2011). "Full-scan high resolution accurate mass spectrometry (HRMS) in regulated bioanalysis: LC-HRMS for the quantitation of prednisone and prednisolone in human plasma." *J Chromatogr B* **879**(27): 2919–2927.

Heudi O, Barteau S, Zimmer D, et al. (2008). "Towards absolute quantification of therapeutic monoclonal antibody in serum by LC-MS/MS using isotope-labeled antibody standard and protein cleavage isotope dilution mass spectrometry." *Anal Chem* **80**(11): 4200–4207.

Hu Q, Noll RJ, Li H, et al. (2005). "The Orbitrap: a new mass spectrometer." *J Mass Spectrom* **40**(4): 430–443.

Huang M-Q, Lin Z and Weng N (2013). "Applications of high-resolution MS in bioanalysis." *Bioanalysis* **5**(10): 1269–1276.

Jemal M (2000). "High-throughput quantitative bioanalysis by LC/MS/MS." *Biomed Chromatogr* **14**(6): 422–429.

Jenkins R, Duggan J, Aubry A-F, et al. (2014). "Recommendations for validation of LC-MS/MS bioanalytical methods for protein biotherapeutics." *AAPS J* **17**(1): 1–16.

Josephs JL (2012). "HRMS: current usage, future directions and the promise of integration with unified data streams suited to post-acquisition mining." *Bioanalysis* **4**(5): 471–476.

Kaufmann A, Butcher P, Maden K, et al. (2010). "Comprehensive comparison of liquid chromatography selectivity as provided by two types of liquid chromatography detectors (high resolution mass spectrometry and tandem mass spectrometry): Where is the crossover point?." *Anal Chim Acta* **673**(1): 60–72.

Korfmacher W (2011). "High-resolution mass spectrometry will dramatically change our drug-discovery bioanalysis procedures." *Bioanalysis* **3**(11): 1169–1171.

Korfmacher WA (2005). "Foundation review: principles and applications of LC-MS in new drug discovery." *Drug Discov Today* **10**(20): 1357–1367.

Lange V, Picotti P, Domon B, et al. (2008). "Selected reaction monitoring for quantitative proteomics: a tutorial." *Molec Syst Biol* **4**: 222.

Leader B, Baca QJ and Golan DE (2008). "Protein therapeutics: a summary and pharmacological classification." *Nat Rev Drug Discov* **7**(1): 21–39.

Li H, Ortiz R, Tran L, et al. (2012). "General LC-MS/MS method approach to quantify therapeutic monoclonal antibodies using a common whole antibody internal standard with application to preclinical studies." *Anal Chem* **84**(3): 1267–1273.

Marshall AG, Hendrickson CL and Jackson GS (1998). "Fourier transform ion cyclotron resonance mass spectrometry: a primer." *Mass Spectrom Rev* **17**(1): 1–35.

Ong S-E, Blagoev B, Kratchmarova I, et al. (2002). "Stable isotope labeling by amino acids in cell culture, silac, as a simple and accurate approach to expression proteomics." *Molec Cell Proteomics* **1**(5): 376–386.

Pillay V, Gan HK and Scott AM (2011). "Antibodies in oncology." *New Biotechnol* **28**(5): 518–529.

Ramanathan R, Jemal M, Ramagiri S, et al. (2011). "It is time for a paradigm shift in drug discovery bioanalysis: from SRM to HRMS." *J Mass Spectrom* **46**(6): 595–601.

Ramanathan R and Korfmacher W (2012). "The emergence of high-resolution MS as the premier analytical tool in the pharmaceutical bioanalysis arena." *Bioanalysis* **4**(5): 467–469.

Rochat B, Kottelat E and McMullen J (2012). "The future key role of LC high-resolution-MS analyses in clinical laboratories: a focus on quantification." *Bioanalysis* **4**(24): 2939–2958.

Ruan Q, Ji QC, Arnold ME, , et al. (2011). "Strategy and its implications of protein bioanalysis utilizing high-resolution mass spectrometric detection of intact protein." *Anal Chem* **83**(23): 8937–8944.

Scott AM, Wolchok JD and Old LJ (2012). "Antibody therapy of cancer." *Nat Rev Cancer* **12**(4): 278–287.

Voelker T, Wang H, Irish M, et al. (2013). "Method development and validation of six bile acids for regulated bioanalysis: improving selectivity and sensitivity." *Bioanalysis* **5**(10): 1229–1248.

Voynov V, Caravella JA and Dimitrov DS (2012). Therapeutic proteins, in *Therapeutic Proteins, Methods in Molecular Biology*, Humana Press, Clifton, NJ, Vol. **899**: pp. 1–26.

Wang H and Bennett P (2013). "Performance assessment of microflow LC combined with high-resolution MS in bioanalysis." *Bioanalysis* **5**(10): 1249–1267.

Wang H, Cao H and Wang Y (2010). "Quantification of n^2-carboxymethyl-2′-deoxyguanosine in calf-thymus DNA and cultured human kidney epithelial cells by capillary HPLC-tandem mass spectrometry coupled with stable isotope-dilution method." *Chem Res Toxicol* **23**(1): 74–81.

Wang Y, Qu Y, Bellows CL, et al. (2012). "Simultaneous quantification of davalintide, a novel amylin-mimetic peptide, and its active metabolite in beagle and rat plasma by online SPE and LC-MS/MS." *Bioanalysis* **4**(17): 2141–2152.

Zheng J, Mehl J, Zhu Y, et al. (2014). "Application and challenges in using LC-MS assays for absolute quantitative analysis of therapeutic proteins in drug discovery." *Bioanalysis* **6**(6): 859–879.

Zhou S, Song Q, Tang Y, et al. (2005). "Critical review of development, validation, and transfer for high throughput bioanalytical LC-MS/MS methods." *Curr Pharm Anal* **1**(1): 3–14.

3

Current Methods for the Characterization of Posttranslational Modifications in Therapeutic Proteins Using Orbitrap Mass Spectrometry

Zhiqi Hao[1], Qiuting Hong[2], Fan Zhang[4], Shiaw-Lin Wu[3,4], and Patrick Bennett[5]

[1] Thermo Fisher Scientific, San Jose, CA, USA
[2] Eurofins Lancaster Laboratories, Inc., Lancaster, PA, USA
[3] BioAnalytix Inc, Cambridge, MA, USA
[4] Northeastern University, Boston, MA, USA
[5] PPD Bioanalytical Lab, Richmond, VA, USA

3.1 Introduction

Protein-based therapeutics were introduced in the 1920s, when the first therapeutic protein, insulin, was purified from animal pancreases to treat diabetes. However, the availability, cost, and immunogenicity largely precluded the use of animal-derived proteins for medical treatments for a long period of time. Today, most of the protein therapeutics are recombinant proteins made by recombinant DNA technology, which not only allows mass production of proteins but also enables engineered proteins with well-defined characteristics and high level of consistency. Since the first introduction of recombinant therapeutic proteins in the 1980s, the recombinant human insulin, protein-based therapeutics have been highly successful in clinics (Pavlou and Reichert 2004). Recombinant monoclonal antibodies (mAbs) and their derivatives have become the fastest growing class of human therapeutics and drug candidates because of their long circulation lifetime, high selectivity in binding to the targeted antigen, low levels of toxicity, and the ability to invoke immune cell effector responses. There has been a dramatic increase of mAb-based drug development in the past two decades. These mAb-based pharmaceuticals include "naked" antibodies, radioimmunoconjugates, antibody–drug conjugates (ADCs), bispecific antibodies, Fab fragments, Fc-fusion proteins/peptides, and immunocytokines (Beck et al. 2010, 2013). The fast development of protein therapeutics has resulted in an increasing demand for protein characterization techniques.

Unlike the traditional drugs that are chemically synthesized homogenous small molecules, therapeutic proteins are heterogeneous, complex macromolecules that are produced by living cells. Various posttranslational modifications (PTM) and degradations on amino acid residues (i.e., oxidation, deamidation, glycation, glycosylation, formation of disulfide bonds, C-terminal lysine processing) could generate minor variants and cause heterogeneity. Since modifications such as these can potentially change the efficacy and safety profiles of a drug, characterization of site-specific modifications is an important part of protein therapeutics development and manufacturing process control.

Several different techniques are frequently applied to study protein modifications and minor variants. Most of these techniques are chromatography or electrophoresis methods aiming at resolving the antibody main isoform from its minor variants. Size exclusion chromatography (SEC), sodium dodecyl sulfate-gel electrophoresis (SDS-PAGE), and capillary electrophoresis-sodium dodecyl sulfate (CE-SDS) separate proteins on the basis of the protein size, while isoelectric focusing (IEF) and ion exchange chromatography (IEX) are on the basis of the global charge state. Hydrophobic interaction chromatography (HIC) and reversed-phase high-performance liquid chromatography (RP-HPLC) are based on the protein hydrophobicity (Little et al. 2006, Yang et al. 2007, Rustandi et al. 2008, Lu et al. 2013). Most of these techniques, although they may be quantitative, do not provide information on the type, nor on the site of sequence modifications. Recent advancements in mass spectrometry (MS) over the past two decades have enabled scientists to address the biophysical and biochemical properties of proteins by providing sequence and structure-related information. MS-based methodologies have become one of the most highly utilized analytical techniques from early development to the manufacturing of protein-based pharmaceutics, providing information on primary sequences, the type and site of PTM, as well as

Protein Analysis using Mass Spectrometry: Accelerating Protein Biotherapeutics from Lab to Patient, First Edition.
Edited by Mike S. Lee and Qin C. Ji.

higher order structures and conformations (Bondarenko et al. 2009, Zhang et al. 2009, Chen 2013).

MS characterization of protein can be performed either at intact protein level or at peptide level. Although dissociation of intact proteins into smaller fragment ions using tandem spectrometry (MS/MS) was reported to provide information of site-specific modifications for small proteins, MS/MS of intact proteins of over 20–30 kDa in size is challenging (Kelleher et al. 1999, McLafferty et al. 2007). The challenge of intact protein analysis is mainly because of reduced solubility and poor separation of intact proteins under typical LC conditions, as well as the relatively low sensitivity of mass spectrometers for intact proteins. Moreover, as the size of protein increases, the tertiary structure of proteins becomes more difficult to disrupt, thereby limiting the MS/MS fragmentation efficiency (Steen and Mann 2004, McLafferty et al. 2007). Therefore, MS characterization of residue modification is usually performed at the peptide level using a bottom-up approach, in which the protein is enzymatically digested into peptides followed by RP-HPLC separation and MS analysis (Lewis et al. 1994, Chelius et al. 2008). Other separation techniques can also be coupled to MS for protein characterization, such as capillary electrophoresis coupled to mass spectrometry (CE-MS), SEC-MS and ion mobility mass spectrometry (IM-MS) (Chen et al. 2011, Pioch et al. 2012). Besides MS measurement of peptide ions, MS/MS experiments, whereby precursor peptides are fragmented followed by detection of the product ions, are performed to generate sequence information of the peptides (Chen et al. 2011, Tao et al. 2011). For data analysis, experimental MS/MS spectra are compared with the predicted, *in silico* generated fragmentation patterns of peptides.

An MS instrument usually consists of ion source and optics, a mass analyzer, a detector, and the data processing electronics. Some mass analyzers, such as the ion trap, Orbitrap™ analyzer, and the ion cyclotron resonance (ICR) mass analyzers separate ions based on their *m/z*-dependent oscillation frequency. Quadrupoles separate ions based on *m/z*-dependent stability, while time-of-flight (TOF) analyzers are based on flight time separation. A review of different types of mass analyzers is out of the scope of this chapter; the readers are therefore referred to other review articles (Gross 2004, Yates 2004, Han et al. 2008). This chapter is focused on the current Orbitrap MS methods for characterization of sequence modifications in therapeutic proteins.

Since its launch in 2005, Orbitrap MS has become established as one of the most powerful tools for protein identification and characterization. Orbitrap borrowed several important principles from other previously existing mass analyzers, including the image current detection from Fourier transform-ion cyclotron resonance (FT-ICR), the pulsed injection and the use of electrostatic field from TOF analyzer, and the ion trapping in precisely defined electrode structures as practiced in radiofrequency (RF) ion traps (Zubarev and Makarov 2013). Compared to the ultrahigh-resolution FT-ICR, Orbitrap provides comparable resolving power and mass accuracy without the need for large and expensive superconductive magnets. Orbitrap also provides a wider dynamic range of mass accuracy than TOF and lower space charge effects than RF ion trap (Makarov et al. 2006a, b).

Some key features of the Orbitrap mass analyzer include high resolving power (up to 500,000 in serial instruments), high mass accuracy (<3 ppm), high sensitivity, and wide dynamic range (over 5e3 for intrascan and over 5e5 for interscan dynamic range) (Makarov et al. 2006a, b, Hao et al. 2012, Michalski et al. 2012). One of the critical technologies, the automatic gain control (AGC), ensures that Orbitrap is always filled with the optimum number of ions (the AGC target value) for all scan functions (Makarov et al. 2006a, b). In practice, this is achieved by estimating from a short prescan the time needed to fill the ions into the trap [ion injection time (IT)] based on the ion flux and the AGC target value. The use of AGC ensures that the key features of Orbitrap instrument, including high resolving power, high mass accuracy, high sensitivity, wide dynamic range as well as quantitation linearity, are achieved to the optimum level. Another advantage of the latest Orbitrap instrumentation is that three different fragmentation techniques, collision-induced dissociation (CID), higher-energy collision dissociation (HCD), and electron transfer dissociation (ETD), are available in the same instrument. The combined use of complementary dissociation methods have been proven to improve the number of identified peptides and PTMs (Shen et al. 2011, Guthals et al. 2013).

In the following two sections, we describe the current development of Orbitrap methods for the characterization of protein modifications to support the discovery and development of protein therapeutics in the pharmaceutical industry. The first section is focused on routine peptide sequencing and PTM analysis using higher-energy collision dissociation approach. Data presented in this section was generated with a Q Exactive™ instrument using a standard reference mAb, NISTmAb, supplied by the National Institute of Standards and Technology. The second section presents a novel approach designed to identify PTMs that are labile to collision dissociation. Results presented in the second section were obtained from a study of tissue plasminogen activator (TPA) and a recombinant TPA (TNK) with an Orbitrap Fusion™ Tribrid™ instrument. The advantages of the combined use of multiple fragmentation mechanisms in an intelligent way are presented and discussed.

Recommendations for best practice are provided with special attention paid to key MS parameters. The approaches described here are generally applicable to other therapeutic proteins.

3.2 Characterization of PTMs Using Higher-Energy Collision Dissociation

In MS/MS experiments, peptide precursor ions are selected and fragmented to generate product ions for detection. Figure 3.1 illustrates the ion pairs that can be generated from backbone bond fragmentation when MS/MS is performed on a peptide. With CID, which is by far the most widely used fragmentation technique for peptide sequencing, the major product ions are b and y fragment ions. From the spectrum of the fragment ions, identity of a peptide and PTMs can be obtained. The sites of PTMs can be located if near-complete ion series from peptide backbone fragmentation can be detected. When only limited variety of ions from backbone fragmentation can be detected, localization of PTMs is challenging although in some cases it can be achieved when knowledge of potential PTM sites is available. Two types of collision dissociation, CID and HCD, are available in Orbitrap instrument. CID is performed in the quadrupole linear ion trap, while HCD is performed in a separate multipole collision cell. Despite the name, the collision energy of HCD is typically in the regime of low-energy collision (less than 1000 eV) but higher than the energy in CID. CID offers high sensitivity and fast scanning speed. However, fragment ions of lower than 25–30% mass-to-charge (m/z) value of that of the precursor ions are not trapped efficiently (Olsen et al. 2007). HCD does not suffer from the low mass cutoff spectrum while it shows similar fragmentation pattern with CID (Olsen et al. 2007). Detection of HCD fragment ions are usually in the Orbitrap, generating high-resolution, high mass accuracy MS/MS spectra, which lead to accurate and confident peptide identification. CID has been established as a powerful and robust method for peptide MS/MS, while routine application of HCD is still relatively new in the pharmaceutical industry. In this section, the application of HCD on peptide sequencing and PTM analysis is presented using NISTmAb as a model protein.

Typically, a data-dependent MS/MS instrument method is used for peptide sequencing and PTM localization. Each scan cycle of the method includes a full MS survey scan followed by a series of MS/MS, each for one of the several most intense ions determined by the previous survey scan. Dynamic exclusion is usually enabled to cover the duration of LC elution peak (usually 20–30 s) so that in the next scan cycle the instrument will choose another set of ions for MS/MS rather than repeating the same set of ions as the previous scan cycle. If the instrument scan rate is fast enough, then dynamic exclusion will allow low-abundant species with PTMs selected for MS/MS. The new-generation Orbitrap instruments are capable of performing 10 (the Q Exactive instrument), or 20 (the Q Exactive HF instrument) data-dependent MS/MS in 1 s along with the survey full scan, which allows identification of very-low-abundant protein modifications, sequence variants, or impurities. The resolution settings for the survey scan and MS/MS scan of a bottom-up data-dependent experiment is usually 60,000–70,000 and 15,000–17,500, respectively. AGC target values for survey scan and MS/MS scans are recommended by the instrument software method editor. For Q Exactive instrument series, target values of 1–3e6 for full MS survey scan and 1–2e5 for MS/MS scan are the routine choices. To ensure a good quality MS/MS spectrum for low-abundant peptides, maximum IT for MS/MS scan is set to at least 100 ms, sometimes even higher, to allow enough ions filled into trap when the ion flux is low. The actual IT for ion injection, which is determined by instrument control

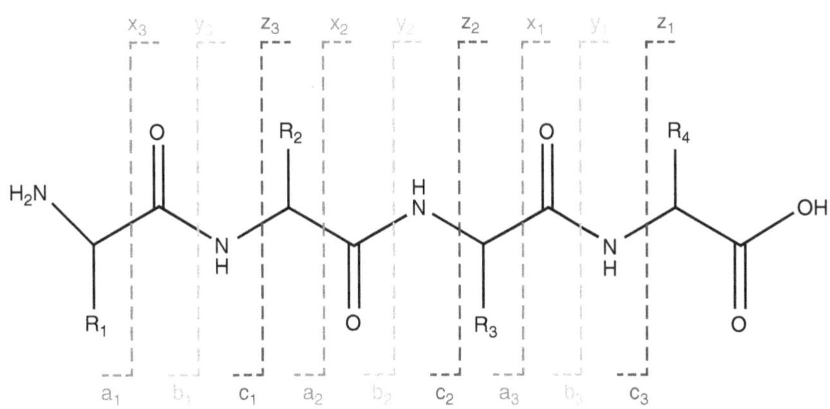

Figure 3.1 The most commonly used nomenclature for peptide fragmentation.

software based on the AGC setting and the ion flux, is much shorter when the ion flux is strong. Therefore, the maximum IT is reached only when the ion flux is low. A normalized collision energy of 27 is typically used for fragmentation of peptides. Besides the MS parameters, ionization conditions can affect spectrum quality. The temperature of the heated electrospray ionization (HESI) source probe, gas flow, and spray voltage are flow rate-dependent and are usually optimized during tuning of the instrument based on the spray stability and signal intensity.

Tryptic digest of NISTmAb was analyzed using a 60-min data-dependent top 10 LC–MS/MS method on a Q Exactive quadrupole-Orbitrap instrument. In each scan cycle, 10 most intense precursor ions were selected for HCD with dynamic exclusion. The raw data were analyzed using PepFinder 2.0 software for peptide and site-specific PTM identification. Besides 100% sequence coverage for both light and heavy chain, a list of PTMs and variants were identified including oxidation, deamidation, glycation, C-terminal Lys truncation, and N-terminal pyroglutamine. Some of the PTMs were less than 0.1% in abundance compared to the unmodified peptides species (data not shown). The following is a detailed discussion on the characterization of two common PTMs, oxidation and deamidation.

3.2.1 Oxidation

Oxidation is a common nonenzymatic modification of protein, which can occur during the manufacture process and storage. The amino acid methionine (Met) and cysteine (Cys) are most susceptible to oxidation, while oxidations of tryptophan (Trp), histidine (His), and tyrosine (Tyr) are also observed. However, Cys oxidation is less observed in proteins because free thiol groups in Cys are prone to form disulfide bonds. Oxidation of Met yields Met sulfoxide, in some extreme cases Met sulfone (double oxidation), causing a mass increase of 16 and 32 Da, respectively. Met oxidation also makes the hydrophobic site chain more polar, leading to change in liquid chromatography profile. It has been observed that proteins that contain oxidized Met elute later on weak cation exchange column but earlier on HIC and RP stationary phases (Liu et al. 2008a, b). Met oxidation is readily detected and was found to increase in a couple of circumstances, such as long-term storage, incubation with oxidants, exposure to UV light, and elevated temperature (Lam et al. 1997, Chumsae et al. 2007, Liu et al. 2008a, b). Oxidation of Trp may yield several different products (e.g., 5-hydroxy-Trp (Δmass = +16 Da), Kynurenine (Δmass = +4 Da), and *N*-formylkynurenine (Δmass = +32 Da)) (Yang et al. 2007). His oxidation could also yield multiple products but mainly oxo-His. Both Try and His oxidation are believed to be metal-catalyzed or free-radical-mediated (Ji et al. 2009). Oxidation on these two residues is usually very slow and low abundant, but is found to increase when protein is exposed to catalyst or oxidants.

In therapeutic mAbs, oxidation of Met and Trp are the most common chemical modifications that occur during purification, formulation, and storage processes (Chumsae et al. 2007, Yang et al. 2007, Ji et al. 2009). Met oxidation in Fc could decrease bioactivity and stability of IgGs (Wang et al. 2011, Hmiel et al. 2015). Trp oxidation in the CDR can be a significant issue and has been shown to correlate with activity loss (Qi et al. 2009, Hmiel et al. 2015). Therefore, the sites of oxidation need to be characterized and their abundance relative to the unmodified peptide should be monitored for product quality and safety assurance. Shown in Figure 3.2 is the characterization of Met oxidation in NISTmAb. HCD of the oxidized peptide yielded the same fragment ions as the wild type except for those that contain the Met residue (e.g., b4 and y15 fragment ions), which shows an increase of 15.99 Da. The level of oxidation at this site is only about 1% relative to the nonmodified peptide based on the ratio of peak areas of the extracted ion chromatogram (data not shown). As shown here, AGC technology allowed accumulation of a sufficient number of ions within the defined time period, providing high-quality spectrum for the characterization of low-abundant PTMs.

3.2.2 Deamidation

Deamidation of asparagine (Asp) residue is another common nonenzymatic modification of proteins, which can significantly impact protein structure and function. Identifying and monitoring the level of deamidation are important in product characterization during protein therapeutics development and production. It is found that protein local structure and external factors, such as temperature or pH, play important roles in the rate of deamidation (Patel and Borchardt 1990, Xie and Schowen 1999). Asp residue followed by glycine (Gly) was found to be the most susceptible site for deamidation (Stephenson and Clarke 1989, Tonie Wright and Urry 1991, Liu et al. 2009). Increased deamidation level was observed for proteins under basic pH conditions (Patel and Borchardt 1990, Song et al. 2001). At neutral and basic pH, deamidation proceeds via formation of a five-member ring aspartyl succinimide (Asu) intermediate (Vlasak et al. 2009). The unstable Asu is hydrolyzed into a mixture of two isomers, isoAsp and Asp. The ratio between isoAsp and Asp in short peptides is usually 3:1 to 4:1. However, as a natural amino acid, Asp can be predominant in proteins. When deamidation occurs at

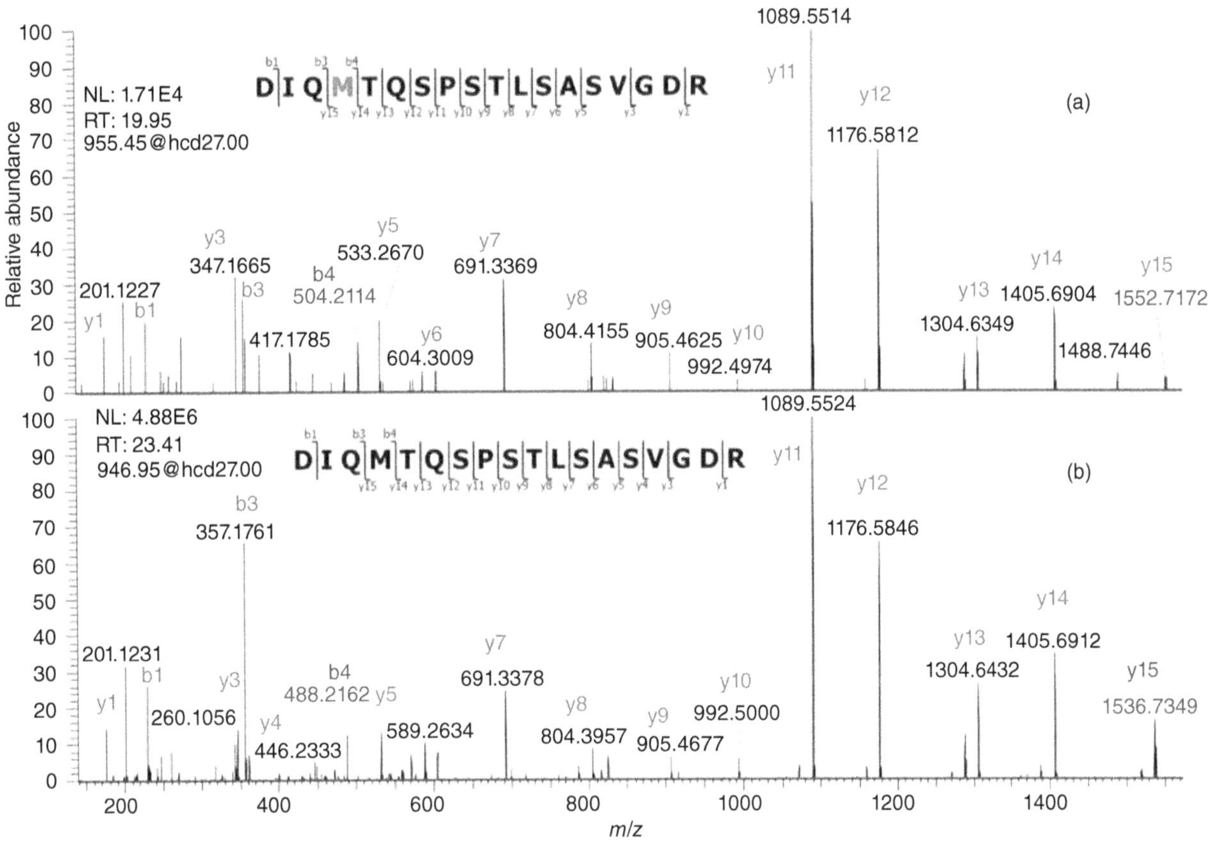

Figure 3.2 Characterization of methionine oxidation in NISTmAb. HCD spectra of the oxidized peptide at Met residue (a), and its unmodified peptide, DIQMTQSPSTLSASVGDR (b). The 15.99 Da mass shift on b4 and y15 fragment ions indicates oxidation at the methionine residue.

acidic pH, Asp is usually the only observed product, presumably produced from direct hydrolysis of the side chain amide (Ren et al. 2009). The hydrolysis of Asu into Asp and isoAsp occurs during sample handling and, especially, under conventional proteolytic digestion condition when prolonged exposure to alkaline pH is required. The result is, therefore, the increased amount of Asp and isoAsp and decreased amount of Asu in the sample. Thus, to accurately characterize deamidation level in drug substance during drug development, special care must be taken to reduce artifactual overdeamidation during sample preparation. A high-fidelity peptide mapping procedure for the study of succinimide in mAbs was developed using reduction under denaturing conditions at pH 5.0, and digestion of antibody with porcine trypsin at pH 7.0. This experimental approach allows correct identification and quantitation of relative amounts of Asu, isoAsp, and Asp (Yu et al. 2011).

Deamidation causes a mass increase of 0.984 Da, which results in an overlapped isotope profile between the non-deamidated and the deamidated species of the same peptide. As shown in Figure 3.3, the monoisotopic mass of the deamidated peptide is 7.4 ppm different from the mass of the second isotope of the same peptide without deamidation. Differentiation of these two peaks would require MS resolving power of above 280 K (at *m/z* 200), which is not an option in a standard Q Exactive instrument. Thus, if these two peptides are not chromatographically separated, an isotope profile that consists of merged peaks will be obtained in the MS spectrum. Therefore, for confident identification and accurate quantification of deamidation by MS technology, a good LC separation is necessary of the deamidated peptide from the same peptide without deamidation. Shown in Figure 3.4 are the HCD spectra of a NISTmAb peptide with and without deamidation. A mass increase of 0.98 Da was observed for fragment ions that contain Asp11, for example, y6–y10 ions, indicating deamidation at Asp11, rather than Asp12.

The data-dependent HCD method presented here provides informative and high-quality spectra for peptide sequencing. For peptides containing PTMs that are not labile, this approach usually generates complete or near complete fragment ion series providing sufficient site-specific information to localize PTMs. However, CID or HCD does impose some limitations in identifying

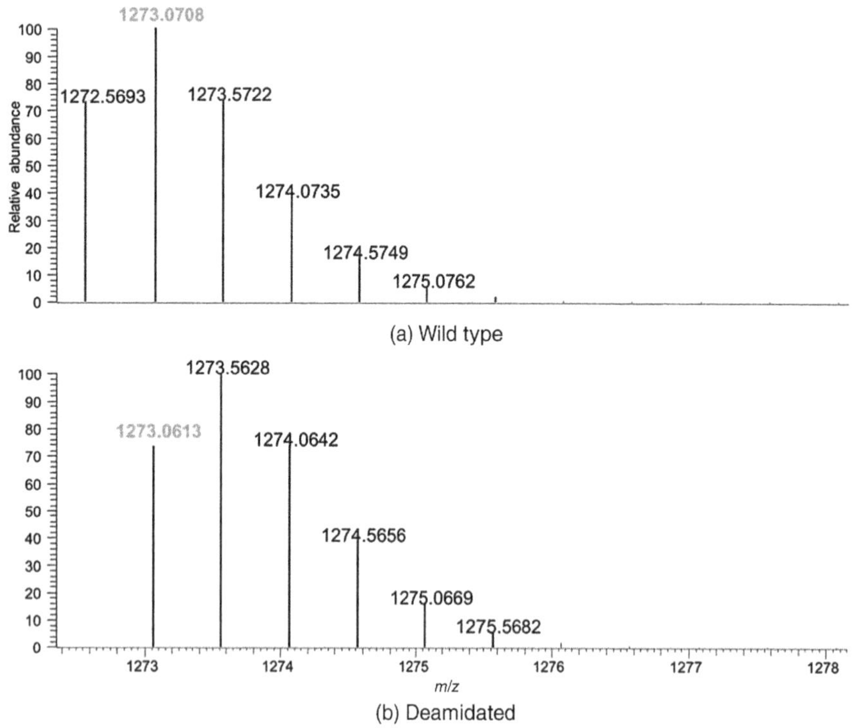

Figure 3.3 Simulated isotopic profile of the deamidated and nondeamidated version of NISTmAb peptide, GFYPSDIAVEWESNGQPENNYK ($[M+2H]^{2+}$). The isotopic peaks are too close in mass between the two peptides to be separated by a standard Q Exactive. The separation of the two species by chromatography is necessary, especially for accurate quantitation of relative abundance.

and localizing labile PTMs. PTMs of certain residues of a peptide can redirect the sites of preferred cleavage during CID or HCD, often resulting in cleaving the modifying moiety and leaving the peptide backbone intact. Examples of these "MS/MS labile" PTMs include glycosylation, phosphorylation, sulfonation, and nitrosylation (Mirza et al. 1995, Mann and Jensen 2003, Mikesh et al. 2006). Loss of these labile PTMs during CID or HCD results in spectra lacking product ions with the labile PTM moiety retained, therefore, reduced success rates in sequence determination and PTM localization. An alternative dissociation method, ETD, is advantageous for characterization of labile PTMs because ETD preserves PTMs so that peptide sequence information can be obtained. In the following section, the application of ETD on characterization of labile PTMs is described using glycopeptides as examples.

3.3 Application of Electron Transfer Dissociation to the Characterization of Labile PTMs

ETD fragments peptides by transferring an electron from a radical anion to a protonated peptide. This rapid neutralization of the charged site by an electron leads to the generation of a radical, which in turn induces fragmentation of the peptide backbone, causing cleavage of the Cα–N bond. ETD generates complementary c- and z-type ions instead of the typical b- and y-type ions observed in CID or HCD (Figure 3.1) (Syka et al. 2004, Mikesh et al. 2006). The two types of fragmentation methods have different preference with regard to *m/z* values and amino acids. CID or HCD works best for doubly or triply charged peptides with less charge density (higher *m/z*), while ETD performs the best for peptides with more than two charge and of higher charge density (lower *m/z*). The peptide bond at the aminoterminal end of proline is the preferred dissociation site under CID, whereas dissociation at the aminoterminal end of proline is almost impossible with ETD because two bonds have to be broken to generate fragment ions. Due to the complementary nature of CID (or HCD) and ETD, the combined use of both methods has been shown to provide much more identification than a single fragmentation method alone (Molina et al. 2008, Swaney et al. 2008, Frese et al. 2011).

One of the major advantages of ETD in MS-based protein characterization is its ability to localize the exact site of PTMs that can be missed by CID or HCD (Good et al. 2007, Wiesner et al. 2008). Although ETD was initially described as an advantageous tool for phosphoproteomics, it has been applied to many other labile PTMs, including those that are important in protein therapeutics: O- or, N-linked glycosylation and glycation. ETD has also been applied to differentiate Asp and isoAsp (Kim and Pandey 2012). In some cases, such as glycation, O-GlcNAc and isoAsp, ETD can be used as the primary fragmentation method. In the case of N- and O-linked glycopeptides, a combined use of CID (or HCD) with ETD is recommended.

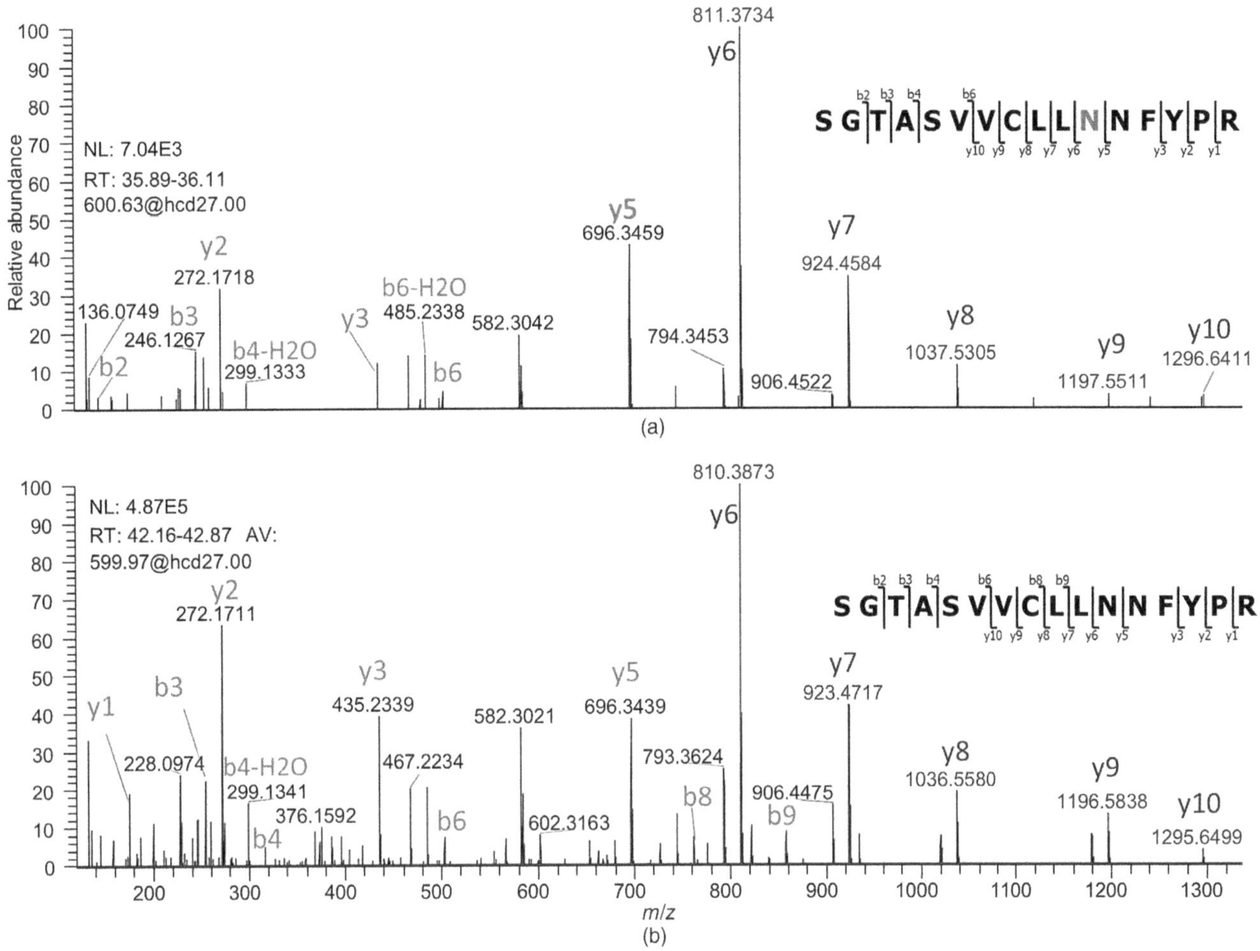

Figure 3.4 Characterization of deamidation in NISTmAb. HCD spectra of the nondeamidated peptide (b), SGTASVVCLL**N**NFYPR, and the deamidated species (a). HCD fragmentation of the deamidated peptide yields the same fragment ions as the nondeamidated species except for ions that contain the Asp11 residue, which has a mass increase of 0.98 Da. The cysteine residue on this peptide is reduced and alkylated.

CID or HCD often leads to the loss of glycan from peptides, as it preferentially fragments glycosidic and/or peptide–glycan bonds but not peptide bonds, thus provides compositional and some structural information of the oligosaccharides part of glycopeptides. ETD, on the other hand, produces a more complete fragmentation of peptide backbone while keeps the glycan attached, providing information of peptide sequence and the site of the glycan attachment. It has been demonstrated that using both ETD and CID, it is possible to simultaneously elucidate both the glycan structure and peptide sequence of N- and O-linked glycopeptides (Hogan et al. 2005, Catalina et al. 2007, Scott et al. 2011).

3.3.1 Performing ETD Experiments in Orbitrap Instruments

ETD technology has been most actively developed in ion trap mass spectrometer although developments with respect to ion/ion chemistry have been introduced to QTOF mass spectrometers as well. As a fragmentation method, ETD has lower fragmentation efficiency than that of CID or HCD, especially for peptide precursor ions with lower charge density (higher *m/z*). This inefficiency of ETD is due, at least in part, to nonspecific interactions between ions produced by ETD. The use of additional energy (supplemental activation, SA) has proved to improve the dissociation of the charge reduced precursor ions, generating more informative ETD spectrum (Swaney et al. 2007).

ETD is available as an option for all the hybrid and tribrid Orbitrap instruments as one of the fragmentation methods. For any data-dependent LC–MS/MS experiment, ETD can be selected as the only or one of the fragmentation methods. The resulting fragment ions from ETD can be detected either in ion trap or in Orbitrap analyzer. A few ETD-specific instrument parameters are important for optimized ETD performance. The AGC value for anion ETD reagent ions is usually set to 1–2e5. The maximum IT of 50–100 ms for ETD reagent ion injection is recommended. The ETD reaction time is set to 50–100 ms for doubly charged precursor ions. For a sample of peptide mixtures, a charge-dependent ETD reaction time is recommended where the instrument software determines ETD reaction time based on the precursor charge state, with shorter reaction time for higher charged precursors. In this way, ETD overreaction is avoided to limit the production of internal fragments and the neutralization of singly charged fragments. The fragmentation efficiency of ETD can be improved by using SA, especially for samples containing mostly doubly charged peptides, or large peptides with low charge density ($m/z > 1000$). A normalized SA energy of 15–20 is usually enough to improve the dissociation of the charge reduced precursor ions. Before any ETD experiment, the signal intensity of ETD reagent anion should be evaluated and a normalized intensity of at least e6 level is optimum for an LC–MS/MS experiment.

ETD can be used with CID or/and HCD in a number of ways. One can carry out experiments in the way that for each peptide precursor, an alternating CID and ETD is performed. This strategy produces both ETD and CID spectra for each peptide, providing the most information while spending unnecessary instrument time in the cases when only one of the methods can provide enough information. A more intelligent strategy involves conditional selection of either ETD or CID during the LC–MS/MS run based on certain preset parameters related to physicochemical properties of peptide ions. This strategy chooses the more appropriate method for each peptide ion while making the best use of instrument time. Examples of this strategy include data-dependent decision tree method and product-ion-dependent ETD method. In the most recently developed Orbitrap Fusion instrument, the method editor was designed to provide more flexibility to setting up intelligent methods. Besides signal intensity from the previous survey scan, some physicochemical properties of peptide ions such as charge state and m/z of precursor or product ions can be used to make intelligent decision. These properties can also be prioritized to best fit the purpose of the analysis when there are more than one levels of decision-making. For example, a data-dependent HCD product-dependent ETD (HCDpdETD) method performs regular data-dependent HCD experiment and triggers ETD only on certain peptides that produce specified ions during HCD. As described in the following section, this method can be applied to glycopeptides by specifying in the method editor those characteristic oxonium product ions generated by HCD of glycopeptides. Once those ions are detected in a HCD spectrum, ETD is triggered on the same peptide precursor. A pair of spectra is obtained for each glycopeptide, an ETD spectrum that contains ions from peptide backbone bond fragmentation and a HCD spectrum that contains ions from glycan fragmentation. Combined information from both spectra allows more comprehensive elucidation of glycopeptide structures. Besides triggering ETD by HCD product ions, one can also choose to give priority to high charge density (low m/z) precursor ions to increase the success rate of ETD experiments, thus the likelihood of identifying more glycopeptides in the sample.

3.3.2 Structure Elucidation of Glycopeptides Using Multiple Fragmentation Mechanisms in Orbitrap Instruments

Glycosylation is one of the most common posttranslational modifications of therapeutic proteins. More than one-third of approved biotherapeutics and many in clinical trials are glycoproteins. The presence and nature of the oligosaccharides are known to significantly modulate yield, bioactivity, solubility, stability, immunogenicity, and clearance rate from circulation (Rudd and Dwek 1997, Arnold et al. 2007, Durocher and Butler 2009). The N-glycans attached to Asn297 of Fc region of therapeutic recombinant antibodies and fusion proteins of immunoglobulin G1 (IgG1) are critical to the activation of downstream effector mechanisms. Besides the presence of core glycans at the Fc regions, the N-linked sites can be at the variable regions of either heavy chains or/and light chains. Cetuximab has an N-linked glycan at Asn88 of the heavy chain variable region. Some Fc-fusion therapeutics proteins, such as TNFRII-Fc, CD2-Fc, and CTLA4-Fc, contain glycosylation modifications in the fusion portions, in addition to their Fc glycans (Qian et al. 2007, Jefferis 2009). Many nonimmunoproteins such as growth factors, cytokines, hormones, and therapeutic enzymes, are also glycoproteins. Glycosylation is known to be important for their biosynthesis, secretion, metabolic fate, biological activity, and circulatory half-life (Ulloa-Aguirre et al. 1999, Van Patten et al. 2007). For example, erythropoietin (EPO) has three N-linked and one O-linked glycan chains. Removal of either two or all three sites results in poor product secretion (Egrie et al. 1993).

Protein glycosylation biosynthesis is a complex pathway catalyzed by several different types of glycosidase and transglycosidase, leading to a large heterogeneity in the glycoforms. At least 10 relatively abundant glycan compositions are readily detected on the human IgG1 Fc region (Wuhrer et al. 2007, Hong et al. 2013). Each glycan composition may also contain several isomers, which are difficult to separate using LC. The glycoform heterogeneity makes characterization of protein glycosylation challenging. Many proteins contain multiple glycosylation sites, which further complicates the characterization.

Three major types of glycosylation have been found in protein therapeutics: N-linked glycosylation at Asp, O-linked glycosylation at serine (Ser) and threonine (Thr), and O-GlcNAc at Ser and Thr. One standard glycosylation analysis is to characterize the released glycans, in which glycans were cleaved from protein and enriched for analysis. This technique yields good sensitivity for the glycans but provides no information on protein or sites of glycosylation. Characterization of glycopeptide is challenging because the glycosidic bonds are relatively more labile than the peptide bonds. As a result, standard CID and HCD experiments yield informative glycan fragments but very limited information on peptide sequence (Hogan et al. 2005, Catalina et al. 2007). ETD fragmentation of glycopeptides, on the other hand, produces mainly c and z fragment ions from the peptide backbone while keeping the glycans intact. It has been demonstrated that using both ETD and CID, it is possible to simultaneously elucidate both the glycan structure and peptide sequence of N- and O-linked glycopeptides (Scott et al. 2011, Zhao et al. 2011, Zhu et al. 2013). In the following, we present a recently developed LC–MS method for glycoprotein characterization using Orbitrap Fusion instrument, in which HCD is applied as the major fragmentation technique for peptide identification while additional ETD fragmentations are triggered automatically when glycopeptides are eluted from column. We have successfully applied this method to characterize glycopeptides from tissue plasminogen activator (TPA or PLAT). TPA is a serine protease found on endothelial cells to catalyze the conversion of plasminogen to plasmin and is used as a medical treatment for embolic or thrombotic stroke.

Samples analyzed in this study were TPA and G-Tenecteplase (TNK), which is a recombinant TPA molecule with minor sequence changes mainly at the glycosylation sites: T103N that introduces an N-glycosylation site, N117Q that removes an N-glycosylation site, and KHRR (296–299) to AAAA. I-TNK is a recombinant variant, which shares the same primary structure with G-TNK. All three proteins were reduced, alkylated, and digested with trypsin for LC–MS analysis using a data-dependent top-10 HCDpdETD method with an Orbitrap Fusion mass spectrometer. This method primarily acquired MS/MS spectra using HCD with ETD triggered only when glycopeptides were eluted from column. Collision dissociation of glycopeptides typically yields abundant carbohydrate oxonium ions such as HexHexNAc (366.14 *m/z*), HexNAc (204.08 *m/z*), Neu5Ac (292.09 *m/z*), and HexNAc fragment (138.05 *m/z*), which can be used as the diagnostic fragment ions of glycopeptides. If any of these diagnostic glycan fragment ions were detected in the HCD MS/MS experiment, then a subsequent ETD fragmentation was triggered on the same precursor ion. Therefore, this HCDpdETD method generates a pair of HCD and ETD spectra for each glycopeptide that yields structure information for both the peptide sequence and the glycan composition. The raw data files were analyzed using PepFinder 2.0 software for identification and quantification of known and unknown modifications. Peak areas of related peptide ions under their extracted ion chromatograms (XIC) are used for relative quantification of modified peptides. A mass tolerance of 5 ppm was used to ensure accurate identification.

The Orbitrap MS survey scans were acquired at a resolution of 120,000 (at *m/z* 200) with an AGC target of 4e5, and a maximum IT of 60 ms. The HCD MS/MS spectra were acquired in the Orbitrap at a resolution of 30,000 (at *m/z* 200) with an AGC target of 5e4, a maximum IT of 120 ms and a normalized collision energy of 30. ETD MS/MS spectra were acquired in the linear ion trap at the rapid scan rate with an AGC target of 1e4, and a maximum IT of 500 ms. Ions of the highest charge state were prioritized for data-dependent MS/MS followed by ions of the lowest *m/z* value. This way, the data-dependent priority was given to peptide precursor ions with higher charge density, which resulted in good ETD spectra for glycopeptide sequencing and glycan localization. ETD activation time was charge dependent and the actual time was determined by instrument control software based on the standard calibration as suggested by manufacturer.

Using this method, a total of four glycosylation sites were identified, three of which are over 99% glycosylated. N448 was glycosylated in all three samples, while N103 was detected in I-TNK and G-TNK and N117 only in TPA. The fourth glycosylation site, N184, was identified only in I-TNK and only 19% of this site was glycosylated (data not shown). I-TNK has an additional glycosylation site (N184) even though it shares the same amino acid sequence as G-TNK, suggesting a different manufacturing process. An example of a glycopeptide identified using HCDpdETD, G-TNK peptide C441-R449 with glycosylation on N448, is shown in Figure 3.5. HCD of this glycopeptide produced abundant fragments that show the loss of monosaccharides from the glycopeptide. HCD also produced abundant carbohydrate oxonium ions such as 366.14 (HexHexNAc), 204.08 (HexNAc), and 292.09 (Neu5Ac), which were used as the diagnostic fragment ions to trigger ETD MS/MS experiments. ETD of this peptide, on the other hand, produced c and z ions from the peptide backbone cleavage. The identification of multiple c and z ions containing glycan, c8, z4, z5, z6, z7, and z8, clearly located the glycosylation site. The calculated relative abundance of this glycoform was 0.52%

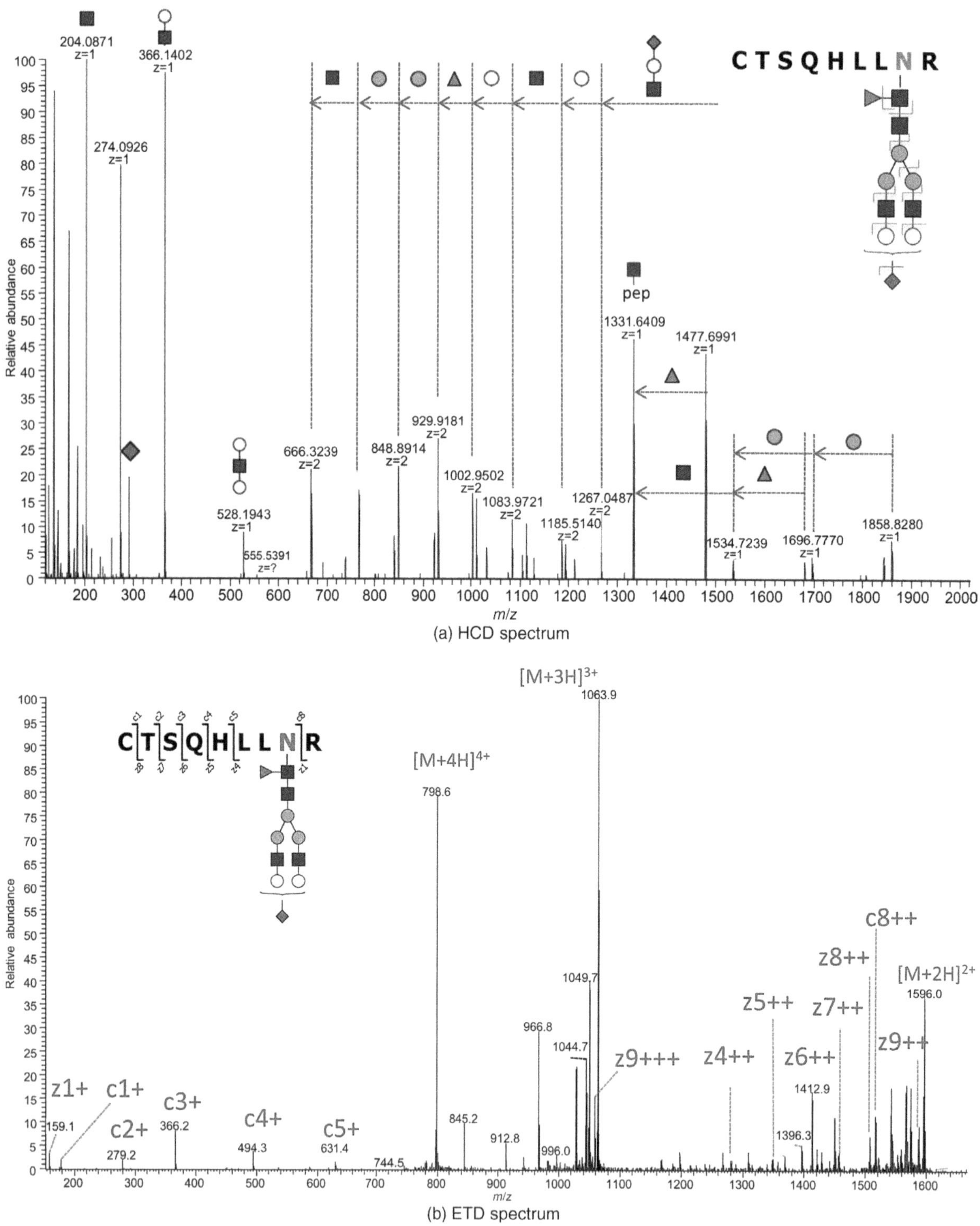

Figure 3.5 Characterization of glycopeptides, C441–R449, with glycosylation on N448, using HCDpdETD. HCD enables identification of glycan composition (a), and ETD enables peptide back bond sequencing (b). Symbol key: light circles = galactose (Gal); dark circles = mannose (Man); blank circles = hexose (Hex); squares = *N*-acetylglucosamine (GlcNAc); triangles = fucose (Fuc); diamonds = *N*-acetylneuraminic acid (Neu5Ac).

using the ratio of peak area of this peptide to the sum of the peak areas of the unmodified form plus all other glycoforms of this peptide.

The combined use of HCD and ETD fragmentation is a powerful approach for glycopeptide structure elucidation. Table 3.1 contains a list of all the N448 glycoforms

Table 3.1 Identified N448 glycoforms and their relative abundance in the three samples.

N448 Glycoform	TPA	I-TNK	G-TNK
N448 + A2G2F	**6.41%**	**5.40%**	**3.23%**
N448 + A2S1G0	5.18%	2.57%	<1%
N448 + A2S1G0F	<1%	<1%	1.79%
N448 + A2S1G1F	**23.11%**	**16.86%**	**14.43%**
N448 + A2S2F	**37.96%**	**35.34%**	**37.59%**
N448 + A3G3F	<1%	1.29%	<1%
N448 + A2Sg1S1F	1.32%	<1%	<1%
N448 + A3S1G2F	1.59%	2.48%	<1%
N448 + A3S2G0	1.43%	<1%	<1%
N448 + A3S2G1F	**5.19%**	**7.00%**	**5.04%**
N448 + A4S2G2F	<1%	<1%	2.20%
N448 + A4S1G3F	<1%	1.16%	<1%
N448 + A3S3F	**9.33%**	**11.61%**	**16.50%**
N448 + A4S3G1F	1.17%	6.55%	2.62%
N448 + A4S4F	1.67%	7.20%	6.51%

Only those with relative abundance higher than 1% in at least one of the samples are included. The five major glycoforms are highlighted in bold. Abbreviations for glycan structure: Antenna A, core fucose (Fuc) F, mannose (Man) M, galactose (Gal) G, *N*-acetyl neuraminic acid (NANA) S, *N*-glycolyl neuraminic acid (NGNA) Sg.

with relative abundance higher than 1% in the three samples. The relative abundance and identity of the various glycoforms on N448 were consistent among all three samples. Most glycans on this site contain sialic acid. The identity of the glycoforms on N103 is similar between I-TNK and G-TNK, but the profiles of relative abundance are markedly different. The glycoforms on N117 are primarily high mannose species, which is different from the glycans identified on any of the other sites. Glycosylation on N184 was only detected for the I-TNK sample with all of the glycans containing sialic acid (data not shown). The cycle time of this data-dependent method was not affected by the additional ETD experiments because they are performed in parallel in the ion trap while the spectra for survey scan and HCD were acquired in Orbitrap analyzer. For peptides other than glycopeptide, standard data-dependent HCD experiments were performed. One hundred percent sequence coverage and identification of various other sequence modifications were achieved (data not shown).

3.4 Conclusion

In the past decade, hundreds of papers have been published using LC–MS-based approaches to characterize protein modifications. One of the major challenges is that many modifications are two or three orders of magnitudes lower in abundance compared to their wild types. The other challenge is that fragmentation methods do not always generate information-rich spectra for structure elucidation. Orbitrap MS is one of the most advanced MS technologies for bottom-up protein characterization due to its high sensitivity and wide dynamic range for peptide analysis. With recent improvements in scan speed, it is now possible to sequence deeply into low-abundant modifications without extended chromatography run time. The AGC technology allows accumulation of low-abundant ions inside the trap, providing a unique way to enrich the selected precursor ion so that a high-quality spectrum is generated for low-abundant peptides with PTMs. ETD as a relatively new, alternative fragmentation method has demonstrated advantages in analyzing labile PTMs. It generates informative spectrum to sequence peptide and locate PTMs by cleaving at peptide backbone bond while keeping the labile PTMs intact. Both CID (or HCD) and ETD can be carried out in one experiment to maximize the sequence coverage and/or to produce orthogonal information for structure elucidation. Although CID has been the routinely used fragmentation method, ETD will be employed more often for analyzing PTMs in the future. The current and emerging Orbitrap methods presented in this chapter are universal and can be applied to the analysis of other protein therapeutics, including more sophisticated antibody derivatives such as ADCs, bi- and multispecific antibodies, as well as biosimilars.

Acknowledgment

The authors thank Alexander Makarov, Rosa Viner and David Horn at Thermo Fisher Scientific, Li Tao and Guodong Chen at Bristol-Myers Squibb for their critical review of this manuscript.

References

Arnold JN, Wormald MR, Sim RB, et al. (2007). "The impact of glycosylation on the biological function and structure of human immunoglobulins." *Annu Rev Immunol* **25**(1): 21–50.

Beck A, Wagner-Rousset E, Ayoub D, et al. (2013). "Characterization of therapeutic antibodies and related products." *Anal Chem* **85**(2): 715–736.

Beck A, Wurch T, Bailly C, et al. (2010). "Strategies and challenges for the next generation of therapeutic antibodies." *Nat Rev Immunol* **10**(5): 345–352.

Bondarenko PV, Second TP, Zabrouskov V, et al. (2009). "Mass measurement and top-down HPLC/MS analysis of intact monoclonal antibodies on a hybrid linear quadrupole ion trap-Orbitrap mass spectrometer." *J Am Soc Mass Spectrom* **20**(8): 1415–1424.

Catalina MI, Koeleman CA, Deelder AM, et al. (2007). "Electron transfer dissociation of N-glycopeptides: loss of the entire N-glycosylated asparagine side chain." *Rapid Commun Mass Spectrom* **21**(6): 1053–1061.

Chelius D, Xiao G, Nichols AC, et al. (2008). "Automated tryptic digestion procedure for HPLC/MS/MS peptide mapping of immunoglobulin gamma antibodies in pharmaceutics." *J Pharm Biomed Anal* **47**(2): 285–294.

Chen G (2013). *Characterization of Protein Therapeutics Using Mass Spectrometry*, Springer, ISBN 978-1-4419-7862-2.

Chen G, Warrack BM, Goodenough AK, et al. (2011). "Characterization of protein therapeutics by mass spectrometry: recent developments and future directions." *Drug Discov Today* **16**(1–2): 58–64.

Chumsae C, Gaza-Bulseco G, Sun J, et al. (2007). "Comparison of methionine oxidation in thermal stability and chemically stressed samples of a fully human monoclonal antibody." *J Chromatogr B* **850**(1–2): 285–294.

Durocher Y and Butler M (2009). "Expression systems for therapeutic glycoprotein production." *Curr Opin Biotechnol* **20**(6): 700–707.

Egrie JC, Grant JR, Gillies DK, et al. (1993). "S7.7 The role of carbohydrate on the biological activity of erythropoietin." *Glycoconjugate J* **10**(4): 263–263.

Frese CK, Altelaar AF, Hennrich ML, et al. (2011). "Improved peptide identification by targeted fragmentation using CID, HCD and ETD on an LTQ-Orbitrap Velos." *J Proteome Res* **10**(5): 2377–2388.

Good DM, Wirtala M, McAlister GC, et al. (2007). "Performance characteristics of electron transfer dissociation mass spectrometry." *Mol Cell Proteomics* **6**(11): 1942–1951.

Gross JH (2004). *Mass Spectrometry: A Textbook*, Springer Science & Business Media, ISBN 978-3-540-36756-7.

Guthals A, Clauser KR, Frank AM, et al. (2013). "Sequencing-grade de novo analysis of MS/MS triplets (CID/HCD/ETD) from overlapping peptides." *J Proteome Res* **12**(6): 2846–2857.

Han X, Aslanian A and Yates JR, III (2008). "Mass spectrometry for proteomics." *Curr Opin Chem Biol* **12**(5): 483–490.

Hao Z, Zhang Y, Eliuk S, et al. (2012). "A quadrupole-Orbitrap hybrid mass spectrometer offers highest benchtop performance for in-depth analysis of complex proteomes." Thermo Fisher Scientific, Application note 552.

Hmiel L, Brorson K and Boyne M, II (2015). "Post-translational structural modifications of immunoglobulin G and their effect on biological activity." *Anal Bioanal Chem* **407**(1): 79–94.

Hogan JM, Pitteri SJ, Chrisman PA, et al. (2005). "Complementary structural information from a tryptic N-linked glycopeptide via electron transfer ion/ion reactions and collision-induced dissociation." *J Proteome Res* **4**(2): 628–632.

Hong Q, Lebrilla CB, Miyamoto S, et al. (2013). "Absolute quantitation of immunoglobulin g and its glycoforms using multiple reaction monitoring." *Anal Chem* **85**(18): 8585–8593.

Jefferis R (2009). "Glycosylation as a strategy to improve antibody-based therapeutics." *Nat Rev Drug Discov* **8**(3): 226–234.

Ji JA, Zhang B, Cheng W, et al. (2009). "Methionine, tryptophan, and histidine oxidation in a model protein, PTH: mechanisms and stabilization." *J Pharm Sci* **98**(12): 4485–4500.

Kelleher NL, Lin HY, Valaskovic GA, et al. (1999). "Top down versus bottom up protein characterization by tandem high-resolution mass spectrometry." *J Am Chem Soc* **121**(4): 806–812.

Kim MS and Pandey A (2012). "Electron transfer dissociation mass spectrometry in proteomics." *Proteomics* **12**(4–5): 530–542.

Lam XM, Yang JY and Cleland JL (1997). "Antioxidants for prevention of methionine oxidation in recombinant monoclonal antibody HER2." *J Pharm Sci* **86**(11): 1250–1255.

Lewis DA, Guzzetta AW, Hancock WS, et al. (1994). "Characterization of humanized anti-TAC, an antibody directed against the interleukin 2 receptor, using electrospray ionization mass spectrometry by direct infusion, LC/MS, and MS/MS." *Anal Chem* **66**(5): 585–595.

Little MJ, Paquette DM and Roos PK (2006). "Electrophoresis of pharmaceutical proteins: status quo." *Electrophoresis* **27**(12): 2477–2485.

Liu H, Gaza-Bulseco G, Faldu D, et al. (2008a). "Heterogeneity of monoclonal antibodies." *J Pharm Sci* **97**(7): 2426–2447.

Liu H, Gaza-Bulseco G, Xiang T, et al. (2008b). "Structural effect of deglycosylation and methionine oxidation on a recombinant monoclonal antibody." *Mol Immunol* **45**(3): 701–708.

Liu YD, van Enk JZ and Flynn GC (2009). "Human antibody Fc deamidation in vivo." *Biologicals* **37**(5): 313–322.

Lu C, Liu D, Liu H, et al. (2013). "Characterization of monoclonal antibody size variants containing extra light chains." *MAbs* **5**(1): 102–113.

Makarov A, Denisov E, Kholomeev A, et al. (2006a). "Performance evaluation of a hybrid linear ion TRAP/Orbitrap mass spectrometer." *Anal Chem* **78**(7): 2113–2120.

Makarov A, Denisov E, Lange O, et al. (2006b). "Dynamic range of mass accuracy in LTQ Orbitrap hybrid mass spectrometer." *J Am Soc Mass Spectrom* **17**(7): 977–982.

Mann M and Jensen ON (2003). "Proteomic analysis of post-translational modifications." *Nat Biotechnol* **21**(3): 255–261.

McLafferty FW, Breuker K, Jin M, et al. (2007). "Top-down MS, a powerful complement to the high capabilities of proteolysis proteomics." *FEBS J* **274**(24): 6256–6268.

Michalski A, Damoc E, Lange O, et al. (2012). "Ultra high resolution linear ion trap Orbitrap mass spectrometer (Orbitrap Elite) facilitates top down LC MS/MS and versatile peptide fragmentation modes." *Mol Cell Proteomics* **11**(3): O111.013698.

Mikesh LM, Ueberheide B, Chi A, et al. (2006). "The utility of ETD mass spectrometry in proteomic analysis." *Biochim Biophys Acta* **1764**(12): 1811–1822.

Mirza UA, Chait BT and Lander HM (1995). "Monitoring reactions of nitric oxide with peptides and proteins by electrospray ionization-mass spectrometry." *J Biol Chem* **270**(29): 17185–17188.

Molina H, Matthiesen R, Kandasamy K, et al. (2008). "Comprehensive comparison of collision induced dissociation and electron transfer dissociation." *Anal Chem* **80**(13): 4825–4835.

Olsen JV, Macek B, Lange O, et al. (2007). "Higher-energy C-trap dissociation for peptide modification analysis." *Nat Methods* **4**(9): 709–712.

Patel K and Borchardt R (1990). "Chemical pathways of peptide degradation. II. Kinetics of deamidation of an asparaginyl residue in a model hexapeptide." *Pharm Res* **7**(7): 703–711.

Pavlou AK and Reichert JM (2004). "Recombinant protein therapeutics—success rates, market trends and values to 2010." *Nat Biotechnol* **22**(12): 1513–1519.

Pioch M, Bunz SC and Neususs C (2012). "Capillary electrophoresis/mass spectrometry relevant to pharmaceutical and biotechnological applications." *Electrophoresis* **33**(11): 1517–1530.

Qi P, Volkin DB, Zhao H, et al. (2009). "Characterization of the photodegradation of a human IgG1 monoclonal antibody formulated as a high-concentration liquid dosage form." *J Pharm Sci* **98**(9): 3117–3130.

Qian J, Liu T, Yang L, et al. (2007). "Structural characterization of N-linked oligosaccharides on monoclonal antibody cetuximab by the combination of orthogonal matrix-assisted laser desorption/ionization hybrid quadrupole–quadrupole time-of-flight tandem mass spectrometry and sequential enzymatic digestion." *Anal Biochem* **364**(1): 8–18.

Ren D, Pipes GD, Liu D, et al. (2009). "An improved trypsin digestion method minimizes digestion-induced modifications on proteins." *Anal Biochem* **392**(1): 12–21.

Rudd PM and Dwek RA (1997). "Glycosylation: heterogeneity and the 3D structure of proteins." *Crit Rev Biochem Mol Biol* **32**(1): 1–100.

Rustandi RR, Washabaugh MW and Wang Y (2008). "Applications of CE SDS gel in development of biopharmaceutical antibody-based products." *Electrophoresis* **29**(17): 3612–3620.

Scott NE, Parker BL, Connolly AM, et al. (2011). "Simultaneous glycan-peptide characterization using hydrophilic interaction chromatography and parallel fragmentation by CID, higher energy collisional dissociation, and electron transfer dissociation MS applied to the N-linked glycoproteome of *Campylobacter jejuni*." *Mol Cell Proteomics* **10**(2): M000031-MCP000201.

Shen Y, Tolić N, Purvine SO, et al. (2011). "Improving collision induced dissociation (CID), high energy collision dissociation (HCD), and electron transfer dissociation (ETD) Fourier transform MS/MS degradome–peptidome identifications using high accuracy mass information." *J Proteome Res* **11**(2): 668–677.

Song Y, Schowen RL, Borchardt RT, et al. (2001). "Effect of 'pH' on the rate of asparagine deamidation in polymeric formulations: 'pH'–rate profile." *J Pharm Sci* **90**(2): 141–156.

Steen H and Mann M (2004). "The ABC's (and XYZ's) of peptide sequencing." *Nat Rev Mol Cell Biol* **5**(9): 699–711.

Stephenson RC and Clarke S (1989). "Succinimide formation from aspartyl and asparaginyl peptides as a model for the spontaneous degradation of proteins." *J Biol Chem* **264**(11): 6164–6170.

Swaney DL, McAlister GC and Coon JJ (2008). "Decision tree-driven tandem mass spectrometry for shotgun proteomics." *Nat Methods* **5**(11): 959–964.

Swaney DL, McAlister GC, Wirtala M, et al. (2007). "Supplemental activation method for high-efficiency electron-transfer dissociation of doubly protonated peptide precursors." *Anal Chem* **79**(2): 477–485.

Syka JEP, Coon JJ, Schroeder MJ, et al. (2004). "Peptide and protein sequence analysis by electron transfer dissociation mass spectrometry." *Proc Natl Acad Sci USA* **101**(26): 9528–9533.

Tao L, Ackerman M, Wu W, et al. (2011). Characterization of impurities and degradants in protein therapeutics by mass spectrometry. *Characterization of Impurities and Degradants Using Mass Spectrometry*, John Wiley & Sons, Inc., pp. 391–426.

Tonie Wright H and Urry DW (1991). "Nonenzymatic Deamidation of Asparaginyl and Glutaminyl Residues in Protein." *Crit Rev Biochem Mol Biol* **26**(1): 1–52.

Ulloa-Aguirre A, Timossi C, Damian-Matsumura P, et al. (1999). "Role of glycosylation in function of follicle-stimulating hormone." *Endocrine* **11**(3): 205–215.

Van Patten SM, Hughes H, Huff MR, et al. (2007). "Effect of mannose chain length on targeting of glucocerebrosidase for enzyme replacement therapy of Gaucher disease." *Glycobiology* **17**(5): 467–478.

Vlasak J, Bussat MC, Wang S, et al. (2009). "Identification and characterization of asparagine deamidation in the light chain CDR1 of a humanized IgG1 antibody." *Anal Biochem* **392**(2): 145–154.

Wang W, Vlasak J, Li Y, et al. (2011). "Impact of methionine oxidation in human IgG1 Fc on serum half-life of monoclonal antibodies." *Mol Immunol* **48**(6–7): 860–866.

Wiesner J, Premsler T and Sickmann A (2008). "Application of electron transfer dissociation (ETD) for the analysis of posttranslational modifications." *Proteomics* **8**(21): 4466–4483.

Wuhrer M, Stam JC, van de Geijn FE, et al. (2007). "Glycosylation profiling of immunoglobulin G (IgG) subclasses from human serum." *Proteomics* **7**(22): 4070–4081.

Xie M and Schowen RL (1999). "Secondary structure and protein deamidation." *J Pharm Sci* **88**(1): 8–13.

Yang J, Wang S, Liu J, et al. (2007a). "Determination of tryptophan oxidation of monoclonal antibody by reversed phase high performance liquid chromatography." *J Chromatogr A* **1156**(1–2): 174–182.

Yates JR (2004). "Mass spectral analysis in proteomics." *Annu Rev Biophys Biomol Struct* **33**(1): 297–316.

Yu XC, Joe K, Zhang Y, et al. (2011). "Accurate determination of succinimide degradation products using high fidelity trypsin digestion peptide map analysis." *Anal Chem* **83**(15): 5912–5919.

Zhang Z, Pan H and Chen X (2009). "Mass spectrometry for structural characterization of therapeutic antibodies." *Mass Spectrom Rev* **28**(1): 147–176.

Zhao P, Viner R, Teo CF, et al. (2011). "Combining high-energy C-trap dissociation and electron transfer dissociation for protein O-GlcNAc modification site assignment." *J Proteome Res* **10**(9): 4088–4104.

Zhu Z, Hua D, Clark DF, et al. (2013). "GlycoPep detector: a tool for assigning mass spectrometry data of N-linked glycopeptides on the basis of their electron transfer dissociation spectra." *Anal Chem* **85**(10): 5023–5032.

Zubarev RA and Makarov A (2013). "Orbitrap mass spectrometry." *Anal Chem* **85**(11): 5288–5296.

4

Macro- to Micromolecular Quantitation of Proteins and Peptides by Mass Spectrometry

Suma Ramagiri, Brigitte Simons, and Laura Baker

SCIEX, Framingham, MA, USA

4.1 Introduction

The importance of biotherapeutics as a class of drugs has increased significantly over recent years due to their enormous potential to treat a wide array of human diseases ranging from autoimmune and inflammatory diseases to cancer, cardiovascular diseases, and rare genetic disorders. These highly promising therapeutic agents, including very small peptide chains, such as insulin, up to much larger proteins, such as antibodies and novel Fc-like fusion proteins, are extremely attractive as drug candidates because of their low toxicity and high specificity. Thus, these compounds continue to fill the preclinical and clinical pipelines of many pharmaceutical companies. The rapid growth of biotherapeutics is a good indicator of its success, with the global market valued at around US$199.7 billion in 2013 and projected to grow by 13.5% through 2020 (Smith 2013). The number of clinically approved protein and peptide therapies has jumped to over 170 products (Zhong et al. 2011) with 350 antibody-based therapies that currently await clinical trials (Reichert 2013), making biotherapeutics the fastest growing class of drugs in the last decade. With increased industry interest and investment and rising demand from the medical community for these unique, targeted therapies, there is a growing requirement to develop high-throughput analytical techniques to expand biotherapeutic product lines.

To overcome regulatory hurdles and to advance to clinical trials, biopharmaceutical drug development and discovery requires careful metabolic monitoring of a candidate drug, a process that necessitates accurate quantitation during pharmacokinetic (PK), toxicokinetic (TK), bioequivalence, and clinical drug monitoring studies – all of which are conducted in complex biological matrices (blood, plasma, or urine). With this rapid growth in biotherapies comes increased demand for an analytical platform that is flexible, robust, and easily integrated into preexisting drug development workflows. Widely used for small-molecule drug development, liquid chromatography–tandem mass spectrometry (LC/MS/MS) has positively impacted bioanalysis applications due to recent technological developments that comply with regulatory expectations and overcome analyte detection limitations. Presented here, we demonstrate how key mass spectrometry technologies can coalesce into straightforward, accurate, extremely sensitive, and, most importantly, high-throughput quantitative solutions. Already considered the gold standard for quantitation in other areas of bioanalytical quantitation such as proteomics, antidoping, forensics, and clinical chemistry (Hopfgartner and Bourgogne 2003, Shi et al. 2012), LC/MS/MS is poised to replace and outperform other techniques for biotherapeutic analysis. The current standard conventions for protein and peptide quantitation are based on the ligand binding assay (LBA), such as the enzyme-linked immunosorbent assay (ELISA) or on UV detection of individual peptides using high-pressure liquid chromatography (HPLC) separations. LBAs rely on immunoaffinity detection of a unique epitope on the protein or peptide of interest, and due to the highly specific nature of the antibody-based detection, they offer high sensitivity, although the dynamic range is limited to just one or two orders of magnitude (Ezan and Bitsch 2009). Because antibodies unique to the target analyte require a lengthy development process, assay development can often be time-consuming and expensive; in addition, LBA results are often plagued by interferences and high background from antibody cross-reactivity (van den Broek et al. 2013). UV detection and quantitation of peptides is commonly used for peptide mapping, and this analytical method can be useful after extensive sample preparation and cleanup. UV detection with HPLC also does not require the expense and time commitment of antibody production, but the applicability of this method narrows as the complexity of the sample matrix increases.

Protein Analysis using Mass Spectrometry: Accelerating Protein Biotherapeutics from Lab to Patient, First Edition.
Edited by Mike S. Lee and Qin C. Ji.

Herein, we present an extensive resource composed of peptide quantitation studies performed on mass spectrometry instruments. Also, we illustrate how sensitive and selective detection can be achieved even in the presence of high background noise. To meet bioanalytical quantitative standards and assay validation parameters, peptide bioanalysis must be sensitive enough to achieve the standard benchmarks for excellence in accuracy and precision. Due to the inherent complexities of biological samples, bioanalysis is often negatively impacted by high background noise and interfering peaks. Section 4.2 illustrates how realizing superb analyte selectivity – even in biological samples with numerous, highly abundant, endogenous proteins – is driven by innovative tools such as multiple reaction monitoring cubed (MRM^3).

Workflow and SelexION™ Differential Ion Separation Technology. Advances in high-resolution mass spectrometry are detailed in Section 4.3, which highlights targeted workflows on the high-resolution, accurate-mass spectrometry (HRMS) system that extend the sensitivity and selectivity of quantitative assays due to the narrow extraction widths and high-resolution time-of-flight (TOF) data. Lastly, Section 4.4 investigates the software tools available for robust peptide quantitation workflows that give researchers intuitive tools to automate the complex, multistep calculations for peak-area quantitation.

Each section and experiment featured in this resource includes an overview of the *key challenges*, *benefits*, and *features* of the bioanalytical technique presented. In this way, the technique of mass spectrometry can be put into context with other bioanalytical tools and help provide insights into its many advantages. LC/MS/MS analysis offers many attractive features to support biopharmaceutical drug development; however, the integration of LC/MS/MS into the biopharmaceutical workflow has been slow in spite of its dominance in the small-molecule laboratory. Widely accepted and easily validated, the LBA technique remains a popular method for protein and peptide bioanalysis due to its relatively lower investment in infrastructure and ease of implementation into the high-throughput environment. Yet, even LBA methods have their drawbacks, and straightforward LC/MS/MS alternatives are sought that can support the operational challenges of accelerating the further development of biotherapies.

4.2 Key Challenges of Peptide Bioanalysis

To understand why the pharmaceutical industry has been hesitant to fully embrace LC/MS/MS strategies for peptide quantitation, the complexities and challenges of the workflow must be fully appreciated. (For a summary of excellent reviews on LC/MS/MS protein and peptide quantitation, see Table 4.1). For both protein and peptide quantitation, calibration curves based on standards are used to calculate concentration values for unknowns in biological samples; in addition, the amassed data must be stringent enough to meet the rigorous benchmarks prescribed by the USFDA (FDA 2001). For therapeutic peptides, the workflow is more straightforward because proteolysis is omitted, and the intact peptide can be directly quantitated by MS/MS after relatively limited sample preparation (Figure 4.1). There is appreciably much more complexity when evaluating larger molecular weight biotherapeutics (>10 kDa), which are not always suitable in their entirety for direct MS/MS analysis. Therefore, bioanalysis of larger proteins and antibodies is based on quantitation of a small portion of the protein, typically a signature peptide released through tryptic digestion with an m/z ratio that is determined to be unique from all other peptides in the digest mixture. When coupled with the inclusion of a stable-isotope-labeled (SIL) internal standard, the response ratio of the released signature peptide to the SIL internal standard reveals a concentration representative of the intact protein. To build this multifaceted process into the framework of regulated bioanalysis is an extremely challenging prospect, which provides unique insights into why LC/MS/MS quantitation of biopharmaceuticals has been slow to gain acceptance in the GLP laboratory.

Evaluation of LC/MS/MS bioanalysis reveals that the major challenges for accurate and precise quantitation lie primarily in the realm of sample preparation, which includes (i) the lengthy and extensive workflows for the production of signature peptides and (ii) the diminishing accuracy of quantitative measurements in highly complex biological samples due to background interferences. Because the multistep reduction/alkylation/digestion process generates a more complex mixture than the starting sample, bioanalysis of low-level therapeutic peptides can be extremely challenging. Therefore, achieving LLOQs in the low ng/mL range is highly dependent at this time on the optimization of sample preparation steps (Bischoff et al. 2013). The numerous competing background peptides are a major consideration in sample preparation, which typically requires enrichment and semipurification of the analyte, and thus, introduces additional complexity to the workflow (Bischoff et al. 2013). The potential for variable peptide release during digestion of the target protein is a significant concern from a regulatory perspective. If digestion conditions are not well controlled, then irregular signature peptide release can have a lasting impact on the overall data quality (van den Broek et al. 2013). To overcome these drawbacks, strategies such as condensing sample prep steps

Table 4.1 Selected citations for further reading on protein and peptide LC/MS/MS methodologies.

Title	Article highlights	Citation
"Analysis of biopharmaceutical proteins in biological matrices by LC-MS/MS I. Sample preparation"	• Sample-preparation aspects for quantifying biopharmaceutical proteins in body-derived fluids by LC/MS/MS • Enrichment at the peptide level after proteolytic digestion • Chemical derivatization of peptides for enhancing ionization efficiency • Automation of the entire analytical procedure for routine applications in pharmacokinetic and clinical studies	Bischoff et al. 2013
"Analysis of biopharmaceutical proteins in biological matrices by LC-MS/MS II. LC-MS/MS analysis"	• Overview of selected reaction monitoring (SRM) strategies for quantifying peptides in biological matrices. Selection of signature peptides and internal standards • Selectivity improvements using MS^3 and differential mobility spectrometry (DMS) • Quantitative LC/MS analysis with low- and high-resolution MS • Data-independent acquisition (DIA) for collection of all data in a single analysis	Hopfgartner et al. 2013
"Bioanalytical LC-MS/MS of protein-based biopharmaceuticals"	• Overview of topics relating to the bioanalysis of biopharmaceutical proteins in biological matrices • Compares alternative quantitative methodology, such as ligand binding assays (LBAs), to mass-spectrometry-based platforms • Review of practical aspects of the seven "critical factors" for protein sample preparation • Special focus on the quantitation of monoclonal antibodies in serum and plasma • Advances in selectivity, including high-resolution mass spectrometry	van den Broek et al. 2013
"Liquid chromatography coupled with tandem mass spectrometry for the bioanalysis of proteins in drug development: Practical considerations in assay development and validation"	• Approaches for overcoming operational challenges due to complex sample preparation • Development and validation of a fast, simple, and reliable LC/MS/MS peptide quantitation method that fits into current pharmaceutical workflows • Recommendations for validating quantitative methods based on surrogate peptides	Liu et al. 2013

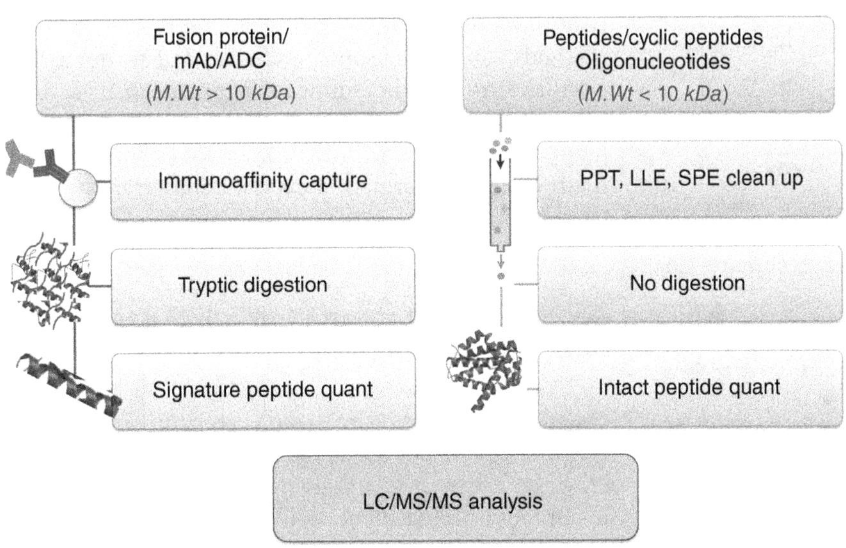

Figure 4.1 Peptide and protein bioanalytical workflow strategies. Protein quantitation typically involves a tryptic digestion step, which is omitted during peptide bioanalysis, thereby simplifying the process.

and digestion optimization can lead to more straightforward method development with wider regulatory appeal (Liu et al. 2013). And to that end, as advances in technology deliver exceedingly more sensitive and selective mass spectrometry workflows for direct quantitation in the subpicomolar range (Hopfgartner et al. 2013), sample preparation protocols can be further streamlined and simplified. Such a mass-spectrometry-based workflow would rely less on intricate sample enrichment and baseline reduction protocols, which will help propel this versatile and reliable MS methodology firmly into the domain of regulated biotherapeutic quantitation.

Following are some of the key challenges in bioanalytical quantification of biotherapeutic drugs such as peptides and proteins:

- *Limited quantitation range* – Analytical range of ELISA-based method is less than two orders of magnitude; at least three orders of magnitude is desired in bioanalysis. Poor MS/MS sensitivity combined with often poor selectivity can compromise the desired lower limits of quantitation (LLOQ).
- *Impaired sensitivity in complex matrices* – Very low-level peptide detection (sub-pg/mL) can be suppressed by high background and competing ions in biological samples. The best, previously reported LOQ is 100 pg/mL; extended-release pharmacokinetic studies demand lower levels of detection.
- *Low specificity* – Complex biological matrices hamper data resolution and require sophisticated sample preparation and/or advanced instrumentation.
- *Coeluting multiple charge interference* – Limits accurate quantification and also peak integration at LOQ levels. *Isobaric interferences* will limit selectivity and specificity of the assay and cause issues for accurate identification during bioanalytical method development process.
- *Poor data quality* – Precision and accuracy can be compromised at low peptide levels and may lead to results that are below accepted bioanalytical standards.
- *Reduced recovery, low sensitivity* – The adsorptive properties and/or polarity of peptides can compromise recovery, and interferences from biological matrices can negatively impact sensitivity and selectivity.
- *Physico-chemical proprieties* of peptides – Nonspecific binding, poor solubility, and complex charge-state envelopes require a highly versatile technique for quantification.
- *Limited MRM selectivity* – MRM approaches and efficient ultra high pressure liquid chromatography (UHPLC) separations may not provide adequate signal-to-noise ratios at LLOQ due to isobaric interferences or high baseline noise.
- *Systematic measurement errors* – Especially for ultralow level quantitation, measurement errors have a significant effect on data accuracy and precision.
- *Poorly fragmenting peptides* – Cyclic fragments often fragment poorly resulting in few product ions for analysis.

4.2.1 Key Benefits of the LC/MS/MS Peptide Quantitation Workflow

While LBAs may be primarily used in industry at this time, LC/MS/MS techniques provide many potential benefits that are grounded in the direct evaluation of the analyte's chemical nature, rather than indirect signals that stem from an immunological interaction. Quantitative data obtained by LC/MS/MS methodology correlates well with LBA-derived concentrations (van den Broek et al. 2013). Unlike LBA assays that require specific antibodies for each analyte, mass spectrometry platforms have universal applicability and provide a single technique for a diverse range of analytes. All proteins and peptides can be evaluated by LC/MS/MS without exception, and other biomolecules such as lipids and carbohydrates and other molecules of interest to the purification process can be identified. LBAs are generally more limited in their applicability because of autoantibody cross-reactivity and the restricted availability of commercial kits for every protein of interest (Hopfgartner et al. 2013). Nonspecific binding and molecular class limitations are surpassed with LC/MS/MS, which can even meet the challenge of quantitating highly homologous isoforms that are impossible to distinguish using immunoaffinity techniques. Because of the need to provide selective data in a background of highly abundant, endogenous proteins, low-level biomolecule quantitation is analogous to finding a needle in a haystack; yet LC/MS/MS is able to deliver quantitative data with excellent accuracy and precision over a wide linear dynamic range, often over three to four orders of magnitude (van den Broek et al. 2013). In addition, in contrast to the repeated expense and time-consuming nature of antibody production, LC/MS/MS methods can be developed and validated within a relatively shorter amount of time for multiple targets at once. This advantage along with the inherent flexibility and high selectivity of LC/MS/MS provides an attractive method for biopharmaceutical quantitation in the regulated laboratory.

4.3 Key Features of LC/MS/MS-Based Peptide Quantitation

Ongoing optimization of sample preparation steps will continue to enhance the LC/MS/MS quantitation process. However, the most significant gains in protein and peptide quantitation will likely be realized through technological innovations in mass spectrometry instrumentation.

A focus on improvements with sensitivity and selectivity for the detection of very low levels of proteins and peptides in very complex backgrounds may deliver high-performance instruments that can rapidly and simultaneously measure multiple analytes – and power the pharmaceutical discovery and development process into the future.

4.3.1 Sensitivity

Biopharmaceuticals are very potent, highly targeted therapies that are administered in low concentration doses and exhibit a narrow therapeutic range. Often found at circulating levels in the sub-ng/mL range, detection of biotherapies typically requires very highly sensitive methods. The enhancement of ionization efficiency and ion transmission has made it possible to detect drugs and metabolites in the subfemtogram levels (Thompson 2012). New technologies, such as the IonDrive™ QJet Ion Guide, underpin the sensitivity enhancements in the quadrupole/linear ion trap (QTRAP) and Triple Quad MS systems and help to transmit more ions to the detector through improved collisional focusing of ions. Heating and desolvation improvements in the IonDrive™ Turbo V Source and increased size and improved design of the aperture release more ions into the instrument. To fully detect the augmented signal, improvements to the dynamic range of the detector allow for accurate ion counting; the high energy conversion dynode (HED) detection system measures high ion signals without saturation to produce a linear dynamic range of over six orders of magnitude. These technologies are pivotal for providing continued improvements to sensitive bioanalysis.

4.3.2 Selectivity

Even if the pinnacle of sensitivity is reached, researchers will still be faced with the challenges of separating low levels of pharmaceutically active biomolecules from the highly complex biological matrix, where every endogenous compound can potentially interfere with the target signal. On the sample prep side, several strategies exist for the selective removal of competing background ions as well as enrichment of the analyte fraction. However, the required time and the potential for sample loss with additional cleanup steps make this approach much less appealing. Currently, advances in MS selectivity are focused on methods that provide an additional degree of separation subsequent to the entrance to the MS or post-MS/MS selection to help improve separation capacity in highly complex biological matrices. To maximize instrument performance when detecting low-level analytes masked by high background, the QTRAP system offers MRM^3 scans and the SelexION™ Differential Mobility Separation Device for improved peak shapes and signal-to-noise ratios during protein and peptide quantitation.

4.3.2.1 MRM^3

Peak measurements obtained by multiple reaction monitoring (MRM) scans are occasionally challenged by interferences and overlapping peaks that cannot be removed without further, more elaborate sample cleanup. To provide additional specificity, the technique of MRM^3 can be applied using the QTRAP system of instruments – extremely sensitive, hybrid triple quadrupole instruments with a linear ion trap for further fragmentation of the primary product ions. Quantitation of the secondary product ions is usually not affected by competing or overlapping ions, which are filtered out in previous MRM selection steps (Figure 4.2). This reduction in baseline results in improved peak shape, higher signal-to-noise ratios, and superior LLOQs. The QTRAP system is powered by eQ™ Electronics for scan speeds that are fast enough to be compatible with fast LC flow rates; and these instruments are equipped with single frequency excitation for highest selectivity of the product ion prior to secondary fragmentation. The Linear Accelerator™ Trap Electrodes provide 100-fold more sensitivity for the detection of low-level secondary fragments resulting from the use of MRM^3 to resolve issues of high background noise.

4.3.2.2 Differential Mobility Spectrometry (DMS)

In some cases, if secondary product ions are not specific enough or are too low for MRM^3 to be used, or method development time is too limited for prolonged MRM^3 development, then additional selectivity can be gained through differential mobility spectrometry (DMS). This technique selects ions of interest based on their inherent mobility difference between a set of planar plates with high and low energy fields applied, where coeluting interferences can be tuned out prior to analyte entrance into the mass spectrometer. Along with SelexION™ Differential Ion Separation Technology for quickly resolving isobaric species and single and multiple charge state interferences on a timescale compatible with UHPLC and multiple MRM acquisitions, thus providing an additional, orthogonal level of separation for difficult-to-address overlapping peaks (Figure 4.3).

4.3.3 High-Resolution Accurate-Mass Spectrometry

Improvements to selectivity can also be gained through high-resolution mass spectrometry using an instrument such as a high-resolution, accurate-mass TOF mass spectrometer (the TripleTOF® system), which combines qualitative exploration and high resolution on a single accurate-mass platform. When using the high-resolution multiple reaction monitoring (MRMHR) workflow, the

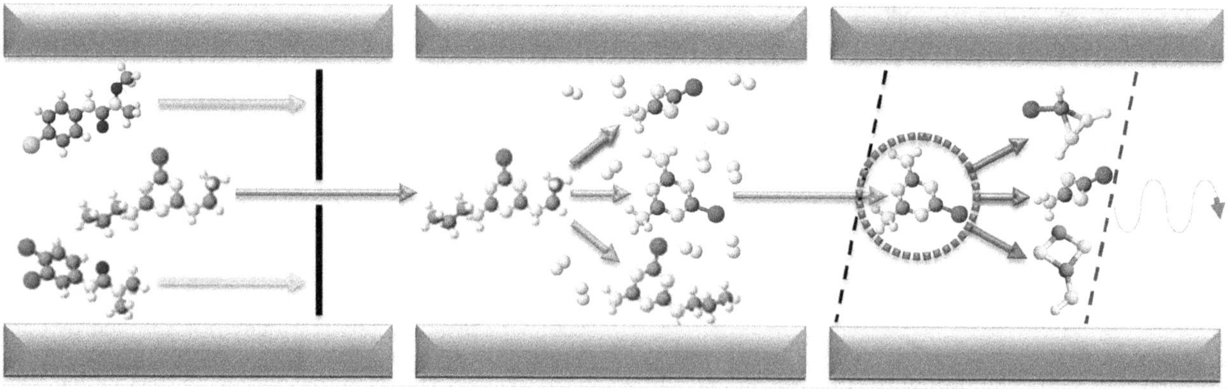

Figure 4.2 MRM^3 scan description: selection of the precursor ion is made in Q1, followed by fragmentation in Q2, then the residual precursor ion and resulting fragments are trapped in the Q3 linear ion trap (LIT) for a designated fill time. A specific fragment ion is selected for further fragmentation, it is isolated in Q3, and then this ion is fragmented to form second-generation fragment ions. These ions are rapidly scanned out of the LIT and are used as the analytical signal for the MRM^3 experiment. This can result in LOQs almost an order of magnitude lower than analogous MRM experiments through reduced background and the elimination of interferences. Their fast acquisition rates enable MRM^3 experiments to be combined with higher throughput chromatography, resulting in greater efficiencies.

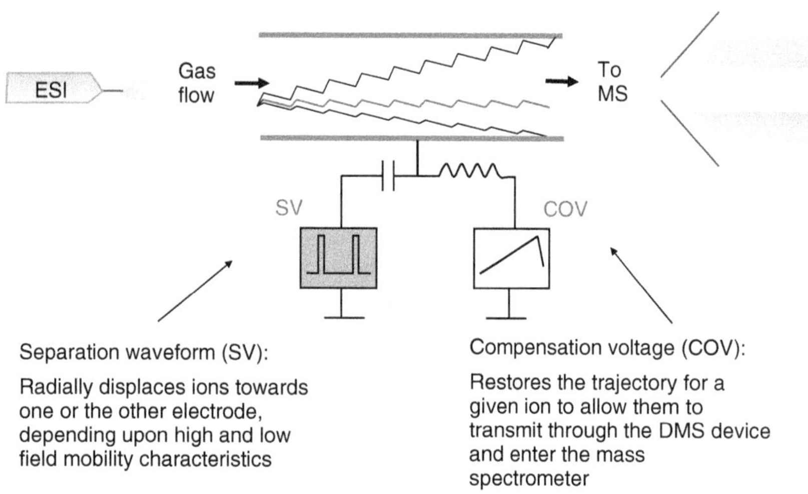

Figure 4.3 SelexION™ technology (differential mobility separation (DMS)): The DMS cell is easily mounted without any tools required. The DMS cell consists of two plates (planar geometry) where the gas flows through the cell transporting ion toward the MS orifice. An asymmetric waveform is applied that alternates between high- and low-field energy, which results in a separation voltage (SV). The ions move back and forth between the plates as they are dragged toward the exit of the cell by gas flow, and will have a net drift toward one of the electrodes based on their high and low field mobility. The compensation voltage (COV), which is a small DC offset applied to one of the electrodes, is optimized for the ion of interest and ensures transmission through the cell. The COV can be considered a filtering voltage. Individual peptides can be tuned with specific COV values and hence separated from other background and interfering peaks.

TOF analyzer detects all the fragments from the precursor at high resolution and high mass accuracy. Using narrower extraction widths than the unit resolution of triple quadrupole-based experiments, difficult separations between background peaks and analytes can now be achieved and improved to such an extent that minimal interferences are observed. When fragment ions are extracted at these narrow extraction widths, analytes can be detected at higher specificity and at accurate mass in complex matrices (Figure 4.4).

4.3.4 Software

The evaluation of protein and peptide quantitation results can often be time-consuming and repetitive and relies on manual peak identification and data integration – a process that does not lend itself well to the high-throughput environment. Comprehensive, powerful, and easy-to-use software solutions such as MultiQuant™ Software and DiscoveryQuant™ Software have been developed that enable the simultaneous and automated processing of multiple analytes. Not only do these software packages rapidly

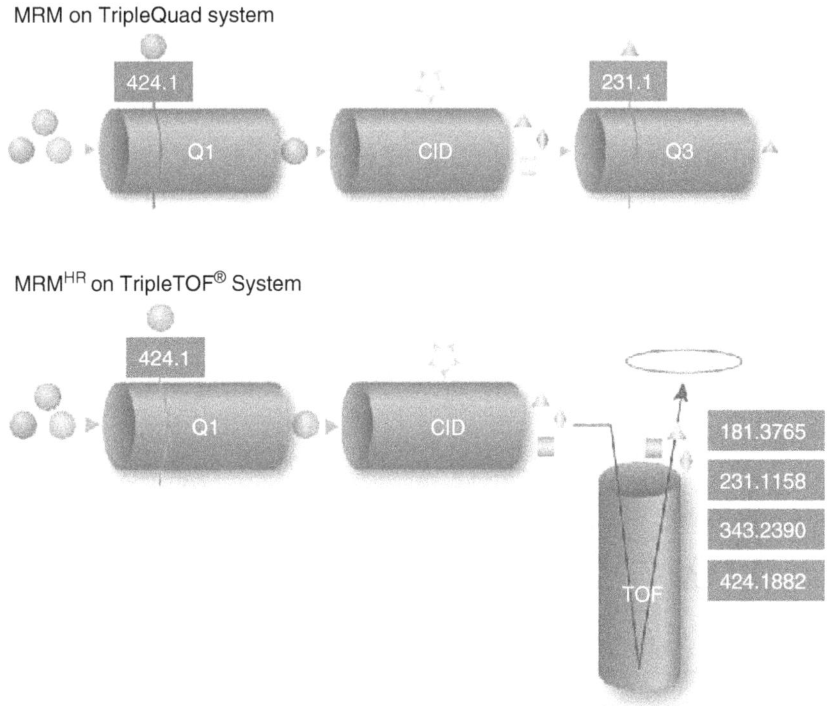

Figure 4.4 Schematic overview of MRM^{HR} acquisition technique on a TripleTOF® System compared to standard triple quadrupole mass spectrometer. The high-resolution multiple reaction monitoring (MRM^{HR}) method carried out on a fast-scanning high-resolution mass spectrometer allows for accurate quantification due to the acquisition of a large number of data points per peak and wide linear dynamic range.

process MS scans and data, but the software also supports improved data integrity and security, unique audit trail functionality for improved regulatory compliance, and an embedded digital link to the Watson laboratory information management system (LIMS) for increased confidence in data safety.

4.4 Advantages of the Diversity of Mass Spectrometry Systems

We primarily focus on experiments conducted on two types of hybrid triple quadrupole instruments: (i) high-resolution, accurate-mass spectrometry quadrupole time-of-flight (QTOF) system and (ii) the Triple Quad or the QTRAP systems. Each platform has distinct advantages. The high-resolution, accurate-mass spectrometry system is uniquely suited to qualitative discovery (as well as quantitation) due to the underlying acquisition of a full spectrum of secondary fragments at high resolution, while the QTRAP system and its augmented ion generation, transmission, and detection works best for applications requiring high sensitivity and expanded linear ranges. The QTRAP system is fully accepted for regulated bioanalysis at the Phase 1 level and above, but the high-resolution, accurate-mass spectrometry system dominates in ease of method development and nontargeted analysis during drug discovery protocols. In the event that one application demands the benefits and strengths of an alternative MS platform, the transfer of methods is easy and intuitive; the two MS systems have identical source and collision cell designs based on the innovative LINAC® collision cell, which allows for seamless coordination of quantitative data with qualitative analysis (Figure 4.5).

4.5 Perspectives for the Future

As instrumental technological innovations surpass the quantitative limitations imposed by biological sample complexity, LC/MS/MS biopharmaceutical quantitation will become more fully established as a routine methodology in the regulated laboratory. Time-consuming and complicated sample preparation steps will evolve to become better suited to the automated requirements of the MS-based bioanalytical workflow, and sample extraction procedures are likely to become more highly selective to achieve the sensitivities required for monitoring subpicomolar concentrations of biotherapeutic agents. Highly sensitive methods based on the enhanced MS ionization efficiency and transmission have yielded promising results on the QTRAP system and have produced sufficient LLOQs for low-level biomolecule quantitation needed for PK and TK studies. In addition, distinct gains with the use of DMS and MRM^3 add an additional layer of selectivity and remove hard-to-separate background and leading to better signal-to-noise

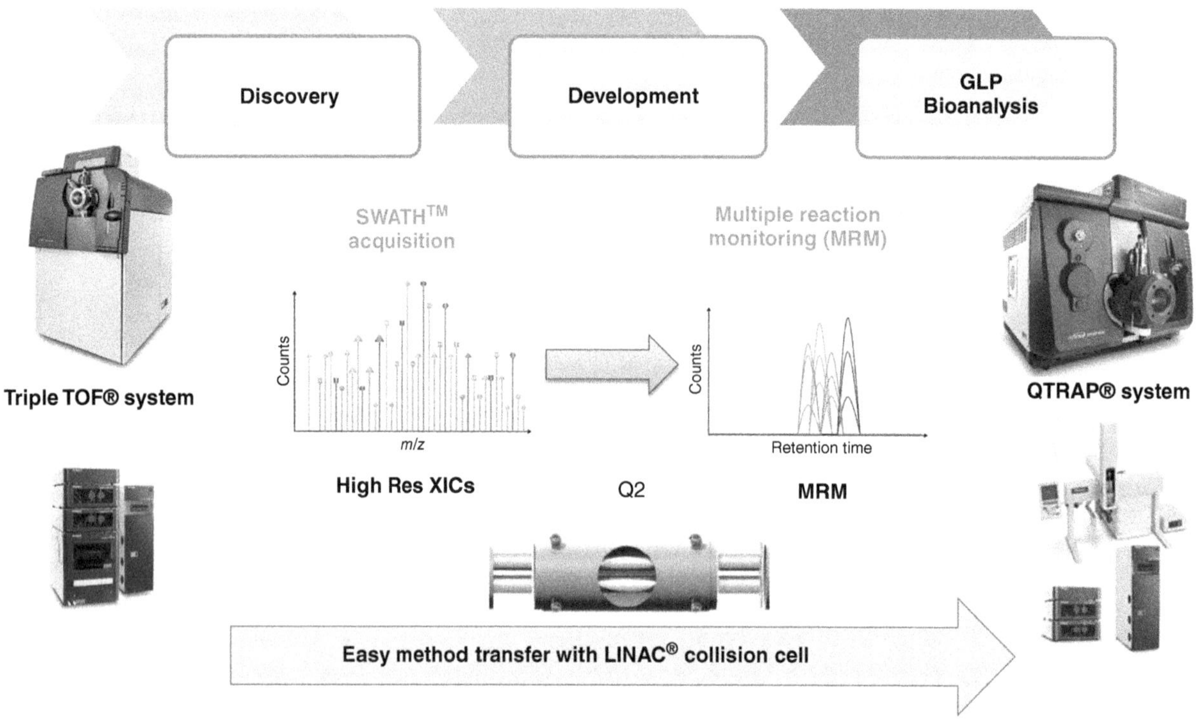

Figure 4.5 Continuity of workflows between TripleTOF to QTRAP. From product characterization during research and development process to biotransformation and bioanalysis during PK/PD analysis in preclinical and clinical studies.

parameters. The potential of high-resolution mass spectrometry to measure intact, high-molecular-weight biomolecules will gain significant interest as technological advances push TOF sensitivities toward those of the hybrid linear ion trap instruments. By reducing the need for additional sample preparation steps with enhanced MS detection and selectivity capacities, LC/MS/MS techniques will become more closely aligned with the high-throughput workflows necessary for regulated bioanalysis.

References

Bischoff R, Bronsema KJ, van de Merbel NC, et al. (2013). "Analysis of biopharmaceutical proteins in biological matrices by LC-MS/MS I. Sample preparation." *TrAC Trends Anal Chem* **48**: 41–51.

Ezan E and Bitsch F (2009). "Critical comparison of MS and immunoassays for the bioanalysis of therapeutic antibodies." *Bioanalysis* **1**(8): 1375–1388.

FDA (2001). "*Guidance for industry; bioanalytical method validation." US Department of Health and Human Services, Food and Drug Administration, Center for Drug Evaluation and Research,* Rockville, MD.

Hopfgartner G and Bourgogne E (2003). "Quantitative high-throughput analysis of drugs in biological matrices by mass spectrometry." *Mass Spectrom Rev* **22**(3): 195–214.

Hopfgartner G, Lesur A and Varesio E (2013). "Analysis of biopharmaceutical proteins in biological matrices by LC-MS/MS II. LC-MS/MS analysis." *TrAC Trends Anal Chem* **48**: 52–61.

Liu G, Ji QC, Dodge R, et al. (2013). "Liquid chromatography coupled with tandem mass spectrometry for the bioanalysis of proteins in drug development: Practical considerations in assay development and validation." *J Chromatogr A* **1284**: 155–162.

Reichert JM (2013). "Which are the antibodies to watch in 2013?" *mAbs* **5**(1): 1–4.

Shi T, Su D, Liu T, et al. (2012). "Advancing the sensitivity of selected reaction monitoring-based targeted quantitative proteomics." *Proteomics* **12**(8): 1074–1092.

Smith S (2013). "Biopharmaceuticals – a global market overview." Reportbuyer.com, http://www.prweb.com/releases/2013/11/prweb11314067.htm.

Thompson B (2012). "Driving high sensitivity in biomolecular MS." *Genet Eng Biotechnol News* **32**(20). http://www.genengnews.com/gen-articles/driving-high-sensitivity-in-biomolecular-ms/4603/?kwrd=high%4620sensitivity%4620in%4620biomolecular%4620MS.

van den Broek I, Niessen WMA and van Dongen WD (2013). "Bioanalytical LC–MS/MS of protein-based biopharmaceuticals." *J Chromatogr B* **929**: 161–179.

Zhong X, Peter N, Corbo M, et al. (2011). Recent advances in biotherapeutics drug discovery and development, in *Drug Discovery and Development – Present and Future,* Kapetanović I (Ed.), InTech, Croatia.

5

Peptide and Protein Bioanalysis Using Integrated Column-to-Source Technology for High-Flow Nanospray

Shane R. Needham[1] *and Gary A. Valaskovic*[2]

[1] *Alturas Analytics, Moscow, ID, USA*
[2] *New Objective Inc., Woburn, MA, USA*

5.1 Introduction – LC–MS Has Enabled the Field of Protein Biomarker Discovery

Molecular biomarkers of disease and treatment efficacy, comprised of genes, proteins, peptides, and metabolites, present unique analytical challenges as they are required for the development and success of personalized molecular medicine (Duffy and Crown 2008, Overdevest et al. 2009, Heger and Kellogg 2014). Furthermore, the confluence of advances in genomics, mass spectrometry, and separation science places mass spectrometry-based sciences at a unique time. These advances combined with an increased focus on the patient appear to have positioned mass spectrometry to make significant and sustained contributions to fundamental life science and applied human health.

Mass spectrometry has played a central role in the fundamental development of global, qualitative proteomics. (Answering the question: "What proteins are present in my sample?"). In particular, electrospray ionization mass spectrometry (ESI-MS), coupled to nanobore liquid chromatography (nLC) has revolutionized the identification and analysis of proteins and peptides present in cells, tissues, and biological fluids (Haynes and Yates 2000, Pandey and Mann 2000, Washburn et al. 2001). The previous technology for protein characterization, Edman sequencing, required picomoles of purified material (as single proteins/peptides) analyzed over many hours (approx. 0.5 h per amino acid) of time. nLC-MS, stands in stark contrast, as it is capable of the shotgun identification of many thousands of proteins present in complex mixtures obtained from subfemtomole (10^{-15}) samples in approximately 1 h (Hebert et al. 2014).

This facile tool for protein identification has rapidly evolved into the key platform for the discovery of protein and peptide biomarkers. (Answering the question: "Which proteins manifest, characterize, or reflect the disease/treatment state?"). Nanobore LC–ESI-MS platforms for qualitative proteomics and biomarker discovery are essentially identical. The biomarker discovery platform typically employs methods of "relative quantitation" in which protein abundance between disease (or treatment) and nondisease (or nontreatment) populations are compared through a combination of bioinformatic and statistical strategies (Haynes and Yates 2000, Meng et al. 2007, Yates and Washburn 2013). The desired result from such comparisons is a list of target protein(s) or peptides of high relevance characterizing the disease state, and/or the efficacy of medical treatment.

For such biomarkers to translate to clinical utility through the process of validation, the analytical question shifts from qualitative to quantitative (Carr and Anderson 2008, Boja et al. 2011, Anderson 2012). The analytical methodology must be suitable to answer the question: How much of my desired protein target, or targets, are present? This analytical methodology is, of course, the realm of quantitative bioanalysis by mass spectrometry.

Qualitative LC–MS systems that are used to identify proteins, and discover protein biomarkers, differ from the systems typically employed in bioanalytical quantitation. Different sample preparation and chromatographic formats are used, which ultimately translates into different laboratory workflows. Complex sample mixtures that result from biomarker discovery activities place a high emphasis on sensitivity and extreme chromatographic performance in terms of separating power and peak capacity (long columns, using slow gradients). Generally, these formats are not amenable to high-throughput assays. Throughput for biomarker discovery activities is typically measured in tens of samples analyzed per day.

Quantitative bioanalytical assays also place a high emphasis on sensitivity, though on a limited number of

Protein Analysis using Mass Spectrometry: Accelerating Protein Biotherapeutics from Lab to Patient, First Edition.
Edited by Mike S. Lee and Qin C. Ji.

analytes, with a high degree of selectivity. Triple-quadrupole mass spectrometry is typically the preferred platform. LC performance, however, is usually targeted and measured in terms of throughput rather than separation power and peak capacity (short columns, using fast gradients). Throughput is relatively high with bioanalytical quantitation applications and typically generates 200 to more than 1000 samples per day.

To be successful in the translation of biomarker discoveries to clinical utility, it is of keen interest to combine the desirable characteristics of the traditional qualitative proteomics workflow within the quantitative bioanalytical landscape. Sensitivity must be preserved, as proteins are present in biofluids (e.g., blood, plasma, urine, spinal fluid) over a concentration range of up to 11 orders of magnitude (Anderson and Anderson 2002). It is highly probable that biomarkers of interest are present at low-to-sub picogram-per-mL levels (Anderson 2010). Chromatographic separation quality must also be preserved, as endogenous protein biomarkers, and the chemically digested peptides that are the analytical surrogate for those proteins (Addona et al. 2009) suffer from high levels of mass interference and, thus, result in a strictly limited selectivity for tandem mass spectrometry (MS/MS) (Sherman et al. 2009). LC–MS technology developments (sample preparation, chromatography, ionization, mass analysis) that bridge and translate these requirements will be key for success.

5.2 Integration of Miniaturized LC with Nanospray ESI-MS Is a Key for Success

LC–MS has experienced significant technical evolution having established trends toward decreasing column diameter, lower flow rates, and smaller column packing particle sizes (Lee and Kerns 1999). State-of-the-art LC has evolved from 4.6-mm-inside-diameter columns (ID) with 1 mL/min flow rates to 1–2-mm-ID columns that operate at less than 200 µL/min. Smaller micro- (0.2–0.3 mm ID) and nanoscale (<0.2 mm ID) column formats that operate at 10 and 0.3 µL/min, respectively, have strong application-specific roles, particularly when high sensitivity is required and/or sample volumes are strictly limited (Tomer et al. 1994).

There are many driving factors in the trend to use smaller diameter columns. Primary benefits in the switch to these smaller columns include reduced solvent consumption, improved cleanliness of the mass spectrometer source/inlet/vacuum system, and perhaps most importantly, reduced sample injection volume. A 0.3-mm-ID column, on average, will consume one-twentieth and one-hundredth the solvent required for 2- and 4.6-mm-ID columns, respectively. Less solvent consumption reduces both purchase and chemical waste disposal costs; small diameter columns are truly "green." The low flow rates associated with small ID columns achieve the same cleanliness goals as the typical postcolumn LC divert valve approach (Wang et al. 2000). Less mobile phase flowing through the ESI source, and presented to the front-end inlet of the mass spectrometer, translates directly into a cleaner MS and reduces the maintenance expense and increases the MS acquisition time.

The reduced injection volume required by small ID columns, and the resulting relative concentration advantage when using reverse-phase gradient elution HPLC, results in clear experimental advantages. As shown in Figure 5.1, the concentration of analyte on the head of the column, for injections of a fixed volume, increases as the column diameter decreases (Tomer et al. 1994). A 0.3-mm-ID column has a 40-fold concentration advantage compared to a 2.1-mm-ID column, while a 75-µm-ID column has a nearly 800-fold advantage. This concentration benefit is often described in the literature as an "increase in (column) sensitivity." However, this concentration benefit is perhaps more correctly viewed as an effective *sampling advantage*. Micro- and nanobore columns enable the handling and analysis of much smaller absolute sample sizes and/or concentration of trace components present in larger sample volumes. Reduced sampling volume has an underappreciated workflow benefit. This benefit is unrelated to sensitivity. To illustrate the advantages of smaller ID columns and reduced sampling volume, consider the following example. A 100 µL sample is sufficient for nearly twenty 5 µL injections with microscale LC, compared with only two 40 µL injections using a conventional LC column. Thus, the reasons to implement 0.3-mm-ID (and smaller) columns are as compelling as the motivation to switch from 4.6-mm-ID columns to 2.1-mm-ID columns. Certainly, it is difficult to envision modern quantitative bioanalytical laboratories migrating back to the routine use of 4.6 mm columns.

Much attention has been given to microsampling strategies such as dried blood spots (De Jesús and Chace 2012, Ji et al. 2012, Meesters and Hooff 2013) (DBS) or capillary sampling (Bowen et al. 2013, Nilsson et al. 2013, Spreadborough et al. 2013) as methodologies suitable for use in discovery bioanalytical, regulated bioanalytical, and clinical laboratory settings. The analysis of reduced sample volumes (typically less than 10–20 µL of blood or plasma) obtained from microsampling benefits from miniaturized LC formats (Rainville et al. 2011, Arnold and Needham 2013). The combination of microsampling and micro-LC, for example, enables serial sampling from small animal models such as mice (Rahavendran et al. 2012) while maintaining a low limit of quantitation.

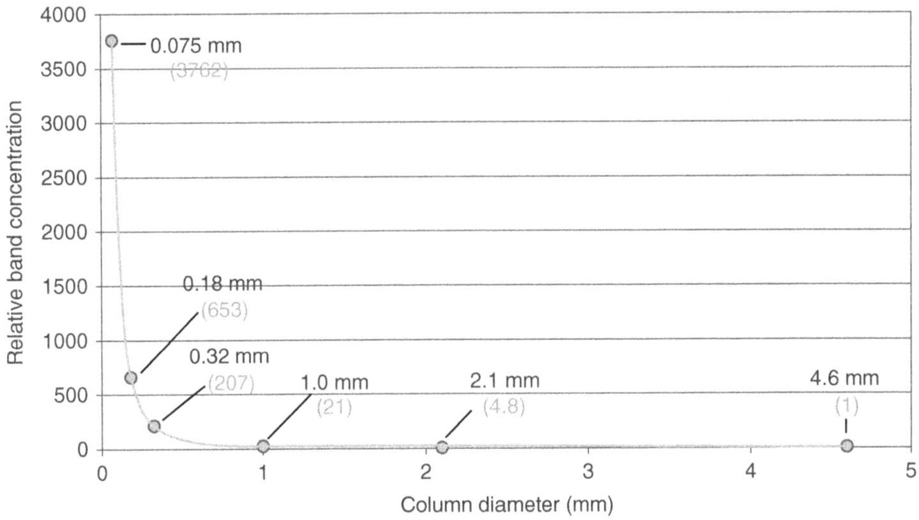

Figure 5.1 Theoretical relative concentration enrichment versus column inside diameter for a fixed injection volume. Values in parentheses are the relative concentrations normalized to 1 for a 4.6-mm-ID column.

Along with the compelling advantages provided by miniaturized sampling and separation, detection by ESI-MS must also fit the paradigm of miniaturization. A driving factor to trend to lower flow rates for ESI-MS is an improvement in ionization efficiency. Early work on ESI response demonstrated an increase in proportional MS signal-to-noise ratio (S/N), as the mobile phase flow rate of ESI is reduced (Kebarle and Tang 1993, Juraschek et al. 1998, Geronmanos et al. 2000). Thus, a system that provides optimal chromatographic benefits and optimal flow rates to the MS is an ideal situation for high-throughput, high-sensitivity LC–MS bioanalysis. Nanobore chromatography, using LC columns having a typical ID of 150 μm or less, is conveniently coupled to mass spectrometry via nanospray ionization, a low-flow variant of ESI (Wood et al. 2003). Nanospray provides a highly efficient means to transport liquid ions to gas phase ions with little or no added thermal energy. The low flow rates (10–500 nL/min) associated with nanospray result in the generation of submicrometer droplets and generate a maximal surface area-to-volume ratio of column effluent (Wilm and Mann 1994). Such high surface area translates directly to high ionization efficiency (Juraschek et al. 1999, Cech and Enke 2000, Valaskovic et al. 2006). The low thermal requirement for nanospray is an added benefit as it is directly compatible with high-molecular-weight, thermally sensitive peptides and proteins (Wilm and Mann 1995, Valaskovic et al. 1996). Additional advantages at ultralow flow rates (<100 nL/min) include reduced ion suppression (Hatsis et al. 2009), a trend toward equimolar response (Valaskovic et al. 2006), linear ionization response (Wickremsinhe et al. 2006), and improved S/N (Zhou et al. 2012, 2013).

In addition to the smaller droplets afforded by low flow rates, the overall physical size of the ESI plume also scales with flow rate. Conventional flow (mL/min) ESI-MS ESI generates an aerosol plume that is centimeters in diameter. Compared with the (sub) millimeter MS inlet, only a small fraction of the plume is actually sampled (Schneider et al. 2005, 2006). Under conventional conditions, the MS inlet is an atmospheric pressure flow splitter, with greater than 99% of the generated plume flowing to waste. High-flow nanospray and microspray, with a (sub) μL/min flow rate, generates an aerosol plume that is on the same dimensional scale as the MS inlet (Valaskovic et al. 2004, Schneider et al. 2006) (Figure 5.2). This gain in apparent sensitivity is best thought of in terms of utilization and sampling efficiency (Juraschek et al. 1998). Zero postcolumn waste directly translates into the ability to handle microscale samples with complete efficiency. In some cases, especially for the detection of peptides in acidic conditions, ultralow flow rates can translate into higher ion-current (Geronmanos et al. 2000, Tang et al. 2004, Valaskovic et al. 2006), and this is, of course, an analyte and matrix-dependent observation.

5.3 Micro- and Nano-LC Are Well Suited for Quantitative Bioanalysis

The combination of micro- or nano-LC combined with nanospray ionization would appear to yield a nearly ideal platform for high-sensitivity bioanalysis by LC–MS/MS. In theory and in the hands of experts, nanospray ionization does indeed provide for an ideal platform. However,

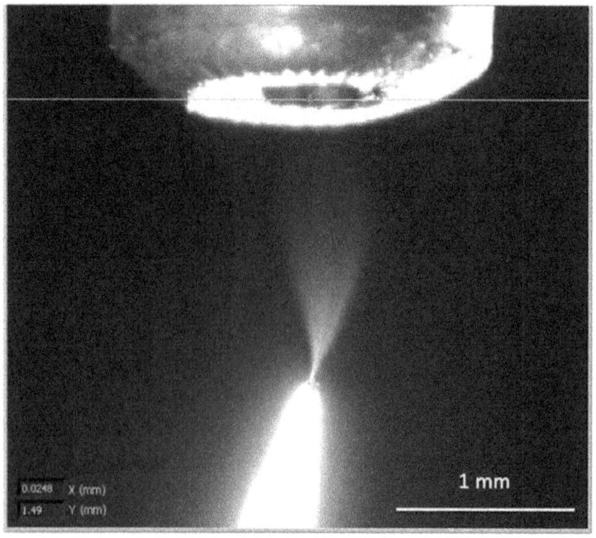

Figure 5.2 Magnified view of a 15 μm-ID nanospray emitter at 1800V operating on a Thermo Scientific LTQ™ mass spectrometer equipped with a New Objective Digital PicoView™ nanospray source. Mobile phase 30% ACN, 0.1% formic acid, flowing at approx. 300 nL/min. Distance between the MS inlet and the emitter is 1.5 mm.

experimental challenges must be taken into consideration when such a platform is transitioned into the mainstream analytical laboratories (i.e., high flow rates, conventional chromatography, and ESI). Thus, assembly and use of a high-performance micro- or nano-LC system is presently in the domain of the expert user. Challenges in pre- and postcolumn plumbing, sample preparation of complex matrices, and mass spectrometer tuning requires significant training and experience with method development.

A common misconception is that miniaturized LC is "too slow" for use in bioanalysis applications that require short run times and high throughput. The casual observation of the mobile phase flow rate (0.3–10 μL/min) would suggest this is the case. However, it is important to realize that the linear velocity of analyte through a micro- or nanobore column is *identical* to a conventional column. This misperception appears to have two different origins: Early low-flow LC pumps suffered from significant gradient delay. Ten or more minutes were required to move mobile phase from the piston of the pump through the autosampler and onto the column. Second, much of the application literature for miniaturized LC formats feature qualitative peptide analysis using injection-to-injection cycle times measured in hours or days. These long cycle times are necessary due to the complexity of the sample (trace-mixture analysis) faced in qualitative discovery experiments. The long cycle times feature shallow gradients and long LC (≥25 cm) columns; both essential to achieve sufficient chromatographic peak capacity for the mass spectrometer to effectively sample the mixture (Hsieh et al. 2013, Hebert et al. 2014).

Attention to the overall plumbing and connection scheme, starting at the outlet of the HPLC pump, is required (Table 5.1). Critical factors include total system volume, the injection volume, and pre- and postcolumn volumes. Precolumn volume is clearly important as the gradient delay time from pump, through the autosampler, and onto the column, places the ultimate limit on the minimum injection cycle time. It is important to note that a 0.3-mm-ID column system, with a careful choice of connection tubing (from 20 to 50 μm ID), can yield a competitive relationship. System volume does not need to be a limiting factor for injection cycle time.

Controlling and reducing postcolumn volume is perhaps the most critical parameter to maintain the quality of LC peak shape. Unlike precolumn volume, which adds to the injection dispersion prior to sample stacking on the head of the column, postcolumn volume adds directly to the dispersion of the eluting peak. Integration of the

Table 5.1 Representative flow rate, system volume, delay time, and flow rate-to-volume ratio for differing scales of HPLC column ID.

Column ID (mm)	Flow (μL/min)	Tubing ID (μm)	Volume (μL/cm)	Length (cm)	Volume (μL)	Delay (s)	Flow/volume
4.6	1000	0.010″ (250 μm)	0.507	60	30.4	1.8	32.9
2.1	200	0.007″ (175 μm)	0.248	60	14.9	4.5	13.4
0.3	10	0.0016″ (50 μm)	0.020	60	1.2	7.1	8.5
0.3	10	0.0008″ (20 μm)	0.003	60	0.2	1.1	53.8
0.075	0.25	0.0016″ (50 μm)	0.020	60	1.2	282.2	0.2
0.075	0.25	0.0008″ (20 μm)	0.003	60	0.2	44.6	1.3

Note that a 0.3-μm-ID-column system plumbed with 20-μm-ID tubing can have a competitive flow-to-volume ratio with conventional LC formats. This parameter places an ultimate limit that dictates injection cycle time on an LC–MS system. A 60 cm total length of tubing volume is assumed.

ionization emitter into the LC column body eliminates postcolumn volume. Indeed, the technology of the so-called "packed-emitter" column (PicoFrit™ column, Figure 5.3a) has been the *de facto* standard in qualitative proteomics for the past 20 years (Andren et al. 1994, Emmett and Caprioli 1994). As shown in Figure 5.3, the complete elimination of postcolumn volume yields top chromatographic peak shape. For reference, the total column volume was 450 nL/min, and the volume of a typical eluting peak was approximately 30 nL. The postcolumn volume was only on the order of 3% of the column volume. Even this relatively insignificant postcolumn volume, which is generated by a 4 cm length of 20 μm ID tubing (12 nL), compromises peak shape.

5.4 Demonstrating Packed-Emitter Columns Are Suitable for Bioanalysis

Traditional nanospray methods that feature the use of packed-column emitters offer an exceptionally high degree of performance. For optimal performance, nanospray sources are typically equipped with precision translation stages for optimal MS inlet capture, specialized means for the application of high-voltage precolumn, and specialized zero dead volume connections. Such tools are indeed high performance, but are not typically perceived to be easy-to-use, particularly from the nonexpert, nonpractitioner. Requirements for training are high relative to conventional LC/MS-based methods; and thus, these methods are typically taught in graduate-level programs. Recent research and development efforts have focused on translating such methods to easier-to-use formats that incorporate a high degree of integration. The elements of emitter size and shape, emitter position, application of applied voltage, zero-dead-volume connections, and temperature control are especially critical for success. Thus, these elements are combined to provide high performance together with ease of use and robustness. Figure 5.4 highlights two examples of commercially available approaches to such integration. The first is the integration of a nano-/microbore LC column directly inside a conventional ESI electrode/probe assembly (Figure 5.4a the PicoFuze™ column). The second is a dedicated package developed specifically for high-performance

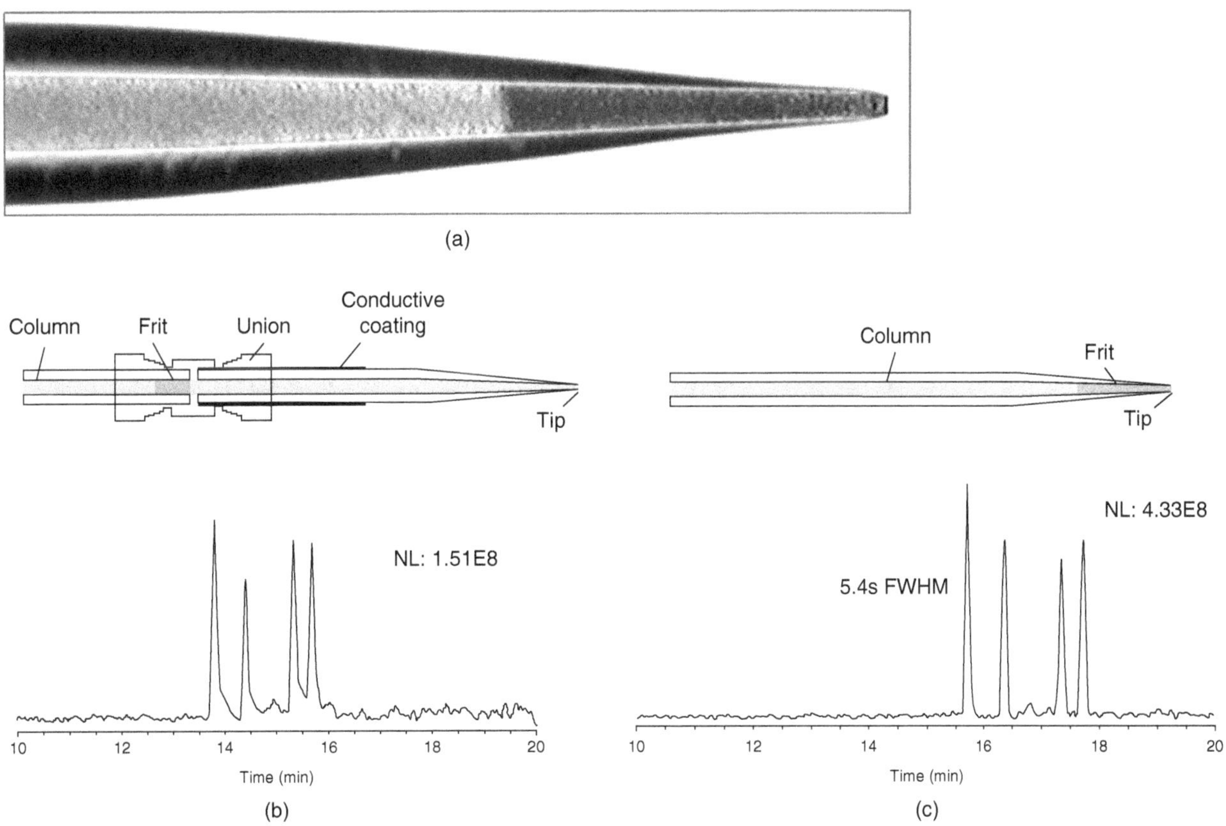

Figure 5.3 (a) Photomicrograph of a 75 μm, 3.5 μm C18, packed emitter column (PicoFrit™, New Objective Inc.) and (b) reconstructed base peak chromatograms (ion trap mass spectrometry) comparing identically prepared conventional nanobore (left) column and packed emitter nanobore (right) column. 25 Femtomole on-column injection of four angiotensin peptides eluted with a linear acetonitrile–water gradient (0.1% formic acid). Note the improved signal and elimination of tailing for the packed emitter column.

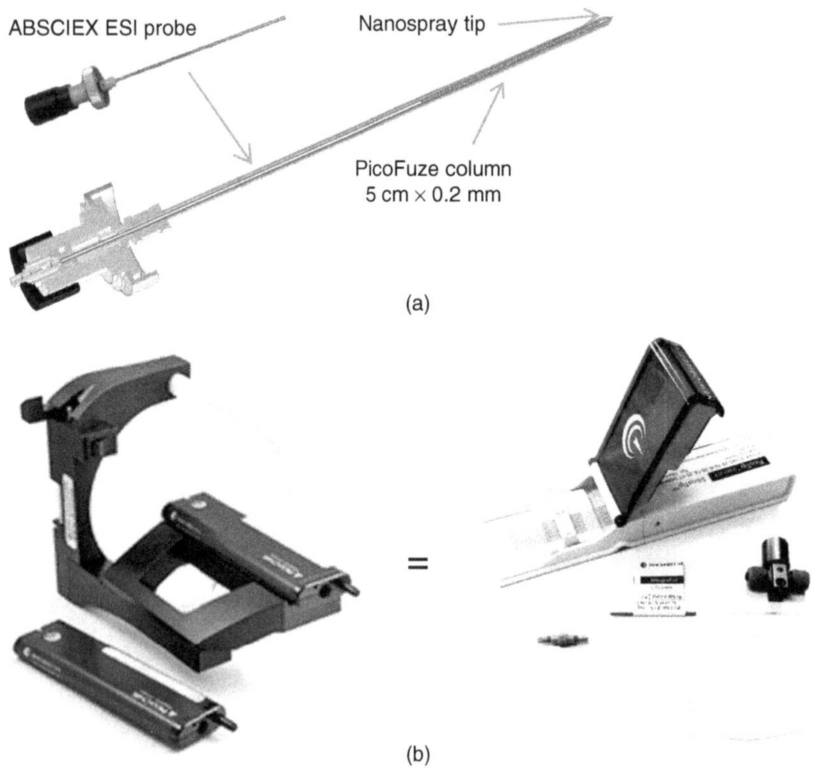

Figure 5.4 (a) Photo and sectional view of an AB SCIEX ESI electrode modified to hold a PicoFuze™ column, directly inside the needle assembly. The column dimensions are 5 cm × 0.2 mm ID. (b) The PicoChip™ system (left) for easy-to-use integrated nanospray. The PicoChip™ and PicoFuze™ column systems replace all of the components (ESI emitter, coupling union, column, HV contact, and transfer line) shown on the right-hand side of Figure 5.5(b).

nano- and microspray (Figure 5.4b the PicoChip™ column). Both technologies are deeply rooted in the traditional fused-silica packed emitters used successfully in qualitative proteomics. Note that each integrated device replaces a host of fittings, tubing, connections, and, importantly, the ESI emitter. Advantages of this integrated approach include ease of use, new high-voltage contact with each column, control or elimination of pre- and postcolumn volumes, a new ESI spray assembly with each column replacement, and preservation of LC performance.

Figure 5.5 illustrates the representative SRM chromatograms obtained from sample injections (1 μL injection) of a tryptic peptide of human monoamine oxidase B (MAO-B) on three different columns: 2 mm conventional, 0.3 mm microflow, and 0.2-mm-ID PicoFuze™, each packed with identical C18 media (ProntoSil, Bischoff Chromatography) to a bed length of 5 cm. Note that the signal response and reduction in peak width is improved with the PicoFuze™ column compared to the microflow and conventional column data.

A peptide quantitation experiment that featured the use of PicoFuze™ columns with human plasma matrix samples was used to test this integrated concept. A triple quadrupole MS (5600, AB SCIEX) was connected to a microflow HPLC and autosampler (MicroLC, Eksigent Technologies). The conventional ESI electrode was replaced with a 0.2-mm-ID × 5 cm PicoFuze™ column (New Objective, Inc.) packed with 3 μm ProntoSil C18 (Bischoff Chromatography). The "surrogate peptide" quantitation method of Olah and coworkers (Ouyang et al. 2012) was adopted for the analysis of human MAO-B on a triple quadrupole LC–MS/MS system. Briefly, in this surrogate approach, one or more peptides, resulting from enzymatic tryptic digestion of the target protein residing in a pellet precipitate, and meeting the validation criteria for genetic and analytical specificity, are selected as the analyte for targeted quantitative measure. Isotopically labeled synthetic peptide, of the same amino acid sequence as the surrogate peptide, is preferably used as the analytical internal standard (IS) for absolute quantitation.

Conventional bioanalytical protocols were followed to establish analytical figures of merit (accuracy and precision). Human MAO-B was extracted from human plasma using organic precipitation and "pellet digestion" using trypsin (Ouyang et al. 2012). Following the surrogate peptide selection scheme, isotopically labeled (stable label) peptide was spiked into calibration and QC samples over a range from 1 to 100 μg/mL. Eight standard curve points and five quality control levels were established (LLOQ, LQC, MQC, HQC, and ULOQ). A summary of the validation data is presented in Table 5.2. Plots of the analyte-to-IS response ratio for the surrogate,

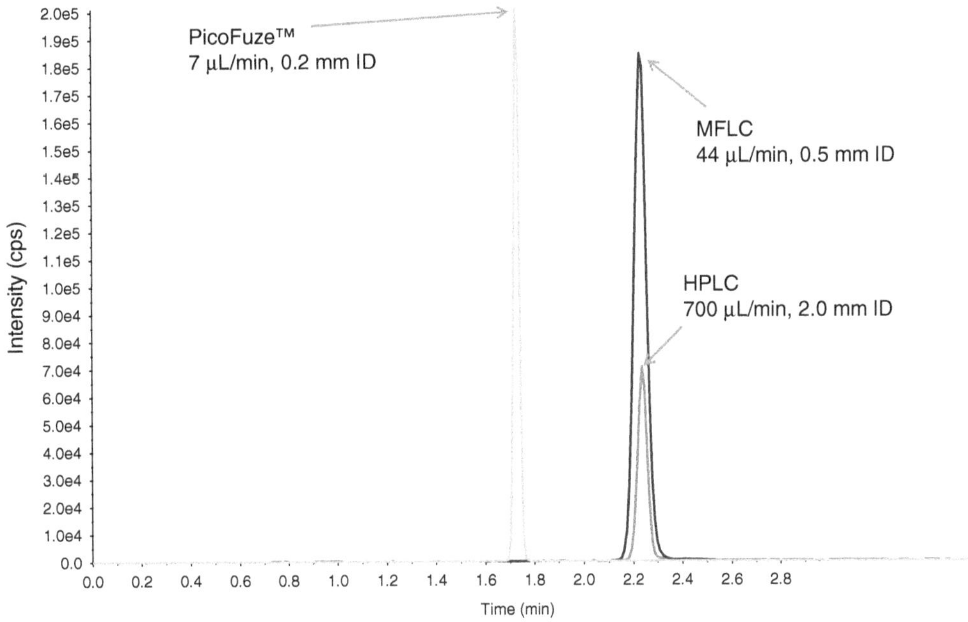

Figure 5.5 Representative gradient elution SRM chromatograms for a 1 μL injection of MOAB digest (standard surrogate peptide), at a concentration of ng/mL. Each column was packed with identical 3 μm, C18 packing material. Column diameter and column flow rates are shown, each column was 5 cm in length.

Table 5.2 Summary of validation for the surrogate peptide of MAOB

Surrogate peptide accuracy/precision			
Sample name	**Concentration (μg/mL)**	**Average accuracy (%)**	**%CV**
LLOQ	1.00	118	19.5
LQC	3.00	93.9	22.6
MQC	15.0	86.6	13.9
HQC	80.0	86.8	23.8
ULOQ	100	94.6	8.9

Average accuracy and %CV are shown for lower limit of quantitation (LLOQ), lower QC (LQC), mid QC (MQC), high QC (HQC), and upper limit of quantitation (ULOQ).

and a confirmatory peptide versus concentration are shown in Figure 5.6. The data was fitted to a quadratic curve and the value of r was 0.9906 and 0.9913 calculated for the surrogate and confirmatory peptides, respectively. While the accuracy and % CV data do not meet stringent acceptance criteria (better than 15% CV), it should be kept in mind that this is first-in-kind data, and much about the overall protocol is not optimized (i.e., digestion conditions, recovery methods, ESI tuning conditions). Additional validation studies of this technology are underway in the author's respective laboratories. The fundamental advantages of small injection volume, lower solvent consumption, and reduced costs will be beneficial to bioanalytical environment.

5.5 Future Outlook

The history of analytical measurements by LC–MS/MS has repeatedly followed the path from qualitative analysis to quantitative analysis (Lee and Kerns 1999). The need to shift the analytical question from which analytes are present in a sample to the quantity of analytes present is largely driven by the importance of the measurement. These measurements will have an impact on the diagnosis, evaluation, and treatment of human health and are clearly important questions that require quantitative bioanalytical measurements. The analytical advantages miniaturized LC and ESI have empowered qualitative proteomics and metabolomics over the past 20 years. These methods and, of course, analytical trends toward miniaturization and integration will likely continue to be adopted into critical quantitative measurements in the years ahead. Continuous improvements to instrumentation and methods will pave the way to a new era of mass spectrometry in the clinical laboratory (Kushnir et al. 2013). The hallmarks of such an era will likely feature analytical platforms that are easy to use, high performing, and easily adaptable to a clinical laboratory workflow (Chace et al. 2015).

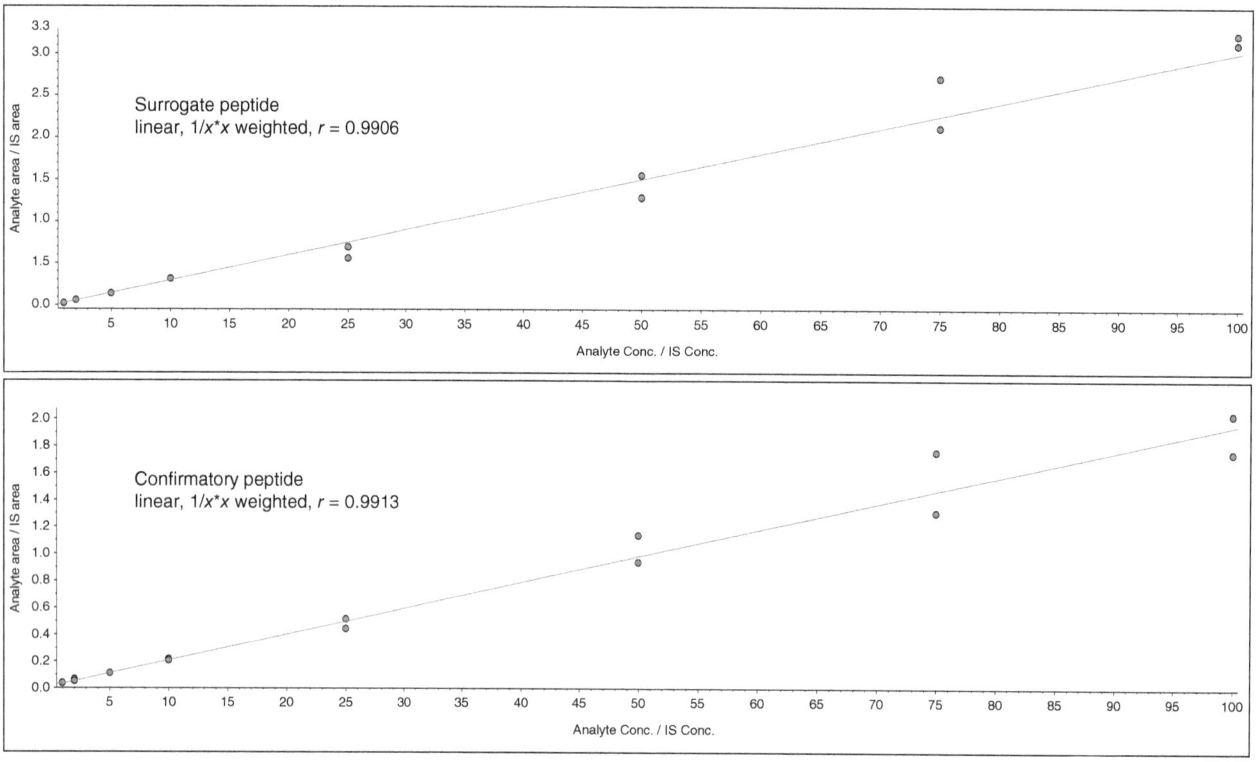

Figure 5.6 Peak area ratio (analyte/internal standard) for the surrogate peptide and confirmatory peptide for the MAOB digest. The data were plotted and fitted to a quadratic curve, with a calculated *r* value as shown.

References

Addona TA, Abbatiello SE, Schilling B, et al. (2009). "Multi-site assessment of the precision and reproducibility of multiple reaction monitoring-based measurements of proteins in plasma." *Nat Biotechnol* **27**(7): 633–641.

Anderson L (2012). "Within sight of a rational pipeline for development of protein diagnostics." *Clin Chem* **58**(1): 28–30.

Anderson NL (2010). "The clinical plasma proteome: a survey of clinical assays for proteins in plasma and serum." *Clin Chem* **56**(2): 177–185.

Anderson NL and Anderson NG (2002). "The human plasma proteome: history, character, and diagnostic prospects." *Mol Cell Proteomics* **1**(11): 845–867.

Andren PE, Emmett MR and Caprioli RM (1994). "Micro-electrospray: zeptomole/attomole per microliter sensitivity for peptides." *J Am Soc Mass Spectrom* **5**(9): 867–869.

Arnold DW and Needham SR (2013). "Micro-LC-MS/MS: The future of bioanalysis." *Bioanalysis* **5**(11): 1329–1331.

Boja ES, Jortani SA, Ritchie J, et al. (2011). "The journey to regulation of protein-based multiplex quantitative assays." *Clin Chem* **57**(4): 560–567.

Bowen CL, Licea-Perez H, Karlinsey MZ, et al. (2013). "A novel approach to capillary plasma microsampling for quantitative bioanalysis." *Bioanalysis* **5**(9): 1131–1135.

Carr SA and Anderson L (2008). "Protein quantitation through targeted mass spectrometry: the way out of biomarker purgatory?" *Clin Chem* **54**(11): 1749–1752.

Cech NB and Enke CG (2000). "Relating electrospray ionization response to nonpolar character of small peptides." *Anal Chem* **72**(13): 2717–2723.

Chace DH, Valaskovic GA, Lee MS, et al. (2015). "Multiple metabolite profiles and LC-MS: applications in screening and diagnostics," *Advanced LC-MS Applications in Metabolomics*, Future Science Ltd, London, pp. 22–39.

De Jesús VR and Chace DH (2012). "Letter to the editor: commentary on the history and quantitative nature of filter paper used in blood collection devices." *Bioanalysis* **4**(6): 645–647.

Duffy MJ and Crown J (2008). "A personalized approach to cancer treatment: how biomarkers can help." *Clin Chem* **54**(11): 1770–1779.

Emmett MR and Caprioli RM (1994). "Micro-electrospray mass spectrometry: ultra-high-sensitivity analysis of peptides and proteins." *J Am Soc Mass Spectrom* **5**(7): 605–613.

Geronmanos S, Freckleton G and Tempst P (2000). "Tuning of an electrospray ionization source for maximum peptide-ion transmission into a mass spectrometer." *Anal Chem* **72**: 777–790.

Hatsis P, Valaskovic G and Wu JT (2009). "Online nanoelectrospray/high-field asymmetric waveform ion mobility spectrometry as a potential tool for discovery pharmaceutical bioanalysis." *Rapid Commun Mass Spectrom* **23**(23): 3736–3742.

Haynes PA and Yates JR, 3rd (2000). "Proteome profiling-pitfalls and progress." *Yeast* **17**(2): 81–87.

Hebert AS, Richards AL, Bailey DJ, et al. (2014). "The one hour yeast proteome." *Mol Cell Proteomics* **13**(1): 339–347.

Heger NE and Kellogg MD (2014). "Personalized therapeutics and companion diagnostics: a new paradigm in testing and treatment." *Clin Chem* **60**(5): 795–796.

Hsieh EJ, Bereman MS, Durand S, et al. (2013). "Effects of column and gradient lengths on peak capacity and peptide identification in nanoflow LC-MS/MS of complex proteomic samples." *J Am Soc Mass Spectrom* **24**(1): 148–153.

Ji QC, Liu G, D'Arienzo CJ, et al. (2012). "What is next for dried blood spots?" *Bioanalysis* **4**(16): 2059–2065.

Juraschek R, Dulcks T and Karas M (1999). "Nanoelectrospray – more than just a minimized-flow electrospray ionization source." *J Am Soc Mass Spectrom* **10**: 300–308.

Juraschek R, Schmidt A, Karas M, et al. (1998). "Dependence of electrospray ionization efficiency on axial spray modes." *Adv Mass Spectrom* **14**: 1–15.

Kebarle P and Tang L (1993). "From ions in solution to ions in the gas phase: the mechanism of electrospray ionization mass spectrometry." *Anal Chem* **65**: 972A–986A.

Kushnir MM, Rockwood AL and Bergquist J (2013). "LC-MS/MS in clinical laboratories." *Bioanalysis* **5**(1): 5–6.

Lee MS and Kerns EH (1999). "LC/MS applications in drug development." *Mass Spectrom Rev* **18**(3–4): 187–279.

Meesters RJW and Hooff GP (2013). "State-of-the-art dried blood spot analysis: an overview of recent advances and future trends." *Bioanalysis* **5**(17): 2187–2208.

Meng F, Wiener MC, Sachs JR, et al. (2007). "Quantitative analysis of complex peptide mixtures using FTMS and differential mass spectrometry." *J Am Soc Mass Spectrom* **18**(2): 226–233.

Nilsson LB, Ahnoff M and Jonsson O (2013). "Capillary microsampling in the regulatory environment: validation and use of bioanalytical capillary microsampling methods." *Bioanalysis* **5**(6): 731–738.

Ouyang Z, Furlong MT, Wu S, et al. (2012). "Pellet digestion: a simple and efficient sample preparation technique for LC-MS/MS quantification of large therapeutic proteins in plasma." *Bioanalysis* **4**(1): 17–28.

Overdevest JB, Theodorescu D and Lee JK (2009). "Utilizing the molecular gateway: the path to personalized cancer management." *Clin Chem* **55**(4): 684–697.

Pandey A and Mann M (2000). "Proteomics to study genes and genomes." *Nature* **405**(6788): 837–846.

Rahavendran SV, Vekich S, Skor H, et al. (2012). "Discovery pharmacokinetic studies in mice using serial microsampling, dried blood spots and microbore LC–MS/MS." *Bioanalysis* **4**(9): 1077–1095.

Rainville PD, Smith NW, Wilson ID, et al. (2011). "Addressing the challenge of limited sample volumes in in vitro studies with capillary-scale microfluidic LC–MS/MS." *Bioanalysis* **3**(8): 873–882.

Schneider BB, Guo X, Fell LM, et al. (2005). "Stable gradient nanoflow LC-MS." *J Am Soc Mass Spectrom* **16**(9): 1545–1551.

Schneider BB, Javaheri H and Covey TR (2006). "Ion sampling effects under conditions of total solvent consumption." *Rapid Commun Mass Spectrom* **20**(10): 1538–1544.

Sherman J, McKay MJ, Ashman K, et al. (2009). "How specific is my SRM? The issue of precursor and product ion redundancy." *Proteomics* **9**(5): 1120–1123.

Spreadborough MJ, Day J, Jackson-Addie K, et al. (2013). "Bioanalytical implementation of plasma capillary microsampling: small hurdles, large gains." *Bioanalysis* **5**(12): 1485–1489.

Tang K, Page JS and Smith RD (2004). "Charge competition and the linear dynamic range of detection in electrospray ionization mass spectrometry." *J Am Soc Mass Spectrom* **15**(10): 1416–1423.

Tomer KB, Moseley MA, Deterding LJ, et al. (1994). "Capillary liquid chromatography/mass spectrometry." *Mass Spectrom Rev.* **13**: 431–457.

Valaskovic GA, Kelleher NL and McLafferty FW (1996). "Attomole protein characterization by capillary electrophoresis-mass spectrometry." *Science* **273**(5279): 1199–1202.

Valaskovic GA, Murphy JP, 3rd and Lee MS (2004). "Automated orthogonal control system for electrospray ionization." *J Am Soc Mass Spectrom* **15**(8): 1201–1215.

Valaskovic GA, Utley L, Lee MS, et al. (2006). "Ultra-low flow nanospray for the normalization of conventional liquid chromatography/mass spectrometry through equimolar response: standard-free quantitative estimation of metabolite levels in drug discovery." *Rapid Commun Mass Spectrom* **20**(7): 1087–1096.

Wang Z, Hop CE, Leung KH, et al. (2000). "Determination of in vitro permeability of drug candidates through a caco-2 cell monolayer by liquid chromatography/tandem mass spectrometry." *J Mass Spectrom* **35**(1): 71–76.

Washburn MP, Wolters D and Yates JR, 3rd (2001). "Large-scale analysis of the yeast proteome by multidimensional protein identification technology." *Nat Biotechnol* **19**(3): 242–247.

Wickremsinhe ER, Singh G, Ackermann BL, et al. (2006). "A review of nanoelectrospray ionization applications for drug metabolism and pharmacokinetics." *Curr Drug Metab* **7**(8): 913–928.

Wilm MS and Mann M (1994). "Electrospray and Taylor-cone theory, Dole's beam of macromolecules at last?" *Int J Mass Spectrom Ion Processes* **136**: 167–180.

Wilm MS and Mann M (1995). "Analytical properties of the nanoelectrospray ion source." *Anal Chem* **68**: 1–8.

Wood TD, Moy MA, Dolan AR, et al. (2003). "Miniaturization of electrospray ionization mass spectrometry." *Appl Spec Rev* **38**: 187–244.

Yates JR and Washburn MP (2013). "Quantitative proteomics." *Anal Chem* **85**(19): 8881.

Zhou F, Lu Y, Ficarro SB, et al. (2013). "Genome-scale proteome quantification by deep SEQ mass spectrometry." *Nat Commun* **4**: 2171.

Zhou F, Lu Y, Ficarro SB, et al. (2012). "Nanoflow low pressure high peak capacity single dimension LC-MS/MS platform for high-throughput, in-depth analysis of mammalian proteomes." *Anal Chem* **84**(11): 5133–5139.

6

Targeting the Right Protein Isoform: Mass Spectrometry-Based Proteomic Characterization of Alternative Splice Variants

Jiang Wu

Shire Pharmaceuticals, Lexington, MA, USA

6.1 Introduction

Human genes are transcribed into precursor-messenger ribonucleic acids (pre-mRNAs) and undergo various posttranscriptional modifications. One of the posttranscriptional modifications, alternative splicing, removes introns from the pre-mRNA sequences and stitches their surrounding exons together in different combinations to produce various mature mRNAs with different sequences. Therefore, alternative splicing is a regulated cellular process by which a single gene is capable of generating multiple forms of mature mRNAs and gives rise to a variety of protein splice variants that differ in their amino acid sequences and possibly biological functions (Pajares et al. 2007, Carling et al. 2009, Kelemen et al. 2013, Tavares et al. 2014a, b).

Alternative splicing significantly increases the diversity of mRNAs expressed from the genome and is a major source of cell-specific proteomic variation in humans. Several genome-wide analyses indicate that more than half of human genes present alternative spliced forms and 90% of protein functions are affected by alternative splicing mechanisms. Numerous evidences have shown that individual splice variants may regulate vastly different or even opposite cellular functions. Therefore, studies simply reporting up- or down-regulation of protein or mRNA expression without detecting and quantifying the individual splice variants are not sufficient because the transcripts and encoded proteins are mixtures of functionally different isoforms. Understanding relative expression and structure–function relationship of the splice isoforms is essential for the discovery and development of more specific therapeutics and biomarkers.

6.2 Alternative Splicing and Human Diseases

In recent years, expression pattern and functional studies of gene splice variants in normal or disease conditions have emerged as a rapidly growing area of research (Carling et al. 2009, Power et al. 2009, Ning and Nesvizhskii 2010, Omenn et al. 2010, Blair and Zi 2011, Lallous et al. 2013, Liu et al. 2013, Omenn et al. 2013, Sheynkman et al. 2013, Tang et al. 2013, Zhang et al. 2013, Boja and Rodriguez 2014, Lisitsa et al. 2014, Menon et al. 2014, Omenn 2014, Omenn et al. 2014, Tavares et al. 2014a, 2014b). Aberrant regulation of alternative splicing of pre-mRNAs leads to translation of proteins with changed sequences, structures, abundance, cellular localization as well as susceptibility to proteasome degradation, and, thus, unsuitable for the normal cellular functions. In fact, it is well recognized that abnormal alternative splicing is correlated with pathophysiological change and many major human disorders. A specific splicing isoform may be a more specific drug target for certain type of disease or a biomarker for diagnosis and stratification of patients.

Alternative splicing regulates heart development, blood coagulation, lipid metabolism, and metabolic pathways. For instance, alternative splicing of 3-hydroxy-3-methylglutaryl coenzyme A reductase (HMGCR) and low-density lipoprotein receptor (LDLR) may suppress their protein activities for cholesterol production and uptake (Zhu et al. 2007, Burkhardt et al. 2008). Proprotein convertase subtilisin/kexin type 9 (PCSK9) is involved in cholesterol biosynthesis and receptor-mediated uptake through alternative splicing (Schmidt et al. 2008). In neurological and muscle diseases, alternative splicing

Protein Analysis using Mass Spectrometry: Accelerating Protein Biotherapeutics from Lab to Patient, First Edition.
Edited by Mike S. Lee and Qin C. Ji.

attributes to the changes of topology, solubility, and signal peptides of integral membrane proteins. Kcnq2 is a brain-derived gene involved in neuronal M current and has two splice variants. The long variant is preferentially expressed in differentiated neurons, whereas the short transcript is prominent in fetal brain undifferentiated neuroblastoma cells and brain tumors (Smith et al. 2001). Dysregulation of alternative splicing is a hallmark of cancer. As an example, RON tyrosine kinase gene can generate a constitutively active kinase due to the skipping of an alternative exon (Eckerich et al. 2009, Moon et al. 2012, Wang et al. 2012). Manipulation of this splicing event aiming to change tumor progression has been a subject of study in various laboratories. As shown here, splice variant C of the osteopontin (OPN) has been implicated in metastasis and progression of a variety of cancers (Gimba and Tilli 2013). Sevcik reported the BRCA1 splice variant Delta14–15 with partial deletion of a regulatory serine-containing domain impairs the deoxyribonucleic acid (DNA) repair capacity in MCF-7 cells (Sevcik et al. 2012). A recent paper by Skalka et al. demonstrated that a splice variant form of carboxypeptidase E activates the Wnt signaling pathway, whereas the full-length canonical variant form is an inhibitor of Wnt/B-catenin signaling (Lertwittayapon et al. 2012). The splice variant C of the serine/threonine kinase Nek2 is involved in breast cancer development; specific inhibition of this isoform may have potential therapeutic benefit for some types of human breast tumors (Liu et al. 2012). Splice variants as potential diagnostic or prognostic biomarkers have been reviewed extensively (Pajares et al. 2007, Blair and Zi 2011, Omenn et al. 2013, 2014).

6.3 Identification of Splice Variant Proteins

Many therapeutic targets are composed of multiple splice variants. Understanding relative expression, structure, and function of individual splice variant proteins is crucial to designing specific therapeutics targeting the particular disease-relevant isoform or discovering disease-specific biomarkers. Unfortunately, current knowledge of splicing is primarily derived from RNA transcripts, with very little known about the tissue- and disease-expression level, three-dimensional structures, and functional differences of the proteins (Pajares et al. 2007, Carling et al. 2009, Omenn et al. 2010, 2013, 2014, Kelemen et al. 2013, Omenn 2014, Tavares et al. 2014a, 2014b). The knowledge gap is largely due to the technical challenges to characterize the splicing variants at proteome level (Ning and Nesvizhskii 2010, Wu et al. 2012, Sheynkman et al. 2013, Tavares et al. 2014a, 2014b).

In this chapter, recent developments in the identification of global splice variant proteins utilizing integrated transcriptomic and proteomic approaches are reviewed. Immunoaffinity mass spectrometry-based proteomic techniques to identify and quantify tissue- and disease-specific expression of protein splicing variants are described. Characterization of these splice isoforms would help the pharmaceutical industry to refine the drug target/biomarker strategy and make go/no-go decisions.

6.3.1 Global Profiling of Splicing Variant Proteins

Several approaches have been employed to interrogate splice variants and compare their expression profiles in normal and disease states. The general approach was to search proteomic data against databases that contain splice variants followed by confirmation of the translation of a spliced sequence by detecting (splice junction) peptide(s) unique to that isoform. However, this method can only identify splicing variant proteins whose sequences are already in the database entry.

Driven by the advances in the next-generation sequencing (e.g., RNA-Seq) and the development of various bioinformatics tools, the high-throughput transcriptomic and proteomic data offer new perspectives to probe protein splice variants in different tissues, organisms, and cells (Ning and Nesvizhskii 2010, Sheynkman et al. 2013, Zhang et al. 2013, Menon et al. 2014, Omenn 2014, Omenn et al. 2014, Tavares et al. 2014a, 2014b). Figure 6.1 is a typical workflow for the proteome-wide splice variant identification (Sheynkman et al. 2013, Tavares et al. 2014a, 2014b). The key is to construct mRNA data containing splice junction sequences, which are translated into analogous polypeptide sequences for mass spectrometry identification of novel splice variant peptides. One way is to utilize computational algorithms to construct putative mRNA database assembling all theoretical exon–exon combinations. RNA-Seq, on the other hand, enables rapid experimental analysis of sequences and expression levels of many alternative splice variants at the transcript level. These methods expand proteome databases to include entries for putative or experimentally confirmed splice variants. Subsequent search of the mass spectra against the database allows identification of the splice variant peptides.

Sheynkman et al. (2013) described a representative application of cell-specific RNA-Seq data to identify new peptides that result from alternative splicing. In their experiments, both RNA-Seq and proteomic profiling data were collected in parallel from the same Jurkat cells. Total RNA was extracted from the cells using a standard Trizol protocol, where 80 million reads of the longest

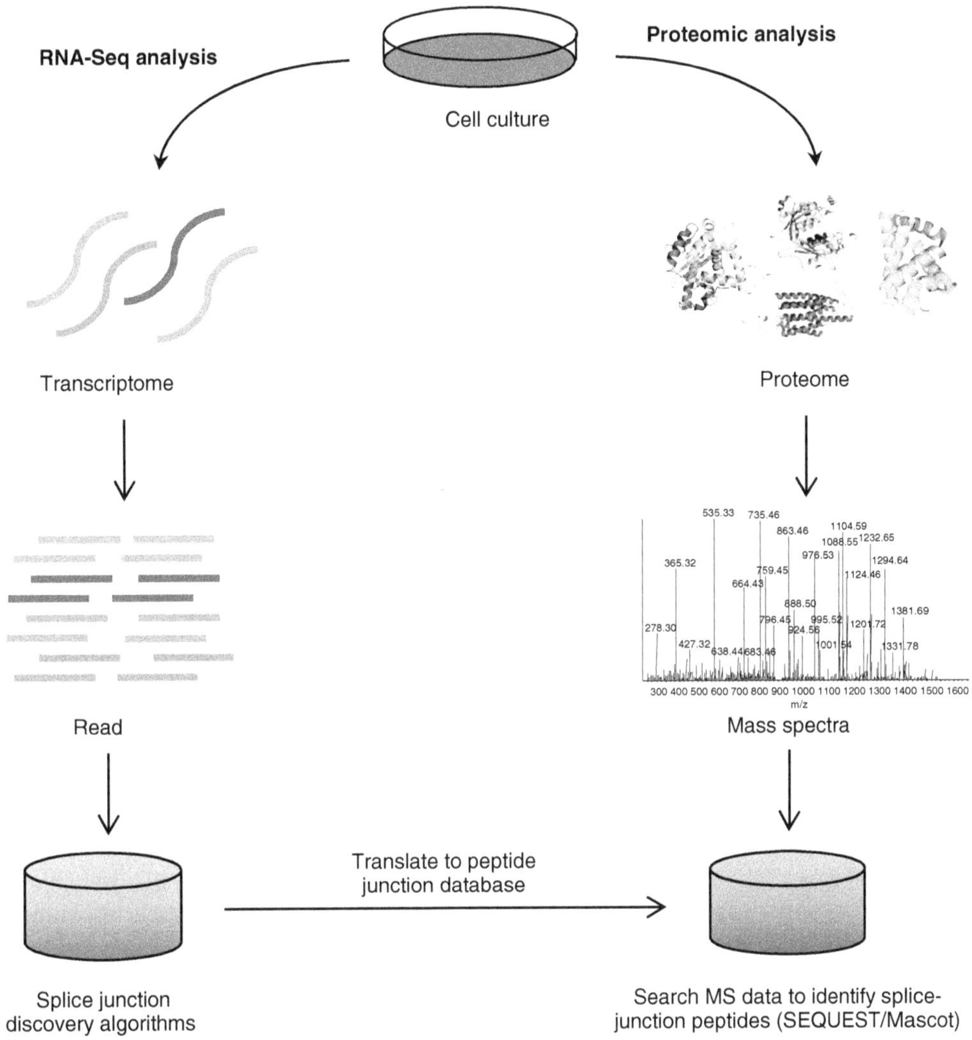

Figure 6.1 Transcriptomic and proteomic data collection workflow for splice-junction peptide identification.

RNA-Seq read type available on the Illumina platform were analyzed to provide a comprehensive RNA-Seq data set so that alternative splice forms at transcript level can be sensitively detected and assembled in a splice-junction database. The splice-junction mRNA sequences were converted into a customized polypeptide sequence database for searching against mass spec data. For proteomic profiling, cellular proteins were extracted using sodium dodecyl sulfate-containing buffer and were subsequently digested by trypsin. The resulting peptides were fractionated by liquid chromatography using high pH mobile phase. The peptide fractions were subjected to nano-LC–MS/MS analysis to maximize the identification of splice-junction peptides. The raw mass spectra were searched against human database that contains the splice-junction peptide sequences. A total of 12,873 transcripts mapped to 12,873 genes and 6810 proteins were identified, of which 57 novel splice junction peptides were identified with high confidence (1% false-discovery rate, FDR) that were absent in the UniProt/TrEMBL database (Sheynkman et al. 2013).

Using similar approaches, Omenn's group described in a series of reports the identification and quantification of splice variant peptides and proteins in mouse models of human KrasG12D activation/Ink4a/Arf deletion-caused pancreatic cancer, human Her2/neu (ERBB2)-induced breast cancer, and human ERBB2+ cancer cell lines (Menon et al. 2009, Menon and Omenn 2010, Liu et al. 2013). They were able to identify 420 splice isoforms in plasma for the pancreatic cancer model and 608 splice isoforms in mammary tissue for the breast cancer model, with striking differential expression between the tumor-bearing and normal mice in each case.

6.3.2 Characterization of Relative Expression of Protein Splice Variants

While profiling of novel splice variants offers rich information about the disease mechanism and signaling pathway at proteome level, a current issue is the challenge to identify globally low abundance of protein variants due to the huge dynamic range of protein expressions. Targeted in-depth characterization of a particular protein to identify its tissue- and disease-specific splice variant(s) would be an attractive option.

Recently, tremendous progress has been made in the application of mass spectrometry for quantification of proteins by measuring their surrogate peptides originating from proteolytic digestion (Meng and Veenstra 2011, Boja and Rodriguez 2012, Liebler and Zimmerman 2013, Yassine et al. 2013). With the introduction of stable-isotope-labeled peptides as internal standards and multiple-reaction monitoring (MRM) for targeted acquisition, multiple proteins or protein isoforms can simultaneously be quantified with extremely high specificity. In conjunction with immunoaffinity capture, such methods enable detection and quantitation of low ng/mL concentrations of the targeted proteins. While this approach has widely been utilized in biomarker identification, its application to splice variant characterization is still limited (Wu et al. 2012).

To measure the relative abundances of the protein splice variants, it is critical to enrich all splice variant forms without biases. This enrichment can be achieved by capture of all isoforms by antibodies, receptors, binding proteins, or chemical probes that are capable of recognizing and pooling down all isoforms. Second, to identify a particular splice variant, one must carefully select proteolytic enzymes to generate splice-junction peptide(s) with reasonable size, sequence specificity, and ionization efficiency to be identified by mass spectrometry. A general flowchart of immunocapture and mass spectrometric identification/quantitation of splice variant proteins is shown in Figure 6.2.

Bone morphogenetic protein (BMP) family members participate in organ regeneration through autocrine and paracrine actions. Alternative splicing of the BMP1 gene gives rise to at least two splice variants, the shorter form BMP1–1 and the longer form BMP1–3; the latter has not been confirmed on the protein level. To understand the expression and function of BMP1–3 isoform in circulation, plasma samples from healthy volunteers and patients with chronic kidney diseases were subjected to heparin affinity chromatography to enrich for proteins specific for bone and cartilage, the majority of which are known to have heparin binding domains (Grgurevic et al. 2007, 2011). The bound proteins were eluted from the column and precipitated with saturated ammonium sulfate. The protein pellet was subjected to SDS-PAGE separation, the respective bands were excised, digested with trypsin, and analyzed by LC–MS. Proteomic analysis identified the endogenous BMP1–3 protein isoform and demonstrated that it circulates as an active enzyme (does not contain the prodomain). Administration of recombinant BMP1–3 increased renal fibrosis. In contrast, inhibition of circulating BMP1–3 with a neutralizing antibody reduced renal fibrosis, suggesting that this pathway may be a therapeutic target for chronic kidney diseases.

We recently described an elegant use of an activity-based chemical probe to explore the tissue-specific distribution of various isoforms (not necessarily splice variants) of AMP-activated protein kinase (AMPK) (Wu et al. 2013). The central significance of AMPK as a metabolic regulator makes it an attractive target for drug action in diseases of energy imbalance, such as obesity, diabetes, and metabolic syndrome. Mammalian AMPK is a heterotrimer composed of α-, β-, and γ-subunits (Figure 6.3). Each subunit exists as either two (α1, α2; β1, β2) or three (γ1, γ2, γ3) isoforms, giving rise to a total of 12 possible AMPK α/β/γ-heterotrimers. Different types of cells, tissues, and animal species express distinct combinations of the subunit isoforms with specialized functions. In the development of tissue-specific therapeutic interventions and selecting preclinical species, understanding the relative abundance of AMPK isoforms in different tissues and species is likely to strengthen correlations of isoform distributions with disease pathology.

Efforts to capture tissue AMPK using different antibodies yielded inconsistent results, as the screened antibody panels failed to capture the subunits. Instead, AMPK was captured from lysates of a range of cells and tissues using an activity-based chemical probe (ActivX ATP probe), which covalently attaches desthiobiotin to conserved lysyl residues in the ATP-binding sites of protein kinases, including AMPK heterotrimer, for affinity enrichment (Figure 6.4). Affinity-captured proteins were fractionated and subsequently identified by mass spectrometry. Relative abundance of the subunit isoforms was assessed by comparing the "spectral counts" of the isoform-specific peptides, an approach widely used in label-free protein quantitation in biological samples.

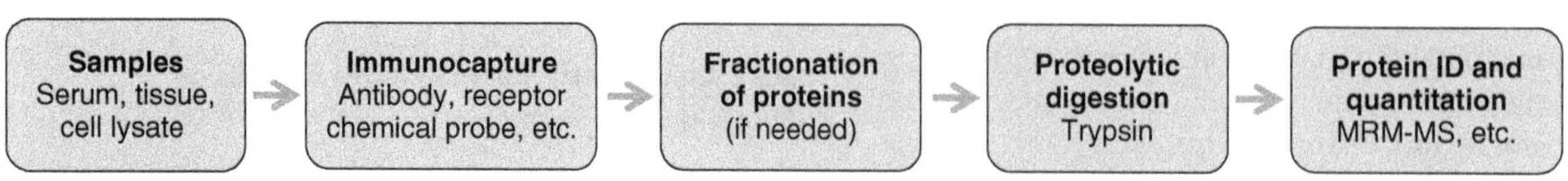

Figure 6.2 Flowchart of immunocapture and mass spectrometric identification/quantitation of splice variant proteins.

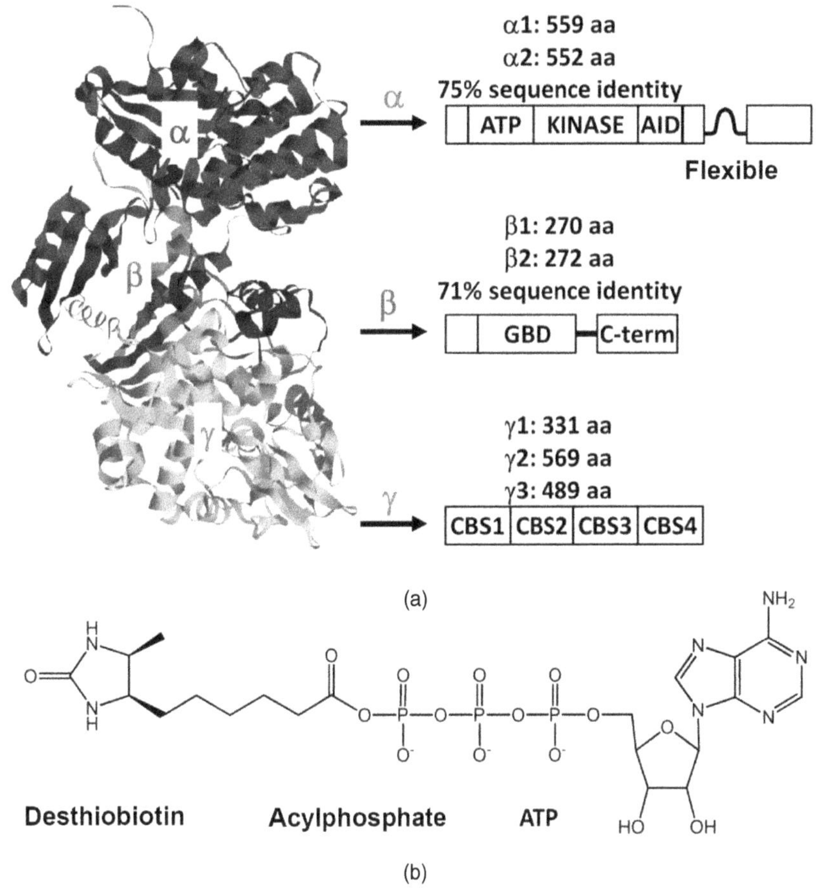

Figure 6.3 (a) Primary basis for diversity of the AMPK heterotrimer. The ribbon diagram of a mammalian AMPK structure is from PDB accession 2Y94. Sequence information refers to canonical UniProtKB entries for human AMPK polypeptides. Abbreviations: AID, autoinhibitory domain; GBD, glycogen-binding domain. (b) Structure of the ActivX ATP probe designed to couple desthiobiotin to conserved Lys in the ATP-binding loop of a protein kinase. (*Source*: Wu et al. 2013. Reproduced with permission of The American Society For Biochemistry and Molecular Biology.)

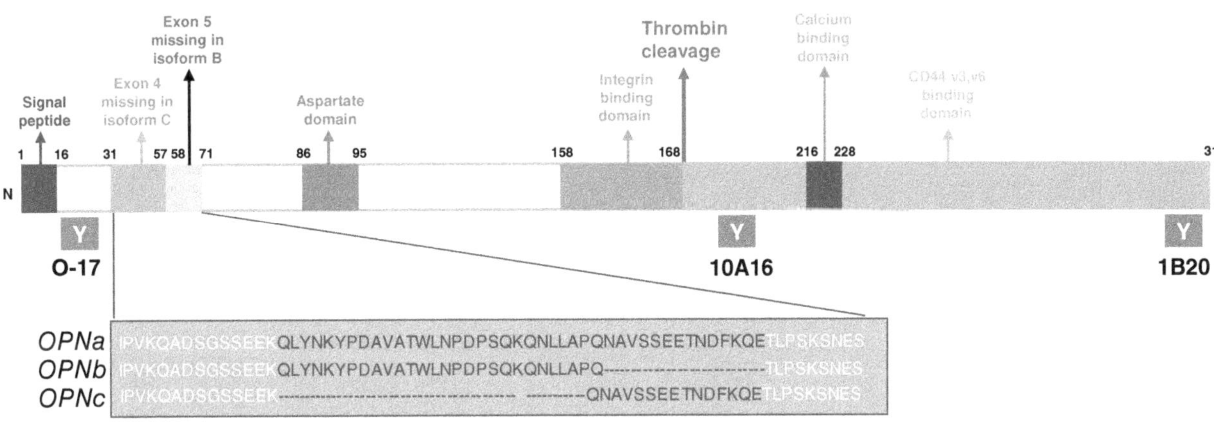

Figure 6.4 Schematic representation of the domain structure of osteopontin (OPN). The inset shows amino acid depletion of OPNb and OPNc variants at exon 5 and exon 4, respectively. The epitope regions of the antibodies used for immunocapture are also marked. (*Source*: Wu et al. 2012. Reproduced with permission of Taylor & Francis.)

Table 6.1 provides a summary of AMPK isoforms identified in different cell lines (HEK293, HepG2), tissues (heart, liver hepatocytes, skeletal muscle) of the different species (human, rat, dog). In agreement with mRNA analysis, proteomic results indicated that the predominant AMPK isoform in the liver of both diabetic patients and healthy individuals is α1β2γ1, but that dog and rat livers mainly contain the α1β1γ1 and α2β1γ1isoforms, respectively. In addition, the data support the tissue-specific expression of AMPK isoforms, the major isoforms in human liver, heart, and skeletal muscle tissues are different, dominated by α1β2γ1, α2β1γ1/2, and α2β2γ1, respectively. Taken together, the species- and tissue-specific AMPK complexes imply that pharmaceutical

Table 6.1 AMPK subunit isoforms identified in different species and tissues/cells (spectral counts were used to measure relative levels of isoform expression).

Species	Tissue/cell	Isoform	Spectral counts	Major AMPK isoform	Species	Tissue/cell	Isoform	Spectral counts	Major AMPK isoform	Species	Tissue/cell	Isoform	Spectral counts	Major AMPK isoform
Human	HEK293	α1	26	α1β1γ1	Human	Healthy Hepa-1	α1	65	α1β2γ1	Rat	Rat Hepa	α1	6	α2β1γ1
	HEK293	α2	20			Healthy Hepa-1	β1	6			Rat Hepa	α2	15	
	HEK293	β1	12			Healthy Hepa-1	β2	92			Rat Hepa	β1	18	
	HEK293	β2	9			Healthy Hepa-1	γ1	208			Rat Hepa	γ1	25	
	HEK293	γ1	33			Healthy Hepa-2	α1	68	α1β2γ1		Rat SM	α1	26	α2β2γ1
	HEK293	γ2	1			Healthy Hepa-2	β1	5			Rat SM	α2	166	
	HepG2	α1	48	α1β1γ1		Healthy Hepa-2	β2	45			Rat SM	β1	7	
	HepG2	α2	5			Healthy Hepa-2	γ1	85			Rat SM	β2	217	
	HepG2	β1	48			Diabetic Hepa-1	α1	37	α1β2γ1		Rat SM	γ1	212	
	HepG2	β2	29			Diabetic Hepa-1	β1	8			Rat Heart	α1	28	α2β1/2γ1
	HepG2	γ1	129			Diabetic Hepa-1	β2	77			Rat Heart	α2	98	
	HepG2 + α2β2γ3	α1	38	n/a		Diabetic Hepa-1	γ1	185			Rat Heart	β1	70	
	HepG2 + α2β2γ3	α2	33			Diabetic Hepa-2	α1	68	α1β2γ1		Rat Heart	β2	61	
	HepG2 + α2β2γ3	β1	15			Diabetic Hepa-2	β1	6			Rat Heart	γ1	106	
	HepG2 + α2β2γ3	β2	35			Diabetic Hepa-2	β2	139			Rat Heart	γ2	20	
	HepG2 + α2β2γ3	γ1	64			Diabetic Hepa-2	γ1	265		Dog	Dog Hepa-1st	α1	49	α1β1γ1

HepG2 + α2β2γ3	γ3	14			Human SM	α1	15	α2β2γ1		Dog Hepa-1st	α2	16
	Human SM	α2	71			Dog Hepa-1st	β1	40				
	Human SM	β1	19			Dog Hepa-1st	β2	1				
	Human SM	β2	64			Dog Hepa-1st	γ1	93				
	Human SM	γ1	94			Dog Hepa-2nd	α1	5	α1β1γ1			
	Human SM	γ2	5			Dog Hepa-2nd	β1	9				
	Human Heart	α2	4	α2β1γ1/2		Dog Hepa-2nd	γ1	43				
	Human Heart	β1	25			Dog Hepa-2nd	γ2	8				
	Human Heart	β2	3									
	Human Heart	γ1	20									
	Human Heart	γ2	24									

activation of AMPK could have different effects in different species and different tissues. These results can be used to guide the design of tissue-specific therapeutics and selection of animal species for preclinical toxicity and efficacy studies.

In principle, the strategy based on activity-based chemical probes is also suitable to identify and quantify splice variants or other types of isoforms of other therapeutically important enzyme classes, provided that antibodies to capture the isoforms are not readily available. This approach may include serine hydrolases, cysteine proteases, protein phosphatases, glycosidases, ubiquitin-conjugating and hydrolyzing enzymes, proteasomes, histone deacetylases, and cytochrome P450s (Rix and Superti-Furga 2009, Simon and Cravatt 2010, Chang et al. 2013, Niphakis and Cravatt 2014).

6.3.3 Quantitation of Splice Variants by MRM-MS

Affinity capture coupled with MRM-MS has become an attractive tool for the quantitation of endogenous biomarkers and therapeutic targets by measuring their proteolytic peptides as surrogates and using stable-isotope-labeled peptides as internal standards. This approach combines the selective enrichment of targeted proteins with the high specificity and sensitivity of MRM detection and allows for accurate, multiplexed protein quantitation (Meng and Veenstra 2011, Boja and Rodriguez 2012, Liebler and Zimmerman 2013, Yassine et al. 2013). However, unlike biomarker quantitation where multiple peptides are available as surrogates, usually only one splice-unique peptide can be selected as surrogate in the quantitation of the splice variants using this approach.

OPN is a secreted glycoprotein implicated in the metastasis and progression of numerous cancers (Gimba and Tilli 2013). Elevated OPN levels in plasma and tissues are correlated with disease progression and poor survival, supporting the note that OPN can be an attractive diagnostic biomarker or a therapeutic target. Three splice variants (OPNa, OPNb, and OPNc) were found in humans, with OPNb and OPNc lacking the exon 5 and exon 4, respectively (Figure 6.4). However, the relative expression of the individual isoforms at protein level and their respective roles in cancer progression were not well characterized due to the lack of reliable assays that are capable of differentiating these highly similar isoforms.

To investigate cancer-specific OPN isoform(s), we developed an immunoaffinity-based MRM-MS method to simultaneously quantify OPNa, OPNb, and OPNc isoforms in plasma from healthy and non-small-cell lung cancer (NSCLC) subjects (Wu et al. 2012). Central to the method are (i) a panel of antibodies that recognize different regions of the OPN sequences was used to capture all OPN isoforms in plasma, eliminating biases against any specific isoform; (ii) immobilized trypsin was used to promote the efficient digestion and quantitative generation of the splice-junction peptides; (iii) the unique splice-junction peptides generated from OPNa, b, and c were used as surrogates for quantification of the respective OPN isoforms and their stable-isotope-labeled counterparts were used as internal standard (Table 6.2); (iv) phosphatase was introduced to restore the native peptides because the surrogate peptides are originally partially phosphorylated; and (v) the resulting splice-junction peptides were quantified by nano-LC–MRM-MS to enhance sensitivity.

Simultaneous quantitation of all three OPN isoforms was achieved at pg/mL concentrations in the plasma of 10 healthy individuals and 10 NSCLC patients (Figure 6.5). Recovery for OPNa peptide was between 78% and 112% over a concentration range of 10–50 ng/mL. A close correlation between the ELISA and MRM-MS assays was observed for the measurement of total OPN concentration in plasma, suggesting that the MRM-MS is a reliable alternative for OPN quantitation in plasma.

Table 6.2 A list of the OPN isoform-specific peptides and their precursor and transition ions used in the MRM-MS experiment.

OPN isoform	OPN peptide	Precursor ion (*m/z*)	Transition ion (*m/z*)
OPNa	QNLLAPQNAVSSEETNDFK[$^{13}C_6$,$^{15}N_2$]	1057.5	1573.8 (y14), 1644.9 (y15)
	QNLLAPQNAVSSEETNDFK	1053.5	1565.8 (y14), 1636.7 (y15)
OPNb	QNLLAPQTLPSK[$^{13}C_6$,$^{15}N_2$]	659.4	778.5 (y7), 849.6 (y8)
	QNLLAPQTLPSK	655.4	770.5 (y7), 841.6 (y8)
OPNc	QNAVSSEETNDFK[$^{13}C_6$,$^{15}N_2$]	738.8	977.4 (y8), 1064.5 (y9)
	QNAVSSEETNDFK	734.8	969.4 (y8), 1056.5 (y9)

Source: Wu et al. 2012. Reproduced with permission of Taylor and Francis.

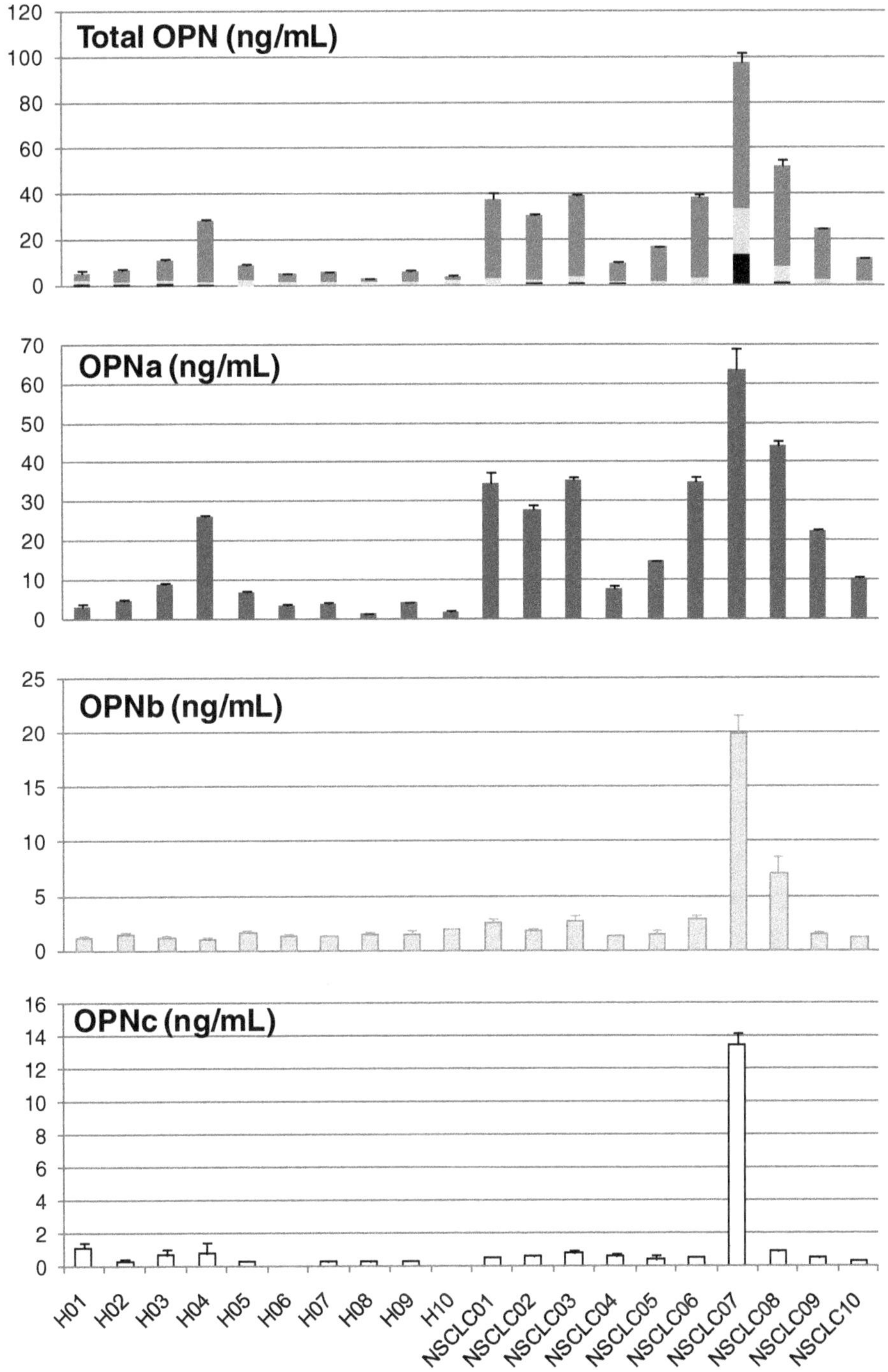

Figure 6.5 Plasma concentration of OPN isoforms in healthy individuals and NSCLC patients determined by immunoaffinity MRM-MS. (*Source*: Wu et al. 2012. Reproduced with permission of Taylor & Francis.)

An important conclusion from this study is that a wide range of concentrations of all three OPN isoforms were present in plasma from healthy individuals, suggesting that none of the splice variants were cancer specific. However, OPNa, the major isoform in healthy and NSCLC plasma, was substantially elevated in NSCLC patients. OPNb and OPNc were also detected albeit at much lower levels. Strikingly, the data demonstrated the circulating OPNc did not correlate with the presence of the disease. While local overexpression of OPNc in NSCLC tumors remains a possibility, the results suggest that therapeutic strategy targeting OPNc inhibition would have to take its endogenous expression levels into account.

6.4 Conclusion

The advances of proteomics in conjunction with high-throughput transcriptomics provide new perspectives to discover novel alternative splice variants in different cell lines, tissues, and organisms. Targeted proteomics integrating immunoaffinity capture and quantitative mass spectrometry (label-free or MRM) have been used to interrogate peptides derived from multiple splice variants to quickly screen for tissue- and disease-specific splice variants in a fast and cost-effective manner. While relative expression levels of the splice variants at mRNA and protein levels alone offer valuable information, understanding their functional roles in normal and pathophysiological conditions is more critical toward targeted drug and biomarker discovery to improve disease diagnosis and treatment.

References

Blair CA and Zi X (2011). "Potential molecular targeting of splice variants for cancer treatment." *Indian J Exp Biol* **49**(11): 836–839.

Boja ES and Rodriguez H (2012). "Mass spectrometry-based targeted quantitative proteomics: achieving sensitive and reproducible detection of proteins." *Proteomics* **12**(8): 1093–1110.

Boja ES and Rodriguez H (2014). "Proteogenomic convergence for understanding cancer pathways and networks." *Clin Proteomics* **11**(1): 22.

Burkhardt R, Kenny EE, Lowe JK, et al. (2008). "Common SNPs in HMGCR in micronesians and whites associated with LDL-cholesterol levels affect alternative splicing of exon13." *Arterioscler Thromb Vasc Biol* **28**(11): 2078–2084.

Carling PJ, Elliott DJ, Robson CN, et al. (2009). "Targeting alternative splicing in prostate oncology." *Discov Med* **8**(41): 74–80.

Chang JW, Cognetta AB, 3rd, Niphakis MJ, et al. (2013). "Proteome-wide reactivity profiling identifies diverse carbamate chemotypes tuned for serine hydrolase inhibition." *ACS Chem Biol* **8**(7): 1590–1599.

Eckerich C, Schulte A, Martens T, et al. (2009). "RON receptor tyrosine kinase in human gliomas: expression, function, and identification of a novel soluble splice variant." *J Neurochem* **109**(4): 969–980.

Gimba ER and Tilli TM (2013). "Human osteopontin splicing isoforms: known roles, potential clinical applications and activated signaling pathways." *Cancer Lett* **331**(1): 11–17.

Grgurevic L, Macek B, Durdevic D, et al. (2007). "Detection of bone and cartilage-related proteins in plasma of patients with a bone fracture using liquid chromatography-mass spectrometry." *Int Orthop* **31**(6): 743–751.

Grgurevic L, Macek B, Healy DR, et al. (2011). "Circulating bone morphogenetic protein 1–3 isoform increases renal fibrosis." *J Am Soc Nephrol* **22**(4): 681–692.

Kelemen O, Convertini P, Zhang Z, et al. (2013). "Function of alternative splicing." *Gene* **514**(1): 1–30.

Lallous N, Dalal K, Cherkasov A, et al. (2013). "Targeting alternative sites on the androgen receptor to treat castration-resistant prostate cancer." *Int J Mol Sci* **14**(6): 12496–12519.

Lertwittayapon T, Tencomnao T and Santiyanont R (2012). "Inhibitory effect of alternatively spliced RAGEv1 on the expression of NF-kB and TNF-alpha in hepatocellular carcinoma cells." *Genet Mol Res* **11**(2): 1712–1720.

Liebler DC and Zimmerman LJ (2013). "Targeted quantitation of proteins by mass spectrometry." *Biochemistry* **52**(22): 3797–3806.

Lisitsa A, Moshkovskii S, Chernobrovkin A, et al. (2014). "Profiling proteoforms: promising follow-up of proteomics for biomarker discovery." *Expert Rev Proteomics* **11**(1): 121–129.

Liu S, Im H, Bairoch A, et al. (2013). "A chromosome-centric human proteome project (C-HPP) to characterize the sets of proteins encoded in chromosome 17." *J Proteome Res* **12**(1): 45–57.

Liu Z, Wang Y, Wang S, et al. (2012). "Nek2C functions as a tumor promoter in human breast tumorigenesis." *Int J Mol Med* **30**(4): 775–782.

Meng Z and Veenstra TD (2011). "Targeted mass spectrometry approaches for protein biomarker verification." *J Proteomics* **74**(12): 2650–2659.

Menon R, Im H, Zhang EY, et al. (2014). "Distinct splice variants and pathway enrichment in the cell-line models of aggressive human breast cancer subtypes." *J Proteome Res* **13**(1): 212–227.

Menon R and Omenn GS (2010). "Proteomic characterization of novel alternative splice variant proteins in human epidermal growth factor receptor 2/neu-induced breast cancers." *Cancer Res* **70**(9): 3440–3449.

Menon R, Zhang Q, Zhang Y, et al. (2009). "Identification of novel alternative splice isoforms of circulating proteins in a mouse model of human pancreatic cancer." *Cancer Res* **69**(1): 300–309.

Moon H, Cho S, Yang X, et al. (2012). "Identification of novel splicing variants from RON proto-oncogene pre-mRNA." *Oncol Rep* **28**(6): 2217–2220.

Ning K and Nesvizhskii AI (2010). "The utility of mass spectrometry-based proteomic data for validation of novel alternative splice forms reconstructed from RNA-Seq data: a preliminary assessment." *BMC Bioinformatics* **11** (Suppl 11): S14.

Niphakis MJ and Cravatt BF (2014). "Enzyme inhibitor discovery by activity-based protein profiling." *Annu Rev Biochem* **83**: 341–377.

Omenn GS (2014). "Plasma proteomics, the Human Proteome Project, and cancer-associated alternative splice variant proteins." *Biochim Biophys Acta* **1844**(5): 866–873.

Omenn GS, Guan Y and Menon R (2014). "A new class of protein cancer biomarker candidates: differentially expressed splice variants of ERBB2 (HER2/neu) and ERBB1 (EGFR) in breast cancer cell lines." *J Proteomics* **107**: 103–112.

Omenn GS, Menon R and Zhang Y (2013). "Innovations in proteomic profiling of cancers: alternative splice variants as a new class of cancer biomarker candidates and bridging of proteomics with structural biology." *J Proteomics* **90**: 28–37.

Omenn GS, Yocum AK and Menon R (2010). "Alternative splice variants, a new class of protein cancer biomarker candidates: findings in pancreatic cancer and breast cancer with systems biology implications." *Dis Markers* **28**(4): 241–251.

Pajares MJ, Ezponda T, Catena R, et al. (2007). "Alternative splicing: an emerging topic in molecular and clinical oncology." *Lancet Oncol* **8**(4): 349–357.

Power KA, McRedmond JP, de Stefani A, et al. (2009). "High-throughput proteomics detection of novel splice isoforms in human platelets." *PLoS One* **4**(3): e5001.

Rix U and Superti-Furga G (2009). "Target profiling of small molecules by chemical proteomics." *Nat Chem Biol* **5**(9): 616–624.

Schmidt RJ, Zhang Y, Zhao Y, et al. (2008). "A novel splicing variant of proprotein convertase subtilisin/kexin type 9." *DNA Cell Biol* **27**(4): 183–189.

Sevcik J, Falk M, Kleiblova P, et al. (2012). "The BRCA1 alternative splicing variant Delta14–15 with an in-frame deletion of part of the regulatory serine-containing domain (SCD) impairs the DNA repair capacity in MCF-7 cells." *Cell Signal* **24**(5): 1023–1030.

Sheynkman GM, Shortreed MR, Frey BL, et al. (2013). "Discovery and mass spectrometric analysis of novel splice-junction peptides using RNA-Seq." *Mol Cell Proteomics* **12**(8): 2341–2353.

Simon GM and Cravatt BF (2010). "Activity-based proteomics of enzyme superfamilies: serine hydrolases as a case study." *J Biol Chem* **285**(15): 11051–11055.

Smith JS, Iannotti CA, Dargis P, et al. (2001). "Differential expression of kcnq2 splice variants: implications to m current function during neuronal development." *J Neurosci* **21**(4): 1096–1103.

Tang JY, Lee JC, Hou MF, et al. (2013). "Alternative splicing for diseases, cancers, drugs, and databases." *Scientific World J* **2013**: 703568.

Tavares R, de Miranda Scherer N, Pauletti BA, et al. (2014a). "SpliceProt: a protein sequence repository of predicted human splice variants." *Proteomics* **14**(2–3): 181–185.

Tavares R, Scherer NM, Ferreira CG, et al. (2014b). "Splice variants in the proteome: a promising and challenging field to targeted drug discovery." *Drug Discov Today* **20**: 353–360.

Wang D, Lao WF, Kuang YY, et al. (2012). "A novel variant of the RON receptor tyrosine kinase derived from colorectal carcinoma cells which lacks tyrosine phosphorylation but induces cell migration." *Exp Cell Res* **318**(20): 2548–2558.

Wu J, Pungaliya P, Kraynov E, et al. (2012). "Identification and quantification of osteopontin splice variants in the plasma of lung cancer patients using immunoaffinity capture and targeted mass spectrometry." *Biomarkers* **17**(2): 125–133.

Wu J, Puppala D, Feng X, et al. (2013). "Chemoproteomic analysis of intertissue and interspecies isoform diversity of AMP-activated protein kinase (AMPK)." *J Biol Chem* **288**(50): 35904–35912.

Yassine H, Borges CR, Schaab MR, et al. (2013). "Mass spectrometric immunoassay and MRM as targeted MS-based quantitative approaches in biomarker development: potential applications to cardiovascular disease and diabetes." *Proteomics Clin Appl* **7**(7–8): 528–540.

Zhang EY, Cristofanilli M, Robertson F, et al. (2013). "Genome wide proteomics of ERBB2 and EGFR and other oncogenic pathways in inflammatory breast cancer." *J Proteome Res* **12**(6): 2805–2817.

Zhu H, Tucker HM, Grear KE, et al. (2007). "A common polymorphism decreases low-density lipoprotein receptor exon 12 splicing efficiency and associates with increased cholesterol." *Hum Mol Genet* **16**(14): 1765–1772.

7

The Application of Immunoaffinity-Based Mass Spectrometry to Characterize Protein Biomarkers and Biotherapeutics

Bradley L. Ackermann and Michael J. Berna

Eli Lilly and Company, Indianapolis, IN, USA

7.1 Introduction

7.1.1 The Importance of Protein Measurement

At the molecular level, protein interactions are responsible for driving function in almost all biological systems. Not surprisingly, the study of proteins permeates the life sciences, and protein measurement is essential to understanding biology. This perspective is certainly true in the pharmaceutical industry where protein characterization, profiling, and quantitative analysis are involved at every stage of drug discovery and development. Since almost all drug targets are proteins, protein identification and profiling are needed for target identification and preclinical discovery. Further, with the recent explosion of biotherapeutic drugs, protein quantification is required from the earliest stages of discovery to understand target biology, perform pharmacokinetic (PK) analysis, and to understand biotransformation of peptide and protein biotherapeutics. Similarly, the measurement of protein biomarkers is important for the clinical translation of new therapeutic agents. Biomarkers are routinely used to assess pharmacodynamic (PD) endpoints such as target engagement and dose selection. Biomarkers may also be used to understand disease progression and in certain cases to enable patient selection. These last two applications form the basis for using protein biomarkers as diagnostic agents.

Mass spectrometry (MS) is widely recognized for its contributions to the analysis and characterization of proteins and is alone in its ability to perform structural characterization while serving as a selective detector for quantitative analysis. In addition, MS may be coupled to chromatography to allow trace multianalyte analysis in complex biological matrices. These combined attributes have made MS an indispensable tool for the development of biotherapeutic agents since both quantitative and qualitative information is needed to understand their PK and biotransformation. A powerful example is found by the introduction of antibody–drug conjugates (ADC) as targeted agents for chemotherapy (Kaur et al. 2013) as the absence of MS would limit our understanding and ability to optimize these complex molecules. An equally compelling case can be made for the importance of MS in protein biomarker discovery. The best example is the field of proteomics, which is built almost entirely on MS technology.

In light of these accolades, it might surprise some to learn that clinical quantification of protein biomarkers by MS remains quite limited in its application. The same observation can be made for clinical bioanalysis of biotherapeutics, a stark contrast from small molecule bioanalysis, a field dominated by liquid chromatography tandem mass spectrometry (LC/MS/MS). While several logistical reasons may contribute to this trend, a primary technical reason in both cases is the need for greater sensitivity. Using current LC/MS/MS technology with conventional isolation methods, lower limits of quantification (LLOQ) in the range of 1–10 ng/mL are accessible in human plasma, albeit with relatively large sample volumes and considerable sample preparation (Keshishian et al. 2007). Unfortunately, most clinically relevant protein biomarkers in human plasma or serum circulate in the low to mid pg/mL range. Similar limitations occur for investigating biotherapeutic PK since clinical doses tend to be lower than the doses given during preclinical assessment. A further complication is the need to digest the target protein to yield a surrogate peptide (bottom-up analysis) owing to comparatively greater signals observed for peptides than proteins by electrospray ionization (ESI). These factors along with the competition for ionization from coextracted matrix molecules (ion suppression) and the signal dilution introduced by multiple charging combine to limit LC/MS/MS sensitivity for proteins.

Protein Analysis using Mass Spectrometry: Accelerating Protein Biotherapeutics from Lab to Patient, First Edition.
Edited by Mike S. Lee and Qin C. Ji.

7.1.2 Ligand Binding Assays

Protein measurement in clinical specimens is currently dominated by ligand binding assays (LBAs), with the exception of solid tissue samples where immunohistochemical staining is performed. The first LBA format was radioimmunoassay (RIA), introduced in Yalow and Berson (1960) for the determination of insulin. To alleviate issues associated with handling radioactivity, enzymes were eventually conjugated to capture antibodies to permit colorimetric detection. Accordingly, these methods are referred to as enzyme immunoassay (EIA). Two EIA-based methods were simultaneously published in 1971, each using a competitive format (Engvall and Perlmann 1971, Van Weemen and Schuurs 1971). Because the method of Engvall and Perlmann incorporated antibody conjugation to a solid substrate to facilitate wash steps, their method was given the name ELISA standing for enzyme-linked immunosorbent assay (ELISA).

Today, most ELISA applications utilize a sandwich format, involving separate antibodies having unique binding epitopes for capture and detection. Among several possible detection formats, absorbance and chemiluminescence are the most popular. Sandwich assays, which have superior selectivity to EIA, are used throughout drug discovery and development for both biomarker and biotherapeutic applications. Applications range from the use of research use only (RUO) kits for discovery applications (Bowsher et al. 2012) to highly validated ELISA methods for clinical investigation (Sloan et al. 2012). Advantages to ELISA methods are that they are relatively simple to run, do not require expensive instrumentation, employ a multiplexed detection format, and have sufficient sensitivity to analyze protein biomarkers (mid to low pg/mL range). PK studies for protein biotherapeutics also employ ELISA for the same reasons. A further advantage is the ability for ELISA to detect proteins in their active conformation, an important consideration for biotherapeutics.

Despite these advantages, LBA methods have significant shortcomings. Chief among these is the limited specificity of antibody reagents, which too often confounds the interpretation of biomarker results. In the case of PK measurement, ELISA assays frequently fail to distinguish a protein therapeutic from its metabolites depending on the epitopes available for measurement. Indeed, the limitations of LBA methods have been well documented (Hoofnagle and Wener 2009, Becker and Hoofnagle 2012). In addition to limited specificity, other issues include excessive time and cost for reagent preparation, susceptibility to matrix effects, poor interlab reproducibility, and concerns about precision and accuracy. It is also important to note that the vast majority of clinical ELISA measurements occur as single analyte assays, despite the commercial availability of several multiplexed LBA platforms (Leng et al. 2008). Uptake of multiplexed methods for clinical analysis has been limited, due in part to difficulty associated with the need to simultaneously optimize multiple analytes under a common set of assay conditions.

7.1.3 The Introduction of Hybrid IA-MS Methods

During the past decade, there has been a clear trend toward the use of targeted methods for proteomic analysis. While several reasons exist for this trend, the most concise answer is that selective detection results in improved sensitivity and grants access to proteins of lower abundance. Targeted methods frequently use triple quadrupole mass spectrometers with multiple reaction monitoring (MRM) detection, which as pointed out earlier, allows plasma proteins to be quantified in the low ng/mL range (Keshishian et al. 2007). In order to quantify proteins in the pg/mL range, the practice of combining immunoaffinity capture with MS detection or immunoaffinity-mass spectrometry (IA-MS) was introduced (Nelson et al. 1995, Anderson et al. 2004, Ackermann and Berna 2007). These methods are sometimes referred to as *hybrid* techniques, because they embody a combination of LBA and MS (Stevenson et al. 2013). The advantage of using IA capture is straightforward to understand since antibody reagents are able to selectively enrich target proteins as much as 1000-fold (Whiteaker et al. 2007). Thus, sub-ng/mL assays can readily be produced, with less sample volume and less sample preparation. For example, IA capture methods obviate the need for a deliberate step to remove highly abundant plasma proteins common to conventional targeted methods. Although IA-MS methods are still not as sensitive as the most sensitive ELISA methods, there are a variety of advantages of IA-MS derived from the improved specificity of MS detection, the use of stable isotopes for improved precision, and the superior facility for analyzing protein mixtures. In addition, the versatility introduced by MS detection allows detailed structural information to be obtained for low-abundance proteins.

The goal of this chapter is to introduce the most common formats for applying IA-MS techniques and show their collective utility for both biomarker analysis and biotherapeutic development. Select examples are cited to describe major application areas without an attempt to be comprehensive in our treatment. In addition to providing examples involving biomarkers and biotherapeutics, general information on IA-MS techniques is provided to inform the reader on important options and considerations in this rapidly growing field.

7.2 Overview of IA-MS Methods

7.2.1 Classification of IA-MS Methods

At a high level, IA-MS methods can be classified into two groups according to whether the antibody reagents used target an intact protein or a surrogate peptide produced by proteolytic digestion during sample processing. A schematic diagram of these two options is shown in Figure 7.1. Figure 7.1(a) depicts the situation involving IA capture of an intact target, whereas Figure 7.1(b) illustrates IA capture by an antibody targeted against a surrogate peptide. This latter methodology, pioneered by Leigh Anderson, has been given the name SISCAPA, which stands for Stable Isotope Standard Capture with Anti-Peptide Antibodies (Anderson et al. 2004). Recently, a third configuration, which incorporates both anti-protein and anti-peptide antibodies, was introduced (Figure 7.1c) (Neubert et al. 2013). This approach, often referred to as sequential IP, allows for unparalleled specificity and sensitivity. The triangles shown in Figure 7.1 indicate the addition of an internal standard, which is typically a stable-isotope-labeled (SIL) version of a peptide or protein. Two arrows are associated with SIL-peptide internal standard addition reflecting the fact that addition can occur before, after, or simultaneously with enzymatic digestion. It should also be mentioned that although digestion is common, it is not always used. Reduction and alkylation (R/A) is also optional, because in many cases it is not needed to achieve adequate digestion.

Several options accompany these IA-MS formats. Samples are not restricted to plasma or serum but encompass a variety of sample types including cell lysates

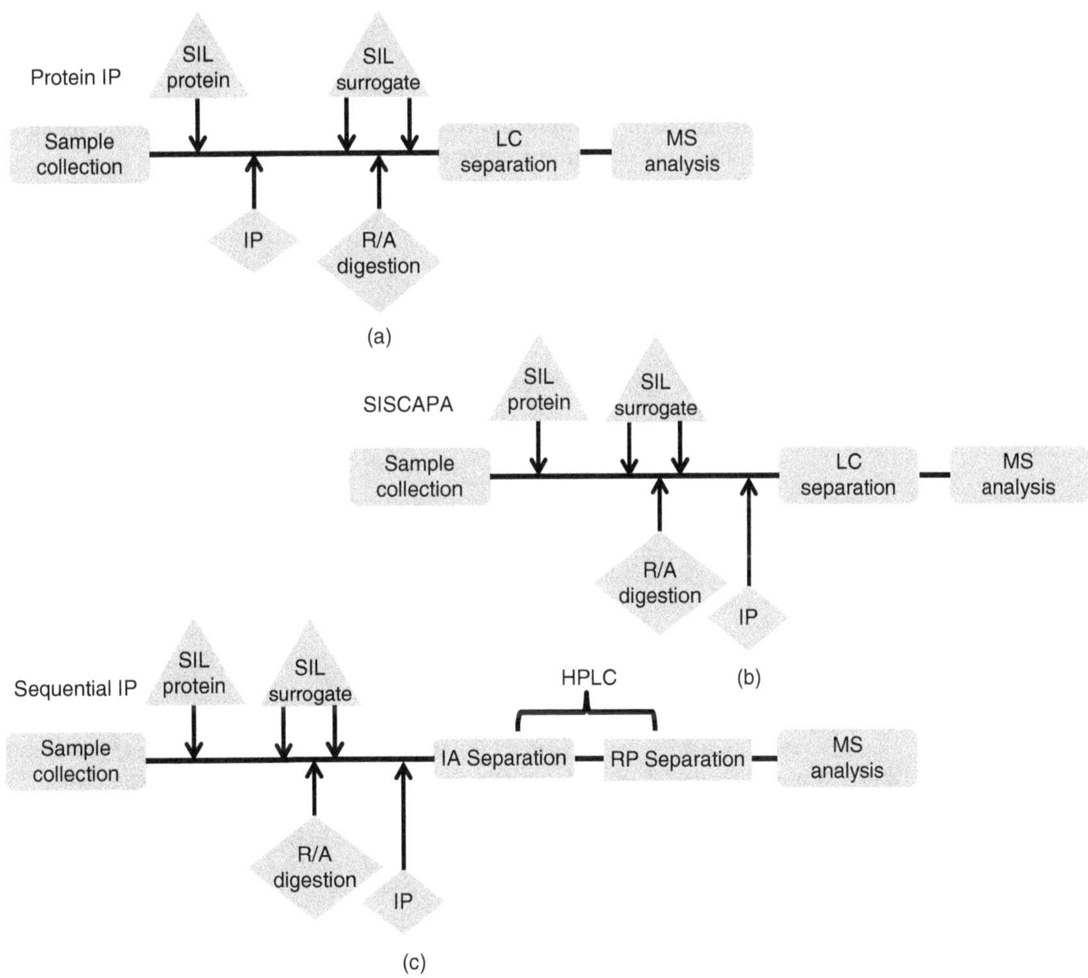

Figure 7.1 Schematic illustration of the basic IA-MS flow schemes: (a) anti-protein capture, (b) anti-surrrogate peptide capture, and (c) sequential IP. Internal standards, usually present as their SIL forms, can exist as either labeled proteins or peptides. Several formats can be used for immunoprecipitation (IP). In the case of sequential IP, purification of the surrogate peptide occurs on-line using an immunoaffinity column. Reduction/alkylation (R/A), typically used to enhance proteolytic digestion, is optional and may occur before or after the addition of the SIL peptide internal standard.

(Gerber et al. 2003) to solid tissue samples (Neubert et al. 2012). In addition, a number of formats exist for IA capture. While magnetic beads are a popular immunoprecipitation (IP) format, several options have been used including agarose beads, direct antibody absorption onto 96-well polystyrene plates (Berna and Ackermann 2009) solution phase capture, and capture using pipette tips filled with a porous monolithic substrate containing the capture antibody (Krastins et al. 2013). Along with these options, which are considered off-line, on-line methods that couple column-based IA capture with LC/MS have also been published (Dufield and Radabaugh 2012).

An expanded classification of IA-MS methods appears in Table 7.1, in which, IA-MS formats are categorized by the *analytical target* for the assay, since this delineation permits an application-centric discussion of IA-MS methods. For our discussion, peptides are considered to be less than 10 kDa, since molecules smaller than this molecular weight are viable candidates for intact detection without enzymatic digestion to yield a surrogate peptide (i.e., top-down). In contrast, protein applications (>10 kDa) involve bottom-up analysis since proteolytic digestion to yield a peptide surrogate is generally required for sensitivity. As illustrated in Figure 7.1, biomarker applications can involve IA capture of an intact protein (Figure 7.1a) or a peptide surrogate (Figure 7.1b). These applications are classified separately in Table 7.1 in relation to protein biomarkers. Sequential IP methods (Figure 7.1c) are not considered as a separate category but are grouped with SISCAPA in Table 7.1 and in subsequent discussion.

Protein biotherapeutics are classified separately from biomarkers in Table 7.1 owing to distinct differences between these applications. Protein biotherapeutic applications are divided into two categories in Table 7.1. The first deals with nonantibody biotherapeutic molecules of both peptide and protein origin. As a class, biotherapeutic molecules tend to have unique biophysical properties arising from the common practice of covalently attaching molecules such as polyethylene glycol (PEG) to enhance their PK. No distinction will be made on molecular weight since the protein and peptide applications tend to be similar. These molecules are, however, considered separately from the final two categories in Table 7.1, which addresses therapeutic constructs derived from monoclonal antibodies (mAbs), an area that has expanded greatly in recent years. Owing to their complexity, ADCs are discussed separately and listed as a unique category in Table 7.1.

Table 7.1 Classification of immunoaffinity mass spectrometry methods.

Application	Target for IA-capture	Comment	References
Biomarker *peptide*	Intact peptide	Most applications use top-down analysis (no enzymatic digestion)	Berna and Ackermann (2009), Krastins et al. (2013), Berna et al. 2006, Chappell et al. (2014), Oran et al. (2014), Oe et al. (2006), Berna et al. (2008), Wolf et al. (2001), Miyachi et al. (2013), Chappell et al. (2012), Kumar et al. (2010), Oran et al. (2011)
Biomarker *protein*	Intact protein	Increasingly being applied to limited protein mixtures involving a common epitope. Provides useful bridge to ELISA as common reagents are used	Berna and Ackermann (2009), Krastins et al. (2013), Callipo et al. (2010), Ocaña and Neubert (2010), Umberger et al. (2012), Niederkofler et al. (2001), Kiernan et al. (2002), Oran et al. (2010), Berna et al. (2007), Zhen et al. (2007), McAvoy et al. (2014), Torsetnes et al. (2013), Torsetnes et al. (2014a), Torsetnes et al. (2014b), Lin et al. (2013), Nicol et al. (2008)
Biomarker *protein*	Surrogate peptide	Useful for multianalyte panels involving diverse protein sets. Also can be used to enhance surrogate peptide sensitivity (sequential IP)	Becker and Hoofnagle (2012), Anderson et al. (2004), Whiteaker et al. (2007), Neubert et al. (2013), Whiteaker et al. (2010), Whiteaker et al. (2012), Razavi et al. (2013), Kuhn et al. (2009), Hoofnagle et al. (2008), Addona et al. (2009), Schoenherr et al. (2012), Palandra et al. (2013)
Biotherapeutic *peptide or protein*	Intact biotherapeutic	Common to use multiple surrogate peptides to track both intact drug and metabolites	Kaspar and Reichert (2013), Dawes et al. (2013), Hess et al. (2012), Thomas et al. (2014), Ezan (2013), Podust et al. (2013), Xu et al. (2010), Wang and Heilig (2012), Hager et al. (2013)
Biotherapeutic *mAb constructs*	Intact mAb	Grants improved sensitivity for PK analysis. Important tool for clinical translation	Chames et al. (2009), Wang et al. (2008), Li et al. (2012), Leary et al. (2013), Furlong et al. (2012), Furlong et al. (2013), Dubois et al. (2008), Fernández Ocaña et al. (2012)
Biotherapeutic *antibody–drug conjugates*	Intact mAb or toxic payload	Useful for PK determination as well as the *in vivo* drug-to-antibody ratio (DAR) distribution for ADCs	Kaur et al. (2013), Jaracz et al. (2005), Chari (2008), Sauerborn and van Dongen (2014), Dere et al. (2013), Xu et al. (2011, 2013), Roberts et al. (2013)

The categories listed in Table 7.1 provide the framework for our discussion of IA-MS applications in the later part of this chapter. In the section that follows, select topics relevant to all IA-MS methods are covered to give the reader appropriate background and to serve as a practical resource.

7.2.2 Stable-Isotope-Labeled Internal Standards

A critical part of all IA-MS assays is the inclusion of an internal standard (IS) to compensate for differences in recovery and/or ionization during LC/MS (Bronsema et al. 2012). While examples of analog internal standards exist (Callipo et al. 2010, Bronsema et al. 2012), a vast majority of analyses are conducted using SIL peptides reflecting the prevalence of solid-phase peptide synthesis techniques. Amino acids containing ^{15}N and/or ^{13}C are generally preferred to deuterated standards to avoid changes in retention time that can occur with deuterated molecules. A disadvantage to peptide internal standards is they do not fully account for protein recovery. Shuford and coworkers extensively studied factors affecting accuracy when using peptide SIL-IS and showed that both peptide production and stability are important factors when using trypsin (Shuford et al. 2012). Further, they recommend adding the SIL-IS concurrently with typsin and using modified trypsin to avoid chymotryptic-based instability enhanced by trypsin autocatalysis (Shuford et al. 2012). SIL-IS are often prepared with extended sequences to track digestion (Figure 7.1). Although finding increased use, the true utility of such *extended* or *winged* peptides is protein dependent and must be established on a case-by-case basis (Barnidge et al. 2004, Ocaña and Neubert 2010).

Protein internal standards, though harder to procure, have been shown to yield improved analytical performance (Brun et al. 2007). Protein internal standards can be divided into two groups: SIL proteins and protein analogs. SIL proteins are prepared by recombinant techniques using SILAC technology, which stands for stable isotope labeling by amino acids in cell culture (Ong et al. 2002). The second approach involves using a protein having close structural similarity to the target. A popular subset of this approach involves chemical derivatization where a known minor chemical modification is introduced during sample processing (Bystrom et al. 2011). A good example is the differential dimethyl labeling strategy outlined by Ji et al. (2009). In this paper, cynomolgus serum samples from monkeys dosed with a mAb were digested with trypsin. Standards of the mAb were also digested. Upon completion, the study samples were treated with d_0 formaldehyde, while the standards were treated with d_2 formaldehyde. Reductive amination of both sample sets ultimately prepared dimethyl labels at lysine residues in each case. Accordingly, each lysine residue (or N-terminus) produced a net shift of 4 Da to provide an *in situ* internal standard for each tryptic peptide monitored.

7.2.3 IA Capture Formats

As cited above, beads, either polymer-coated magnetic or agarose based, are the most popular form of IP used with IA-MS techniques. Beads are available from several vendors providing various coupling techniques for antibody attachment. Attachment can occur via covalent or through noncovalent interactions. Protein A/G and biotin–streptavidin are common methods for noncovalent attachment. Magnetic beads are simple means for bead collection during processing steps and have been used with various automation strategies (Whiteaker et al. 2010, 2012).

An alternative format to beads is a method known as mass spectrometric immunoassay or MSIA (Nelson et al. 1995). This technology uses a porous monolithic phase loaded into pipette tips as a support for antibody attachment. Plasma sample volumes of up to 1 mL may be loaded through successive aspirate and dispense cycles to concentrate the target of interest onto the phase. This technique is readily automated and purports to have less nonspecific binding than bead-based methods. To date, several clinical applications have been published seeking to address the issue of protein isoform heterogeneity in clinical protein biomarker applications (Krastins et al. 2013).

By analogy to ELISA, we and others have shown that it is possible to perform IA capture directly in polystyrene microtiter plates (Berna and Ackermann 2009, Umberger et al. 2012). In addition to lowering expense, plate capture simplifies sample preparation and allows IA-MS to utilize equipment standard to LBA such as plate washers. We performed a comparison between plates and magnetic beads, which showed that comparable results can be obtained (Umberger et al. 2012). The chief limitation to plates is capacity, since sample volumes greater than 200 μL cannot be used. As IA-MS sensitivity continues to approach that obtained by ELISA, we believe that plate-based capture will become the default for many IA-MS applications.

A final format for IA-MS capture that will be mentioned here is on-line IA-MS methods. As will be discussed in greater detail, antibodies can be bound inside through pores of polymeric beads packed to make a chromatographic column. Using antibodies with sufficient affinity, it is possible to capture a target peptide on-line in the order of seconds, making IA capture commensurate with the time frame of a chromatographic method. After loading and washing, the target peptide is typically eluted by lowering the pH and detected using conventional LC/MS/MS methodologies. While there are limitations on the types of applications suitable for this format, on-line IA-MS methods provide tremendous advantages in terms of sample throughput and minimized sample processing. Readers interested in this technology are referred to a chapter by Dufield serving as the definitive reference for this approach (Dufield and Radabaugh 2012).

An example of on-line IA-MS from our laboratory involved the analysis of a neoepitope of type II collagen (NET2C) in urine (Berna et al. 2006). Specific proteolytic fragments of type II collagen, a major component of joint tissue, have been identified as biomarkers of osteoarthritis. A major focus of osteoarthritis research is on therapies to impede joint degradation, and thus, biomarkers of cartilage degradation are important for diagnosis and treatment. NET2C is produced by collagenases (e.g., matrix metalloproteinases) overexpressed in osteoarthritic cartilage. In human urine NETC exists predominantly as a 45-amino-acid peptide, although in rats, NET2C fragments are shorter, ranging in length from 14 to 21 amino acids. Because these peptides are concentrated in urine, urine represents the matrix of choice. To support various pharmacology models, a method to quantify a 14-mer rat NET2C fragment was implemented in our laboratory. The instrumental design for on-line analysis appears in Figure 7.2. Urine aliquots of 0.5 mL spiked with an SIL internal standard were injected onto an IA column prepared using 9A4 antibody covalently attached to Poros™ beads (Applied Biosystems). Following loading, water was introduced at 1.75 mL/min for 1.2 min to wash urea and other urinary matrix components to waste. Glacial acetic acid (5% v/v in water) flowing at 1.75 mL/min was then used to elute the bound material from the IA column allowing it to concentrate on an Aquasil C18 trap column. Using a second HPLC system, reversed-phase gradient elution at 0.5 mL/min was used to back-flush the trap column and deliver the analyte to narrow-bore C18 analytical column for separation and MS analysis. In this configuration, the trap column was needed to bridge the flow rates used for IA capture and LC/MS/MS. Using the instrumental format shown in Figure 7.2, complete IA-MS analysis occurred on the chromatographic timescale with a total run time of less than 7 min/sample. This throughput enabled multiple rat pharmacology models depicting OA disease progression to be investigated.

7.2.4 Liquid Chromatography

Numerous LC formats have been applied to targeted protein analysis incorporating various stationary phases, column diameters, flow rates, and even multidimensional configurations. Because a full review of this subject is beyond the scope of this chapter, our discussion will emphasize the factors having the greatest influence on assay sensitivity with appropriate consideration given to practical issues such as robustness and throughput.

Because ESI is a concentration-dependent ionization technique, greater signal can be derived by using smaller diameter columns as the bands produced are more concentrated. This effect, which scales as the inverse square of the column internal diameter, is plotted in Figure 7.3.

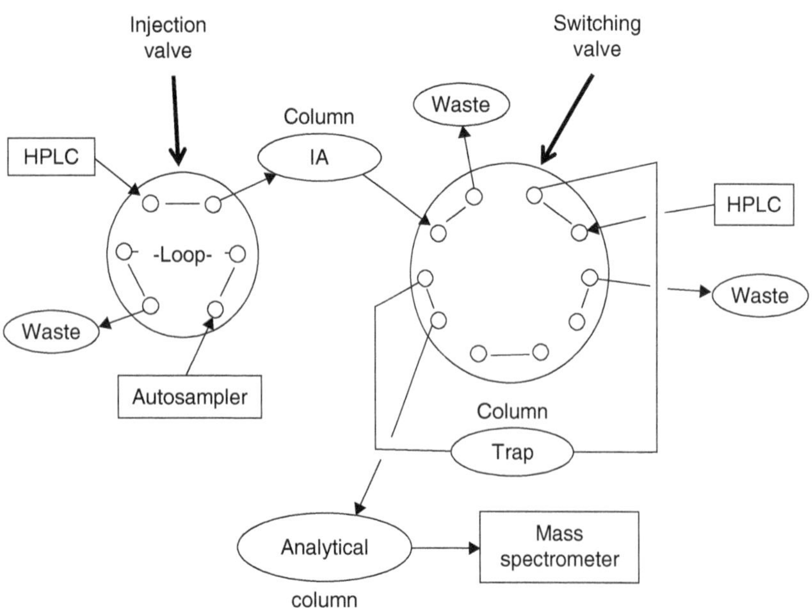

Figure 7.2 Diagram of apparatus used for on-line IA-MS of NET-2C in rat urine. At $t = 0$, 0.5 mL rat urine spiked with internal standard is injected onto IA column (40 °C) at a flow rate of 1.75 mL/min and diverted to waste. At $t = 1.7$ min, the IA column is eluted by lowering the pH of the mobile phase. The peptides are sent to an intermediate trap column (Aquasil C18, 3 × 30 mm) using the same flow rate. At $t = 4.3$ min, the trap is eluted and sent to the analytical column (CapCell-Pak C18 MG, 2 × 35 mm, 50 °C) for gradient elution at 0.5 mL/min. The retention time of NET-2C under these conditions was 5.28 min.

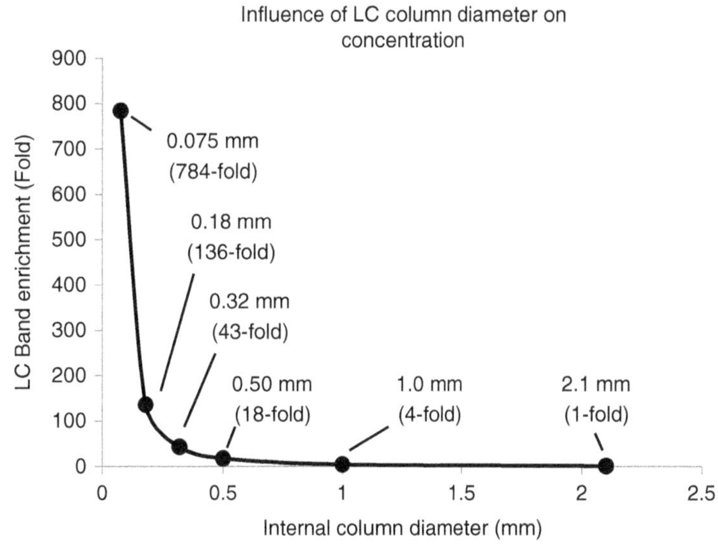

Figure 7.3 Plot that illustrates the increase in LC band concentration obtained using smaller i.d. columns. This phenomenon has a proportional effect on MS sensitivity due to the known concentration dependence of ESI.

Nano-LC is the smallest of four generally recognized columns regimes differentiated by column bore and flow rate. These classifications are defined in Table 7.2. As indicated by Figure 7.3, nano-LC (typically 75 μm id) provides the greatest relative sensitivity. While this effect has been unequivocally demonstrated in practice, it is important to recognize several practical limitations to nano-LC such as high susceptibility to extra-column band broadening, excessive run times attributed to long gradient delays, finite capacity, and limited robustness compared to larger bore columns. To overcome these issues, MS instrument vendors have introduced dedicated nano-LC interfaces to improve robustness and speed (Brennen et al. 2007, Chappell et al. 2014).

In order to capitalize on the impact of reduced column diameter for biological samples, two factors must be addressed. The first relates to injection volume. As shown in Figure 7.3, a fourfold gain can be achieved by switching from a narrow-bore (2.1 mm id) to a microbore (1.0 mm id) column. Unfortunately, while 50 μL can readily be injected onto a narrow-bore column, this volume is not practical for a 1.0 mm column since optimal injection volume also scales inversely with the square of column id. To address this issue, we and others have routinely incorporated a trap column used with a switching value to concentrate the solutes from a large injection prior to back-flushing the analyte onto the smaller bore analytical column for gradient elution. A diagram for typical back-flush configuration appears in Figure 7.4. Such two-dimensional LC methods may be used with each of the column sizes listed in Table 7.2 to accommodate large injection volumes as column diameters are reduced.

The second issue relates to capacity as column overloading occurs more readily as the bore decreases. Given our desire to measure proteins at trace levels, overloading is hardly an issue for the analyte. In contrast, matrix components, which do not change in abundance, can completely overwhelm an analysis, either by producing interfering peaks or by creating ion suppression, if not removed by extraction. The higher enrichment efficiency of protein/peptide analytes from biological matrices using IA compared to conventional extraction techniques is at the heart of explaining why IA methods are vital to maximizing LC/MS sensitivity for proteins. Generally, it

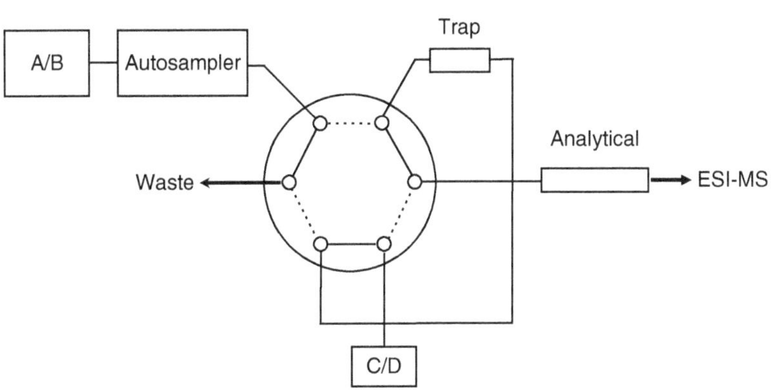

Figure 7.4 Trap and elute column switching diagram. Valve is shown in the elute position using a back-flush configuration. (*Source*: Berna et al. 2007. Reproduced with permission of American Chemical Society.)

Table 7.2 Classification of liquid chromatography formats and the associated column diameters and flow rates.

Format	Column i.d. (mm)	Flow rate (μL/min)	Typical application
Nano	0.05–0.12	0.05–1.0	Global proteomics
Capillary	0.18–0.32	1.0–20	Global and targeted proteomics
Micro	0.50–1.0	20–200	Trace-level quantification
Narrow	1.0–2.1	200–1000	Higher throughput quantification

is not possible to quantify protein/peptide levels below 100 pg/mL using conventional extraction methods, even from large sample volumes (e.g., 1 mL), due to the inability to effectively remove matrix components. As discussed herein, low pg/mL quantification limits have now been achieved in a number of examples using IA-MS.

Interested readers are referred to a recent review by Lassman and Fernandez-Metzler (2014) on LC technology for quantitative LC/MS/MS analysis of proteins. In their review, specific emphasis was given to the recent trend of using *low flow* methods, such as microbore LC, to balance the desire for sensitivity with the need for robustness and throughput. We expect this trend to continue, particularly as LC-vendors offer improved options for performing chromatography in the intermediate flow ranges defined by capillary and micro-LC. Concurrently, investment will also be needed in improved ESI source design, since current ion sources target the flow extremes and have neglected micro-LC and the mid-range characterized as *low flow*.

7.2.5 MS Detection

By far the most common instrument platform to date used for IA-MS operation is the triple quadrupole mass spectrometer (TQMS). TQMS instruments derive their selectivity by selecting the molecular ion for the surrogate peptide of interest, inducing fragmentation in the mass spectrometer through controlled collision with a target gas, and monitoring select fragment ions produced indicative of the peptide sequence. This methodology is referred to as selective reaction monitoring (SRM) or by the interchangeable term MRM. It is important to note here that MRM is not unique to TQMS but can also be performed using ion trapping devices as well as time-of-flight (ToF) mass analyzers.

Although TQMS remains the top choice for targeted protein analysis, an increasing trend has been the use of high-resolution/accurate mass (HR/AM) detection as an alternative way to achieve analyte selectivity in complex biological matrices. Two MS platforms are generally used for this work: quadrupole time-of-flight (Q-ToF) and Orbitraps. While Q-ToF mass spectrometers permit faster data acquisition rates and improved transmission of high *m/z* ions (Han et al. 2008), Orbitrap instruments are finding increased use for biotherapeutics and biomarkers, owing to unprecedented mass resolution attainable on the chromatographic timescale and flexibility in the modes of operation (Gallien and Domon 2014). Because many IA-MS applications tend to be limited by chemical background (i.e., not signal), this technology helps to close the sensitivity gap with respect to MS versus ELISA. Increasing examples of the use of Orbitrap-MS for quantification are indeed being represented in the literature (Wong et al. 2011, Rochat et al. 2013).

Despite the widespread use of ESI, it bears mentioning that a number of IA-MS applications have been reported using matrix-assisted laser desorption ionization (MALDI). In the case of biomarkers, MALDI-ToF applications have shown considerable impact including use for validated clinical assays (Razavi et al. 2013). The fact that MALDI cannot be used with on-line chromatography limits its overall utility. In addition, it is generally less sensitive than ESI. The main advantage of MALDI is that it is amenable for rapid throughput applications. Notable examples come from IA-MS work involving MSIA sample preparation (Niederkofler et al. 2001, Kiernan et al. 2002, Oran et al. 2014).

7.3 IA-MS Applications – Biomarkers

7.3.1 Peptide Biomarkers

Peptide biomarkers, defined here as being less than 10 kDa in molecular weight, play an important role in numerous biological processes and often serve as markers for various disease states. An exciting development regarding MS-based peptide quantification is the prevalence of top-down methods. In fact, most reported applications in this space have achieved sufficient sensitivity without resorting to enzymatic digestion to yield a surrogate peptide. In addition to simplifying LC/MS sample preparation, top-down methods allow direct examination of the intact peptide and its related forms resulting from metabolism, posttranslational modification, or other processes.

Sensitivity remains the key challenge for MS-based peptide analysis since bioactive peptides, such as the incretin hormones, typically circulate at low pg/mL concentrations in clinically accessible fluids. Not surprisingly, LBA methods have dominated this application area. In recent years, LC/MS has been used to analyze bioactive peptides; however, larger sample volumes (0.1–1 mL) are typically required. Although outside the scope of this chapter, it should be mentioned that several LC/MS/MS assays have been reported for bioactive peptides using conventional extraction methods and not IA enrichment (Siskos et al. 2009, Lame et al. 2011, Chambers et al. 2013, Chen et al. 2013). These methods, which have clearly benefited from continuous advancements in MS technology, have a distinct advantage since antibody reagents are not required. The downside is that the large sample volumes and extensive sample preparation required make these methods impractical for most clinical applications. Moreover, conventional methods simply cannot compete with the detection limits possible by IA-MS, in part due to their inability to adequately remove matrix-related ion suppression.

When compared to sandwich ELISA, several advantages to IA-MS may be cited for peptides. For instance, method development is accelerated because it can proceed as soon as a viable capture antibody is identified. In contrast, preparing two high-quality antibodies that do not compete to develop a sandwich ELISA assay can take an indefinite period. A further advantage is derived from the enhanced specificity of MS, which can be used to simultaneously measure multiple forms of a given peptide provided an antibody to a common epitope is used. In contrast, LBA methods require protracted development times to ensure that the antibody reagents used have sufficient specificity at both termini to ensure that the active form of the peptide is measured. Moreover, if successful, separate assays must be built to measure different forms of the peptide (e.g., active vs total). Although multiplexed LBA methods exist, they have not been widely adopted owing to issues related to flexibility and performance.

The advantages cited for IA-MS are nicely illustrated by a recent publication by Chappell and coworkers, which describes their assay for glucagon-like peptide 1 (GLP-1)-related peptides (Chappell et al. 2014). GLP-1 (7–36 amide) is a peptide hormone secreted by ilium L-cells in the intestinal lumen in response to glucose. This peptide binds to the GLP-1 receptor leading to the secretion of insulin and suppression of glucagon. *In vivo* GLP-1 is rapidly deactivated by dipeptidyl peptidase IV, which cleaves two residues from the N-terminus to produce GLP-1 (9–36 amide). Inhibitors of DPP-IV, which prolong the effects of GLP-1, have shown to be effective in the treatment of type II diabetes and, indeed, several DPP-IV inhibitors are currently on the market.

Chappell et al. succeeded in establishing a clinically validated method for both the active and inactive forms of GLP-1. Their assay used an equal mix of two anti-GLP-1 mAbs targeting the mid-domain and C-terminus covalently attached to magnetic beads. Using a sample volume of 1 mL human plasma, the authors were able to establish an LLOQ of 2 ng/mL for GLP-1 (7–36 amide) and (9–36 amide). Analysis was conducted using nano-LC-ESI in conjunction with MRM detection on a TQMS instrument. The assay was an example of top-down analysis as no enzymatic digestion was used.

When making a comparison to LBA, it must be acknowledged that substantially more plasma was needed for IA-MS (>10-fold). However, this method clearly demonstrates the utility of IA-MS since the critical issue of specificity was removed and analysis of both active and inactive forms occurred in a single assay. In contrast, most GLP-1 ELISA assays have trouble differentiating active from inactive forms, perhaps not surprising given the likelihood of overlapping epitopes.

A top-down example from our own laboratory was the application of IA-MS to quantify β-amyloid peptides in human CSF (Oe et al. 2006). Prior to this time, β-amyloid (Aβ) peptide levels were determined exclusively by ELISA methods and significant interest was expressed about the accuracy of these determinations given the discordance in reported literature values and the relevance of these biomarkers to Alzheimer's disease. Using a biotinylated mAb targeting the mid-domain of the Aβ, an immunoprecipitation method based on 1 mL CSF was developed using streptavidin-coated magnetic beads. A key observation in this work was the importance of basic conditions involving ammonium hydroxide, acetonitrile, and water for recovering Aβ peptides from surfaces. As a consequence, LC–ESI–MS/MS was performed in the negative ion mode followed by MRM detection using a linear quadrupole ion trap mass analyzer.

Figure 7.5 displays LC-MRM mass chromatograms acquired for $A\beta_{1-40}$ and $A\beta_{1-42}$ at the LLOQ (0.2 and 0.4 ng/mL, respectively). The two additional mass chromatograms shown correspond to the uniformly ^{15}N-labeled peptide analogs used as internal standards. Calibration standards for this assay were prepared in rat CSF, which has a different Aβ sequence. Excellent parallelism was demonstrated between the surrogate matrix and human CSF (Oe et al. 2006). This assay was ultimately used to cross-validate the existing ELISA method by performing analysis on a common set of clinical CSF samples. The results from this experiment appear in Figure 7.6 and document the close agreement observed between the two methods.

A second example from our laboratory involved the cardiac hypertrophy biomarker NTproBNP and is an example of a bottom-up approach (Berna et al. 2008).

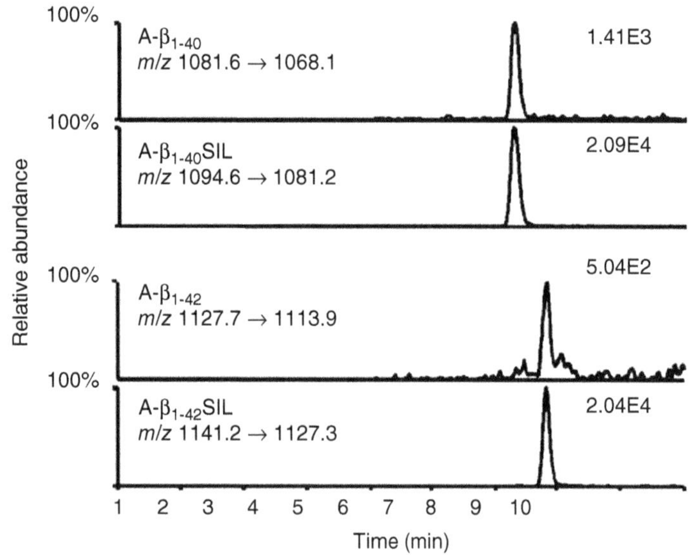

Figure 7.5 SRM extracted ion mass chromatograms from the LC/MS/MS analysis of β-amyloid peptides $A\beta_{1-40}$ and $A\beta_{1-42}$ along with their uniformly ^{15}N-labeled internal standards. The standards shown represent the LLOQ concentrations of 400 and 200 pg/mL, respectively, and were prepared using rat CSF as the surrogate matrix. (*Source*: Oe T 2006. Reproduced with permission of John Wiley & Sons.)

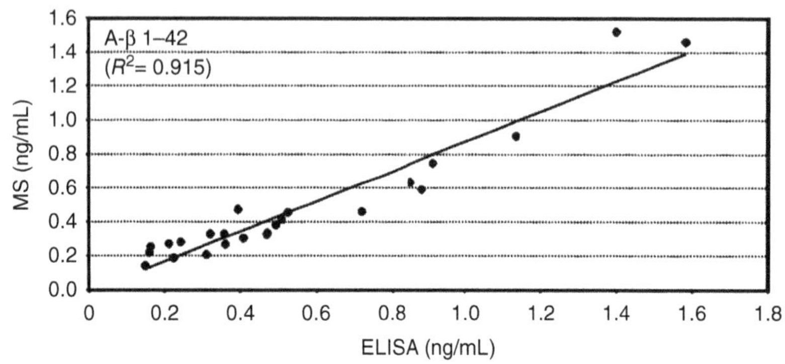

Figure 7.6 Correlation analysis comparing ELISA and IA-MS results for $A\beta_{1-42}$ measured in human CSF. The samples were obtained in a clinical trial and collected by lumbar puncture. (*Source*: Oe T 2006. Reproduced with permission of John Wiley & Sons.)

Previous research had shown that the active hormone brain natriuretic peptide (BNP) is secreted from the cardiac ventricles in response to pressure overload and is believed to mitigate cardiovascular remodeling by promoting vasodilation and natriuresis. In addition, circulating levels of BNP had previously been correlated clinically with left ventricle hypertrophy.

We sought to measure rat natriuretic peptides in order to investigate their utility as preclinical biomarkers of drug-induced cardiac hypertrophy. Owing to known complications associated with measuring the disulfide-linked BNP, we chose to monitor the equimolar NTproBNP as a surrogate. As shown by the diagram in Figure 7.7, NTproBNP is formed simultaneously to BNP during proteolysis of the parent peptide ProBNP. NTproBNP, which is cosecreted with BNP, has also been shown to be a clinically relevant biomarker of cardiac hypertrophy.

The method involved immunoprecipitation from 100 μL rat serum using a mix of rabbit polyclonal anti-NTproBNP antibodies immobilized by a protein-A labeled agarose gel. Following an overnight incubation, the gel was washed, an internal standard was added, and on-gel digestion by trypsin was used to yield the surrogate peptide: Leu-Leu-Glu-Leu-Ile-Arg (LLELIR). To maximize tracking, the internal standard, prepared by solid-phase synthesis, consisted of the entire 50-mer NTproBNP containing a single labeled isoleucine residue $[^{13}C_6\ ^{15}N_1]$ residing within the surrogate peptide. The sensitivity of this assay (LLOQ 100 pg/mL) was attributed in part to the lipophilicity of this peptide surrogate, which enhanced its ESI signal and demonstrated enhanced sensitivity over a top-down approach (data not shown). The MS/MS spectrum for the unlabeled parent peptide is shown in Figure 7.8.

Initial demonstration of the utility of rat NTproBNP as a biomarker of drug-induced cardiac hypertrophy was demonstrated by a 2-week study in which rats ($N = 8$) were treated either with vehicle or an internal compound shown previously to induce cardiac hypertrophy given daily by oral gavage (50 mg/kg). The bar graph in Figure 7.9 reveals that a dramatic increase in cardiac mass (+34%) was associated with a 3.5-fold increase in the NTproBNP.

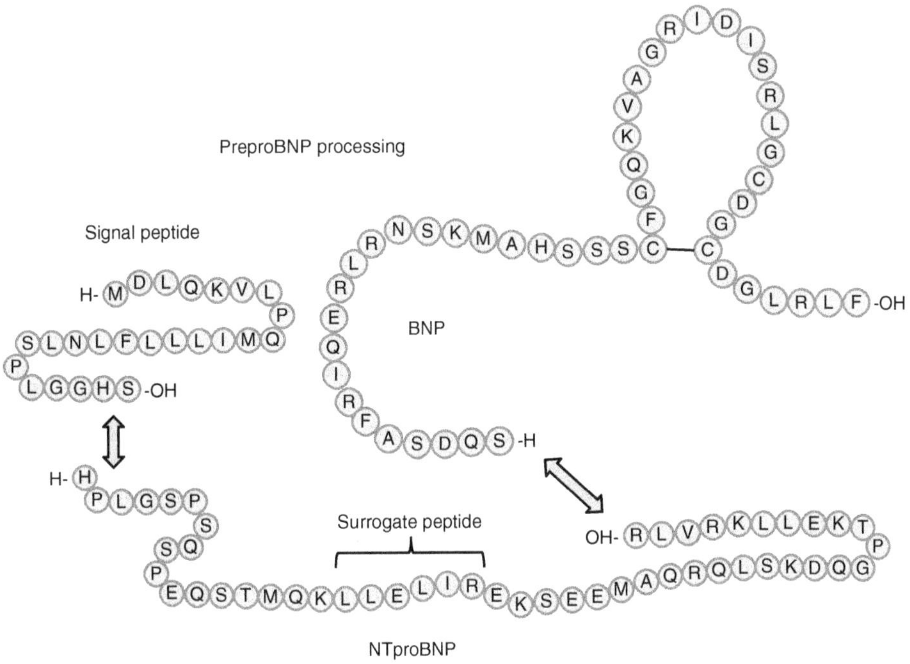

Figure 7.7 Schematic representation showing the production of brain natriuretic peptide (BNP), a biomarker of cardiac hypertrophy. In response to increased pressure, preproBNP is enzymatically cleaved to yield three peptides secreted from cardiomyocytes. The active BNP binds to receptors in the heart producing natriuresis and vasodilation to counteract the pressure load. NTproBNP, also a known biomarker of cardiac hypertrophy, was measured using the surrogate peptide identified on the figure. (*Source*: Berna et al. 2008. Reproduced with permission of American Chemical Society.)

Figure 7.8 Product ion mass spectrum for the tryptic peptide LLELIR showing the predominant *y* and *b* ions. This spectrum was obtained at low collision energy to display the full range of ions produced. The product ion at *m/z* 530.3, which became the base peak after collision energy optimization, was used for SRM. (*Source*: Berna et al. 2008. Reproduced with permission of American Chemical Society.)

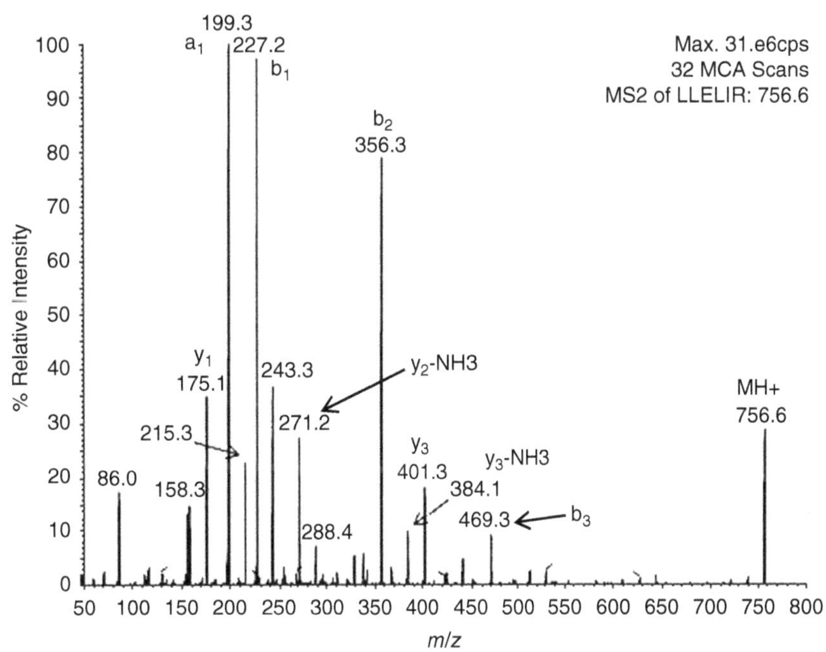

Other relevant examples of peptide biomarker IA-MS have appeared in the literature. Perhaps the earliest example was published by Wolf et al. (2001) and involved the incretin hormone glucose-dependent insulinotropic polypeptide. Similar to GLP1, active GIP1–42 is rapidly degraded by DPP-IV to give GIP3–42. Using a bead-based IP method with a C-terminal antibody, these authors successfully measured basal levels of these peptides in human plasma (LLOQ 20 pg/mL) starting with a volume of 1.9 mL. Amazingly, this work was performed

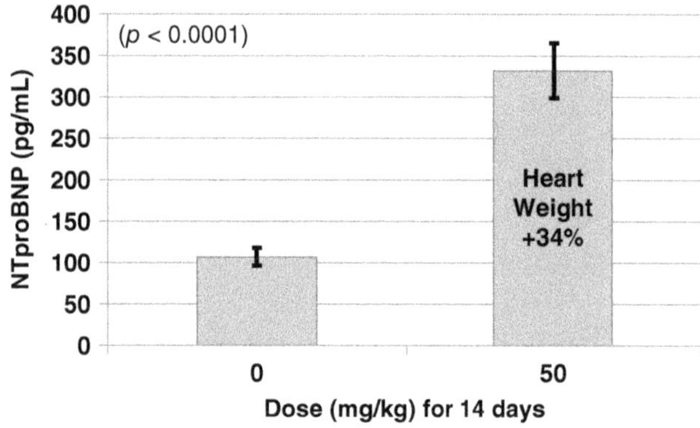

Figure 7.9 NTproBNP concentrations in rat serum were shown to correlate with increases in cardiac weight after treatment with a compound known to induce cardiac hypertrophy. Two groups of Sprague-Dawley rats ($N = 8$) were dosed daily with either vehicle or compound (50 mg/kg) administered by oral gavage. The observed increase (>threefold) was highly significant ($p < 0.0001$) and was accompanied by a 34% mean increase in cardiac weight. (*Source*: Berna et al. 2008. Reproduced with permission of American Chemical Society.)

using a single quadrupole mass spectrometer. It is worth noting that over a decade later a method using conventional extraction appeared in the literature by Miyachi et al. (2013). Their method, based on 0.2 mL plasma, reported an LLOQ of 5 pg/mL. Because this method requires Asp-N digestion and nano-LC (~30 min/sample), the practical utility of this method for clinical analysis is in question.

A second example of IA-MS by Chappell and coworkers bears mentioning. In 2012, their group published a method for *ex vivo* determination of plasma renin activity (PRA) used for clinical development of renin inhibitors (Chappell et al. 2012). The conversion of plasma renin to angiotensin I (Ang I) is the rate-limiting step of the renin–angiotensin–aldosterone system controlling hypertension. Because conversion of Ang I to Ang II is central to raising blood pressure, several existing treatments have focused on this step. Clearly, in order to study the upstream blockade by renin inhibition, Ang I must be reliably measured. These workers used an anti-Ang I antibody to capture the Ang I produced *ex vivo* over a 3-h period during which previously collected plasma samples were incubated at 37 °C. Acetonitrile precipitation was used to release bound Ang I for determination by LC/MS/MS. The results from this method compared favorably to the EIA LBA method previously used for PRA determination.

A further example involving clinical quantification of full-length parathyroid hormone (PTH 1–84) was published by a group at the Mayo Clinic and illustrates the importance of IA-MS to simultaneously deliver sensitivity and specificity (Kumar et al. 2010). Because inactive, C-terminal containing, degradation products account for 90% of circulating PTH, LBA methods lacking sufficient specificity are prone to giving spuriously high results. Using 1 mL serum, an IA-MS method using antibody-labeled polystyrene beads and detection by LC-MRM was prepared using the N-terminal tryptic fragment PTH 1–13 as a surrogate for the active hormone. Bland-Altman analysis derived from a comparison to an immunoassay performed on a Roche Cobas™ biochemical analyzer gave excellent results indicating a mean negative bias for the IA-MS method of only −9.4% (Kumar et al. 2010).

A number of IA-MS peptide applications come from examples involving MSIA extraction. Generally speaking, these applications fall into two types. The first class are MALDI applications (no-LC) involving top-down analysis. These methods grant higher throughput, but at the expense of sensitivity. Examples include analysis of insulin variants (Oran et al. 2011), the cytokine CCL5 (Oran et al. 2010), and β-2 microglobulin (Niederkofler et al. 2001). More recent examples have used bottom-up LC/MS/MS analysis to deliver improved detection limits. This strategy has been extended to β-amyloid peptides, PTH, and insulin-like growth factor 1 (IGF-1) (Krastins et al. 2013).

7.3.2 Protein Biomarkers – Anti-Protein Capture

A majority of reported IA-MS applications have been for protein targets, defined herein as being larger than 10 kDa. Because current MS technology does not provide sufficient sensitivity for intact proteins, a majority of these applications involve bottom-up analysis. As depicted in Figure 7.1, IA-MS analysis for proteins can occur using either anti-protein or anti-surrogate peptide antibodies (or a combination of both). Applications from our laboratory have involved IA capture of the target protein (Figure 1a). One reason is that for most pharmaceutical applications the target of interest is predefined. Hence, the task at hand is to produce selective assays to study pharmacology or understand the fate of the biotherapeutic agent. Moreover, high-affinity mAbs for the target are often available in the research setting to permit IA-MS assay development.

This approach was illustrated by work we conducted to evaluate putative protein biomarkers of cardiotoxicity in preclinical species (Berna et al. 2007, Zhen et al. 2007). One example involved rat myosin light chain 1 (myl3), a 23 kDa isoform of myosin, a protein involved in cardiac muscle contraction. Although previous literature had implicated myl3 as a candidate biomarker of cardiac necrosis, internal work was inhibited by the lack of antibodies for ELISA development. Using a biotinylated version of a commercially available mAb against rat myl3, an IP method was made using streptavidin-coated magnetic

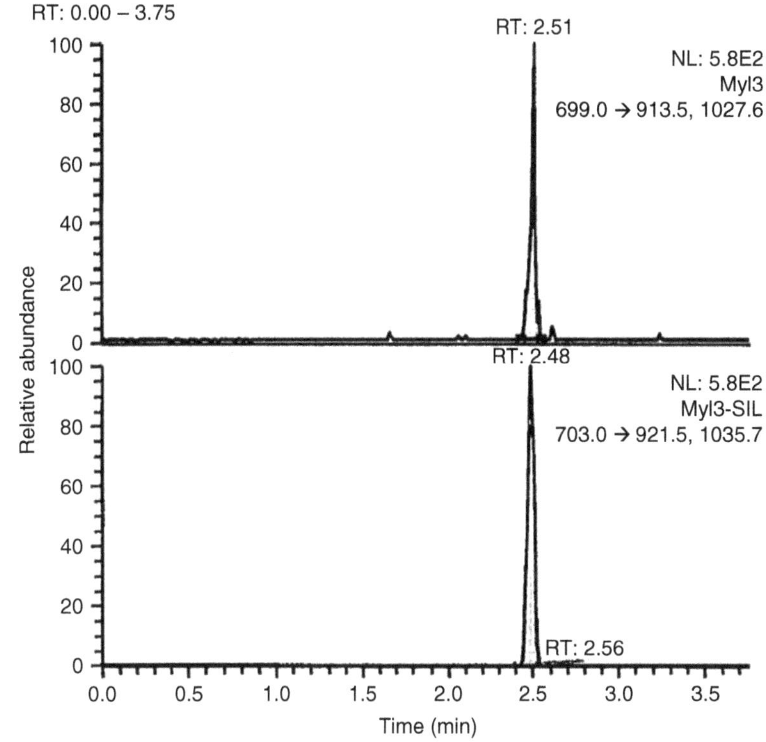

Figure 7.10 SRM extracted ion mass chromatograms for the 13-mer surrogate peptide for Myl3 shown at the assay LLOQ (0.07 nM). The lower trace corresponds to the SIL internal standard present at 0.71 nM. All standards were prepared using immunodepleted rat serum as the surrogate matrix. Sample preparation consisted of using antibody-coated magnetic beads for Myl3 capture followed by on-bead digestion with trypsin to yield the surrogate peptide. (*Source*: Berna et al. 2007. Reproduced with permission of American Chemical Society.)

beads with 100 μL rat serum. On-bead digestion with trypsin, without reduction–alkylation, liberated the 13-mer surrogate peptide: ALGQNPTQAEVLR that was ultimately measured by MS. LC/MS/MS occurred by TQMS in conjunction with narrow-bore LC incorporating the trap-back-flush design depicted in Figure 7.4. Validation of the method, which had a range of 3–150 ng/mL for myl3, was conducted using immunodepleted rat serum as the surrogate matrix for calibration standards and QC samples. Excellent signal was achieved at the LLOQ as shown by the SRM mass chromatograms for the surrogate peptide and its deuterated internal standard appearing in Figure 7.10.

This first application of this assay was to support a study in which rats were given a single subcutaneous dose of 50 mg/kg isoproterenol, an agent known to cause cardiac necrosis. As shown in Figure 7.11, this dose produced a 25-fold increase in myl3 over baseline levels and remained elevated over a 24-h period (Berna et al. 2007). A second example of this strategy involved fatty acid binding protein (fabp3) (Zhen et al. 2007). The method produced provided critical cross-validation data for a LBA method that was being used to study this marker. Ultimately, both myl3 and fabp3 were incorporated into a panel of LBA methods currently used to assess drug-induced cardiotoxicity in preclinical discovery.

To date, several IA-MS applications have been reported using anti-protein capture. Similar to our cardiotoxicity work, the methods tend to be validated assays targeting a single protein. Also, in many cases direct comparison was made to existing LBA methods. Callipo et al. 2010 published an IA-MS assay for carbonic anhydrase (CA-II), a 29 kDa protein, using 20 μL human serum. Using a 2.1-mm-id trap column, a large injection (100 μL) was permitted followed by gradient elution on a microbore column (1 mm id). The tryptic surrogate peptide had a retention time of 4.2 min using these conditions. Also in 2010, an IA-MS method for matrix metalloproteinase 9 (MMP9) in mouse serum was reported (Ocaña and

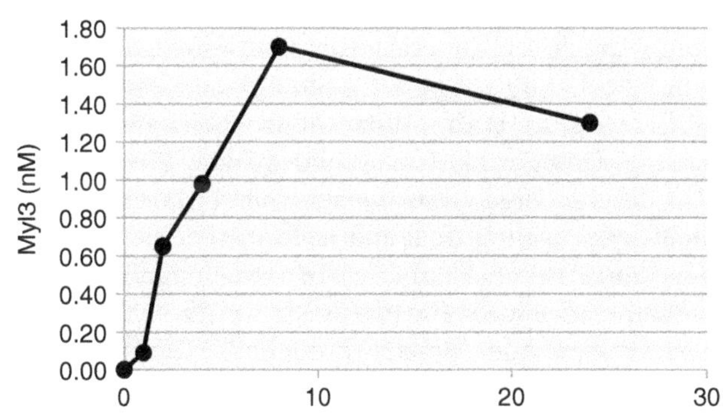

Figure 7.11 Time course observed for Myl3 in rat serum following a 50 mg/kg subcutaneous dose of isoproterenol, a β-adrenergic receptor agonist known to produce cardiac injury at elevated exposures. The results for vehicle group were below the assay LLOQ (0.07 nM) at each time point. (*Source*: Berna et al. 2007. Reproduced with permission of American Chemical Society.)

Neubert 2010). This method, which also used a 20 µL sample volume, incorporated a 96-well robotics platform for performing IP using magnetic beads. LC-MRM analysis of a tryptic surrogate occurred by a 2-D nano-LC method incorporating strong cation exchange as the first dimension followed by reversed-phase C18 (75 µm id) for analysis at a flow rate of 400 nL/min. Interestingly, a poor correlation was observed upon cross-validation with an existing sandwich ELISA method with the latter method producing higher values. While the reasons for this difference were not completely sorted out, this example demonstrates the importance of MS methods for assessing LBA specificity.

In a recent publication, McAvoy et al. (2014) used an anti-protein IA-MS approach to quantify tau, an important biomarker for neurodegeneration. Their method involved bead-based IP using 150 µL of human CSF, followed by trypsin digestion and LC/MS/MS. By selecting a conserved surrogate peptide, the method represented six tau isoforms and thus was a measure of total tau. An important feature of this method, which was successfully cross-validated with ELISA, was the use of a full-length, SIL protein as the internal standard.

Further examples involving anti-protein IA-MS have occurred in the field of clinical oncology. Two applications, recently published by Torsetnes and colleagues, relate to biomarkers used for small-cell lung cancer diagnosis (SCLC): neuron-specific enolase (NSE) (Torsetnes et al. 2013) and progastrin-releasing peptide ProGRP (Torsetnes et al. 2014a, b). While ELISA methods were available in both cases, these authors developed IA-MS methods capable of differentiating the total protein biomarker from specific isoforms. A common analytical approach was used in both methods and involved magnetic bead IP capture from 1 mL human serum. Following trypsin digestion, LC-MRM analysis occurred using a 2-D microbore LC system. In the case of NSE, both homo- and heterodimeric forms were detected. In addition, the LLOQ (38 pg/mL) allowed quantification of NSE levels in both normal individuals and cancer patients (>1 ng/mL). The second marker, ProGRP, was also differentiated from the standard ELISA measurement by providing total and isoform-specific information (Torsetnes et al. 2014a, b). In a recent publication by the same laboratory, the two assays were combined into a panel to provide a greater throughput (Torsetnes et al. 2014a, b). More important, it is believed that the throughput and availability of these two tests will allow early-stage detection of SCLC.

An investigation comparing IA-MS and ELISA for six colon cancer biomarkers in human plasma was recently conducted by Lin et al. (2013). In this study, the capture antibodies available from six commercial ELISA kits were used for anti-protein IA-MS. After comparing the individual methods for tissue inhibitor of metalloproteinase 1 (TIMP1), the remaining five proteins were coextracted as a multiplexed IA-MS panel. The overall results from this investigation showed comparable results in terms of levels, trends, and precision. In addition to the throughput advantage of the IA-MS multiplex, it was noted that MS gave superior discrimination of control versus diseased samples for five of the six proteins studied. The authors commented that an advantage of anti-protein capture, over anti-peptide IP (SISCAPA), is that the antibodies used can be used for subsequent ELISA methods.

Another example of a cancer IA-MS multiplexed assay was published by Nicol et al. (2008). Their study involved a five-plex panel, which included the known tumor biomarker carcinoembryonic antigen (CEA). Results were presented from the analysis of serum collected from lung cancer patients, which indicated an elevation in CEA observed both by IA-MS and ELISA. The method, based on 0.1 mL serum, involved overnight capture by antibodies covalently attached to hydrazide beads. Reduction, alkylation, and trypsin digestion were performed on-bead before SPE cleanup. The extracts were injected onto a 2-D nano-LC system using a 300-µm-id trap column and a 75-µm-id analytical column (both C18). The low ng/mL LLOQ values observed enabled viable quantification and comparison to ELISA for the biomarkers studied. In addition, an interesting comparison was made between IA-MS and a conventional method that used immunosubtraction of abundant serum proteins prior to SPE extraction. The comparison revealed a 150-fold enhancement observed by IA-MS.

Similar to peptides, IA-MS methods for proteins have also been reported using MSIA technology and have included top-down (MALDI) and bottom-up (LC/MS/MS) applications. An example of the former is IGF-1, which was measured in human plasma down to 5 ng/mL with a throughput of 1000 samples per day (Oran et al. 2014). Examples of the latter approach include PCSK9, β-microglobulin, C-reactive protein, apolipoprotein E, and procalcitonin (Krastins et al. 2013).

7.3.3 Protein Biomarkers – Anti-Peptide Capture

Anti-peptide IA-MS (Figure 7.1b) represents a complementary strategy for protein biomarker quantification. This technique, originally introduced by Anderson, goes by the familiar term SISCAPA and represents the largest segment of reported IA-MS applications (Anderson et al. 2004). SISCAPA is closely linked to bottom-up proteomics since similar up-front sample preparation occurs, typically involving reduction, alkylation, and trypsin digestion. Anti-peptide IP was initially developed to follow hypotheses generated by shotgun proteomics in clinically accessible fluids. The process of translating biomarker leads for clinical application is often referred to as *verification*, a term coined by Rifai et al. in a landmark paper addressing the discovery and development of protein biomarkers (Rifai et al. 2006). Owing to its connection to proteomic workflows, SISCAPA methods have frequently been applied using antibody mixtures to target multiple proteins. An important paper highlighting multiplexed SISCAPA was published by Whiteaker et al. detecting as many as 50 surrogate peptides in a single assay (Whiteaker et al. 2012).

A representative example of SISCAPA for clinical work was published by Kuhn et al. and involved the analysis of cardiac troponin I (cTnI), a known marker of cardiac damage, and interleukin-33 (IL-33), a candidate cardiac biomarker (Kuhn et al. 2009). SISCAPA-based LC/MS/MS was used to quantify both biomarkers down to a range of 1–10 ng/mL in human plasma using a volume of 50 µL. Good correlations were established with existing LBA methods. Perhaps the most recognized clinical SISCAPA method was reported by Hoofnagle and coworkers for the cancer biomarker thyroglobulin (Hoofnagle et al. 2008). This protein, which is used to diagnose the recurrence of thyroid cancer, can go undetected in ELISA methods owing to the presence of anti-thyroglobulin antibodies produced by the patient. As a consequence, two ELISA methods are routinely performed to assess the influence of autoantibodies. This approach is not an issue with SISCAPA because of up-front sample digestion prior to capture.

SISCAPA applications have been well represented in oncology, in part due to the National Cancer Institute consortium on Clinical Proteomic Technology Assessment for Cancer (CPTAC). As part of this effort a large interlab comparison was conducted showing the ability of SISCAPA methods to be transferred across laboratories (Addona et al. 2009). A novel SISCAPA cancer application published in 2012 applied IA-MS to quantify the breast tumor markers ER (estrogen receptor) and HER2/neu (human epidermal growth factor 2) (Schoenherr et al. 2012). These proteins, commonly used in cancer diagnosis, were readily quantified from lysate obtained from surgically resected tumor tissue. The data compared favorably to ELISA measurements using dedicated assays and the potential of SISCAPA for multiplexed analysis was discussed. The use of SISCAPA with MALDI-ToF has also been demonstrated and offers the ability for higher throughput in the clinical setting. An example is protein C inhibitor (PCI), which was quantified in the sera of 51 prostate cancer patients involved in a study of cancer recurrence (Razavi et al. 2013). The authors reported that elevated PCI levels correlated with lack of recurrence, a trend not attainable using standard tests for prostate-specific antigen (PSA).

A special category of IA-MS is referred to as *Sequential IP* since it uses both anti-protein and anti-peptide IA-enrichment in the same assay to provide superior sensitivity and selectivity for protein biomarker assays. This format, introduced by Neubert et al. (2013), was demonstrated by the clinical analysis of Nerve Growth Factor (NGF). In this assay, NGF levels were measured at baseline and after administration of the mAb drug Tanezumab®, which targets NGF. The assay, based on 0.6 mL human serum, was validated over a range of 7–450 pg/mL and extended by sample dilution to measure total NGF levels (free plus bound) in the low ng/mL range observed in clinical trials. Initial capture was performed on magnetic beads using a goat anti-NGF polyclonal antibody. After reduction, alkylation, and trypsin digestion, a peptide surrogate was purified by on-line IA-MS using an anti-peptide antibody column (SISCAPA) before quantification by nano-ESI-LC/MS/MS. This latter step was essential to achieve the reported LLOQ since ion suppression and interferences were minimized. Baseline levels were determined in both human and preclinical species owing to sufficient cross-reactivity of the capture polyclonal antibody and the fact the surrogate peptide represented a conserved sequence. Despite the complexity of this approach, the authors emphasized the utility of this IA-MS configuration to produce drug-tolerant assays provided the capture antibody binds to a different epitope as the drug.

A second application of the sequential IP approach was published by the same group for the cytokine IL-21 (Palandra et al. 2013). An IA-MS method was desired for both serum and tissue owing to wide variations in the literature for serum ELISA methods and difficulty applying these methods to tissue. Using a similar analytical configuration to NGF, a validated method was established in human serum and plasma having an LLOQ of 0.78 pg/mL. Despite achieving this sensitivity, human samples were below this level calling into question previous ELISA levels reported to vary from 40 to 800 pg/mL. Successful analysis was conducted in various tissues, which included isolated human T cells, tonsils, and colon samples along with various samples from cynomolgus monkeys.

7.4 IA-MS Applications – Biotherapeutics

7.4.1 Therapeutic Peptides

Peptides and polypeptides are involved in numerous biochemical processes where they can function as agonists or antagonists of cellular receptors, modulators of protein–protein interactions, or competitors/inhibitors of enzyme-mediated reactions, to name a few. Consequently, peptide therapeutics can be used to target a wide variety of disease-associated processes ranging from cancer and cardiovascular disorders to respiratory illnesses and renal failure. This potential served as the impetus for a significant expansion of therapeutic peptide discovery research starting in the late 1990s, which led to an unprecedented number of peptide drug approvals starting in 2010 (Kaspar and Reichert 2013).

Peptide therapeutics face similar bioanalytical challenges to peptide biomarkers, namely the need for high sensitivity and specificity. These requirements are exacerbated by the tendency for peptides to be metabolized into structurally similar metabolites, which may be active or inactive. In the absence of a highly specific assay, it is possible for inactive metabolites to interfere with the measurement of active drug. Conversely, it is also possible for active metabolites to go undetected depending on the specificity of the method used for detection. Although LBA methods play an important role in biotherapeutic peptide analysis, LC/MS has become the method of choice. This is especially true in drug discovery where LBA approaches are challenged to produce antibody pairs within the time frame of a structure–activity-relationship (SAR) screen. During SAR, a large number of peptide variants are synthesized to optimize potency, formulation and *in vivo* stability, and PKs. Oftentimes, a single region in the peptide backbone is sufficiently conserved across variants to provide a single epitope for IA-MS assay development. However, it should be noted that according to the potency and clearance of a therapeutic peptide, LC/MS assays can sometimes be developed without the need for an IA enrichment step (Dawes et al. 2013).

As mentioned above, one of the true advantages of IA-MS is the ability to discretely measure structurally related molecules that would cross-react by LBA analysis. For example, Hess et al. (2012) used a pan-specific anti-insulin antibody to immunoprecipitate (IP) and measure human insulin and several synthetic insulin variants in human samples. Following protein precipitation from human serum, insulin, Humalog®, NovoLog®, Apidra®, Lantus®, and Levemir® were isolated via IP followed by intact (top-down) analysis using LC-SRM. The assay was validated to support clinical and forensic toxicology investigations and was able to measure the insulin analogs as low as 2.6 μU/mL. Owing to similarity between insulin variants, it was necessary to utilize a top-down approach to differentiate the molecules studied. Similarly, Thomas et al. (2014) used IA capture followed by ion mobility, high-resolution MS to measure several insulin analogs to support forensic investigations. The authors' assay was able to sensitively (0.2 to >5 ng/mL) and unambiguously differentiate between human insulin, synthetic insulin variants, and animal insulin analogs.

In addition to SAR support and clinical investigation, IA-MS is starting to see widespread use to monitor therapeutic peptides with time-extension modifications. These modifications, which include glycosylation, PEG conjugation, XTENylation, or Fc fusion, are engineered into therapeutic peptides as a means to increase their *in vivo* exposure by reducing clearance (Ezan 2013, Podust et al. 2013). These modifications can pose a significant challenge to LBA assay development by masking peptide epitopes needed for sandwich ELISA. This issue is exacerbated during SAR screening where large numbers of variants are typically investigated. Similarly, time-extension modifications are challenging for MS because they significantly increase the molecular weight of a peptide, necessitating the use of a bottom-up approach. A further complication is that moving away from intact analysis requires prior knowledge of the molecule's *in vivo* metabolism and biologically active sequences in order to select an appropriate surrogate peptide. Clearly, this can present a significant challenge for MS detection in cases where the tryptic surrogate peptide contains the time-extension modification. In such examples, it is necessary to investigate the use of another protease to generate a more suitable surrogate peptide.

One of the challenges associated with PEG attachment is to find an antibody that recognizes the peptide irrespective of the site of attachment to allow multiple peptide-PEG variants to be compared. One approach to this problem is to use an anti-PEG antibody for capture. This strategy was utilized by Xu et al. to quantify the 40 kDa PEGylated-glucagon-like peptide-1 (GLP-1) agonist, MK-2662, from human serum to support clinical analysis (Xu et al. 2010). Following trypsin digestion, the authors followed the active, N-terminal sequence (1–12) over the range of 2–200 nM. Interestingly, when the results obtained by IA-MS were compared to a second MS analysis using protein precipitation (PPT), the IA-MS concentrations were approximately 15–30% lower. The authors attributed this difference to a lack of specificity when using PPT for sample preparation.

IA-MS has also been applied to the analysis of therapeutic polypeptides (or small proteins) using both intact and bottom-up methods. An example is the use of IA-MS to measure recombinant human methionyl-leptin (rMet-hLeptin), which has the same sequence as human leptin, apart from an N-terminal methionyl residue attributed to expression in *E. coli* (Wang and Heilig 2012). Attempts to use immunoassay formats were unsuccessful due to cross-reactivity with the nearly identical endogenous human leptin (Endo-Leptin). Using an IP method that captured both rMet-hLeptin and Endo-Leptin, the two variants were easily differentiated and quantified by LC-MRM over the range of 15–1000 ng/mL.

In addition to quantitative analysis, MS grants access to structural information not available by LBA methods. However, the sensitivity and speed of LBA assays and its ability to assess functional binding make ELISA an indispensable and complimentary tool. An illustration of this interaction was published by Hager et al. (2013), who combined differential ELISA and IA-MS to investigate the PK and biotransformation of biotherapeutic drugs. An illustration of this approach was provided by the analysis of various fibroblast growth factor 21 (FGF21) variants, which included wild-type FGF21, truncated species, and an Fc-FGF21 fusion protein. ELISA assays were developed to recognize specific areas of the molecule known to be important for target binding since a loss in signal could be interpreted as indicative of proteolysis in that region. This assay was followed up with IA enrichment and MALDI-ToF analysis to identify specific sites of proteolysis.

7.4.2 Therapeutic Antibodies

mAbs have become commonplace in the biotechnology/pharmaceutical industry as treatments for a variety of conditions such as autoimmunity, cardiovascular disease, infectious disease, inflammation, and cancer (Chames et al. 2009). As of 2012, more than 20 mAb were approved by the FDA with hundreds more still under development (Wang et al. 2008, Li et al. 2012). Historically, LBA methods have been used to assess mAb PK. However, more recently, IA-MS has gained popularity due to its high selectivity, reduced requirement for antibody generation, relatively short time for assay development, and the ability to generate structural information often critical to discovery research.

A fundamental need in the development of mAbs is *in vivo* measurement for PK determination. As mentioned earlier, it is often necessary to develop assays for several structurally similar molecules that may differ with respect to their PK parameters and/or metabolic profiles. To address this challenge, Li et al. (2012) developed a generic LC/MS approach utilizing a common automated IP protocol for streamlined sample preparation. This method also incorporated a uniformly SIL mAb internal standard for improved assay performance. The authors validated their approach using four different IgG1 and four IgG2 antibodies and were able to obtain good accuracy and precision over the range of 0.1–15 µg/mL. As a demonstration of their method, rat PK analysis was conducted for eight antibodies using IA-MS and the results obtained were shown to compare favorably to similar analysis performed using LBA.

Leary et al. (2013) evaluated four different LBA formats for their ability to perform routine total IgG measurements to support drug discovery. In addition to ELISA, Meso Scale Discovery™ (MSD), Gyros™, and AlphaLISA™ LBA formats, the authors included IA-LC/MS/MS to see if it could be used as an alternative platform for generic Fc measurements. To test the different formats, the authors prepared mock PK samples in rat serum using a human IgG1 antibody. Based on this comparison, they established that IA-MS compared similarly to the other formats in terms of assay cost per sample, sensitivity, dynamic range, and other requisite assay parameters.

Furlong et al. (2012) reported their "universal" peptide approach for quantifying mAbs in preclinical species. To streamline the assay development required for multiple drug candidates, the authors developed a bottom-up approach in which a single, homologous, surrogate peptide was selected as the basis for a common IA-MS assay applied to multiple mAb variants. An updated version of the assay was published in 2013 which included a second homologous peptide in the mAb light chain to complement the original heavy chain sequence (Furlong et al. 2013). The new assay termed their "dual universal" approach incorporated the light chain surrogate peptide in an attempt to differentiate intact versus degraded forms of the mAb. The assay also included a uniformly stable-isotope-labeled mAb internal standard, similar to the approach by Li et al. (2012) to improve assay accuracy and precision.

In addition to preclinical analyses, IA-MS has been applied to analyze human mAbs under clinical investigation. An example is the work by Dubois et al. (2008) who used immunopurification and LC-MRM to measure the chimeric human:mouse mAb Erbitux® in human serum. To avoid interferences from endogenous human antibodies, the authors utilized the soluble form of the Erbitux® target, epithelial growth factor receptor (sEGFR), to IP the drug from serum. When isolating human mAbs from human matrices, it is necessary to use selective reagents, such as anti-idiotypic antibodies (anti-IDs) or receptor targets during immunopurification to avoid capturing endogenous antibodies that are 3–4 orders higher in concentration. Following IP and trypsin digestion, the authors measured heavy- and light-chain peptides by MRM down to 20 ng/mL (LLOQ for ELISA was 5 ng/mL).

In another example, Ocana and coworkers used IA-MS to quantify a humanized mAb against mucosal addressin cell adhesion molecule (MadCAM), implicated in ulcerative colitis (Fernández Ocaña et al. 2012). In contrast to the Erbitux® example cited above, anti-IDs were used to IP "free" therapeutic mAb (i.e., not bound to target) from human serum. Protein G was used to provide a "total" therapeutic mAb assay (free mAb plus mAb:MadCAM complex). Because this reagent targets the Fc region of the antibody, both free and bound forms were captured. For both assays, a surrogate peptide selected from the complementary determining region (CDR) on the light chain was selected for quantification by LC-MRM. The free assay had a range of 7–450 ng/mL, whereas the total assay covered a range from 0.781 to 50.0 μg/mL. Both methods were applied to a series of clinical samples from ulcerative colitis patients and compared to ELISA methods also measuring free and total drug in the same samples. Excellent correlations were obtained between the methods.

7.4.3 Antibody–Drug Conjugates

ADCs are a relatively new class of biotherapeutics engineered by conjugating a cytotoxic payload via a flexible chemical linker to a mAb. Upon internalization of the ADC into a target cell, it is trafficked to lysosomes where linker cleavage occurs, resulting in payload release and ultimately cell death. The benefit of using ADCs, as opposed to traditional chemotherapeutic agents, is the selectivity by which the cytotoxin is delivered to cancer cells, taking advantage of an antibody's ability to target a receptor that is highly expressed by the tumor (Jaracz et al. 2005, Chari 2008).

It goes without saying that ADCs are highly complex molecules. This complexity arises not only via their interactions with soluble proteins (e.g., albumin) and cell surface receptors but also because of their inherent heterogeneity. This heterogeneity leads to a large number of circulating forms that must be measured in order to characterize an ADC's PK/PD relationship, safety profile, metabolism, and immunogenicity (Sauerborn and van Dongen 2014). As with the other classes of biologic drugs, ADCs have been extensively characterized using LBA approaches (Dere et al. 2013). However, because of the diversity in bioanalytical targets that must be measured, ranging from free small-molecule payload to mAb proteolytic metabolites, IA-MS is ideal for *in vivo* ADC characterization. It should be noted that measurement of "total IgG" for an ADC can be accomplished using the same approaches that were outlined for therapeutic mAbs; however, it should be verified that ADC binding by the affinity capture reagent is not influenced by the conjugated payload.

A principal concern when developing ADCs is the premature release of payload in systemic circulation, which can negatively impact the molecule's efficacy as well as its safety profile. Because premature payload release leads to a change in an ADC's drug-to-antibody ratio (DAR), changes in DAR must be monitored over the course of an *in vivo* study to aid in interpreting the molecule's efficacy and safety data. Unfortunately, it can be difficult to distinguish between molecules with different DAR distributions using immunoassay formats. To address the need for better bioanalytical tools, Xu et al. (2011) implemented bead-based affinity capture and capillary LC/MS to quantitate the DAR distributions of an anti-MUC16 Thiomab-drug conjugate both *in vitro* and *in vivo*. Affinity capture was accomplished using biotinylated target antigen (MUC16 extracellular domain) immobilized to streptavidin-coated magnetic beads. Following IA capture and deglycosylation, the different ADC DAR forms were characterized by intact analysis using Q-ToF MS. The authors followed up this work with a review of different IA-MS and LC-UV approaches that can be used to profile changes in DAR in plasma/serum (Xu et al. 2013). In this more recent work, ADCs prepared using three of the most prevalent conjugation strategies (lysine, site-specific cysteine, and interchain disulfide conjugation) were used to validate their methodology.

In addition to characterizing an ADC's DAR profile and PK, the heterogeneous nature of ADCs makes characterizing their toxicological and immunogenic properties challenging as well. As is true for other large-molecule drug classes, a combination of IA-MS and immunoassay formats is applied to characterize the aforementioned properties. A review of approaches (both IA-MS and LBA) to characterize ADC PKs and immunogenicity was recently published by Sauerborn and van Dongen (2014), while Roberts et al. (2013) focused on strategies to evaluate nonclinical ADC safety for oncology indications.

7.5 Future Direction

In the years ahead, the notion of combining IA sample preparation with MS detection will become commonplace in laboratories studying either protein biomarkers or biotherapeutics. Pharmaceutical laboratories, which tend to be well supplied with MS instrumentation, will increasingly invest in the cross-training necessary to establish a workforce having comparable fluency in MS and LBA methods. This strategy will surely expedite method development and ensure that the proper proteins, including their various forms, are measured with the specificity needed for valid decision-making. As part of this combined interaction, an increased expectation will be for clinical LBA methods to undergo an appropriate cross-validation with IA-MS. While LBA methods will continue to exist as the default for wide-scale clinical analysis, MS methods will be deployed when multiple proteins need to be measured and there will be a concomitant increase in the number of MS-based laboratory developed tests (LDTs) offered by large clinical laboratories for protein biomarkers.

At present, IA-MS methods often require exaggerated sample volumes to provide the sensitivity needed for *in vivo* biological applications. We expect this gap to close over the next decade leading to several implications. One consequence will be that bead-based IP methods will gradually give way to plate-based formats for sample preparation. Further, as sensitivity gains increase, the current molecular weight ceiling for top-down analysis will rise above the current mark of approximately 10–15 kDa. Although top-down methods impose a bottleneck regarding the procurement of protein internal standards, the benefits associated with removal of proteolytic digestion are obvious in terms of time saved and improved analytical performance.

Another manifestation of increased sensitivity will be the avoidance of nano-LC, which presently limits robustness and leads to excessive run times. The current movement toward capillary and micro-LC formats is an encouraging sign of this inevitable change. Higher flow LC methods will still exist owing to relative simplicity and ease of operation.

Based on our current laboratory experience, we are extremely encouraged by the ability of HR/AM methods, to provide unparalleled specificity on the chromatographic timescale. Both MS and HR/AM options will continue to be used with HR/AM guided by matrix complexity. We expect HR/AM technology to improve while becoming easier to deploy and more cost-effective.

In closing, it is interesting to reflect on how the introduction of IA-MS, coupled with improvements in sample preparation and MS technology, has led to an analytical revolution for proteins reminiscent of how LC/MS/MS transformed small-molecule bioanalysis. We anticipate the years ahead to be as equally transformative.

References

Ackermann BL and Berna MJ (2007). "Coupling immunoaffinity techniques with MS for quantitative analysis of low-abundance protein biomarkers." *Expert Rev. Proteomics* **4**(2): 175–186.

Addona TA, Abbatiello SE, Schilling B, et al. (2009). "Multi-site assessment of the precision and reproducibility of multiple reaction monitoring-based measurements of proteins in plasma." *Nat Biotechnol* **27**(7): 633–641.

Anderson NL, Anderson NG, Haines LR, et al. (2004). "Mass spectrometric quantitation of peptides and proteins using stable isotope standards and capture by anti-peptide antibodies (SISCAPA)." *J Proteome Res* **3**(2): 235–244.

Barnidge DR, Hall GD, Stocker JL, et al. (2004). "Evaluation of a cleavable stable isotope labeled synthetic peptide for absolute protein quantification using LC–MS/MS." *J Proteome Res* **3**(3): 658–661.

Becker JO and Hoofnagle AN (2012). "Replacing immunoassays with tryptic digestion-peptide immunoaffinity enrichment and LC–MS/MS." *Bioanalysis* **4**(3): 281–290.

Berna M and Ackermann B (2009). "Increased throughput for low-abundance protein biomarker verification by liquid chromatography/tandem mass spectrometry." *Anal Chem* **81**(10): 3950–3956.

Berna M, Ott L, Engle S, et al. (2008). "Quantification of ntprobnp in rat serum using immunoprecipitation and LC/MS/MS: a biomarker of drug-induced cardiac hypertrophy." *Anal Chem* **80**(3): 561–566.

Berna M, Schmalz C, Duffin K, et al. (2006). "Online immunoaffinity liquid chromatography/tandem mass spectrometry determination of a type II collagen peptide biomarker in rat urine: investigation of the impact of collision-induced dissociation fluctuation on peptide quantitation." *Anal Biochem* **356**(2): 235–243.

Berna MJ, Zhen Y, Watson DE, et al. (2007). "Strategic use of immunoprecipitation and LC/MS/MS for trace-level protein quantification: Myosin light chain 1, a biomarker of cardiac necrosis." *Anal Chem* **79**(11): 4199–4205.

Bowsher RR, Nowatzke W, Sailstad JM, et al. (2012). "Application of commercial research-grade biomarker assays in drug development: is it time to create 'pharmaceutical-grade' kits?" *Bioanalysis* **4**(20): 2427–2430.

Brennen RA, Yin H and Killeen KP (2007). "Microfluidic gradient formation for nanoflow chip LC." *Anal Chem* **79**(24): 9302–9309.

Bronsema KJ, Bischoff R and van de Merbel NC (2012). "Internal standards in the quantitative determination of protein biopharmaceuticals using liquid chromatography coupled to mass spectrometry." *J Chromatogr B* **893–894**: 1–14.

Brun V, Dupuis A, Adrait A, et al. (2007). "Isotope-labeled protein standards: toward absolute quantitative proteomics." *Mol Cell Proteomics* **6**(12): 2139–2149.

Bystrom CE, Sheng S and Clarke NJ (2011). "Narrow mass extraction of time-of-flight data for quantitative analysis of proteins: determination of insulin-like growth factor-1." *Anal Chem* **83**(23): 9005–9010.

Callipo L, Caruso G, Foglia P, et al. (2010). "Immunoprecipitation on magnetic beads and liquid chromatography–tandem mass spectrometry for carbonic anhydrase II quantification in human serum." *Anal Biochem* **400**(2): 195–202.

Chambers EE, Lame ME, Bardsley J, et al. (2013). "High sensitivity LC–MS/MS method for direct quantification of human parathyroid 1–34 (teriparatide) in human plasma." *J Chromatogr B* **938**: 96–104.

Chames P, Van Regenmortel M, Weiss E, et al. (2009). "Therapeutic antibodies: successes, limitations and hopes for the future." *Br J Pharmacol* **157**(2): 220–233.

Chappell DL, Lee AYH, Castro-Perez J, et al. (2014). "An ultrasensitive method for the quantitation of active and inactive GLP-1 in human plasma via immunoaffinity LC–MS/MS." *Bioanalysis* **6**(1): 33–42.

Chappell DL, McAvoy T, Weiss B, et al. (2012). "Development and validation of an ultra-sensitive method for the measurement of plasma renin activity in human plasma via LC–MS/MS." *Bioanalysis* **4**(23): 2843–2850.

Chari RVJ (2008). "Targeted cancer therapy: conferring specificity to cytotoxic drugs." *Acc Chem Res* **41**(1): 98–107.

Chen Z, Caulfield MP, McPhaul MJ, et al. (2013). "Quantitative insulin analysis using liquid chromatography–tandem mass spectrometry in a high-throughput clinical laboratory." *Clin Chem* **59**(9): 1349–1356.

Dawes ML, Gu H, Wang J, et al. (2013). "Development of a validated liquid chromatography tandem mass spectrometry assay for a pegylated adnectin in cynomolgus monkey plasma using protein precipitation and trypsin digestion." *J Chromatogr B* **934**: 1–7.

Dere R, Yi J-H, Lei C, et al. (2013). "Pk assays for antibody–drug conjugates: case study with ado-trastuzumab emtansine." *Bioanalysis* **5**(9): 1025–1040.

Dubois M, Fenaille F, Clement G, et al. (2008). "Immunopurification and mass spectrometric quantification of the active form of a chimeric therapeutic antibody in human serum." *Anal Chem* **80**(5): 1737–1745.

Dufield DR and Radabaugh MR (2012). "Online immunoaffinity LC/MS/MS. A general method to increase sensitivity and specificity: how do you do it and what do you need?" *Methods* **56**(2): 236–245.

Engvall E and Perlmann P (1971). "Enzyme-linked immunosorbent assay (ELISA) quantitative assay of immunoglobulin g." *Immunochemistry* **8**(9): 871–874.

Ezan E (2013). "Pharmacokinetic studies of protein drugs: past, present and future." *Adv Drug Deliv Rev* **65**(8): 1065–1073.

Fernández Ocaña M, James IT, Kabir M, et al. (2012). "Clinical pharmacokinetic assessment of an anti-madcam monoclonal antibody therapeutic by LC-MS/MS." *Anal Chem* **84**(14): 5959–5967.

Furlong MT, Ouyang Z, Wu S, et al. (2012). "A universal surrogate peptide to enable LC-MS/MS bioanalysis of a diversity of human monoclonal antibody and human fc-fusion protein drug candidates in pre-clinical animal studies." *Biomed Chromatogr* **26**(8): 1024–1032.

Furlong MT, Zhao S, Mylott W, et al. (2013). "Dual universal peptide approach to bioanalysis of human monoclonal antibody protein drug candidates in animal studies." *Bioanalysis* **5**(11): 1363–1376.

Gallien S and Domon B (2014). "Quantitative proteomics using the high resolution accurate mass capabilities of the quadrupole-orbitrap mass spectrometer." *Bioanalysis* **6**(16): 2159–2170.

Gerber SA, Rush J, Stemman O, et al. (2003). "Absolute quantification of proteins and phosphoproteins from cell lysates by tandem MS." *Proc Nat Acad Sci* **100**(12): 6940–6945.

Hager T, Spahr C, Xu J, et al. (2013). "Differential enzyme-linked immunosorbent assay and ligand-binding mass spectrometry for analysis of biotransformation of protein therapeutics: application to various fgf21 modalities." *Anal Chem* **85**(5): 2731–2738.

Han X, Aslanian A and Yates Iii JR (2008). "Mass spectrometry for proteomics." *Curr Opin Chem Biol* **12**(5): 483–490.

Hess C, Thomas A, Thevis M, et al. (2012). "Simultaneous determination and validated quantification of human insulin and its synthetic analogues in human blood serum by immunoaffinity purification and liquid chromatography-mass spectrometry." *Anal Bioanal Chem* **404**(6): 1813–1822.

Hoofnagle AN, Becker JO, Wener MH, et al. (2008). "Quantification of thyroglobulin, a low-abundance serum protein, by immunoaffinity peptide enrichment and tandem mass spectrometry." *Clin Chem* **54**(11): 1796–1804.

Hoofnagle AN and Wener MH (2009). "The fundamental flaws of immunoassays and potential solutions using tandem mass spectrometry." *J Immunol Methods* **347**(1–2): 3–11.

Jaracz S, Chen J, Kuznetsova LV, et al. (2005). "Recent advances in tumor-targeting anticancer drug conjugates." *Bioorg Med Chem* **13**(17): 5043–5054.

Ji C, Sadagopan N, Zhang Y, et al. (2009). "A universal strategy for development of a method for absolute quantification of therapeutic monoclonal antibodies in biological matrices using differential dimethyl labeling coupled with ultra performance liquid chromatography-tandem mass spectrometry." *Anal Chem* **81**(22): 9321–9328.

Kaspar AA and Reichert JM (2013). "Future directions for peptide therapeutics development." *Drug Discov Today* **18**(17–18): 807–817.

Kaur S, Xu K, Saad OM, et al. (2013). "Bioanalytical assay strategies for the development of antibody–drug conjugate biotherapeutics." *Bioanalysis* **5**(2): 201–226.

Keshishian H, Addona T, Burgess M, et al. (2007). "Quantitative, multiplexed assays for low abundance proteins in plasma by targeted mass spectrometry and stable isotope dilution." *Mol Cell Proteomics* **6**(12): 2212–2229.

Kiernan UA, Tubbs KA, Nedelkov D, et al. (2002). "Comparative phenotypic analyses of human plasma and urinary retinol binding protein using mass spectrometric immunoassay." *Biochem Biophys Res Commun* **297**(2): 401–405.

Krastins B, Prakash A, Sarracino DA, et al. (2013). "Rapid development of sensitive, high-throughput, quantitative and highly selective mass spectrometric targeted immunoassays for clinically important proteins in human plasma and serum." *Clin Biochem* **46**(6): 399–410.

Kuhn E, Addona T, Keshishian H, et al. (2009). "Developing multiplexed assays for troponin i and interleukin-33 in plasma by peptide immunoaffinity enrichment and targeted mass spectrometry." *Clin Chem* **55**(6): 1108–1117.

Kumar V, Barnidge DR, Chen L-S, et al. (2010). "Quantification of serum 1–84 parathyroid hormone in patients with hyperparathyroidism by immunocapture *in situ* digestion liquid chromatography–tandem mass spectrometry." *Clin Chem* **56**(2): 306–313.

Lame ME, Chambers EE and Blatnik M (2011). "Quantitation of amyloid beta peptides aβ1–38, aβ1–40, and aβ1–42 in human cerebrospinal fluid by ultra-performance liquid chromatography–tandem mass spectrometry." *Anal Biochem* **419**(2): 133–139.

Lassman ME and Fernandez-Metzler C (2014). "Applications of low-flow LC–SRM for the analysis of large molecules in pharmaceutical R&D." *Bioanalysis* **6**(13): 1859–1867.

Leary BA, Lawrence-Henderson R, Mallozzi C, et al. (2013). "Bioanalytical platform comparison using a generic human IgG PK assay format." *J Immunol Methods* **397**(1–2): 28–36.

Leng SX, McElhaney JE, Walston JD, et al. (2008). "ELISA and multiplex technologies for cytokine measurement in inflammation and aging research." *J Gerontol A: Biol Sci Med Sci* **63**(8): 879–884.

Li H, Ortiz R, Tran L, et al. (2012). "General LC-MS/MS method approach to quantify therapeutic monoclonal antibodies using a common whole antibody internal standard with application to preclinical studies." *Anal Chem* **84**(3): 1267–1273.

Lin D, Alborn WE, Slebos RJC, et al. (2013). "Comparison of protein immunoprecipitation-multiple reaction monitoring with ELISA for assay of biomarker candidates in plasma." *J Proteome Res* **12**(12): 5996–6003.

McAvoy T, Lassman ME, Spellman DS, et al. (2014). "Quantification of tau in cerebrospinal fluid by immunoaffinity enrichment and tandem mass spectrometry." *Clin Chem* **60**(4): 683–689.

Miyachi A, Murase T, Yamada Y, et al. (2013). "Quantitative analytical method for determining the levels of gastric inhibitory polypeptides GIP1–42 and GIP3–42 in human plasma using LC–MS/MS/MS." *J Proteome Res* **12**(6): 2690–2699.

Nelson RW, Krone JR, Bieber AL, et al. (1995). "Mass spectrometric immunoassay." *Anal Chem* **67**(7): 1153–1158.

Neubert H, Fountain S, King L, et al. (2012). "Tissue bioanalysis of biotherapeutics and drug targets to support PK/PD." *Bioanalysis* **4**(21): 2589–2604.

Neubert H, Muirhead D, Kabir M, et al. (2013). "Sequential protein and peptide immunoaffinity capture for mass spectrometry-based quantification of total human β-nerve growth factor." *Anal Chem* **85**(3): 1719–1726.

Nicol GR, Han M, Kim J, et al. (2008). "Use of an immunoaffinity-mass spectrometry-based approach for the quantification of protein biomarkers from serum samples of lung cancer patients." *Mol Cell Proteomics* **7**(10): 1974–1982.

Niederkofler EE, Tubbs KA, Gruber K, et al. (2001). "Determination of β-2 microglobulin levels in plasma using a high-throughput mass spectrometric immunoassay system." *Anal Chem* **73**(14): 3294–3299.

Ocaña MF and Neubert H (2010). "An immunoaffinity liquid chromatography–tandem mass spectrometry assay for the quantitation of matrix metalloproteinase 9 in mouse serum." *Anal Biochem* **399**(2): 202–210.

Oe T, Ackermann BL, Inoue K, et al. (2006). "Quantitative analysis of amyloid β peptides in cerebrospinal fluid of Alzheimer's disease patients by immunoaffinity purification and stable isotope dilution liquid chromatography/negative electrospray ionization tandem mass spectrometry." *Rapid Commun Mass Spectrom* **20**(24): 3723–3735.

Ong S-E, Blagoev B, Kratchmarova I, et al. (2002). "Stable isotope labeling by amino acids in cell culture, SILAC, as a simple and accurate approach to expression proteomics." *Mol Cell Proteomics* **1**(5): 376–386.

Oran PE, Jarvis JW, Borges CR, et al. (2011). "Mass spectrometric immunoassay of intact insulin and related variants for population proteomics studies." *PROTEOMICS – Clin Appl* **5**(7–8): 454–459.

Oran PE, Sherma ND, Borges CR, et al. (2010). "Intrapersonal and populational heterogeneity of the chemokine rantes." *Clin Chem* **56**(9): 1432–1441.

Oran PE, Trenchevska O, Nedelkov D, et al. (2014). "Parallel workflow for high-throughput (>1,000 samples/

day) quantitative analysis of human insulin-like growth factor 1 using mass spectrometric immunoassay." *PLoS ONE* **9**(3): e92801.

Palandra J, Finelli A, Zhu M, et al. (2013). "Highly specific and sensitive measurements of human and monkey interleukin 21 using sequential protein and tryptic peptide immunoaffinity LC-MS/MS." *Anal Chem* **85**(11): 5522–5529.

Podust VN, Sim B-C, Kothari D, et al. (2013). "Extension of *in vivo* half-life of biologically active peptides via chemical conjugation to xten protein polymer." *Protein Eng Des Select* **26**(11): 743–753.

Razavi M, Johnson LDS, Lum JJ, et al. (2013). "Quantification of a proteotypic peptide from protein C inhibitor by liquid chromatography–free SISCAPA-MALDI mass spectrometry: application to identification of recurrence of prostate cancer." *Clin Chem* **59**(10): 1514–1522.

Rifai N, Gillette MA and Carr SA (2006). "Protein biomarker discovery and validation: the long and uncertain path to clinical utility." *Nat Biotechnol* **24**(8): 971–983.

Roberts SA, Andrews PA, Blanset D, et al. (2013). "Considerations for the nonclinical safety evaluation of antibody drug conjugates for oncology." *Regulat Toxicol Pharmacol* **67**(3): 382–391.

Rochat B, Peduzzi D, McMullen J, et al. (2013). "Validation of hepcidin quantification in plasma using LC–HRMS and discovery of a new hepcidin isoform." *Bioanalysis* **5**(20): 2509–2520.

Sauerborn M and van Dongen W (2014). "Practical considerations for the pharmacokinetic and immunogenic assessment of antibody–drug conjugates." *BioDrugs* **28**(4): 383–391.

Schoenherr RM, Whiteaker JR, Zhao L, et al. (2012). "Multiplexed quantification of estrogen receptor and HER2/Neu in tissue and cell lysates by peptide immunoaffinity enrichment mass spectrometry." *PROTEOMICS* **12**(8): 1253–1260.

Shuford CM, Sederoff RR, Chiang VL, et al. (2012). "Peptide production and decay rates affect the quantitative accuracy of protein cleavage isotope dilution mass spectrometry (PC-IDMS)." *Mol Cell Proteomics* **11**(9): 814–823.

Siskos AP, Katsila T, Balafas E, et al. (2009). "Simultaneous absolute quantification of the glucose-dependent insulinotropic polypeptides GIP1–42 and GIP3–42 in mouse plasma by LC/ESI-MS/MS: preclinical evaluation of DP-IV inhibitors." *J Proteome Res* **8**(7): 3487–3496.

Sloan JH, Siegel RW, Ivanova-Cox YT, et al. (2012). "A novel high-sensitivity electrochemiluminescence (ECL) sandwich immunoassay for the specific quantitative measurement of plasma glucagon." *Clin Biochem* **45**(18): 1640–1644.

Stevenson L, Garofolo F, DeSilva B, et al. (2013). "2013 white paper on recent issues in bioanalysis: 'hybrid' – the best of LBA and LCMS." *Bioanalysis* **5**(23): 2903–2918.

Thomas A, Schänzer W and Thevis M (2014). "Determination of human insulin and its analogues in human blood using liquid chromatography coupled to ion mobility mass spectrometry (LC-IM-MS)." *Drug Test Anal* **6**(11–12): 1125–1132.

Torsetnes SB, Broughton MN, Paus E, et al. (2014a). "Determining progrp and isoforms in lung and thyroid cancer patient samples: comparing an MS method with a routine clinical immunoassay." *Anal Bioanal Chem* **406**(11): 2733–2738.

Torsetnes SB, Levernæs MS, Broughton MN, et al. (2014b). "Multiplexing determination of small cell lung cancer biomarkers and their isovariants in serum by immunocapture LC-MS/MS." *Anal Chem* **86**(14): 6983–6992.

Torsetnes SB, Løvbak SG, Claus C, et al. (2013). "Immunocapture and LC–MS/MS for selective quantification and differentiation of the isozymes of the biomarker neuron-specific enolase in serum." *J Chromatogr B* **929**: 125–132.

Umberger TC, Cox JM, Berna, MJ and Ackermann, BL (2012). "A direct comparison of sample preparation formats used for immunoaffinity-mass spectrometry based quantification of protein biomarkers." *Proceedings of the 60th ASMS Conference on Mass Spectrometry and Allied Topics Vancouver, BC, May 19–24.*

Van Weemen BK and Schuurs AHWM (1971). "Immunoassay using antigen–enzyme conjugates." *FEBS Lett* **15**(3): 232–236.

Wang W, Wang EQ and Balthasar JP (2008). "Monoclonal antibody pharmacokinetics and pharmacodynamics." *Clin Pharmacol Therapeut* **84**(5): 548–558.

Wang Y and Heilig JS (2012). "Differentiation and quantification of endogenous and recombinant-methionyl human leptin in clinical plasma samples by immunocapture/mass spectrometry." *J Pharm Biomed Anal* **70**: 440–446.

Whiteaker JR, Zhao L, Anderson L, et al. (2010). "An automated and multiplexed method for high throughput peptide immunoaffinity enrichment and multiple reaction monitoring mass spectrometry-based quantification of protein biomarkers." *Mol Cell Proteomics* **9**(1): 184–196.

Whiteaker JR, Zhao L, Lin C, et al. (2012). "Sequential multiplexed analyte quantification using peptide immunoaffinity enrichment coupled to mass spectrometry." *Mol Cell Proteomics* **11**(6).

Whiteaker JR, Zhao L, Zhang HY, et al. (2007). "Antibody-based enrichment of peptides on magnetic beads for mass-spectrometry-based quantification of serum biomarkers." *Anal Biochem* **362**(1): 44–54.

Wolf R, Rosche F, Hoffmann T, et al. (2001). "Immunoprecipitation and liquid chromatographic–mass spectrometric determination of the peptide glucose-dependent insulinotropic polypeptides GIP1–42 and GIP3–42 from human plasma samples: new sensitive method to analyze physiological concentrations of peptide hormones." *J Chromatogr A* **926**(1): 21–27.

Wong RL, Xin B and Olah T (2011). "Optimization of exactive Orbitrap™ acquisition parameters for quantitative bioanalysis." *Bioanalysis* **3**(8): 863–871.

Xu K, Liu L, Dere R, et al. (2013). "Characterization of the drug-to-antibody ratio distribution for antibody–drug conjugates in plasma/serum." *Bioanalysis* **5**(9): 1057–1071.

Xu K, Liu L, Saad OM, et al. (2011). "Characterization of intact antibody–drug conjugates from plasma/serum *in vivo* by affinity capture capillary liquid chromatography–mass spectrometry." *Anal Biochem* **412**(1): 56–66.

Xu Y, Mehl JT, Bakhtiar R, et al. (2010). "Immunoaffinity purification using anti-peg antibody followed by two-dimensional liquid chromatography/tandem mass spectrometry for the quantification of a pegylated therapeutic peptide in human plasma." *Anal Chem* **82**(16): 6877–6886.

Yalow RS and Berson SA (1960). "Immunoassay of endogenous plasma insulin in man." *J Clin Invest* **39**(7): 1157–1175.

Zhen EY, Berna MJ, Jin Z, et al. (2007). "Quantification of heart fatty acid binding protein as a biomarker for drug-induced cardiac and musculoskeletal necroses." *PROTEOMICS – Clin Appl* **1**(7): 661–671.

8

Semiquantification and Isotyping of Antidrug Antibodies by Immunocapture-LC/MS for Immunogenicity Assessment

Jianing Zeng, Hao Jiang, and Linlin Luo

Analytical and Bioanalytical Operations, Bristol-Myers Squibb Co., Princeton, NJ, USA

8.1 Introduction

Immunogenicity refers to immune responses of humans or animals to antigens such as biotherapeutics. In drug development, multiple factors can contribute to the development of immunogenicity against a biotherapeutic, and they can be patient or product related (Vugmeyster et al. 2012, U.S. Department of Health and Human Services 2014, Koren et al. 2002, Ratanji et al. 2014). Patient-specific factors include dosing frequency and amount, administration route, whether immune modulators are used, and patients' immune status. Product-specific factors include the drug's degree of sequence and structure homology with endogenous proteins, pharmacological target, production system, impurities, and degree of aggregation. Immunogenicity of biotherapeutics can cause unwanted immune responses ranging from minor allergic reactions at the injection site to serious adverse effects such as anaphylaxis, cytokine release syndrome, and critical events caused by cross-reactivity with endogenous proteins. Antidrug antibodies (ADAs) generated by the immune system may alter the pharmacokinetic profile of the drug by removing therapeutic proteins from circulation, typically through intracellular degradation of the proteins, thereby, reducing the exposure. On the other hand, ADAs can increase the half-life of a biotherapeutic and, thus, cause prolonged exposure with variable effects on bioactivity. Some ADAs reduce pharmacological activity and drug efficacy by preventing the drug from binding to its target or inhibiting its downstream signaling upon binding due to steric hindrance.

Regulatory agencies request that immunogenicity risk assessments for safety and efficacy be conducted during all clinical stages of biologics development (FDA Draft Guidance for Industry: Assay Development and Validation for Immunogenicity Testing of Therapeutic Protein Products 2016). Currently, ligand binding assays (LBAs) are the established technology for evaluating immunogenicity of therapeutic proteins. The bridging LBA is the most commonly used assay format using the biotinylated drug as the capture reagent and the ruthenylated drug, or other detection-molecule bound to the drug, as the detection agent (Figure 8.1). A tiered approach is recommended to screen, confirm, and titer ADAs in patient samples as illustrated in Figure 8.2. Briefly, during assay validation a screening cut point is established statistically to determine background responses in the targeted population. All the samples with a response above or equal to the "cut point" are subjected to a confirmatory assay to evaluate if the response is drug-related. The confirmed ADA-positive samples are further serially diluted for titer determination. In cases when neutralization evaluation (i.e., the ability of the ADA to neutralize the activity of the drug) is needed, either a binding assay or a functional cell-based assay, which is reflective of the mechanism of the drug, can be used.

Immunogenicity assessment by LBA involves multiple assays, which is time and resource consuming. The assay performance and sensitivity heavily depend on availability and quality of the reagents (such as labeled drugs and control antibodies) and the design of assay format (e.g., Meso Scale Discovery (MSD) bridging, solid-phase extraction with acid dissociation, or affinity capture elution) (Butterfield et al. 2010). When developing a method for measuring ADAs, consideration (e.g., choice of surrogate positive controls) should be given so the method can detect ADAs with high, medium, or low affinities, because the patient's ADA response encompasses antibodies with a potentially wide range of different affinities. One major challenge for LBA development is to ensure its ability to detect ADAs in the presence of the circulating drugs that compete with the labeled drugs for ADA capture and detection. Even though drug tolerance can sometimes be overcome by using samples collected at later time points or after a washout period, these samples

Protein Analysis using Mass Spectrometry: Accelerating Protein Biotherapeutics from Lab to Patient, First Edition.
Edited by Mike S. Lee and Qin C. Ji.

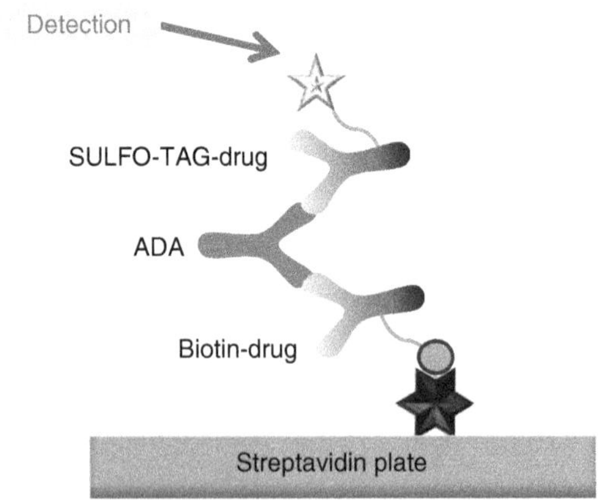

Figure 8.1 Bridging ligand binding assay for immunogenicity assessment.

are not always available due to patient compliance or patient switching to alternate therapies. Other approaches have been developed to overcome drug tolerance issues: the samples can be diluted to decrease the concentration of the circulating drug in the assay by trading sensitivity for increased drug tolerance; the samples can also be pretreated to dissociate the drug–ADA complex or deplete the drug, but this treatment increases variability and cost due to the additional steps and reagents.

A second challenge for LBA development is that sometimes isotyping ADAs is desirable during both nonclinical and clinical drug development for further characterization of the immune response. The conventional bridging assay format is not able to provide ADA isotype information. Some emerging technologies, such as the multiplexing technology from SQI Diagnostics Inc, have been developed for ADA isotyping, but they have not been broadly used due to the complexity of reagent preparation, method development, and high cost. Therefore, a technology for immunogenicity assessment that simultaneously provides capability of multiplexing detection of ADAs and isotype information with sufficient drug tolerance would be a welcome addition to the bioanalytical toolkit.

Liquid chromatography–mass spectrometry (LC–MS) assays have been shown to be orthogonal to traditional LBAs and have many advantages in protein bioanalysis such as unique selectivity and ease of multiplexing. It has been broadly used in proteomic studies

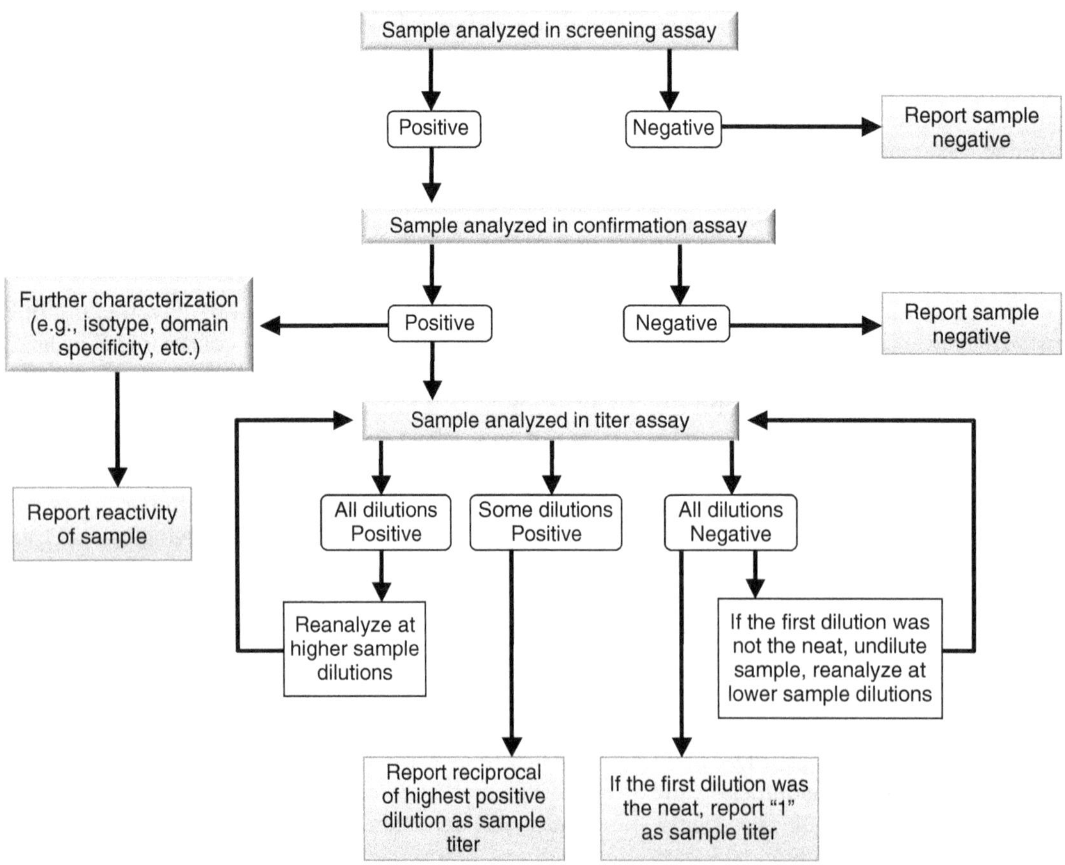

Figure 8.2 Tiered testing paradigm for clinical immunogenicity assessment by LBAs.

and recently has been applied to pharmacokinetic assessments of biologics (Jiang et al. 2013, van den Broek et al. 2013, Chappell et al. 2014, Furlong et al. 2014, Gong et al. 2014, Zhang et al. 2014, Zheng et al. 2014). LC–MS provides good assay selectivity using the multidimensional separation technologies of chromatography and tandem mass spectrometry. It can differentiate a protein from endogenous components (proteins, lipids, etc.) by monitoring unique surrogate peptides derived from the protein. LC–MS is able to simultaneously quantify multiple surrogate peptides using the selected reaction monitoring (SRM) mode due to fast mass scanning (less than 5 ms duty cycle), which allows more than 50 peptides to be measured at a time. In addition, immunocapture recently has been poised to play an increasingly vital role in therapeutic protein quantitation (Kellie and Szapacs 2014) because of its specificity in extracting analyte proteins from the complex serum matrices for the downstream LC–MS analysis. Therefore, with the combination of advantages from both immunocapture and LC–MS techniques, it is feasible to directly measure ADA isotypes within a single analysis.

8.2 Multiplexing Direct Measurement of ADAs by Immunocapture-LC/MS for Immunogenicity Screening, Titering, and Isotyping

Antibodies generated by the immune systems have multiple isotypes, IgG, IgM, IgA, IgE, and IgD, as shown in Figure 8.3. These immunoglobulin isotypes have different *in vivo* half-lives and different concentrations in serum (Brekke and Sandlie 2003). IgG represents approximately 75% of the antibodies in human serum with a half-life of around 20 days. IgM and IgA are about 10% and 15% of the immunoglobulin pool and have a shorter half-life of 5–8 days. IgD and IgE are very scarce in serum. ADAs are the antibodies that the immune system generates after a drug is administered, which have the same constant regions as all other endogenous antibodies but with specific affinity to the drug within their variable regions.

With a traditional bridging LBA (Figure 8.4), besides ADAs, any other serum components that can bind to the drug and bridge two labeled drugs, such as target

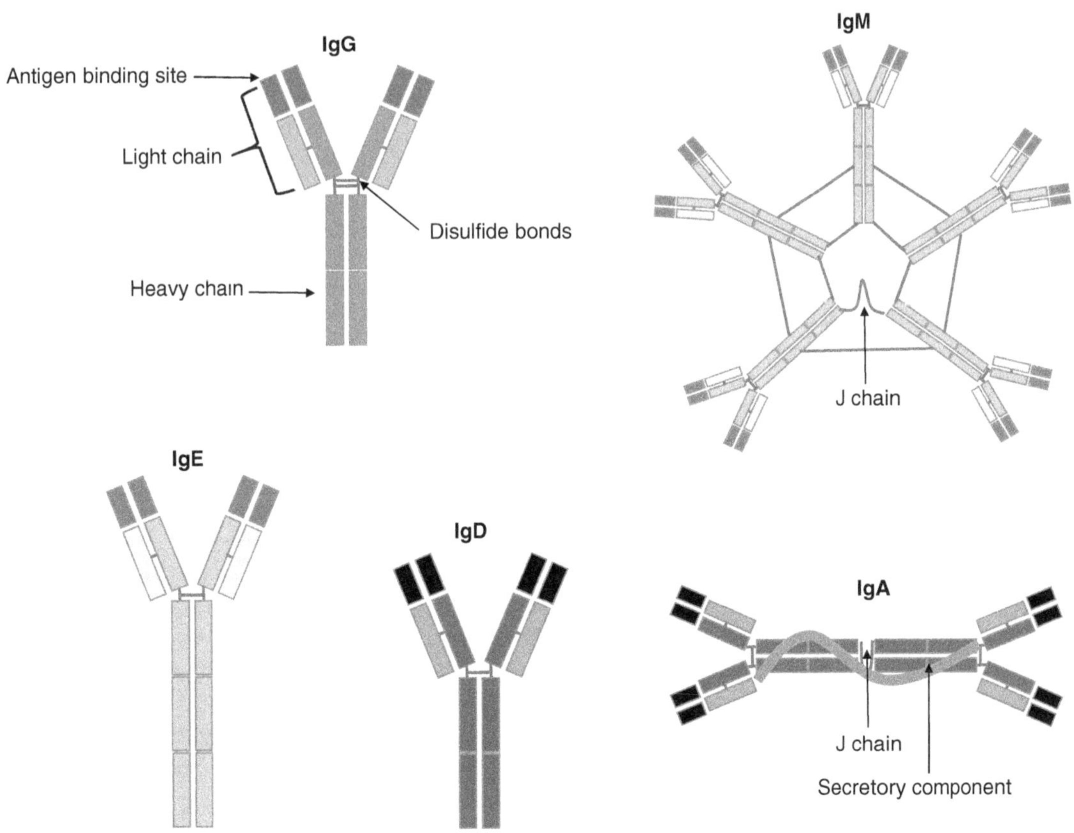

Figure 8.3 Structures of immunoglobulin isotypes.

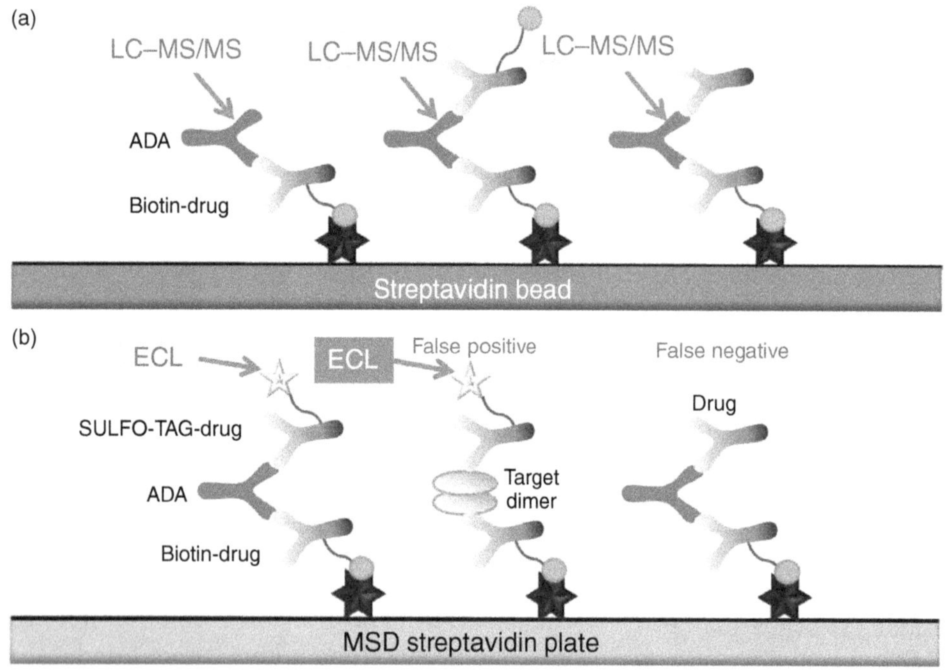

Figure 8.4 Comparison of LC–MS (with streptavidin beads IC) (a) and ligand binding (with MSD streptavidin plates) assays (b) for detection of ADAs in serum.

multimers and nonspecific proteins, will generate positive results in the screening assay, so-called "false positives." So a confirmatory experiment is always required to confirm that the positive results are drug related by spiking extra amount of the unlabeled drug into the samples to compete with the labeled drugs resulting in a decreased detection signal. When the preexisting drug in samples binds to one arm or both arms of ADAs, the ADA–drug complex cannot be captured or detected, which are classified as "false negative" or drug interference.

Multiplexing measurements of ADAs by LC–MS were reported at the 2015 AAPS National Biotechnology Conference (Jiang 2015, Chen et al. 2015). In these case studies, SRM was used to simultaneously quantify unique surrogate peptides from the constant regions of IgG, IgM, IgA, IgE, and IgD after immunocapture with the biotinylated drug (Figure 8.4). The ADAs were captured onto the magnetic beads through the labeled drug (e.g., biotinylated drug bound to streptavidin-coated beads). The captured ADAs were then digested by trypsin before LC–MS analysis of the specific surrogate peptides for each class of Ig. The results were reported as individual concentrations of IgG subclasses, IgM, IgA, IgE, and IgD in the eluate from the magnetic beads coupled with the ADA–drug complex. The LC–MS detection limit in the eluate ranged 5–100 ng/mL for all isotypes. However, the sensitivity of the immunogenicity assay in the method development and validation is not only determined by the LC–MS detection limit but also by the positive control antibodies used to represent the endogenous ADAs. A control antibody with high binding affinity to the drug may generate a high detection signal. So control antibodies, either monoclonal or polyclonal, with different binding affinities are recommended to be used in method development and validation to demonstrate the assay's detection capacity.

It is critical to find a specific surrogate peptide for each isotype (IgG, IgM, IgA, IgD, and IgE). These surrogate peptides are required to be specific for the corresponding immunoglobulins in LC–MS analysis and provide adequate sensitivity. In general, a surrogate peptide from the constant regions of the heavy chain is preferred and it is selected by screening all the peptides derived from a reference immunoglobulin protein after proteolysis. Trypsin is a commonly used protease for proteolysis treatment due to its well-defined specificity to arginine (Arg) or lysine (Lys) residues and reproducible digestion efficiency. Surrogate peptides should be chemically and physiologically stable peptides without allotypic sites or amino acids prone to glycosylation, oxidation, and deamination. Some example surrogate peptides that have been used are listed in Table 8.1 (Chen et al. 2015, 2016). Other specific surrogate peptides can be also used in the LC–MS analysis, as long as they provide adequate sensitivity and specificity. High-resolution accurate mass spectrometry can be considered as an alternative approach to SRM due to its capability of direct monitoring surrogate peptides and obtaining extra confirmatory information with accurate

Table 8.1 List of unique peptides and MRMs for ADA isotopes/subclasses used in the immunocapture-LC/MS assay.

Isotype/subclass	Unique peptide sequence	MRM pairs
Quantitation peptide		
IgG1	GPSVFPLAPSSK	593.83 → 699.40
IgG2	GLPAPIEK	412.75 → 654.38
IgG3	WYVDGVEVHNAK	708.85 → 698.48
IgG4	GLPSSIEK	415.73 → 660.36
IgE	AEWEQK	395.69 → 590.29
IgM	GQPLSPEK	428.23 → 670.38
IgA1, IgA2	YLTWASR	448.73 → 620.32
Confirmation peptide		
IgG1, IgG3	ALPAPIEK	419.76 → 654.38
IgG1, IgG3, IgG4	VVSVLTVLHQDWLNGK	603.34 → 1110.57
IgE	LEVTR	309.18 → 375.24
IgM	VSVFVPPR	450.77 → 615.36
IgA1, IgA2	VAAEDWK	409.71 → 648.30
IgA1	TFTC[CAM]TAAYPESK	688.31 → 765.38

mass while maintaining similar limits of quantification, limits of detection, linear range, and repeatability (Ruan et al. 2011, Herrero et al. 2014, Mekhssian et al. 2014).

In Jiang and Chen's cases, the LC–MS assay directly quantifies ADAs after immunocapture by the immobilized drug on the magnetic beads, which is different from the LBA that only measures the antibodies that are bridged by one capture drug (e.g., biotinylated drug) and one detection drug (e.g., ruthenylated drug) simultaneously. This approach of direct measurement of ADAs by LC–MS is only affected by drug interference at the immunocapture step (Figure 8.4). Any preexisting drug in clinical samples can compete with both labeled drugs binding to the ADAs, causing the bridging assay prone to greater drug interference than the LC–MS assay. Therefore, the LC–MS assay has a lower probability of false-negative results than the bridging LBA. Meanwhile, multimers of the circulating targets or endogenous proteins may also bridge two labeled drugs and then generate false-positive signals in bridging LBA, which will not happen in LC–MS assay due to the direct measurement of the ADA (Figure 8.4).

8.3 Indirect Measurement of ADAs by Quantifying ADA Binding Components

Among biologics, marketed or in development, some constructs are significantly more complex than monoclonal antibodies. Some common modifications to therapeutic proteins can further complicate the evaluation of immunogenicity. For example, as a result of using PEGylation for extending the pharmacokinetic half-life of small therapeutic proteins, the drug remains in circulation longer, which makes drug interference more problematic. In addition, PEG's flexible structure may alter the binding affinity of the drug to ADAs through steric hindrance, and PEG's repeated motif makes it challenging to develop a bridging LBA. A published LC–MS assay for measuring ADAs was reported to overcome the drug interference issue (Neubert et al. 2008) by measuring the drug, a PEGylated human growth hormone analog, hGHA, bound to the ADAs instead of measuring the ADAs directly. An excess amount of the drug was added into the study samples to saturate all binding sites of the ADAs before protein G beads capture of immunoglobulins from the serum samples. The drug bound to the ADAs was then dissociated with guanidine hydrochloride and digested with cyanogen bromide (CNBr). The peptides derived from the drug were quantified by LC-matrix-assisted laser desorption/ionization (MALDI)/MS with stable-isotope-labeled peptides as reference standards. The drug concentration infers the total amount of ADAs including originally drug-bound and unbound ADAs in the samples (Figure 8.5). In this example, only IgG antibodies were captured by protein G beads and measured while other isotypes were not. This approach provides a

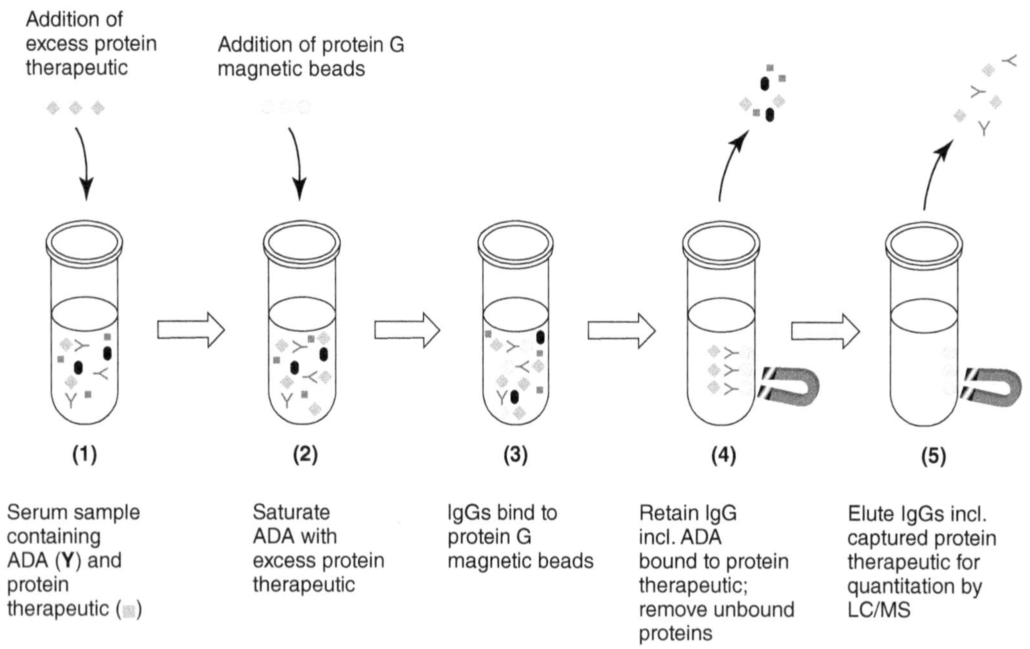

Figure 8.5 Schematic workflow image of the magnetic bead-based immunocapture procedure for sample preparation in the MS immunogenicity assay. (*Source*: Neubert et al. 2008. Reproduced with permission of American Chemical Society.)

complementary methodology that circumvents the drug tolerance issue and has been used to support clinical programs addressing the safety and tolerability of the PEGylated drug.

There have been several approaches used to improve the sensitivity, specificity, and reproducibility in this example that are worth mentioning. The assay uses a MALDI-TOF instrument, instead of a liquid chromatography-triple quadrupole mass spectrometer, which is unconventional. The control antibody was a polyclonal goat anti-hGH antibody. The signal intensity ratio of the peptides derived from the analyte to the one derived from the stable-isotope-labeled peptides (mimicking the N-terminal and C-terminal sequences of the CNBr cleavage product of hGHA) was obtained for calculating the analyte concentration (in fmol) by multiplication with the concentration of the stable-isotope-labeled peptide. These peptides have flanking amino acids at the cleavage sites in order to correct for variability in the digestion step. One advantage of using a control calibrant is that there is no need to do further titration following the initial screening as the initial measurement is semiquantitative. The assay cut point was established by using 24 normal individual cynomolgus serum samples in 3 separate analyses. The assay was able to measure a 50 ng/mL spiked positive control antibody in cynomolgus serum by applying microflow LC–MS. Instead of the traditional trypsin enzyme digestion to generate peptides for mass spectrometry monitoring, CNBr was used to chemically proteolyse the drug, avoiding the steric hindrance of the PEG moiety. These approaches made it possible to evaluate immunogenicity when high drug interference was observed in an LBA.

8.4 Use of LC–MS to Assist in Method Development of Cell-Based Neutralizing Antibody Assays

In addition to using LC–MS to evaluate immunogenicity through direct or indirect measurement of ADAs, the technology has also proved useful in optimizing ADA extraction efficiency and understanding the impact of carryover drug in a cell-based neutralizing antibody (NAb) assay (Jiang et al. 2014, Xu et al. 2015). Functional cell-based assays are the preferred format to evaluate the presence of NAbs. However, cell-based assays are prone to matrix effects and drug interferences (Gupta et al. 2007, Dodge et al. 2009). Growth factors, cytokines, and free drug in a study sample can impact the results of a cell-based NAb assay through the intended signaling pathway. Therefore, sample cleanup to remove these interfering components from samples prior to the cell-based assay is essential. Bead extraction and acid dissociation (BEAD) is an effective approach to purify and enrich the NAbs from study samples and has been demonstrated to improve sensitivity and increase drug tolerance in cell-based assays (Lofgren et al. 2006). However, during the BEAD process, a certain amount of residual drug or matrix components such as endogenous immunoglobulin may also be extracted along with ADAs through one of the two binding arms of antibodies or nonspecific binding to the beads or labware. It is difficult to evaluate the recovery of ADAs and estimate residual drug and nonspecific components mainly due to the interference of competitive binding components.

In our laboratory, we have applied LC–MS for the simultaneous detection of extracted ADAs, residual drug, and endogenous human IgG by quantifying specific surrogate peptides from control antibodies, drug and endogenous immunoglobulins, respectively (Jiang et al. 2014). In this study, the low levels of the residual drug and human IgG in the BEAD eluates indicated that the BEAD procedure efficiently removed the drug and serum components from the serum samples. Meanwhile, the control antibody recovery (~42%) in the BEAD resulted in sufficient ADA to provide an acceptable detection limit for the cell-based assay. This application of LC–MS for the direct and fast measurement of multiple protein analytes in the BEAD eluates takes advantage of the selectivity and multiplexing ability of LC–MS.

8.5 Conclusion and Future Perspectives

In conclusion, the LC–MS technology has a great potential in immunogenicity assessments due to its multiplexing capability and unique selectivity. The applications described herein have demonstrated its capability of mitigating drug interferences and providing ADA isotype information. In addition, LC–MS technology can provide additional information for cell-based NAb assay development. Although LC–MS is often capable of detecting immunoglobulins at low ng/mL levels, the high background responses (specifically IgG1 and IgG2) from serum sometimes prevent developing a highly sensitive assay for ADA measurement if the traditional cut point approach is applied. Understanding the source of the background responses is essential for applying this technology to support clinical studies. Furthermore, the practices of using calibrators for quantification, establishing cut points, and applying the tiered approaches need more data to paint a clearer picture.

References

Brekke OH and Sandlie I (2003). “Therapeutic antibodies for human diseases at the dawn of the twenty-first century.” *Nat Rev Drug Discov* **2**(1): 52–62.

Butterfield AM, Chain JS , Ackermann BL, et al. (2010). “Comparison of assay formats for drug-tolerant immunogenicity testing.” *Bioanalysis* **2**(12): 1961–1969.

Chappell DL, Lassman ME, McAvoy T, et al. (2014). “Quantitation of human peptides and proteins via MS: review of analytically validated assays.” *Bioanalysis* **6**(13): 1843–1857.

Chen LR, Philip D, Norris E, et al. (2015). “Development of LC/MS for ADA analysis.” *AAPS National Biotechnology Conference.*

Chen LZ, Roos D and Philip E (2016). “Development of Immunocapture-LC/MS Assay for Simultaneous ADA Isotyping and Semiquantitation.” *J Immunol Res* **2016**: 7682472.

Dodge, R., C. Daus and D. Yaskanin (2009). “Challenges in developing antidrug antibody screening assays.” *Bioanalysis* **1**(4): 699–704.

Furlong MTC, Titsch W, Xu H, et al. (2014). “An exploratory universal LC-MS/MS assay for bioanalysis of hinge region-stabilized human IgG4 mAbs in clinical studies.” *Bioanalysis* **6**(13): 1747–1758.

Gong C, Zeng J, Akinsanya B, et al. (2014). “Development and validation of an LC-MS/MS assay for the quantitation of a PEGylated anti-CD28 domain antibody in human serum: overcoming interference from antidrug antibodies and soluble target.” *Bioanalysis* **6**(18): 2371–2383.

Gupta S, Indelicato SR , Jethwa V, et al. (2007). “Recommendations for the design, optimization, and qualification of cell-based assays used for the detection of neutralizing antibody responses elicited to biological therapeutics.” *J Immunol Methods* **321**(1–2): 1–18.

Herrero P, Cortes-Francisco N, Borrull F, et al. (2014). “Comparison of triple quadrupole mass spectrometry and Orbitrap high-resolution mass spectrometry in ultrahigh performance liquid chromatography for the determination of veterinary drugs in sewage: benefits and drawbacks.” *J Mass Spectrom* **49**(7): 585–596.

Jiang H (2015). “Using LC-MS for isotyping and quantifying anti-drug antibodies in immunogenicity evaluation.” *AAPS National Biotechnology Conference.*

Jiang H, Xu W, Titsch CA , et al. (2014). “Innovative use of LC-MS/MS for simultaneous quantitation of neutralizing antibody, residual drug, and human immunoglobulin G in immunogenicity assay development.” *Anal Chem* **86**(5): 2673–2680.

Jiang H, Zeng J, Titsch C, et al. (2013). “Fully validated LC-MS/MS assay for the simultaneous quantitation of coadministered therapeutic antibodies in cynomolgus monkey serum.” *Anal Chem* **85**(20): 9859–9867.

Kellie J and Szapacs M (2014). “Quantitative immunocapture MS: current status and challenges in drug discovery.” *Bioanalysis* **6**(18): 2335–2337.

Koren E, Zuckerman LA and Mire-Sluis AR (2002). "Immune responses to therapeutic proteins in humans – clinical significance, assessment and prediction." *Curr Pharm Biotechnol* **3**(4): 349–360.

Lofgren JA, Wala I, Koren E, et al. (2006). "Detection of neutralizing anti-therapeutic protein antibodies in serum or plasma samples containing high levels of the therapeutic protein." *J Immunol Methods* **308**(1–2): 101–108.

Mekhssian K, Mess JN and Garofolo F (2014). "Application of high-resolution MS in the quantification of a therapeutic monoclonal antibody in human plasma." *Bioanalysis* **6**(13): 1767–1779.

Neubert H, Grace C, Rumpel K, et al. (2008). "Assessing immunogenicity in the presence of excess protein therapeutic using immunoprecipitation and quantitative mass spectrometry." *Anal Chem* **80**(18): 6907–6914.

Ratanji KD, Derrick JP, Dearman RJ, et al. (2014). "Immunogenicity of therapeutic proteins: influence of aggregation." *J Immunotoxicol* **11**(2): 99–109.

Ruan Q, Ji QC, Arnold ME, et al. (2011). "Strategy and its implications of protein bioanalysis utilizing high-resolution mass spectrometric detection of intact protein." *Anal Chem* **83**(23): 8937–8944.

U.S. Department of Health and Human Services (2009) F. a. D. A., Center for Drug Evaluation and Research (CDER), Center for Biologics Evaluation and Research (CBER). "Guidance for industry assay development for immunogenicity testing of therapeutic proteins."

U.S. Department of Health and Human Services (2014) F. a. D. A., Center for Drug Evaluation and Research (CDER), Center for Biologics Evaluation and Research (CBER). "Guidance for industry immunogenicity assessment for therapeutic protein products."

van den Broek I, Niessen WM and van Dongen WD (2013). "Bioanalytical LC-MS/MS of protein-based biopharmaceuticals." *J Chromatogr B Analyt Technol Biomed Life Sci* **929**: 161–179.

Vugmeyster Y, Xu X, Theil FP, et al. (2012). "Pharmacokinetics and toxicology of therapeutic proteins: advances and challenges." *World J Biol Chem* **3**(4): 73–92.

Xu W, Jiang H, Titsch C, et al. (2015). "Development and characterization of a pre-treatment procedure to eliminate human monoclonal antibody therapeutic drug and matrix interference in cell-based functional neutralizing antibody assays." *J Immunol Methods* **416**: 94–104.

Zhang YJ, Olah TV and Zeng J (2014). "The integration of ligand binding and LC-MS-based assays into bioanalytical strategies for protein analysis." *Bioanalysis* **6**(13): 1827–1841.

Zheng J, Mehl J, Zhu Y, et al. (2014). "Application and challenges in using LC-MS assays for absolute quantitative analysis of therapeutic proteins in drug discovery." *Bioanalysis* **6**(6): 859–879.

9

Mass Spectrometry-Based Assay for High-Throughput and High-Sensitivity Biomarker Verification

Xuejiang Guo[1,2] *and Keqi Tang*[1]

[1] *Pacific Northwest National Laboratory, Richland, WA, USA*
[2] *Nanjing Medical University, Nanjing, PR China*

9.1 Background

Searching for disease-specific biomarkers has become a major undertaking in the biomedical research field as the effective diagnosis, prognosis, and treatment of many complex human diseases are largely determined by the availability and quality of the biomarkers. A successful biomarker as an indicator to a specific biological or pathological process is usually selected from a large group of candidates by a strict verification and validation process. To be clinically useful, the validated biomarkers must be detectable and quantifiable by the selected testing techniques in their related tissues or body fluids. Due to its easy accessibility, protein biomarkers would ideally be identified in blood plasma or serum. However, most disease-related protein biomarkers in blood exist at very low concentrations (<1 ng/mL) and are "masked" by many nonsignificant species at orders of magnitude higher concentrations. The extreme requirements of measurement sensitivity, dynamic range, and specificity make the method development extremely challenging. The current clinical protein biomarker measurement primarily relies on antibody-based immunoassays, such as ELISA. Although the technique is sensitive and highly specific, the development of high-quality protein antibodies is both expensive and time-consuming. The limited capability of assay multiplexing also makes the measurement an extremely low-throughput one, rendering it impractical when hundreds of thousands of potential biomarkers need to be quantitatively measured across multiple samples.

Assays based on mass spectrometry (MS) have recently shown to be a viable alternative for high throughput and quantitative verification of candidate protein biomarkers. Among them, the triple quadrupole MS-based assay is perhaps the most promising analytical platform. When coupled with liquid chromatography (LC) separation and electrospray ionization (ESI) source, a triple quadrupole mass spectrometer operating in a special selected reaction monitoring (SRM) mode, also known as multiple reaction monitoring (MRM), is capable of quantitatively measuring hundreds of candidate protein biomarkers from a relevant clinical sample in a single analysis. The specificity, reproducibility, and sensitivity could be as good as ELISA. Furthermore, SRM MS can also quantify protein isoforms and posttranslational modifications, for which traditional antibody-based immunoassays often do not exist.

SRM MS-based targeted quantification of proteins is usually based on "bottom-up" method by measuring unique peptides generated by proteolytic enzymes, such as trypsin, which serve as quantitative surrogates for the concentrations of their corresponding proteins in samples. The use of stable-isotope-labeled heavy peptides as the internal standards (IS) has made the SRM MS-based assay a true quantitative technique. In the SRM MS, the peptide precursor ions are first selected by their mass-to-charge ratios (m/z) in the first quadrupole (Q1); after fragmentation of the specific *m*/*z* ions in the collision cell (Q2) by collision-induced dissociation, fragment ions (product ions) with a specific *m*/*z* are selected in the third quadrupole (Q3), which are then transmitted to the detector (Figure 9.1). Each pair of specific *m*/*z* precursor and product ions selected in Q1 and Q3 is defined as a "transition" in an SRM MS assay and the intensity of the specific transition measured by the MS provides the quantitative measurements of the corresponding protein concentration. Once one transition is measured, the mass spectrometer can be quickly switched to measure the next transition. Because both Q1 and Q3 are operated in a very narrow *m*/*z* range and the selection of each transition is highly specific, SRM MS has both higher sensitivity and better selectivity compared to other MS operation methods. The extremely fast measurement of each transition (~10 ms) in SRM MS operation also

Protein Analysis using Mass Spectrometry: Accelerating Protein Biotherapeutics from Lab to Patient, First Edition.
Edited by Mike S. Lee and Qin C. Ji.

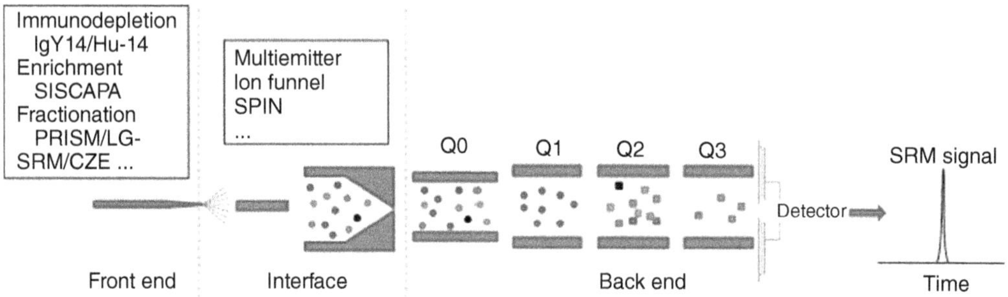

Figure 9.1 Principles and factors governing throughput and sensitivity of biomarker verification.

allows many transitions to be monitored in a single experiment, a potential for the development of multiplexing assays based on a single MS instrument platform (e.g., quantitative measurements of hundreds protein biomarkers in a single experiment). Combined with appropriate IS, the SRM MS can also provide accurate measurements of analyte concentrations with coefficients of variance (CV) similar to immunoassay technique. The effectiveness of SRM MS-based assays for the verification of protein biomarkers in blood is primarily determined by the achievable MS sensitivity, specificity, and dynamic range to quantitatively measure "analytes of interest" in a complex matrix, which can be improved by both the development of advanced sample processing and separation techniques to increase the relative abundance of analytes and separate the analytes of interest from the background interferences in the samples and more sensitive MS instrument.

9.2 Sample Processing Strategies

The abundances of proteins in blood plasma/serum can differ by more than 10 orders of magnitude. The most abundant albumin is present at about 34–54 mg/mL, while the low-abundant cytokines are usually present at the level of pg/mL. The 30 most abundant proteins comprise about 99% of the total protein mass. Most disease-related protein biomarkers in blood plasma/serum are usually of very low abundance. In order to effectively detect and quantify these low-abundant protein biomarkers by MS, a sample depletion step using antigen–antibody affinity columns to remove the most abundant proteins in blood plasma/serum is often used in the sample processing protocols. The commercially available affinity depletion column, such as Immunoglobulin Y14 (IgY14), is very effective in removing 14 most abundant proteins, including albumin, IgG, α1-antitrypsin, Immunoglobulin A (IgA), Immunoglobulin M (IgM), transferrin, haptoglobin, α1-acid glycoprotein, α2-macroglobulin, apolipoprotein A-I, apolipoprotein A-II, fibrinogen, complement C3, and apolipoprotein B in blood plasma/serum. As much as 94% of the total protein mass can be removed from the raw blood plasma/serum by using IgY14 (Sigma–Aldrich, St. Louis, MO) or Hu-14 (Agilent Technologies, Santa Clara, CA) depletion. In addition, SuperMix columns (Sigma–Aldrich, St. Louis, MO) can be further used in tandem with IgY14 or Hu-14 columns to remove moderate abundant proteins in plasma/serum to further improve the detection limit of low-abundance protein biomarkers. Besides the depletion of abundant proteins, antibody-based affinity columns can also be used to enrich the specific proteins or peptides from serum or plasma to improve their MS detection limits. A good example is to use Stable Isotope Standards and Capture by Anti-Peptide Antibodies (SISCAPA) to improve MS quantification of low-abundant plasma/serum proteins (Anderson et al. 2004). An average of 120-fold enrichment of the antigen peptide was demonstrated by the SISCAPA method with the cycle-to-cycle CV near 5%. With the effectiveness of this method, a company, SISCAPA Assay Technologies Inc. was recently formed to develop SISCAPA-based assays.

Following the initial affinity depletion/enrichment, proteolytic enzymes, such as trypsin, are used to digest the remaining proteins in the sample. The resulting peptide mixture is further subjected to LC separations before SRM MS analysis. Due to the extreme complexity of the sample, two-dimensional LC separations are often used to further enrich the targeted peptides and reduce chemical background interferences in the SRM MS measurements. In the first dimensional LC separation, the sample is fractionated into a number of fractions with each fraction containing significantly less number of peptides as compared to the initial peptide mixture, and the relative abundances of the targeted peptides in the corresponding fractions are significantly increased for better detection sensitivity of SRM MS. Several prefractionation methods have been developed by using strong cation exchange (SCX) chromatography and reversed-phase LC, which extend the low limit of quantitation (LLOQ) of SRM MS to low nanogram per milliliter levels in human plasma. Noticeably, Shi et al. recently developed a two-dimensional LC sample processing method, named as high-pressure, high-resolution separations coupled with intelligent selection and multiplexing (PRISM), that further improved the LLOQ of SRM MS to well below 50 pg/mL, making the sensitivity of SRM MS-based protein biomarker verification assay comparable to that of antibody-based immunoassays for the first time. Specifically in PRISM-SRM MS (Shi et al. 2012), as shown in Figure 9.2, human plasma/serum samples are first subjected to immunoaffinity depletion to remove high-abundance proteins followed by tryptic digestion. After sample cleanup, the peptide mixture, spiked with heavy peptide IS corresponding to the selected unique peptide targets related to protein biomarkers, is fractionated by a high-pH reversed-phase LC separation, a key component of PRISM workflow for effective enrichment of the targeted peptides, that has been demonstrated to provide significant advantages

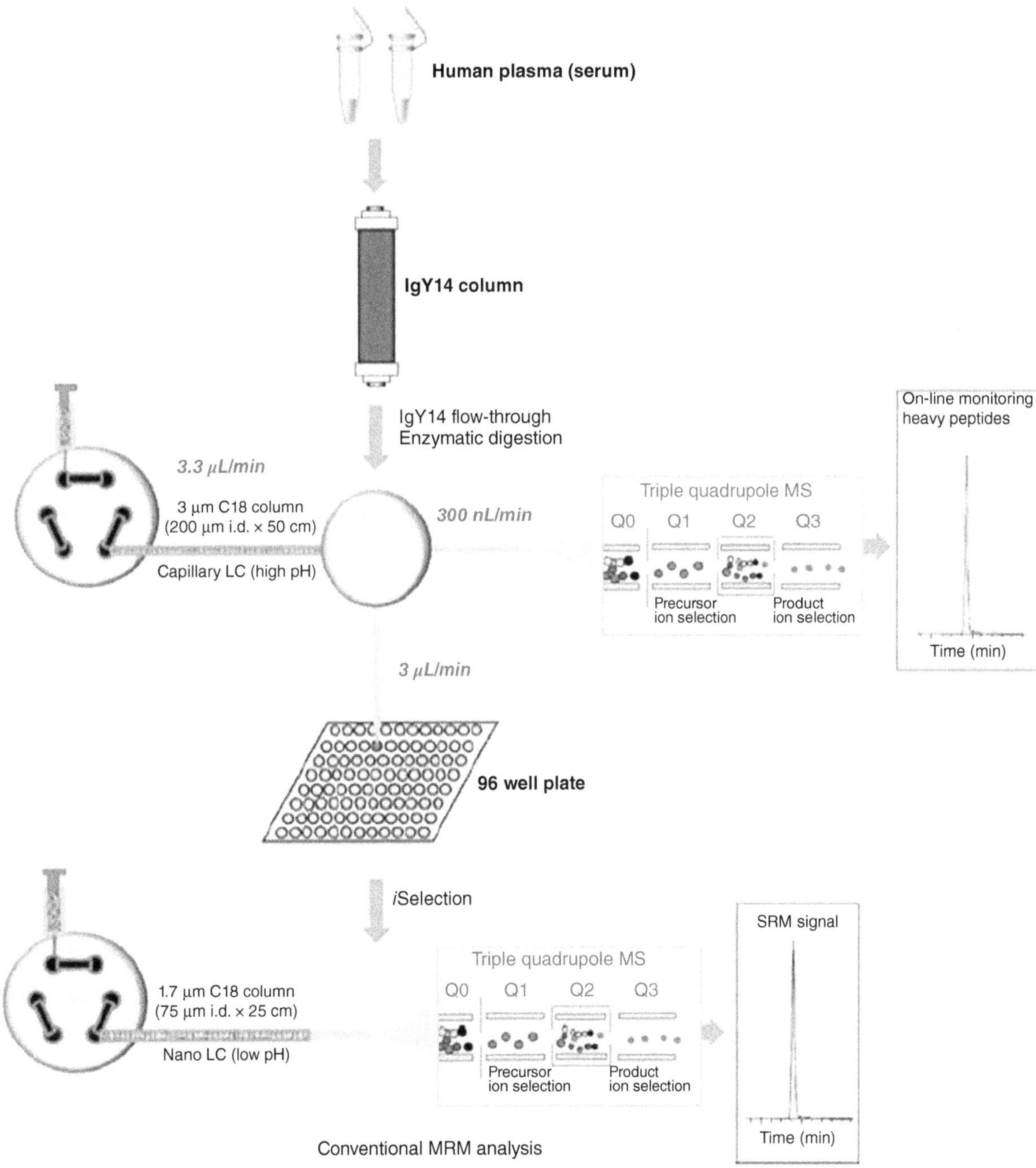

Figure 9.2 The workflow of high-pressure, high-resolution separation with intelligent selection and multiplexing (PRISM)-SRM MS strategy for targeted protein biomarker analysis.

over the conventional SCX chromatography due to its much better resolution. Only the fractions containing specific targeted peptides are subsequently selected (iSelection), based on the elution times of the IS simultaneously monitored by an on-line SRM MS during the sample fractionation, for either direct low-pH reversed-phase LC-SRM MS measurements or fraction multiplexing by pooling multiple fractions together prior to low-pH reversed-phase LC-SRM MS analyses.

The PRISM-SRM MS represents a major advance in the sensitivity of targeted protein biomarker verification. By spiking human prostate-specific antigen (PSA), a clinical biomarker for prostate cancer, together with other protein standards at different concentrations into a female blood plasma sample, a detailed experimental evaluation of PRISM-SRM MS demonstrated that LLOQ below 50 pg/mL could be consistently achieved for all targeted proteins by the workflow. The correlation coefficient for PSA was found to be greater than 0.99 between the measurements of PRISM-SRM MS and immunoassays (Shi et al. 2013a, b). To further test the broad utility of PRISM-SRM MS for targeted protein biomarker verification, the same workflow was also applied for quantitation of anterior gradient 2 (AGR2) in both depleted blood serum and urine. LLOQs of ~130 pg/mL in serum and ~10 pg per 100 µg of total protein mass in urine were obtained, respectively. A correlation coefficient of 0.91 was achieved between the PRISM-SRM MS and ELISA for the measurements of different AGR2 concentrations in urine. The main conclusion of the study was that the urinary AGR2/PSA concentration ratios were found to be significantly differentially expressed between noncancer and cancer, indicating the potential to use the AGR2/PSA ratio to differentiate prostate cancer from noncancer patients (Shi et al. 2014).

To achieve even a lower limit of quantitation in PRISM-SRM MS, different separation techniques were explored to increase the overall peak capacity of the two-dimensional separations in the original PRISM-SRM MS workflow. A long-gradient second-dimension low-pH LC separation was first tested in an attempt to further improve peptide separation before SRM MS quantification (Shi et al. 2013a). A direct comparison of a 5-h gradient LC-SRM MS and a conventional 45-min gradient LC-SRM MS showed an 8- to 100-fold improvement in LLOQ for target proteins in serum using the longer gradient LC separation. Besides improving the PRISM-SRM MS sensitivity, the long-gradient LC-SRM MS also offers higher fraction multiplexing potential as compared to the conventional LC-SRM MS, due to the higher separation resolution, increasing the overall sample analysis throughput. To further overcome the inherent low separation orthogonality between high-pH and low-pH reversed-phase LC separations, an alternative liquid separation technique, such as capillary electrophoresis (CE), was also evaluated for its suitability to replace the second-dimension low-pH separation in PRISM-SRM MS. CE separation is well known for its high resolving power and fast separation speed. CE is also orthogonal to LC separation due to the different separation mechanism. The major limitation of CE-based separation technique is the limited sample loading capacity. To overcome this limitation, a hybrid capillary isotachophoresis (CITP) and capillary zone electrophoresis (CZE) separation was developed to increase the sample loading volume. A recent study by coupling CITP/CZE with SRM MS demonstrated that several microliters of samples can be loaded without degrading CE separation resolution. A linear dynamic range spanning four orders of magnitude with LLOQ well below 50 pM was achieved using CITP/CZE-SRM MS (Wang et al. 2012). To further resolve the mismatching problem in all the existing CE–MS interface designs between the need to use large i.d. separation capillary for large sample loading capacity and small i.d. ESI emitter capillary for stable low flow rate ESI operation, a new sheathless CE–MS was developed that involved the use of a larger i.d. fused silica capillary (360 µm o.d., 100 µm i.d., and 95 cm long) as the separation capillary and a smaller i.d. capillary (90 µm o.d., 20 µm i.d., and ~4 cm long) as the ESI emitter (Wang et al. 2013). The performance evaluation of the new sheathless CITP/CZE-SRM MS demonstrated an LLOQ of 10 pM with measurement reproducibility of the CV <22%. It can be expected that the integration of CITP/CZE into the PRISM-SRM MS workflow would further improve the overall separation peak capacity and LLOQ for the quantitative measurements of targeted protein biomarkers in biofluids.

9.3 Advanced Electrospray Ionization Mass Spectrometry Instrumentation

The detection limit of MS-based assay for measuring protein biomarkers in biomatrix is ultimately determined by the sensitivity of electrospray ionization–mass spectrometry (ESI–MS). In ESI–MS, the peptides eluting from a selected LC separation are ionized by ESI, a soft ionization technique discovered by Fenn et al. in the 1980s to be capable of producing multiply charged intact gas-phase analyte ions from sample solution and measured by the MS for their m/z ratios. The achievable sensitivity of ESI–MS is largely determined by the efficiency of producing gas-phase analyte ions from the charged droplets in electrospray (ionization efficiency) and the effective transfer of the analyte ions from the atmospheric pressure ion source to the high vacuum MS analyzer (ion transmission efficiency). Most of the research and development efforts aiming at improving ESI–MS sensitivity in the last decade have focused on improving the ionization efficiency and the ion transmission efficiency of ESI–MS.

Gas-phase analyte ions in ESI process are generated by four well-known steps including the generation of highly charged droplets from electrospray, charged droplet evaporation, the fission of charged droplet at the Rayleigh limit, and the formation of gas-phase ions via two competing mechanisms described by the charged residue model (Dole et al. 1968) and the ion evaporation model (Iribarne and Thomson 1976). The key to improving analyte ionization efficiency is to produce smaller droplets with a higher charge density so that fewer droplet evaporation-fission cycles are required in the ESI process. At a given sample solution, the initial charged droplet size in the electrospray is largely determined by the liquid flow rate. High ESI efficiency was demonstrated by operating electrospray at tens of nanoliters per minute flow rate. The so-called nanoelectrospray, as proposed by Wilm and Mann (1996), also allows ESI emitter to be placed closer to MS inlet improving the ion transmission efficiency through the ESI–MS interface. To achieve stable electrospray at low nL/min flow rates, both the shape and the size of the emitter are critically important. The traditional stainless steel emitters cannot be machined to a sufficiently small i.d. to establish a stable nanoelectrospray and the fused silica capillary emitters drawn by using laser or flame to a small i.d. at the tip have poor reproducibility and are prone to clogging problem. These challenging problems have been effectively solved by the development of the chemically etched fused silica emitters. The chemically etched fused silica emitter was consistently shown to have taped o.d. and constant i.d. (Figure 9.3). Stable nanoelectrospray at flow rate below 10 nL/min can be easily established by using a 20 μm i.d. etched emitter (Kelly et al. 2006).

While the optimum ionization efficiency can be obtained by operating electrospray in the nanoelectrospray flow rate regime, performing an analysis in the nanoelectrospray ionization (nanoESI) condition is not always practical as commercial LC separations usually operate at optimal flow rate of hundreds of nL/min. To extend the benefits of both robust LC separation at high flow rate and high ESI efficiency at low flow rate, arrays of electrospray emitters have been developed, which splits the high-flow LC eluent into dozens of nanoelectrosprays. For example, capillary-based emitter array comprising 20 individual emitters each operating at 50 nL/min can be used to optimally couple with a capillary LC separation running at 1000 nL/min (Figure 9.4).

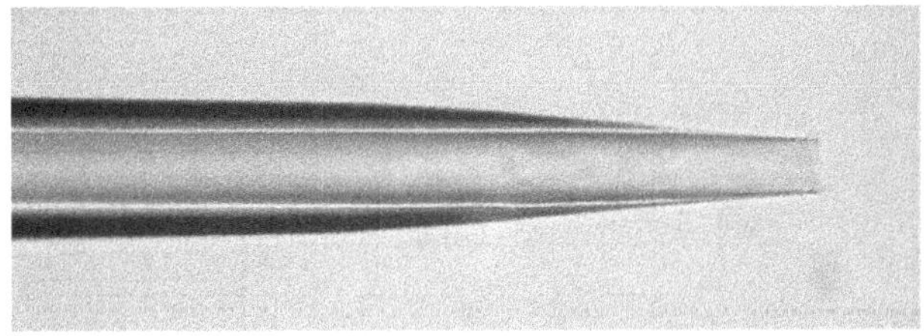

Figure 9.3 A 20 μm i.d. chemically etched nanoelectrospray emitter.

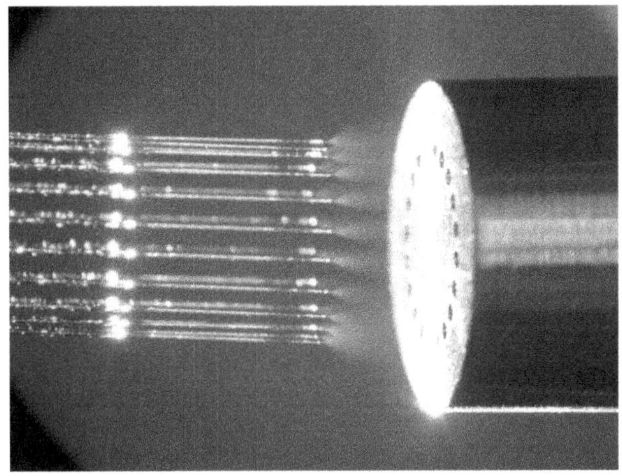

Figure 9.4 Capillary-based multinanoelectrospray source comprised of 20 emitters interfaced with a heated multicapillary MS inlet. The total flow rate is 1 μL/min.

To realize the full advantage of the high ionization efficiency by the nanoelectrospray in high sensitivity ESI–MS, the analyte ions generated in the ESI source have to be effectively transmitted to the MS detector. The majority of the ion loss (>90%) was shown to occur at the ESI–MS interface. The analyte ions are sampled into the first vacuum stage of a mass spectrometer through either a narrow bore heated capillary or a small orifice followed by a small aperture (~1 mm) skimmer in a conventional ESI–MS interface design (Figure 9.5a). The combined ion sampling and ion transmission efficiency for the conventional ESI–MS interface is typically on the order of ~1%, a major ion transmission bottleneck limiting the MS sensitivity. In order to greatly reduce the ion losses at the ESI–MS interface, an electrodynamic ion funnel interface was developed by the Smith laboratory at Pacific Northwest National Laboratory to replace the skimmer in the first MS vacuum stage. A broad range of *m/z* ions exiting the inlet capillary were shown to be focused and transmitted into the second MS vacuum stage with essentially 100% efficiency by using the ion funnel interface. A typical ion funnel consists of a series of ring electrodes with a front section of constant i.d. and a back section that linearly decreases in i.d. to create a

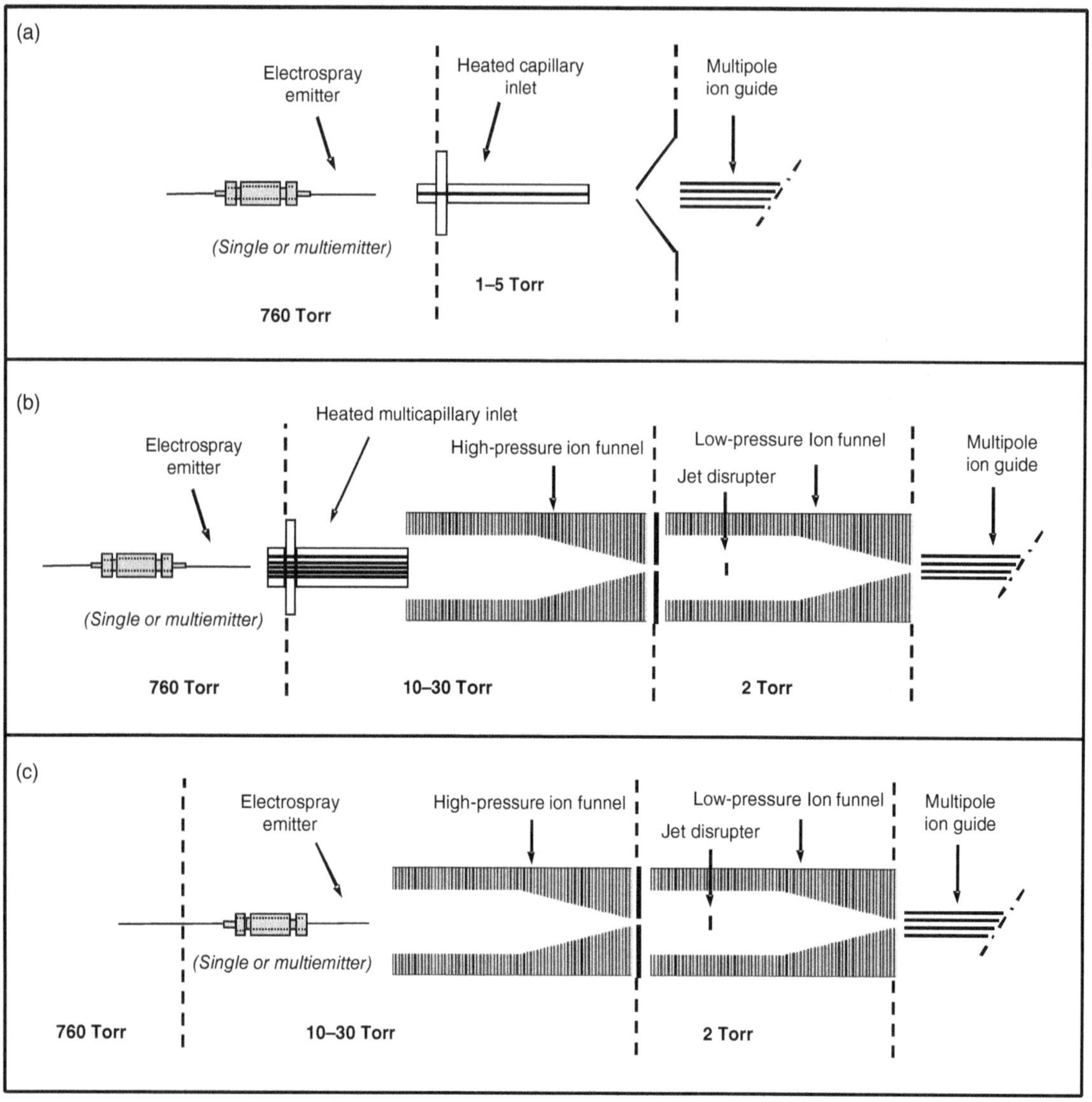

Figure 9.5 Schematic of ESI–MS interfaces: (a) heated capillary-skimmer interface; (b) multicapillary-ion funnel interface; (c) SPIN source interface.

special funnel-shaped ion optics. Both RF and DC voltages are applied to each ion funnel ring electrodes to create a radial RF ion focusing electric field and an axial ion transmission DC electric field. By simply replacing the skimmer with an ion funnel more than 10-fold increase in MS sensitivity was consistently demonstrated on different mass spectrometers (Kelly et al. 2006). To achieve both the high ionization efficiency and high ion transmission efficiency, an ESI emitter array/multicapillary inlet/tandem ion funnel interface was further developed. As shown in Figure 9.5(b), each emitter was aligned with a corresponding capillary inlet in this interface to obtain optimal ion transmission efficiency. A tandem ion funnel design was used to accommodate the significantly increased gas flow by the multicapillary inlet. The experimental evaluation of this interface by using an array of 19 emitters and the geometrically matched 19 heated inlet capillaries showed a sevenfold sensitivity improvement in an LC–MS analysis using a mixture of peptides (Kelly et al. 2008). By coupling the multicapillary/tandem ion funnel interface with SRM MS, the performance evaluation showed over a 70-fold improvement in average peak intensity by using tryptic peptides from targeted proteins spiked into nondepleted mouse plasma over a range of concentrations. The average limit of detections for targeted peptides also improved by 10-fold with notably improved measurement reproducibility (Hossain et al. 2011).

Although the multicapillary/tandem ion funnel interface significantly improves the MS sensitivity, the use of the narrow bore heated capillary as the MS inlet still limits the ion transmission efficiency through the ESI–MS interface. An effective solution to completely eliminate the interface ion loss is to remove the inlet completely and move the ESI source into the first vacuum chamber of the mass spectrometer. Named as subambient pressure ionization with nanoelectrospray (SPIN) as shown in Figure 9.5(c), this new "interfaceless" configuration places the ESI emitter adjacent to a low-capacitance ion funnel in a subambient pressure environment so that the entire electrospray plume can be sampled into the ion funnel and transmitted into the MS analyzer with high efficiency (Page et al. 2008). The detailed experimental evaluation of the SPIN/dual ion funnel interface demonstrated that as much as 50% of ion utilization efficiency could be achieved by operating nanoESI at 50 nL/min flow rate, which essentially implies that one in every two analyte molecules initially in the sample solution was effectively converted to a gas-phase ion and transmitted through the interface into the high vacuum region of the mass spectrometer (Marginean et al. 2010).

9.4 Conclusion

With the recent development of sample preparation and LC–MS interface technologies, MS-based proteomics has achieved significant improvements in sensitivity and throughput. The use of LC-SRM MS-based assays as alternatives to the traditional immunoassays for protein biomarker verification is very promising, especially at the early stages of biomarker development without high-quality antibodies. The LC-SRM MS-based assays also carry the potential to eventually replace the immunoassays in large-scale candidate biomarker verification in which hundreds of the disease-related candidate biomarkers are targeted due to their high multiplexing ability and measurement sensitivity and specificity. However, the current achievable sensitivity of LC-SRM MS may be still not sufficient to effectively quantify many extremely low-abundant proteins in blood plasma/serum. Development of better sample preparation methods and further improvement of LC–MS technologies will certainly help expand the capabilities of LC-SRM MS-based assays in protein biomarker verification and validation.

References

Anderson NL, Anderson NG, Haines LR, et al. (2004). "Mass spectrometric quantitation of peptides and proteins using Stable Isotope Standards and Capture by Anti-Peptide Antibodies (SISCAPA)." *J Proteome Res* **3**(2): 235–244.

Dole M, Mack LL, Hines R, et al. (1968). "Molecular beams of macroions." *J Chem Phys* **49**(5): 2240–2249.

Fenn JB, Mann M, Chin KM, et al. (1989) "Electrospray ionization for mass spectrometry of large biomolecules." *Science*, **246**(4926): 64–71.

Hossain M, Kaleta DT, Robinson EW, et al. (2011) "Enhanced sensitivity for selected reaction monitoring–mass spectrometry-based targeted proteomics using a dual-stage electrodynamic Ion funnel interface." *Mol Cell Proteomics* **10**(2): doi: 10.1074/mcp.M000062-MCP201.

Iribarne JV and Thomson BA (1976). "On the evaporation of small ions from charged droplets." *J Chem Phys* **64**(6): 2287–2294.

Kelly RT, Page JS, Luo Q, et al. (2006). "Chemically etched open tubular and monolithic emitters for nanoelectrospray ionization mass spectrometry." *Anal Chem* **78**(22): 7796–7801.

Kelly RT, Page JS, Zhao R, et al. (2008). "Capillary-based multi nanoelectrospray emitters: improvements in ion transmission efficiency and implementation with capillary reversed-phase LC-ESI-MS." *Anal Chem* **80**(1): 143–149.

Marginean I, Page JS, Tolmachev AV, et al. (2010) "Achieving 50% ionization efficiency in subambient pressure ionization with nanoelectrospray." *Anal Chem* **82**(22): 9344–9349.

Page JS, Tang K, Kelly RT, et al. (2008). "Subambient pressure ionization with nanoelectrospray source and interface for improved sensitivity in mass spectrometry." *Anal Chem* **80**(5): 1800–1805.

Shi T, Fillmore TL, Gao Y, et al. (2013a). "Long-gradient separations coupled with selected reaction monitoring for highly sensitive, large scale targeted protein quantification in a single analysis." *Anal Chem* **85**(19): 9196–9203.

Shi T, Fillmore TL, Sun X, et al. (2012). "Antibody-free, targeted mass-spectrometric approach for quantification of proteins at low picogram per milliliter levels in human plasma/serum." *Proc Natl Acad Sci USA* **109**(38): 15395–15400.

Shi T, Gao Y, Quek SI, et al. (2014). "A highly sensitive targeted mass spectrometric assay for quantification of AGR2 protein in human urine and serum." *J Proteome Res* **13**(2): 875–882.

Shi T, Sun X, Gao Y, et al. (2013b). "Targeted quantification of low ng/mL level proteins in human serum without immunoaffinity depletion." *J Proteome Res* **12**(7): 3353–3361.

Wang C, Lee CS, Smith RD, et al. (2012). "Ultrasensitive sample quantitation via selected reaction monitoring using CITP/CZE-ESI-triple quadrupole MS." *Anal Chem* **84**(23): 10395–10403.

Wang C, Lee CS, Smith RD, et al. (2013). "Capillary isotachophoresis-nanoelectrospray ionization-selected reaction monitoring MS via a novel sheathless interface for high sensitivity sample quantification." *Anal Chem* **85**(15): 7308–7315.

Wilm M and Mann M (1996) "Analytical properties of the nanoelectrospray ion source." *Anal Chem* **68**(1): 1–8.

10

Monitoring Quality of Critical Reagents Used in Ligand Binding Assays with Liquid Chromatography Mass Spectrometry (LC–MS)

Brian Geist, Adrienne Clements-Egan, and Tong-Yuan Yang

Janssen Research & Development, LLC, Spring House, PA, USA

10.1 Introduction

The biopharmaceutical industry is developing protein therapeutics (biologics or biotherapeutics) for a variety of indications in therapeutic areas such as oncology, inflammation, neuroscience, cardiovascular, and metabolism. Various types of biotherapeutic proteins in development include antibodies, cytokines, growth factors/hormones, fusion proteins, peptides, and enzymes. The number of market approvals of biologics has increased exponentially since the original biologic approved for therapeutic use, Humalin®, was launched in 1982 following the fast advancement in recombinant biotechnology. Within the top 100 selling drug products, the percentage of sales generated by biotherapeutics has increased from 17% in 2004 to 39% in 2012 (Evaluate Pharma 2013). In 2012, 18 of the top 50 selling drugs in the world were biologics (therapeutic monoclonal antibodies or recombinant protein/peptide products) (Evaluate Pharma 2013) and in that same year, a record-breaking 14 biologics license applications (BLAs) were submitted to the FDA for approval (Kling 2014). Such a rapid growth of the biotech industry will continue and it is projected by 2018 that 7 out of the top 10 selling pharmaceuticals will be biologics with worldwide sales of approximately $215 billion (Evaluate Pharma 2013). As with all therapeutics, to ensure market approval, a biologic development program must include nonclinical and clinical studies to evaluate and establish a relationship between safety and efficacy of the product. These pivotal studies require a variety of analytical platforms and bioanalytical methods not only to monitor the quality of the biologic but also to analyze and characterize the product in samples from biological matrices (e.g., serum, plasma, body fluids) for characterizing pharmacokinetic (PK), pharmacodynamic (PD), and immunogenicity (IMG) profiles in study subjects.

Ligand binding assays (LBAs), which are often immunoassays in the biotech industry, are the most common bioanalytical platform employed for testing biological study samples. Most LBAs used for PK assessments incorporate drug-specific capture and detection reagents in a sandwich format for analysis, while IMG methods for antidrug antibody (ADA) screening typically employ a "bridging" assay format using the conjugated biologic drug itself as the capture and detection reagent. Most of these bioanalytical LBA methods are developed and validated according to regulatory guidelines and industry best practices, and are subject to regulatory scrutiny (US Food and Drug Administration 2001, European Medicines Agency 2012). The purpose of developing these "regulated" bioanalytical methods for biologics is (i) to accurately quantify the biotherapeutic product in matrix (PK) (DeSilva et al. 2003, Viswanathan et al. 2007), (ii) to reliably detect the formation and progression of host ADA immune response (IMG) (Shankar et al. 2006, 2008), and (iii) to elucidate *in vivo* engagement of drug targets (PD) (Roskos et al. 2011, Lee and Salimi-Moosavi 2012), if applicable. All of these bioanalytical assessments, which are performed throughout the drug development process, are critical to delineating the relationship between safety and efficacy of biologics. Since these LBA methods support the successful application for regulatory approval and postmarketing commitments, it is critical to effectively manage these methods throughout the entire life cycle of the biologic development program.

Regulatory guidance documents, which are based on scientific rationale and stress the need for data integrity, dictate the evaluation of key assay parameters including sensitivity, selectivity, specificity, accuracy, precision, and robustness during the development and implementation of bioanalytical methods used to support the evaluations of biologic drug safety, efficacy, and biocomparability/bioequivalence. These bioanalytical methods

Protein Analysis using Mass Spectrometry: Accelerating Protein Biotherapeutics from Lab to Patient, First Edition.
Edited by Mike S. Lee and Qin C. Ji.

should undergo a robust development and validation process according to currently published industry recommendations and regulatory guidance documents (US Food and Drug Administration 2001, European Medicines Agency 2012, DeSilva et al. 2003, Viswanathan et al. 2007, Shankar et al. 2008, Thway et al. 2010) as their consistent performance during the course of drug development is the foundation for the integrity and interpretation of the bioanalytical data. Therefore, the quality of the critical reagents used in these methods, from initiation of assay development through the final study sample analysis, plays an essential role in (i) the establishment and validation of LBA characteristics and (ii) the successful implementation and optimal performance of the LBA during the entirety of its bioanalytical life cycle (King et al. 2014, O'Hara and Theobald 2013, O'Hara et al. 2012, Staack et al. 2011).

Reagents used in developing LBAs are often complex proteins with large molecular masses that are generated with high selectivity and specificity toward their analyte molecules. O'Hara et al. 2012 had categorized critical reagents used in bioanalytical methods into several categories, such as antibodies, engineered proteins, conjugates, and chemically synthesized molecules (e.g., peptides). A more broadly accepted description of "critical reagents" identifies them as fundamental components used in LBAs whose unique characteristics are essential to defining the quality of the assay performance, such as accuracy and specificity (Staack et al. 2011, Geist et al. 2013). These critical reagents can specifically and selectively bind to their analytes in solution via noncovalent protein–protein interactions. Through these multiple noncovalent interactions (e.g., with the directed immunogenic epitope(s) in an immunoassay) critical reagents bind the target analyte with high affinity (K_D values typically between low nM and pM range) and effectively isolate it from the existing sample matrix. The ability to sustain these unique intermolecular interactions between the reagent and its target analyte enable a selective, sensitive, and reproducible LBA method. However, the introduction of any variable(s) that prevents the reagent from recognizing the analyte (i.e., loss of selectivity), distinguishing the analyte from other unrelated sample components (i.e., loss of specificity), or incorporates specific binding of off-target epitopes (i.e., cross-reactivity) can lead to unexpected LBA performance and impact the integrity and quality of the reported study data (Stubenrauch et al. 2009, Rup and O'Hara 2007).

Most specific LBA reagents used in quantitative assays to support PK assessment fall into one of two categories: (i) drug target or (ii) drug-specific antibody or fragments thereof. Drug-specific monoclonal (mAb) and polyclonal (pAb) antibodies or antigen-binding fragments (Fab) are normally generated against a specific epitope(s) located on the therapeutic protein, such as the complementarity determining region (CDR) of a therapeutic antibody. These specific ADAs (e.g., anti-idiotype (Id) antibodies to therapeutic antibodies) are also most often served as an ADA positive control in IMG assays. The conjugated biologic drug itself is also used in standard bridging format of IMG assays, where they are considered as critical reagents. These critical reagents occasionally can be purchased from commercial sources but are often produced "in-house" by the sponsor as they are generated to the specific biologic drug. LBAs for target engagement (TE) evaluation or biomarker studies more frequently use commercial reagents or kits, which include recombinant protein targets and their antibodies (Lee and Salimi-Moosavi 2012, Zheng et al. 2015). Figure 10.1 shows the various LBA formats used during the development of biologics and highlights where these different types of critical reagents play an important role. The unlabeled source proteins or peptides used to produce the final LBA reagent and its conjugated version are the most critical reagents used in the assay formats and will be the focus of the discussion in this chapter. It should be noted that unlabeled therapeutic drug lots used as the "reference standard" (i.e., in PK assays) or used as reagents in IMG and TE methods are thoroughly characterized and monitored through standard "lot-release" quality control processes (ICH 1997; US FDA Guidance for Industry 1999, 2005). LC–MS is widely used in drug substance/product characterization, which is extensively discussed in Chapters 3 and 16, and will not be discussed here. Other LBA reagents (e.g., coated solid-phase supports, detection enzymes, matrices), while also important to assay performance, are out of the scope of this chapter.

The physicochemical and immunological properties (or quality attributes) of critical reagents used during development provide the foundation for the high degree of selectivity and specificity that enables the robust performance of the LBA during its life cycle. The importance of controlling and monitoring the quality of critical reagents during the life cycle of bioanalytical methods has been recognized by the bioanalytical community and published in numerous industry-wide articles and conference reports (King et al. 2014, O'Hara and Theobald 2013, O'Hara et al. 2012, Geist et al. 2013, Rup and O'Hara 2007, Nicholson et al. 2012). The majority of these critical reagents are produced via recombinant biotechnology or through direct immunization of animals. As with the manufacturing process of biologic drug substances, the development of a controlled production system for critical reagent supply should be considered essential for consistent and efficient bioanalytical

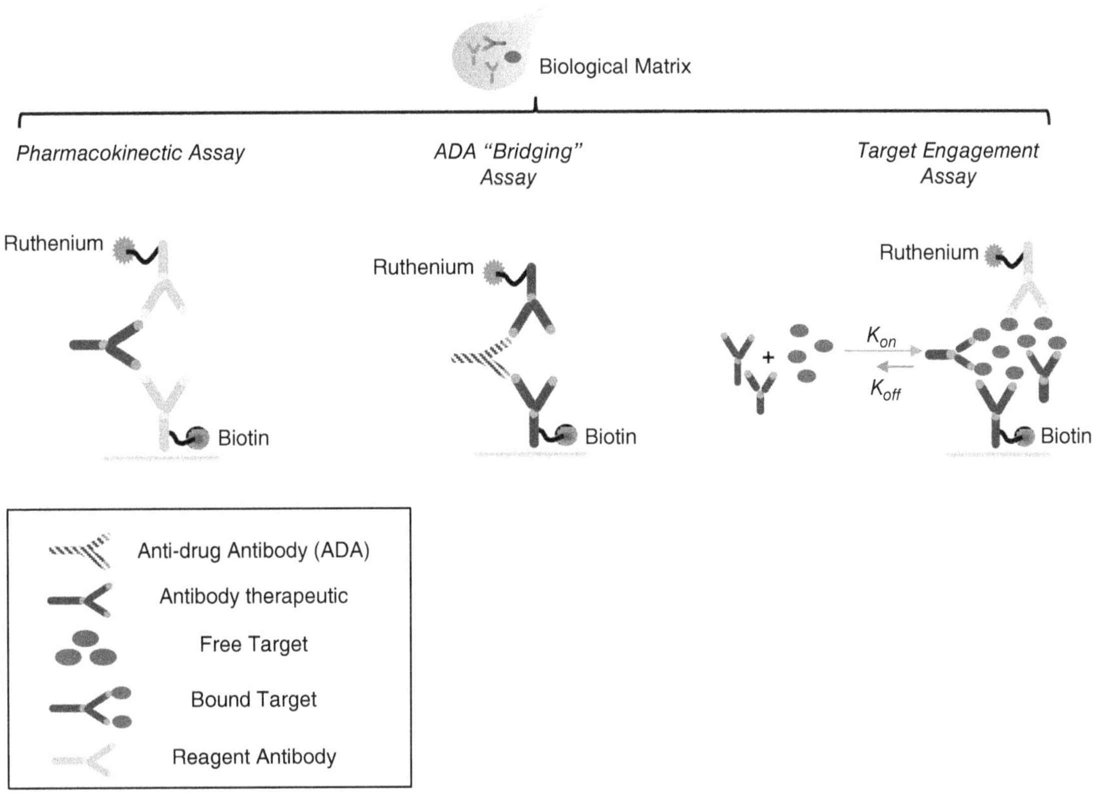

Figure 10.1 Schematic representation of pharmacokinetic, immunogenicity, and target engagement assays typically employed during bioanalysis of biotherapeutics. The schematic above depicts an antibody therapeutic and demonstrates the critical nature of the reagents used during the analysis. For this example, biotin and ruthenium conjugates are shown as capture and detection reagents for a streptavididn based capture method and electrochemilumiscence based detection method. However, a variety of functional labels can be applied, depending on the assay platform. (*Source*: Geist et al. 2013. Reproduced with permission of Future Science.)

operations. Protein reagents derived from transfected or hybridoma cell line systems are susceptible to modifications originating from the producing cells as well as impurities from the culture media and additives. Subtle changes to the medium, culture, or storage conditions can result in unexpected heterogeneity to the final reagent product (Manning et al. 2010, Chirino and Mire-Sluis 2004, Kozlowski and Swann 2006). Alterations in mAb heterogeneity can be attributed to enzymatically produced posttranslational modifications (PTMs including N-linked glycosylation and C-terminal lysine processing) (Sola and Griebenow 2009, Zheng et al. 2011, Houde et al. 2010), nonenzymatic modifications (oxidation, deamidation, glycation, etc.) (Bertolotti-Ciarlet et al. 2009, Wei et al. 2007, Gaza-Bulseco et al. 2008, Vlasak and Ionescu 2008, Rehder et al. 2008, Sinha et al. 2009, Vlasak et al. 2009, Banks et al. 2009, Chelius et al. 2006, Beck et al. 2005), deletions or substitutions to the primary sequence during cell line subcloning and production, and biophysical changes to the molecule (e.g., aggregation state, charge microheterogeneity, hydrophobicity) (Chirino and Mire-Sluis 2004, Xie et al. 2010, Wang et al. 2010, Sahin et al. 2010, Luo et al. 2011). If the assay reagents are purchased from commercial vendors, these modifications may be more prevalent and harder to control from batch to batch and vendor to vendor by end users. It remains important to closely monitor these commercial reagents for batch and vendor consistency due to the routinely undisclosed or proprietary information around the production details and reagent quality control process. Chemical or physical alterations to the critical LBA reagents do not necessarily correlate with changes in biological/immunological activity or functional response of the protein. Furthermore, predicting the propensity for these different modifications to alter the biophysical properties of the protein (e.g., denaturation, aggregation, precipitation) and subsequently the associated LBA performance remains challenging (Luo et al. 2011, Paul et al. 2012). Nevertheless, the potential of any physicochemical alteration to induce changes to the higher order structure of a given critical reagent, possibly impacting the LBA method, should be considered. Several documented findings on the interrelated nature of protein modifications and protein stability and integrity have been published (see Table 10.1). These considerations become indispensable when changes in assay performance are observed and reagent batch testing or troubleshooting is required. In addition to physicochemical and biophysical characterization, demonstration of acceptable biological function and immunoreactivity to the target antigen(s)/or analytes for new batches of critical reagents can help ensure proper protein assembly following production as well as confirm reagent affinity and binding kinetics.

In addition to the challenges involved with production of source proteins, most critical reagents used in LBA methods will also need to be covalently conjugated with other chemical entities to enable appropriate function in the assay. Depending on the LBA format, critical reagents typically require covalent labeling with small chemical or complex protein moieties prior to use within the assay, either to exploit distinct binding relationships for improved assay performance (i.e., biotin-streptavidin/avidin) or for use in specific robust detection systems (i.e., ruthenium chelate, digoxigenin, horseradish peroxidase). Assay performance can also be impacted as a result of batch-to-batch variability of the conjugated critical reagents due to alterations in downstream processing (e.g., changes in purification protocols or conjugation procedures for protein and chemical labels). For example, assay performance may be impacted by inefficient or suboptimal protein:label conjugation (inadequate challenge ratio, inappropriate linker/chemistry, ineffective labeling conditions, etc.) or through epitope masking, where the conjugated label interferes with protein–protein interaction (Acchione et al. 2012).

Due to the importance of critical reagents in LBAs, the reagent generation and selection strategy should be carefully considered. For example, when pAb reagents are utilized, multiple batches obtained through animal-derived immunizations can lead to reagent products that vary in composition. Changes in pAb affinity, avidity, titer, or quality may be impacted by differences in the host animal used, immunization method, and sample collection time points. Furthermore, biophysical characteristics of purified pAb reagents, beyond gross estimations of purity and aggregation state, are not easily determined due to the heterogeneity of the pAb population. For PK methods, pAbs are commonly avoided if possible due to the inconsistency of regenerating similar lots from different animals. It is possible to pool purified pAbs from multiple animals to achieve sufficient amount and consistency for bioanalytical support; however, the limited supply of the source proteins can still be a major concern depending on the stage of drug development. Therefore, it is preferred to use mAbs or *in vitro* purified recombinant proteins in PK method development due to unlimited supply of the source. Depending on the expression systems, batch-to-batch variations may still exist with *in vitro* recombinant technology since these reagents are complex proteins with large molecular weight and PTMs. The relative consistency among batches of mAb reagents can be achieved with adequate analytical quality control process implemented throughout the life cycle of reagent management.

With the inherent variability in the process of generating critical reagents (conjugated and unconjugated) described above, physicochemical and biophysical assessments of these critical reagents, especially at certain stages during the reagent life cycle, can provide great value toward safeguarding the optimal performance of an LBA method and the overall success of a biologic drug development program. Moreover, it becomes important to thoroughly evaluate reagent physicochemical characteristics when there are significant unknown variables involved with the reagent production, such as with reagents obtained from external vendors. These vendors generally will provide insufficient details on the production and quality control process of their reagents to customers due to proprietary protection.

Biochemical and biophysical characterization of the quality attributes for these critical reagents can be analytically challenging due to lack of comprehensive information from the manufacturer. For most LBA critical reagents employed in PK or IMG assay formats, such as anti-Id mAbs produced in hybridoma cell lines, very little may be known about the physicochemical properties of the molecule and the primary amino acid sequence may not have been determined. As mentioned previously, extra considerations should also be taken when the LBA format incorporates a reagent from a commercial source, as reagent screening and selection processes are not controlled by the end user. In addition, changes to the production system and QC process by a vendor are typically unknown to the end user despite the availability of certificates of analyses. Because most qualification procedures for commercial reagents likely vary from vendor to vendor, a core set of relevant physicochemical and biophysical characteristics should be evaluated by the end user to provide quality assurance and batch consistency of the reagent, particularly across multiple commercial batches.

The LBA is usually a quality-indicator for critical reagents; the assay results produced, outside of expected analytical variability, can provide clues to changes in reagent functionality that may cause assay drift if left unaddressed. Analytical assessment of these critical reagents is often limited to acceptable LBA assay performance with very little evaluation of structural or conformational stability. In many situations, simply adjusting the critical reagent working concentration to qualify acceptable performance within the LBA method may indicate a

Table 10.1 Common protein modifications that can impact reagent stability and potentially ligand binding assay (LBA) performance.

Protein/reagent modifications	Targeted residues	Influencing factors	Critical reagent impact	Potential LBA affect	References
Glycosylation	N-linked: Asn; O-linked: Ser, Thr	Production cell line, culture conditions	Altered charge-state and conformation, Fc aggregation	Minimal; ↓ target affinity (Fab-glycoform), aggregation	Sola and Griebenow 2009; Zheng et al. 2011; Houde et al. 2010; Damen et al. 2009
Oxidation	Met, Trp, Cys, His, Tyr	Formulation, temperature, light, pH, storage container	Altered conformation, ↑ unfolding/aggregation	Location dependent; ↓ target affinity, loss of selectivity	Houde et al. 2010; Bertolotti-Ciarlet et al. 2009; Wei et al. 2007; Gaza-Bulseco et al. 2008
Deamidation/ isomerization	Asn/Asp	pH, temperature, buffer, 1° sequence, 2°/3° structure	Altered charge-state, conformation, ↑ fragmentation	Location dependent; ↓ target affinity, loss of selectivity	Vlasak and Ionescu 2008; Robinson 2002; Rehder et al. 2008; Sinha et al. 2009; Vlasak et al. 2009; Hambly et al. 2009
Glycation	Lys, N-term	Cell culture conditions, formulation	Altered charge-state, ↓ pI, acidic species	Minimal; possible change in conformation/target affinity	Goetze et al. 2012; Banks et al. 2009
Pyroglutamate formation	N-term Gln/Glu	Purification process, pH, buffer composition	Increased hydrophobicity	Minimal	Chelius et al. 2006; Beck et al. 2005
C-terminal lysine cleavage	C-term Lys	Production cell line, culture conditions	Altered charge-state and ↑ hydrophobicity	Minimal	Beck et al. 2005
Disulfide rearrangement	Cys S–S	Free –SH, pH, solvent exposure, isoform	Altered conformation, unfolding, aggregation	Decreased target affinity, loss of selectivity/specificity	Liu and May 2012; Hutterer et al. 2013
Hinge region fragmentation	Asp-Xaa, Gly, Xaa-Ser/Cys	Primary sequence, pH, temperature	Fragmentation, unfolding, aggregation	Loss of function, sandwich complex binding interference	Hambly et al. 2009; Cordoba et al. 2005; Gaza-Bulseco and Liu 2008
Primary sequence modification	Any	Production cell line stability, culture conditions	Unknown; location dependent	Varied; possibly severe	Chirino and Mire-Sluis 2004; Kozlowski and Swann 2006; Xie et al. 2010
Protein cross-linking: Biotin	$-NH_2$ (Lys, N-term)	Functional group targeted, molecular incorporation	Altered charge-state or hydrophobicity	Usually minimal; location dependent, change to LBA signal via incorporation ratio	O'Hara et al. 2012; Staack et al. 2011; Acchione et al. 2012
Digoxigenin	–SH (Cys)				
Ruthenium	–COOH (Asp, Glu, C-term)				

Also included are common protein conjugation cross-linkers used in typical LBA assay formats. S–S, disulfide bond; SH, free thiol.

Source: Geist et al. 2013. Reproduced with permission of Future Science.

stressed system that would benefit from a closer investigation into possible changes in the reagent's molecular attributes and properties. Seemingly insignificant changes allowed to progress over an extended length of time that lead to a subtle shift in LBA output may result in a more significant drift throughout the course of method application during biotherapeutic development. Therefore, the implementation of suitable physicochemical and biophysical characterization methods for critical reagents, in addition to standardized critical reagent qualification procedures through LBA analysis, plays an important role in ensuring consistent and optimal assays that produce reliable and high-quality study results. Traditionally, a collection of analytical methods (e.g., size exclusion chromatography (SEC), sodium dodecyl sulfate (SDS)–polyacrylamide gel electrophoresis (PAGE), biomolecular affinity measurement) are employed to characterize critical reagents and the associated conjugates for their purity, concentration, and specificity. Both SEC and SDS–PAGE methods are considered "low-resolution" methods to approximate the intact mass of proteins. Although measurement of affinity kinetics via surface plasmon resonance (SPR) or related technologies can be applied to assess adequate binding activity and function of critical reagents, suitability for use is ultimately dictated through LBA performance. With this in mind, little attention is given to the potential of specific changes in primary structure or physicochemical modifications (e.g., oxidation, deamidation, glycation) to influence the biomolecular interaction of the reagent and target within the assay system. Consequently, issues that arise from poor LBA performance are predominantly viewed as a broader systemic error frequently followed by an intensive troubleshooting effort. Conducting a more detailed molecular evaluation of critical reagents prior to application within the assay may provide a more efficient and prospective, rather than retrospective, approach to LBA development and troubleshooting; and thus reduce the overall cost, time, and resources.

Great advancement has been made in the field of protein-based mass spectrometry (MS) in the last 25 years. The capabilities of MS-based approaches used for the characterization of large biomolecules and intact proteins have advanced significantly following the development of soft ionization techniques such as electrospray ionization (ESI) and matrix-assisted laser desorption/ionization (MALDI) (Siuzdak 1996). ESI techniques, in particular, allow for the ability to generate multiply charged ions of large, more labile biomolecules while minimizing the excessive molecular fragmentation during the ionization process, as can typically be observed with electron or chemical ionization. Another significant advantage to the use of ESI for protein characterization resides in the ability of the technique to be directly interfaced with liquid chromatography (LC), producing samples as electrically charged droplets directly from the liquid phase of the LC effluent within the source of the MS instrument. The electrospray application for most instruments can be performed in both positive and negative ion modes and is tolerant to a wide variety of aqueous and organic mobile phase solvents as well as mobile phase additives (e.g., formic acid, trifluoroacetic acid, ammonium bicarbonate), offering a great deal of flexibility to achieve ideal analytical conditions for distinct compounds. In addition, the continued development into the study of proteomics has resulted in the expertise to produce extremely detailed analyses of protein structure and modifications, largely through peptide digestion and MS sequence analysis. Identification and characterization of protein sequence and structure can be accomplished via several approaches ("top-down" protein analysis vs "bottom-up" peptide analysis following enzymatic digestion) based on instrument capability and the extent of sample manipulation adopted (Li et al. 2011, 2012, Lowenthal et al. 2011, Bondarenko et al. 2009, Zhang et al. 2009). In some cases, the use of less conventional approaches such as "middle-up" protein analysis, where the intact protein undergoes limited digestion into smaller subunits, can deliver the necessary information on protein structure while limiting the complexity of the data analysis (Beck et al. 2013, Chevreux et al. 2011, Zhang et al. 2009, Tipton et al. 2011, Staub et al. 2011). A variety of LC–MS or MS/MS combinations are currently available that couple chromatographic separation and mass analyzers to cover nearly all analytical needs (e.g., sensitivity, resolution, mass accuracy) from the analysis of large intact proteins to small peptides. High-resolution time-of-flight (ToF) and Fourier transform instruments feature a great combination of very good mass accuracy, mass-to-charge (m/z) range, and isotopic resolution well suited for the analysis of intact protein characterization, protein sequence variations, PTMs, and conjugation ratios of labeled proteins. Furthermore, recent progress in MS technology, such as turnkey micro/nanoflow LC, and novel fragmentation techniques (electron transfer dissociation, ETD, electron capture dissociation, ECD), are yielding greater sensitivity and allowing for a wealth of sample information to be collected and analyzed.

With advanced chromatographic techniques, MS analyzers, and "protein-centric" data analysis software, the operation of LC–MS instrumentation and analysis of the generated data has become greatly simplified. The application of LC–MS as an analytical tool for the structural characterization of therapeutic proteins and antibodies has been described well in multiple publications (Bondarenko et al. 2009, Beck et al. 2013, Chevreux et al. 2011, Zhang et al. 2009, Hall et al. 2010, Kaltashov et al. 2012,

Srebalus Barnes and Lim 2007) and Chapters 1–4. LC–MS techniques can be easily implemented in LBA laboratories to monitor critical reagents without extensive training while requiring minimal resources and low cost to the biopharmaceutical companies and contract research organizations (CROs). The primary outcome for critical reagent molecular analysis is to confirm with a high degree of certainty the identity of the protein and define the impurity profile of the source protein and labeled conjugates. Although full sequence determination of most reagents is usually not necessary, some assurance of molecular identity compared to previously produced batches/or lots is strongly encouraged. Therefore, it is imperative to establish a "time zero" characterization profile of a given critical reagent through standard analytical methodologies at the onset of assay development. For critical reagent analysis of mAb/fragment reagents, a middle-up approach provides high-level detail of protein structure at the subunit level while maintaining adequate resolution to detect and identify most PTMs. Complex proteins with tertiary structure, such as mAb reagents, can be derivatized under mild reducing conditions (i.e., dithiothreitol (DTT), tris(2-carboxyethyl)phosphine (TCEP)) and alkylation of free thiols (e.g., iodoacetamide) to break intra- and intermolecular disulfide bonds, resulting in distinct regions or chains of the protein to be analyzed separately with improved mass resolution. These sample treatment procedures only require less than 1 h between preparation and run time. For antibody reagents, variations in the molecular mass of the protein can be separated by domains (e.g., light chain (LC) vs heavy chain (HC); Fab vs Fc) and prioritized based on the probable impact to reagent function in the assay (Figure 10.2). When appropriate, subunit analysis can be performed through limited proteolysis with papain, which generates two Fab and one Fc fragments, and streptococcal immunoglobulin-degrading enzyme (IdeS), which produces one each of $F(ab')_2$ and Fc fragments (Beck et al. 2013, Chevreux et al. 2011). Fab and Fc analysis can provide further information on molecular identity or location of heterogeneity between source lots of a critical reagent to determine if a noted change is relevant. For example, a mass discrepancy within the heavy chains from a qualified and new production lot of reagent mAbs may not affect its function but can be monitored more closely via LBA performance during qualification. Heterogeneity attributed to incomplete C-terminal lysine processing or N-linked glycosylation can introduce additional complexity into the mass spectrum during data analysis but can be alleviated

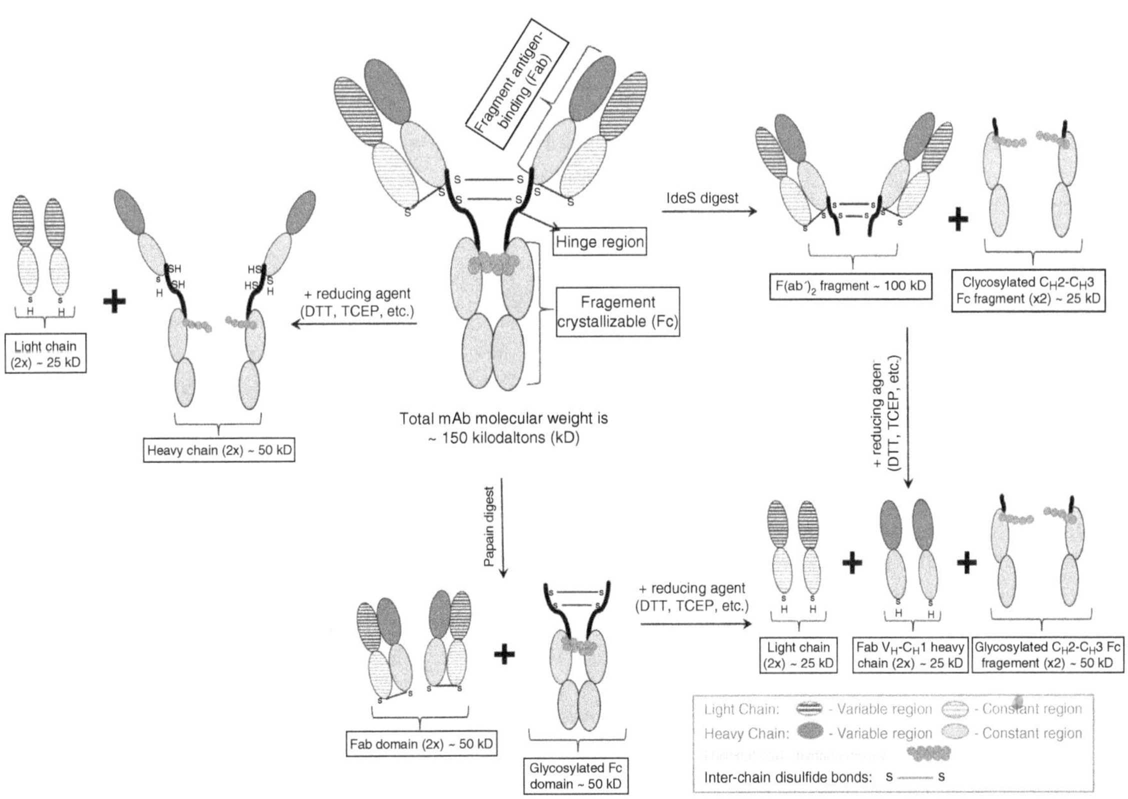

Figure 10.2 A workflow for structural analysis of a mAb LBA critical reagent. Outlined are common derivatization techniques used to help achieve a thorough characterization to determine structural identity between different critical reagent batches. mAb, monoclonal antibody; IdeS, immunoglobulin-degrading *Streptococcus pyogenes;* DTT, dithiothreitol; TCEP, tris(2-carboxyethyl)phosphine. (*Source*: Geist et al. 2013. Reproduced with permission of Future Science.)

through digestion with carboxypeptidase B and an endoglycosidase (e.g., PNGase F) (Beck et al. 2013, Chevreux et al. 2011, Zhang et al. 2009, Kaltashov et al. 2012), respectively. However, the impact of these molecular attributes on the function of the reagent antibody in the LBA is most likely minimal but can be dependent on the assay format. While mass identity of the intact mAb molecule as well as the individual components (e.g., light and heavy chains, Fab/Fc domains) can be accomplished through various analytical techniques, our preferred method is MS.

Analysis of source proteins and their conjugates with LC–MS can provide better resolution to structural details of intact reagents in comparison to conventional biochemical methods. For example, MS analysis can easily discriminate two distinct species of immunoglobulin G (IgG) produced in the same cell line system but containing slight variations in mass that cannot be identified with alternate, conventional biochemical techniques, such as SEC or SDS–PAGE. Selection of the appropriate MS instrument for critical reagent analysis can generally be determined based on the nature of the work being performed (quantitative vs qualitative; peptide vs protein). Since the need for whole protein peptide analysis or *de novo* sequence determination of these LBA reagents would most likely be exhaustive, time-consuming, and unnecessary, tandem MS/MS approaches hold less value unless the information of primary sequences is established for a given critical reagent, for example, recombinant cytokines and molecularly cloned mAbs. Therefore, instrument selection can focus more around MS instruments that provide qualitative value via a broad, accurate mass range and sufficient resolving power for large proteins (e.g., ToF-MS). ToF-MS has become widely accepted in industry as the preferred MS technology for the qualitative mass analysis of macromolecules due to the suitable resolving power (>10,000), high molecular mass accuracy, and broad m/z range required for analysis of large proteins. However, the evolution of tandem quadrupole-ToF mass analyzers (Q-ToF) affords the end user the flexibility of accurate protein characterization via ToF analysis with the option for peptide sequence determination. Accurate molecular mass determinations also often require the use of specific qualitative protein analysis software capable of generating a deconvoluted approximation of the average molecular mass based on the charged data spectrum and identifying PTM or impurity profiles based on the empirically derived native species.

In contrast to the existing regulatory guidance documents and industry best practice white papers outlining the specifications and quality assessments of biotherapeutic products following manufacturer, such assessments are rarely conducted for critical reagents used in LBAs that support preclinical and clinical development of biologics. Moreover, few publications exist on the topic of LC–MS analysis of critical reagents. Geist et al. 2013 published a paper outlining a much needed focus on proactive control of reagent characterization and supply, highlighted some of the benefits to incorporating LC–MS analysis into the routine critical reagent characterization process, and summarized key considerations for the use of LC–MS analysis during the production, characterization, and qualification of critical reagents in LBAs used during the life cycle of bioanalytical method implementation. This chapter is largely centered on the publication from Geist et al and will focus on the use of LC–MS to characterize the most commonly and frequently used critical reagents in LBAs for PK, IMG, and TE/biomarkers (PD) assessment of biologics. In the following section, four case studies are presented to demonstrate (i) the impact of critical reagent quality on assay performance, (ii) the follow-up analytical investigation, and (iii) subsequent strategy to monitor the quality of these reagents. The four examples highlight the importance of monitoring the quality of these critical reagents as well as the benefit of implementing LC–MS from the initial screening after reagent generation through late-stage product development. The primary objective of an effective reagent characterization strategy in a bioanalytical laboratory is the successful management of the LBA during the life cycle of the method. To that end, LC–MS analysis provides a valuable tool for assuring reagent consistency, leading to long-term LBA performance and high-quality data generated and reported.

10.2 Case Study Examples

10.2.1 Case Study #1: Confirmation of Correct Reagent Construct Prior to Use in Development of an LBA Method

The ability to generate and screen affinity-derived panels of reagents, most notably anti-Id mAbs or pAbs, is often considered a key link in the bioanalytical method development for a therapeutic antibody biologic program. The generation of specific anti-Id mAb reagents, in particular, frequently uses recombinant biotechnology for consistent and efficient production. The use of LC–MS for reagent quality monitoring is the most important from the onset of critical reagent generation, or “time-zero” analysis. This “gate-keeping” role can potentially save time and resources prior to evaluation of the reagent within an LBA method by providing a quick approach to assure molecular identity by (i) determining discrepancies between the mass of the protein product from the predicted amino acid sequence of a molecularly cloned critical reagent and (ii) identifying batch-to-batch variability of the protein reagent. If differences in LBA performance are observed without such information, erroneous conclusions about the root cause of reagent differences can be made and significant time might be spent on ineffective assay troubleshooting.

SDS–PAGE and SE-HPLC results on a reproduction of a transiently expressed anti-Id mAb reagent for a PK assay (referred to as "C101") were comparable to the prior batch (data not shown). However, when the first and second batch of the purified C101 proteins were characterized by LC–MS under reduced conditions, the mass of the light chain for the second batch of C101 reagent did not match the expected molecular mass of the protein (based on the DNA sequence) nor did it match the MS profile of the first batch (Figure 10.3). The LC–MS characterization led to the discovery that the wrong light chain construct was coexpressed with the heavy chain construct in the second batch production. It is independently confirmed by affinity measurement that the reagent with the wrong light chain did not bind to its intended analyte (data not shown). The characterization data was shared with the manufacturer, who subsequently made the correct construct.

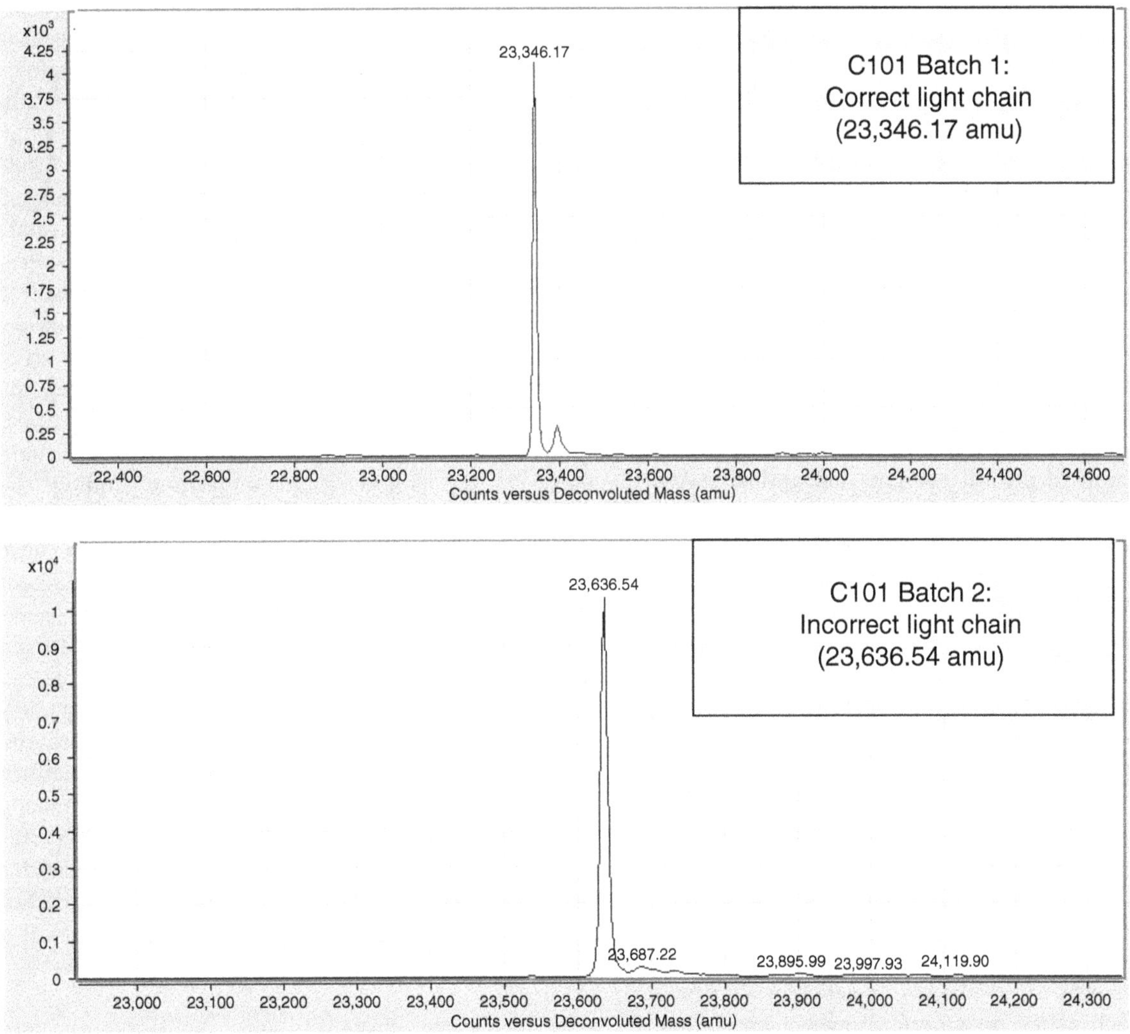

Figure 10.3 Comparison of LC–MS profiles of light chains from two batches of a reagent antibody C101. Batch 1 and 2 of C101 (transiently expressed and purified mAb proteins) were treated with 20 mM dithiothreitol (DTT) to reduce interchain disulfide bonds, allowing for a separate and distinct analysis of heavy and light chain species. The mAb proteins were separated by reversed-phase liquid chromatography on an Intrada WP-RP column (150 mm × 2 mm, 3 μm) and detected via time-of-flight mass spectrometry (ToF-MS) equipped with electrospray ionization (ESI) source. The ToF-MS data was acquired in positive ion mode with an *m/z* acquisition range from 500 to 7000 and the acquired spectral data was subsequently deconvoluted (maximum entropy) with Agilent MassHunter Qualitative Analysis software for molecular structure comparison. The mass of Batch 1 corresponds to the molecular weight of the correct light chain while the mass in Batch 2 corresponds to another light chain construct that was erroneously coexpressed with the correct FM1B11 heavy chain construct.

Without understanding these mass differences early on, the incorrect reagent would have been chemically conjugated and tested in the LBA, at which point incorrect assumptions could have been made about the source of reagent performance variability (e.g., variability could have erroneously been attributed to unknown differences between purification or conjugation protocols). Because LC–MS was performed initially, the other variables were eliminated and subsequent batches were generated correctly, efficiently, and consistently.

10.2.2 Case Study #2: Monitoring the Integrity of the Reagent Cell Line Production System

Typically, most biopharmaceutical companies are using several qualified or validated LBA methods to support the PK, TE/PD, and IMG assessments over multiple biotherapeutic programs. With numerous biologic programs at various stages of development, the need for consistent quality and maintenance of these critical LBA reagents and the cell line systems that produce them is particularly essential. The current case study highlights the value of incorporating LC–MS analysis into routine monitoring of recombinant biotechnology-derived reagents from hybridoma cell lines as it relates to purity and integrity of the overall production system.

During the course of supporting two separate ongoing biologic programs in approximately the same timeframe, new production of critical reagents (all anti-Id mAbs previously developed as capture or detection reagents in the LBA methods, e.g., C390A, C1546A, and C1547A) for two different PK methods was required due to low supply. Each of the reagents was produced from distinct hybridoma cell lines and aliquots of each of the hybridoma cell lines were subsequently stored under liquid nitrogen until needed for new production of the protein reagent. Assay performance in all of the new production batches of these critical reagents was suboptimal and varied from method to method. Initially, each individual assay reagent was investigated separately through various analytical techniques to determine the root cause of the suboptimal LBA performance. Little information was gained from several biophysical analyses with techniques such as SEC or SDS–PAGE (data not shown).

LC–MS analysis did reveal the presence of abnormal species in each of the poor performing reagent source materials, as shown for C390A in Figure 10.4(a) and (b). When compared to prior reagent lots used in the LBA methods under optimally performing conditions, the new production reagent lots were the only samples that appeared to contain either more than one or completely different "antibody-resembling" protein product. The abnormal subspecies in each of the reagent materials derived from the distinct hybridoma clones could be reduced with DTT into apparent heavy and light chains based on molecular weight and could also undergo deglycosylation in the heavy chain following digestion with the endoglycosidase, PNGase F, both characteristic of IgG antibodies. Furthermore, traditional separation of chromatographic peaks observed during the reversed-phase LC gradient and molecular weight determination via SDS–PAGE was also representative of an IgG molecule.

Although the suboptimal assay performance was correlated well with the presence of the mixed species in the reagent source lots, it remained difficult to determine if the subspecies were directly interacting with the specific binding of the molecular complex or indirectly interfering through an alternative mechanism. However, because the effect was observed across two independent biologic programs that were being supported by separately validated LBA methods gives credence to the theory of an indirect mechanism influencing the assay system. While it was not determined where the subspecies originated for the three individual critical reagents, viewing the data collectively exposed an obvious and surprising manifestation. The subspecies prevalent in each of the three source reagents generated from the different hybridoma clones appeared to consist of the same molecular structure (Figure 10.4c). As observed in Figure 10.4(c), the deconvoluted MS profiles of the apparent heavy and light chains of the subspecies in the three samples could nearly overlay. Following this important finding, the results of the analysis were reported back to the cell line manufacturer in hopes of determining the likely cause and possible solution to the problem. After diligent investigation, it was verified that near the time of the new reagent generation there had been a change from the media components historically used during the cell line production process. Since the standard media components were returned to use during reagent generation, no issue has been noted that would resemble the pattern observed in the example above. However, continually monitoring the purity and quality of new reagents after production as well as maintaining open communications with the cell line manufacturer to keep them apprised of any potential situations helps facilitate the efforts if a challenge were to arise.

10.2.3 Case Study #3: Investigation of the Loss of LBA Specificity During Clinical Development

10.2.3.1 Prestudy Investigation

During a PK method development for a therapeutic mAb product in clinical development, the method was developed in a typical sandwich format using a streptavidin-coated solid-phase support and electrochemiluminescent detection system. The critical reagents used in the sandwich format were two distinct hybridoma-derived noncompeting anti-Id mAbs, which served as the biotinylated capture and ruthenium-labeled detection reagents. During the course of assay development, depletion of the supply of the most recently qualified mAb detection reagent source protein (C437A) (i.e., unlabeled antibody) necessitated the generation of a new production batch of the same reagent material for the assay from the same hybridoma cell line bank used to produce the prior source lots. Following production of the new C437A mAb, the reagent was affinity purified and evaluated for molecular characterization using SEC as well as reducing and nonreducing SDS–PAGE. The traditional biochemical and biophysical characterization techniques revealed no distinguishable differences between the new production source material and the previously qualified reagent material. The new C437A source material was subsequently ruthenium-labeled and tested at the validated working concentration for the assay method at 1.25 μg/mL. However, while attempting to qualify the new detection reagent in the assay, it failed to meet the standard LBA acceptance criteria. Following routine LBA

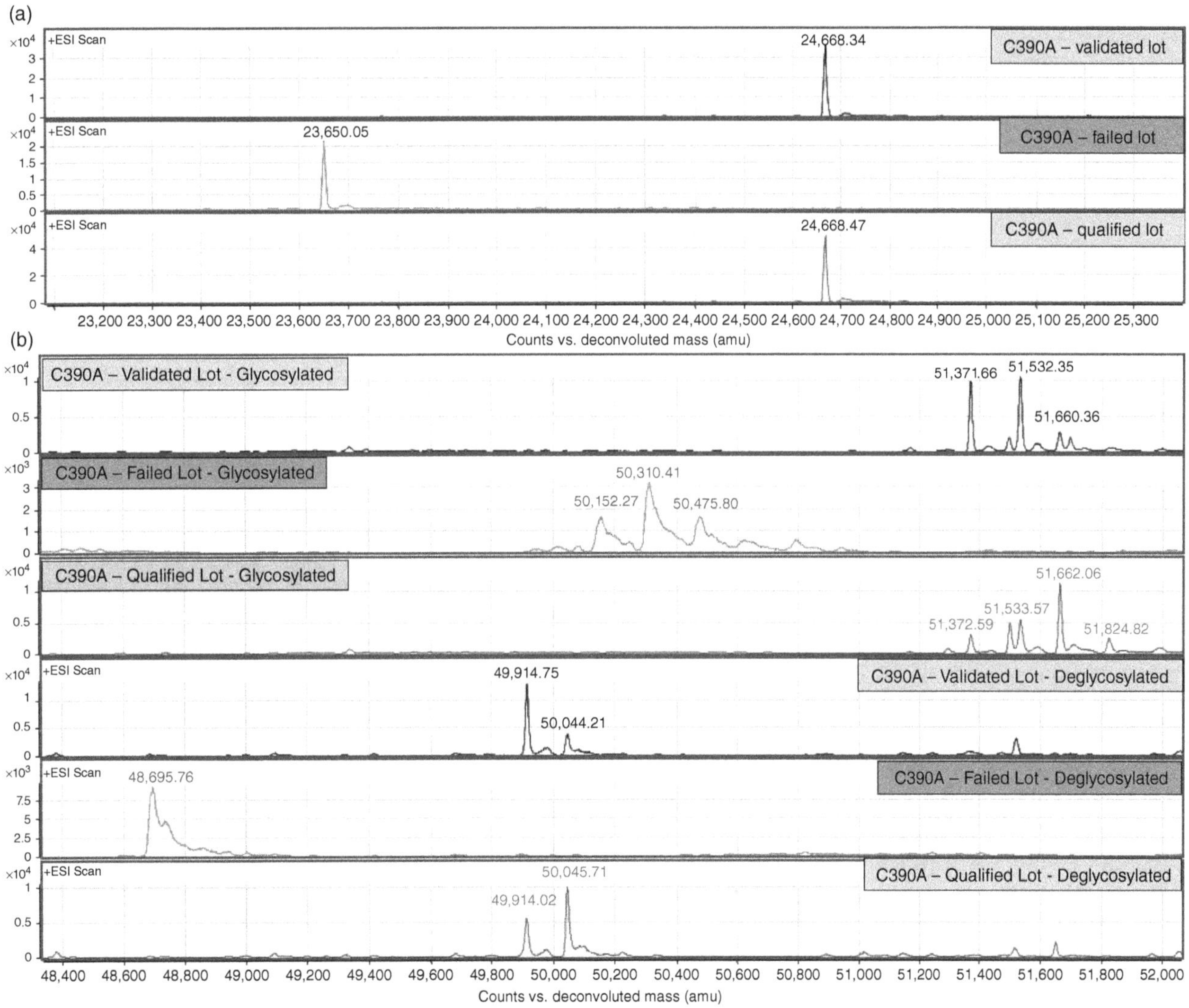

Figure 10.4 LC–MS analysis on reagent samples was performed by reversed-phase liquid chromatography on an Intrada WP-RP column (150 × 2 mm, 3 μm) and detected via time-of-flight mass spectrometry (ToF-MS) in positive ion mode equipped with electrospray ionization (ESI) source. The ToF-MS data was acquired over an acquisition range from 1000 to 3200 *m/z* and the mass spectrum data for each individual reagent was subsequently deconvoluted (maximum entropy) with Agilent MassHunter Qualitative Analysis software for intact molecular comparison. LC–MS analysis comparison between light (a) and heavy (b) chains of qualified and poor performing reagent lots of C390A source protein are shown for reference. Mass shift in heavy chain was not due to glycoform variation. The aligned panels of impurity profiles following LC–MS analysis of poor performing critical reagent source lots are shown separated by light chain (4c-top panel) and heavy chain (4c-bottom panel). Nearly identical impurity profiles were observed for the independently cloned products.

(Continued)

Figure 10.4 Continued

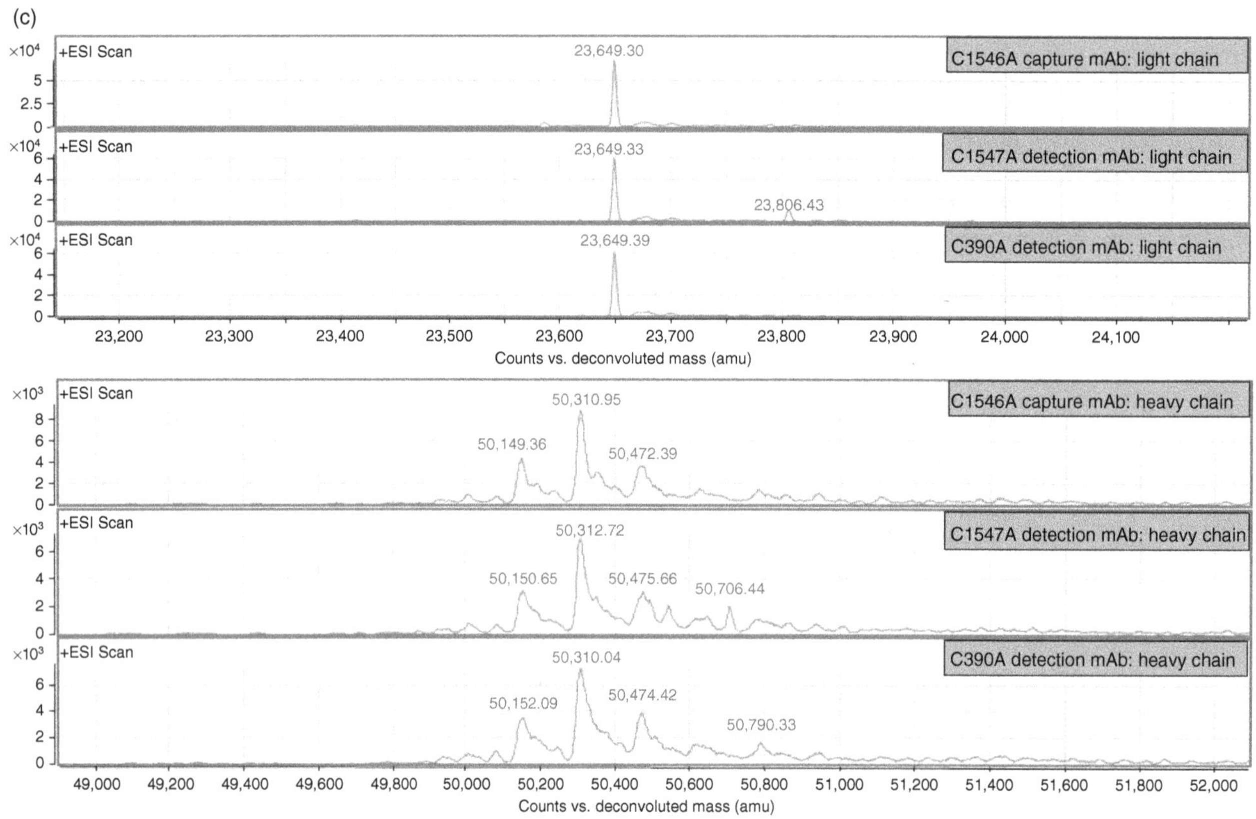

troubleshooting and the inability to move clinical assay development forward without a qualified lot of detection reagent, production of another independent preparation of the C437A mAb source protein reagent was initiated and completed. After similar reagent purification and labeling steps, multiple attempts to qualify the C437A mAb detection reagent in the assay finally resulted in successful qualification in pooled healthy human serum matrix, but only at double the original working concentration (e.g., 1.25 vs 2.5 μg/mL). The qualified reagent was, therefore, considered acceptable for use but noted as a nonideal performing reagent.

With no distinguishable biophysical alteration confirmed from previous analytical results and a clear impact on LBA performance, an investigation into the different source lots of the C437A mAb protein was performed via LC–MS. Structural characterization of the heavy chains from the different lots of C437A source proteins showed the presence of a secondary species with a distinct mass increase of approximately 1 kDa between the qualified and "poorly performing" reagents (Figure 10.5b) that was not clearly distinguishable by conventional HPLC, SEC, and SDS–PAGE methods (Figure 10.5a and c). Within the same lots of reagents, no molecular weight variation in the light chain was observed in the deconvoluted mass spectra (data not shown). In order to ensure the observed increase in mass of the heavy chain was not due to N-linked glycosylation variants, the source proteins were incubated with the endoglycosidase, PNGase F, and reanalyzed. Following the PNGase F digestion, the shifted mass profile in the heavy chain was still evident confirming that the increase in mass could not be attributed to abnormal N-linked glycosylation species (Figure 10.5b). LC–MS evaluation provided a detailed and essential molecular characterization technique to elucidate the problematic components of the assay and expedite the troubleshooting effort. Interestingly, the estimated quantity of the atypical subpopulation present following production of the second resupply of the source lot was much smaller than the amount observed in the first resupply of the failed lot. Several possibilities exist for this observed assay variation including the production of a competing mAb reagent species generated by an incomplete monoclonal hybridoma system or a subspecies generated that lacks adequate specificity for the therapeutic mAb. In both cases, molar ratio titration of the therapeutic to reagent mAb may minimize the impact from the unexpected subspecies. This was evident from the ability of the second resupplied source lot to be qualified, but only through adjusting the reagent concentration to twice the original validated concentration.

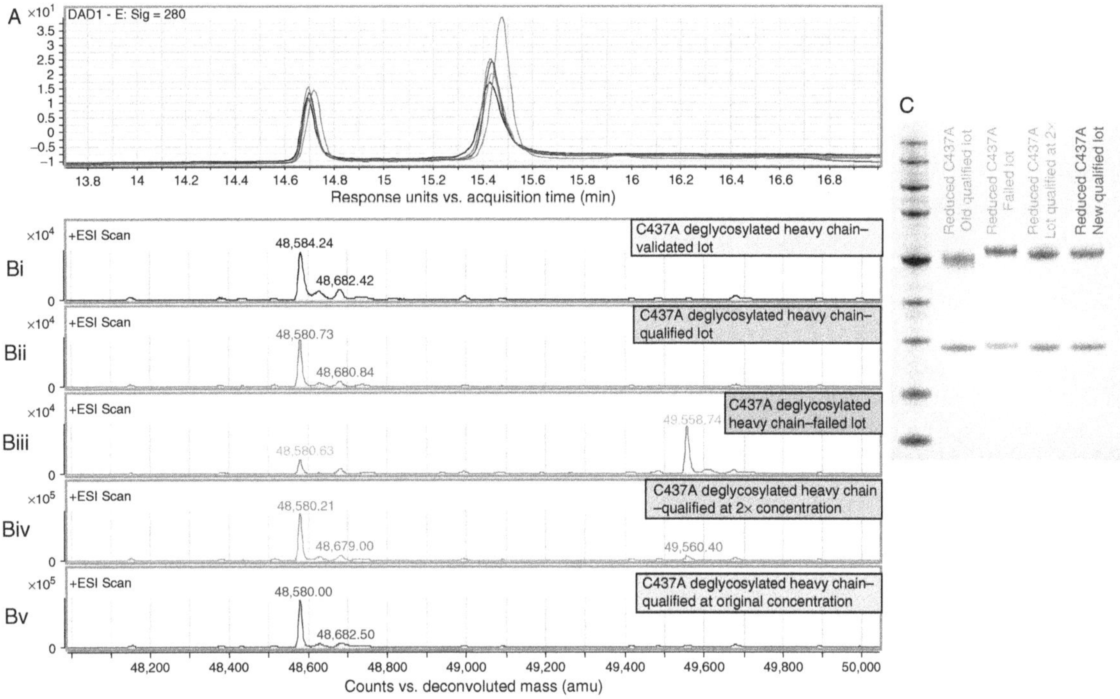

Figure 10.5 LC-MS and SDS-PAGE structural analysis of LBA detection reagent C437A for Case Study #3. The mAb proteins were separated by reversed-phase liquid chromatography on an Intrada WP-RP column (150x2mm, 3μm) and detected via time-of-flight mass spectrometry (TO F-MS) equipped with electrospray ionization (ESI) source. The TOF-MS data was acquired in positive ion mode with an *m/z* acquisition range from 1000 – 3200 and the acquired spectral data was subsequently deconvoluted (maximum entropy) with Agilent MassHunter Qualitative Analysis software for molecular structure comparison.
The deconvoluted mass profiles show five lots of C437A mAb deglycosylated heavy chain (Bi-v). Abnormal reagent species with a significant mass shift is noted for the heavy chain of the "failed lot" (Biii). The same abnormal reagent species is also seen in smaller amounts in the lot qualified at 2-fold the original LBA concentration (Biv). The absorbance trace at 280nm (A) and SDS-PAGE analysis (C) of the reduced mAb show minor differences between the light and heavy chains of the different lots. (*Source*: Geist et al. 2013. Reproduced with permission of Future Science.)

10.2.3.2 In-Study Investigation

During the implementation of another validated PK method, bioanalysis supporting the pharmacokinetic assessment on a biologic was being conducted for two ongoing studies in the late stage of clinical development within a relatively short time frame. Similar to the previous example, a validated immunoassay method was being used that incorporated a typical sandwich format with an electrochemiluminescent detection system. Hybridoma-derived, noncompeting anti-Id mAbs served as the specific antitherapeutic critical reagents within the assay, biotinylated capture mAb (C1415A) and ruthenium-labeled detection mAb (C1414A), and the complex was bound to a streptavidin-coated solid-phase support. Prior to bioanalysis, a resupply of the critical reagent antibodies was required due to depletion of the current lots used during the LBA method validation. The new source lots of both capture and detection reagent mAbs were generated from the preexisting hybridoma cell banks, and affinity purified following standard practices. The reagents were subsequently labeled for the appropriate capture or detection function and qualified in the assay in pooled serum from healthy individuals. Preliminary analytical characterization determined that the new source mAb lots contained no appreciable aggregation (SEC) and were verified for reagent identity by molecular mass confirmation (SDS–PAGE).

However, during the course of bioanalysis for the ongoing clinical studies, an unexpectedly high number of placebo-treated and predose samples from both studies generated quantifiable therapeutic concentrations (31% and 34%, respectively), classified as "false positives" [greater than the lower limit of quantification (LLOQ)]. In-study bioanalysis was stopped and a coordinated, full-scale troubleshooting effort was initiated to determine the root cause of the "false positive" results in the validated LBA. LC–MS analysis was employed to obtain a more detailed comparison of molecular structure and protein identity of the newly produced mAb unlabeled source lots against the previously validated source lots. The results from the LC–MS analysis for both capture and detection reagent mAbs showed clear disparities between the light and heavy chain mass profiles of the original validated lot and the new resupplied source lot.

The new C1415A capture mAb showed a distinct molecular mass shift in both light chain (+65 Da) and heavy chain (approx. −800 Da) with a negligible amount of the original reagent mass species (Figure 10.6a and b). The new detection mAb (C1414A) also showed a shift in the light chain (−156 Da) and heavy chain mass (−390 Da) but did contain a more considerable amount of the original mAb species from the validated reagent lot (Figure 10.7a and b). In both reagents' cases, SDS–PAGE did not detect any dramatic mass differences (Figures 10.6c, d and 10.7c, d).

In parallel with the identification of the abnormal protein species in the new source lots of the mAb reagents, expedited production of replacement batches from the existing hybridoma cell banks with fresh cell culture materials was initiated. LC–MS structural characterization of the replacement source lots of both reagents displayed molecular profiles matching that of the original validated reagents. In order to ensure the heterogeneity observed between the source proteins was not due to N-linked glycosylation of the Fc domain, the mAb reagents were reanalyzed following enzymatic digestion with PNGase F. The glycosylated and deglycosylated heavy chain profiles showed nearly identical mass variation between the new and old source lots. The sample concentrations from the "false positive" predose and placebo controls following reanalysis with the reagents from the replacement lots decreased by an average of 7.2-fold over the two studies (Figure 10.8). Moreover, the total number of samples from predose and placebo-treated subjects that tested as "false positive" in the LBA decreased from 153 to 4 across both studies.

The use of LC–MS to efficiently characterize the molecular variations between the prior and new lots of

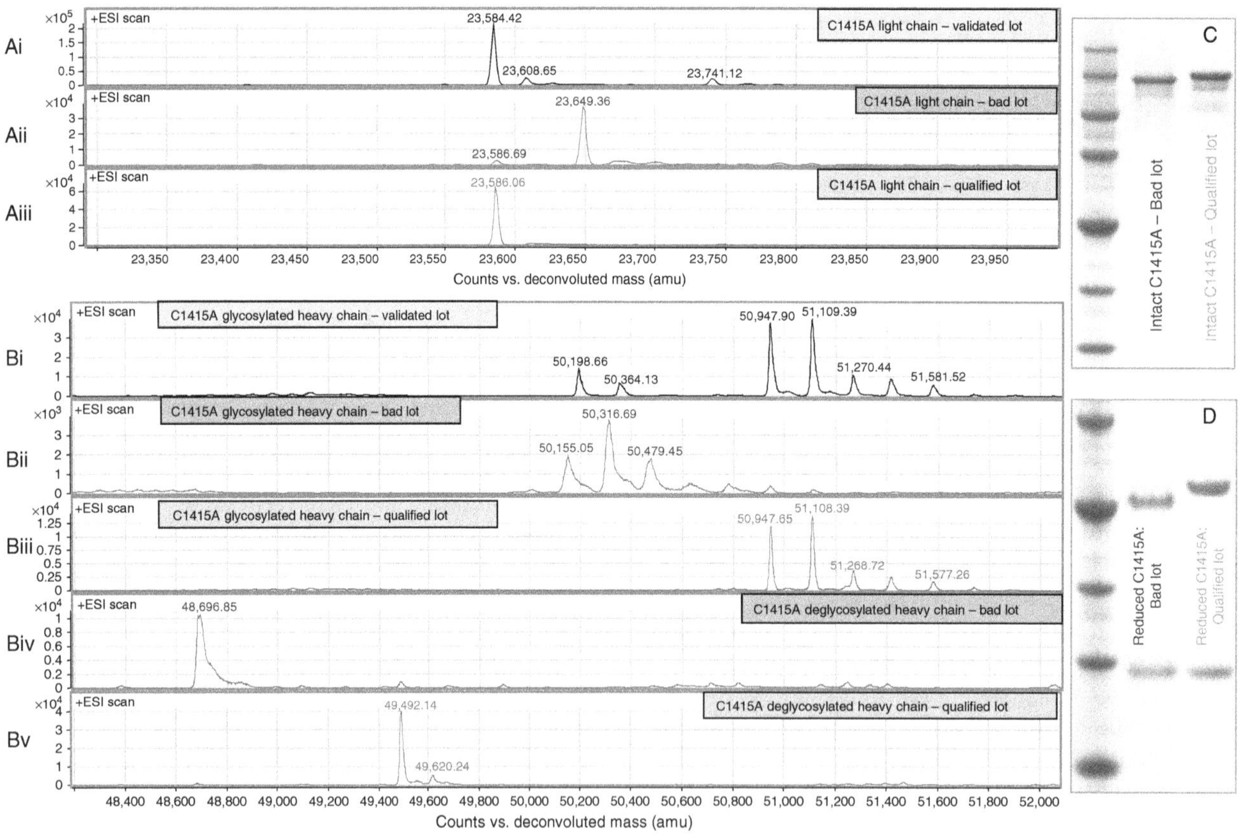

Figure 10.6 LC-MS and SDS-PAGE structural analysis of LBA capture reagent, C1415A. The reduced samples were separated by reversed-phase liquid chromatography on an Intrada WP-RP column (150x2mm, 3μm) and detected via time-of-flight mass spectrometry (TOF-MS) equipped with electrospray ionization (ESI) source. The acquired mass spectrum of the individual heavy and light chain collected for each sample was subsequently deconvoluted into the average intact molecular mass using Agilent MassHunter Qualitative Analysis software. The deconvoluted mass profiles show three lots of C1415A mAb separated by light chain (Ai-iii) and heavy chain glycosylated and deglycosylated (Bi-v). Noticeable mass shifts are seen for both light (Aii) and heavy chains (Bii, Biv) of the "Bad lot". The observed mass difference is unrelated to the glycosylation profile. SDS-PAGE showed small differences in migration for intact (C) and reduced (D) mAb profiles, but confirming the variation in the molecular species of the reagent remained difficult. (*Source:* Geist et al. 2013. Reproduced with permission of Future Science.)

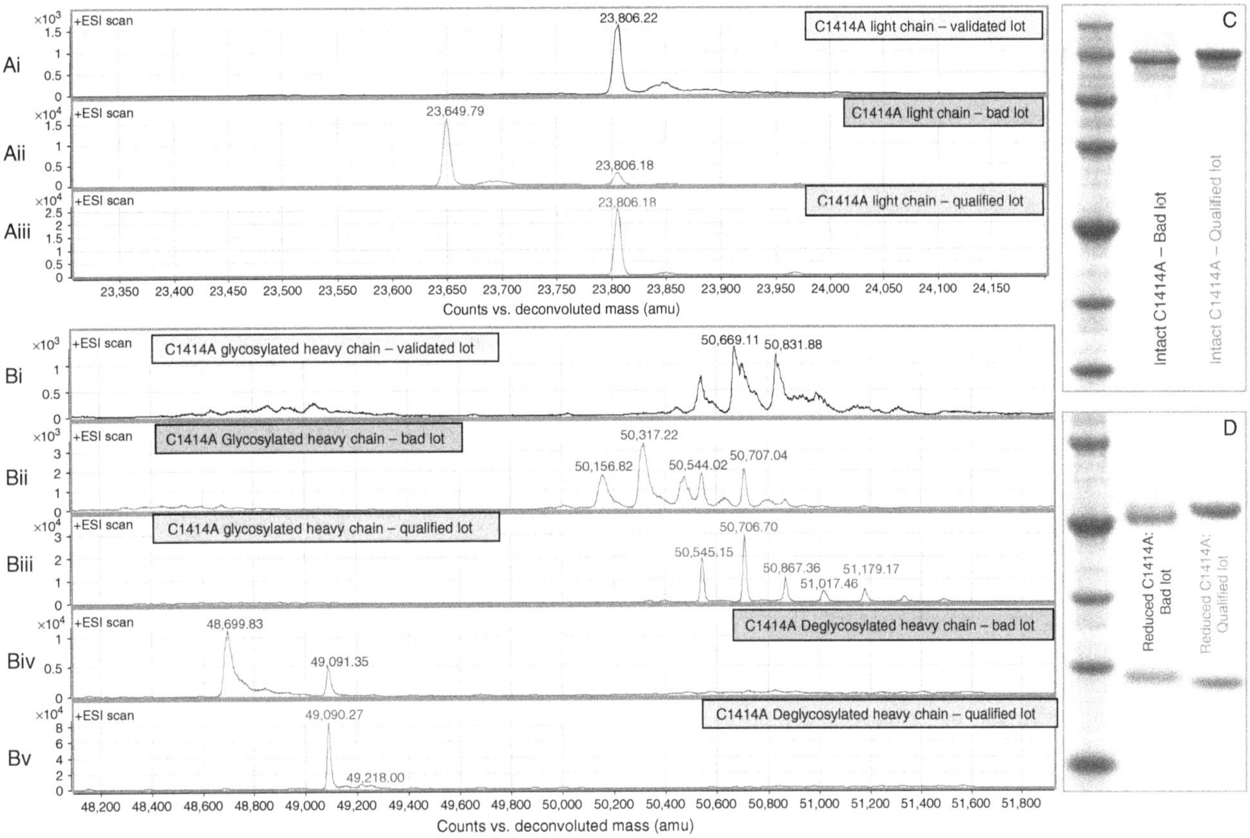

Figure 10.7 LC-MS and SDS-PAGE structural analysis of LBA detection reagent C1414A. The reduced samples were separated by reversed-phase liquid chromatography on an Intrada WP-RP column (150x2mm, 3μm) and detected via time-of-flight mass spectrometry (TOF-MS) equipped with electrospray ionization (ESI) source. The acquired mass spectrum of the individual heavy and light chain collected for each sample was subsequently deconvoluted into the average intact molecular mass using Agilent MassHunter Qualitative Analysis software. The deconvoluted mass profiles show three lots of C1414A mAb separated by light chain (Ai-iii) and heavy chain glycosylated and deglycosylated (Bi-v). Noticeable mass shifts, unrelated to N-glycosylation, were noted for both light (Aii) and heavy chains (Bii, Biv) of the "Bad lot". A small quantity of the primary mAb reagent is still present in the "Bad lot". SDS-PAGE showed small differences in migration for intact (C) and reduced (D) mAb profiles, but confirming the variation in the molecular species of the reagent remained difficult. (*Source:* Geist et al. 2013. Reproduced with permission of Future Science.)

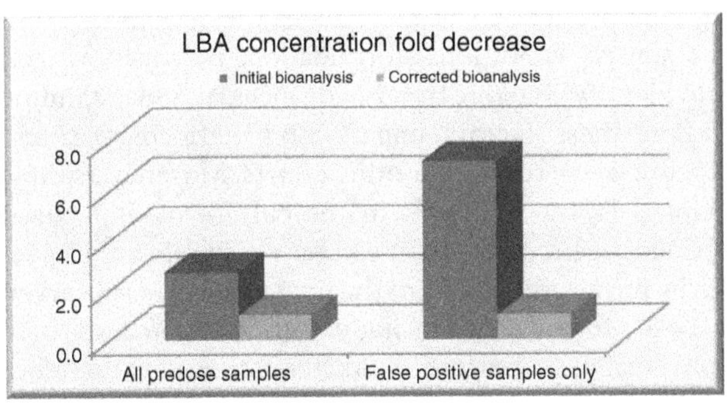

Figure 10.8 Summary of serum drug concentration decrease following reanalysis of predose and placebo "false positives" from Case Study #3. The average concentration from all "false-positive" samples decreased 7.2-fold while the total number of "false positive" samples was reduced from 153 to 4. (Source: Geist et al. 2013. Reproduced with permission of Future Science.)

these critical reagents enabled timely and successful LBA troubleshooting. Whether the mass variation between the reagents was due to the incorporation of an unknown PTM(s) or was attributed to an altered amino acid sequence was not determined. However, because LBA performance has been the principal endpoint to assessing critical reagent function, determination of the root cause was likely extraneous. Furthermore, the therapeutic supported in this case study was being investigated for a rheumatoid arthritis (RA) indication and the ongoing clinical studies were being conducted in diseased patients. There is a well-documented association between RA disease and the presence of elevated rheumatoid factor (endogenous autoantibody response against IgG), which has been shown to produce interference in LBA assay systems (DeForge et al. 2010, Tatarewicz et al. 2010). Therefore, the variation in the source lots of the critical reagents may have only been one factor in the anomalous assay results. In the case of suspected instability of the hybridoma cell line, sponsors may consider to clone the critical reagent sequences and express them in a stably transfected mammalian cell line, such as 293 or CHO cells.

10.2.4 Case Study #4: Monitoring the Incorporation Ratio of Conjugated Critical Reagent Used in LBAs

Chemical moieties are commonly attached to critical LBA protein reagents in order to serve as a molecular handle or as a means to produce absorbance, fluorescence, or luminescence-based assay signals. Conjugated reagents are widely used in LBA methods. After these chemical moieties are conjugated to a protein, it may be important to understand the molar incorporation ratios of these chemicals to the protein reagent. This information is frequently useful to understand lot-to-lot differences in conjugated reagent performance, which can be eliminated by adjustments to protein conjugation protocols. While spectrophotometric methods may be used to determine the overall incorporation ratio of chemical labels to a reagent antibody, LC–MS is an alternative approach that can also provide information on the distribution of conjugated species for a given protein lot.

In this case study, two lots of an electrochemiluminescence detection reagent (i.e., ruthenium-chelate conjugated to an mAb therapeutic) were produced and evaluated in the LBA to assure robustness of an IMG method. When two different lots of the conjugated reagent were tested in the LBA, the second lot led to a much higher assay background signal compared to lot 1, which indicated that the assay was sensitive to lot differences of this reagent. The different lots of the conjugated materials were subjected to LC–MS analysis under reduced conditions in the presence of DTT. Even though the overall conjugation level was similar, MS revealed different proportions of incorporation of the Ru label. For example, the observed overall incorporation ratios of ruthenium chelate were approximately the same at 0.63–0.64 two lots (Figure 10.9). Despite the apparent comparable overall conjugation ratios of the two lots, the proportion of bilabeled protein was greater for lot 2 than for lot 1 (14% for lot 2 vs 6% for lot 1). In this particular instance, the assay diluent was adjusted to eliminate the extreme sensitivity of this assay to the differences in performance of these two reagents. However, an adjustment in the conjugation protocol was a potentially viable approach to assure little reagent conjugate lot-to-lot variability.

10.3 Discussion

Reliable bioanalytical results from robust LBA methods are essential for a successful biologic program, from nonclinical and clinical development through postmarket approval. It is vital to ensure the LBA methods are properly supported and maintained according to established regulatory guidance and industry standards throughout the life cycle of the biologic development program. At the forefront of assuring the optimal performance of the established LBA methods is the consistent production and performance of the critical reagents (source proteins and their conjugates). In this chapter, we have provided case studies demonstrating the significance of applying LC–MS to perform physicochemical and biophysical assessments on critical reagents from initial generation through late-stage bioanalytical use in clinical development.

10.3.1 Keys to Reagent Management

Since the quality attributes of critical reagents can have a direct impact on the performance of LBAs, effective life cycle management of these reagents should include biophysical, biochemical, and immunological characterization prior to and during method development/implementation. Well-characterized quality attributes provide a point of reference for troubleshooting activities during and after the implementation of an LBA method, or even serve a "gate-keeping" function prior to initiation of any method development activities, as demonstrated in Case Study #1. Continued investigations into factors that may impact the structural integrity and stability of critical reagents over time are expected to facilitate consistent reagent production and maintenance from the onset of the LBA development to its implementation. The maintenance of historical results from critical reagent analytical characterization following new production or changes to the system (i.e., cell line source, conditions, formulation, purification) can help identify systemic trends, advance the understanding of how these factors impact LBA performance, and expedite assay troubleshooting efforts. Moreover, simple changes to the materials in the culture medium used during reagent production can have a significant impact on the purity and quality of the final product, as observed in Case Study #2. Capturing empirical "time-zero" data from analytical reagent characterization can therefore provide a reference control to quickly compare old and new reagent profiles. Analytical characterization can also determine the average incorporation ratio of the label:protein in the final conjugated reagent, assess conjugation efficiency, and compare binding affinity of unlabeled and previously conjugated proteins to provide

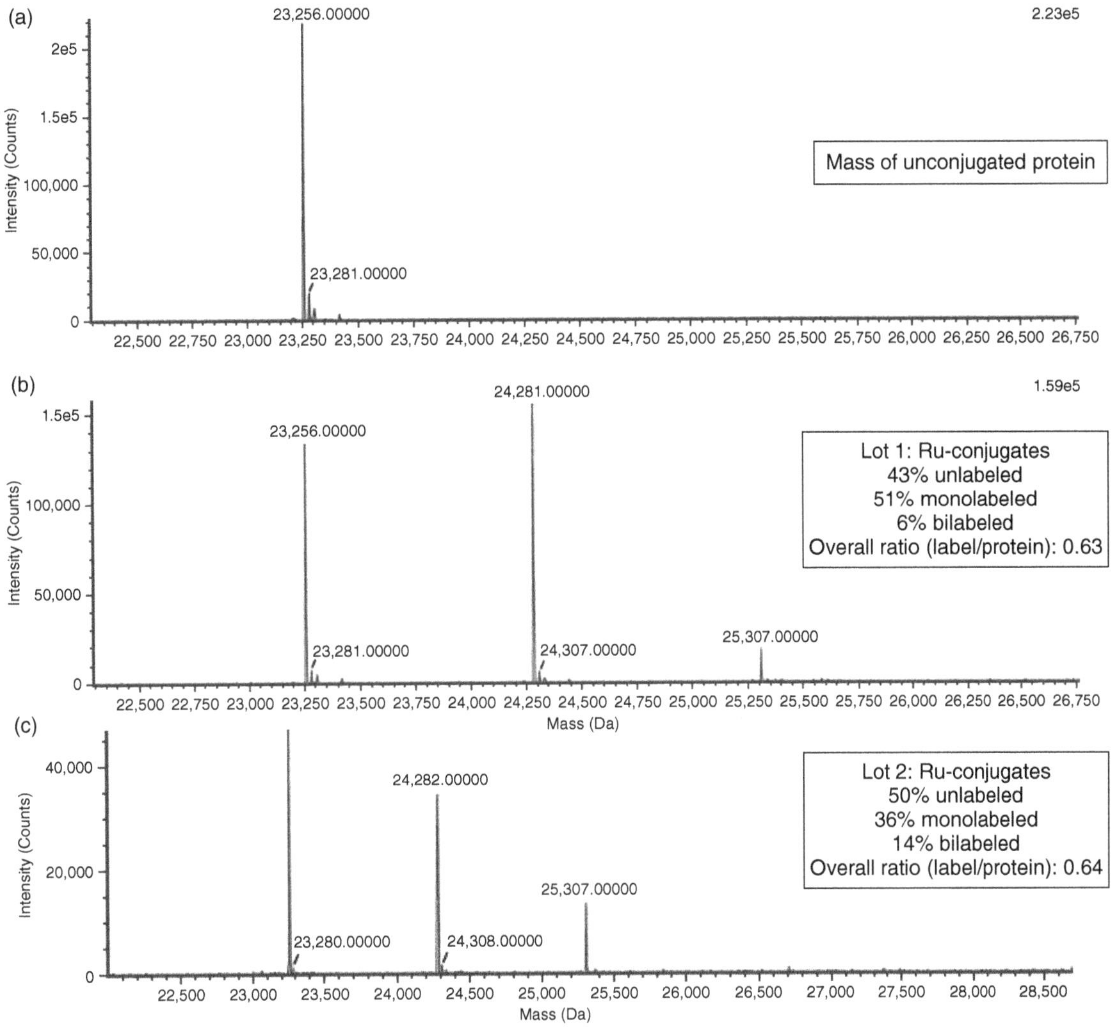

Figure 10.9 Comparison of LC–MS profiles of light chains from two lots of a reagent antibody that was conjugated with Ru-chelate. Lots 1 and 2 of the Ru-conjugate reagent antibody were treated with 20 mM dithiothreitol (DTT) to reduce interchain disulfide bonds, allowing for separate and distinct analysis of heavy and light chain species. The mAb proteins were separated by reversed-phase liquid chromatography and detected via a Waters Xevo G2 Q-ToF equipped with electrospray ionization (ESI) source. The MS data was acquired in positive ion mode with an *m/z* acquisition range from 500 to 2000 and the acquired spectral data was subsequently deconvoluted (maximum entropy) with the Waters Unify Scientific Information System software. (a) Mass of the unlabeled light chain, (b) and (c) proportions of Ru-conjugates for lots 1 and 2, respectively.

assurance that the conjugation process did not impact the specificity of the reagents (Case Study #4). Ideally, this type of analytical monitoring will help maintain the consistency of LBA performance, while reducing the time spent on troubleshooting efforts during LBA development, validation, and bioanalysis.

Commercial reagents are widely used in PD and IMG methods, and more and more biopharmaceutical companies are outsourcing reagent generation process to CROs. It is challenging but even more critical to characterize these commercial or CRO-generated reagents used in LBA method development because most vendors/or CROs will not disclose their detailed production and quality control processes to customers due to proprietary protection. LC–MS represents a powerful technique to monitor the lot-to-lot homogeneity from commercial reagents, which may suffer from variable affinity characteristics or target binding interference. It is important to establish the correlation between observed quality attributes and assay performance of a given commercial reagent in order to effectively manage this external process.

10.3.2 Importance of LC–MS Characterization

During production, purification, or storage of source proteins, conventional protein modifications (oxidation, deamidation, glycosylation, etc.), changes in primary sequences, and process-related impurities can occur. These changes have shown the capacity to affect biophysical properties of protein reagents. Although exhaustive characterization of these protein modifications may not be warranted nor feasible due to the likely minimal impact on the assay, identifying and monitoring these changes from a global perspective may provide insights into the consistency of reagent production and storage stability. Shifts in LC–MS reagent profiles can identify sources of system instability within the reagent management process and help address the factors responsible while also providing valuable information for the selection of optimal reagent production and storage conditions. For example, in Case Study #2, LC–MS characterization was able to identify the presence of production-related IgG-like impurities among different reagent clones, which could not be readily detected by SEC and SDS–PAGE. Although the majority of chemical modifications on reagents have little or no LBA impact, the effect is largely dependent on the location of the reagent structure that is modified. For example, reagent mAb oxidation of susceptible residues (e.g., Met, Trp) located in the light chain may impact epitope binding to a greater degree when compared to oxidation on the Fc domain. When the presence of particular impurities or modifications is noted, further evaluation of the critical reagent through SPR analysis may be applied to confirm that no effect on target binding is observed.

The central goal for physicochemical characterization of critical reagents with LC–MS is to provide a molecular "fingerprint" for each reagent that can be maintained throughout the LBA method life cycle to help monitor biophysical variation at susceptible timepoints in the reagent production process (new lot production, storage, label conjugation, etc.) that might cause suboptimal LBA performance. Mass spectrometric analysis of critical reagents provides the greatest advantage when performed at appropriate timepoints during the course of the reagent life cycle but may depend on the preference of the individual organization. Occasions where molecular characterization is appropriate would include original reagent generation, conjugation, resupply, requalification for added shelf life (expiration date extension), and LBA troubleshooting. With robust MS instruments and sophisticated data analysis software, intact protein or subunit characterization can be performed in a relatively high-throughput manner with limited sample preparation or processing.

Table 10.2 Example of analytical toolkit to evaluate protein reagent quality characteristics for LBAs

Protein reagent quality characteristics (common biochemical/biophysical techniques)	Potential advantages of standard LC–MS methods
Aggregation level (SEC, AUC, SLS/DLS)	NA
Binding activity (Western blot)	NA
Binding kinetics (SPR, ITC, BLI)	NA
Concentration (Spectrophotometry, Western blot, LC–MS)	Accurate concentration may be determined from complex mixture of proteins without the need for immunoreactive reagents (if appropriate reference is available)
Conjugate incorporation ratio (Spectrophotometry, LC–MS)	Conjugates without chromophores can be quantified
	Incorporation efficiency and distribution of conjugate ratios on a protein can be determined
Cross-reactivity (Western blot, SPR, LC–MS)	Unknown interfering species can be identified directly
Formulation buffer (SDS–PAGE, SEC, DLS, LC–MS)	Covalent modifications (PTMs)/degradation products of proteins induced by formulation conditions are more readily detected
Isoelectric point (IEF, cIEF)	NA
Molecular mass/identity (SEC, SDS–PAGE, LC–MS)	Distribution of covalent modifications can be determined
	High resolution "finger printing" of protein reagent profile can be established and compared over batches/lots
Potency (Activity assay)	NA
Purity (SDS–PAGE, SEC, LC–MS)	Non-protein contaminants without chromophores may be elucidated more readily
	Contaminants with overlapping molecular mass and its proportion to that of the critical reagent can be identified

AUC, analytical ultracentrifugation; BLI, biolayer interferometry; SLS/DLS, static/dynamic light scattering; IEF (cIEF), isoelectric focusing/capillary isoelectric focusing; ITC, isothermal titration calorimetry; LC–MS, liquid chromatography–mass spectrometry; SEC, size exclusion chromatography; SPR, surface plasmon resonance; SDS–PAGE: sodium dodecyl sulfate–polyacrylamide gel electrophoresis.

10.3.3 The Analytical Toolbox and a "Fit-for-Purpose" Approach for Reagent Management

While LC–MS may have advantages of detecting mass difference due to microheterogeneity and impurities in protein reagents, it may not be adequate to detect other abnormal behavior of protein reagents in solution, such as aggregation, which can potentially have significant impacts on LBA performance. Monitoring critical reagent physical stability can be accomplished with standard biochemical, biophysical, and immunological methods (e.g., SDS–PAGE and affinity measurement) with applications toward conformational variation, molecular size, and size distribution analysis as referenced in Table 10.2. For example, analysis of aggregation states of critical reagents in solution can be accomplished by various techniques such as SEC, analytical ultracentrifugation (AUC), and static/dynamic light scattering (SLS/DLS), which may be difficult to detect by standard LC–MS methods. The existence of critical reagent aggregates can have unpredictable effects on the binding and function of the aggregated reagent and may increase the susceptibility to chemical modifications. Introducing these nonnative conformations or aggregated species of critical reagents into the assay system can often have aberrant or deleterious effects on LBA performance (Table 10.2). Nevertheless, conventional methods are not always sensitive enough to demonstrate a direct correlation between structural integrity of a critical reagent and optimal performance in LBAs as shown by the case studies. Due to the recent advancement and operational efficiency of the LC–MS platform, LC–MS augments conventional methods in the analytical toolbox to characterize intact protein structure.

Due to the complexity of generating critical protein reagents and conjugation processes, batch-to-batch heterogeneity cannot be avoided due to several factors, for example, potential instability of hybridoma cell clones, change in production conditions, system error in handling expression constructs, and inconsistency of label incorporation during conjugation. While comprehensive analytical characterization and highly controlled manufacturing processes are not practical for generation of critical reagents due to their intended use, cost, and resource limitations, employing LC–MS to monitor lot-to-lot variability, even when the primary sequence is unknown for a reagent produced in a hybridoma cell line, is valuable and time saving. The aforementioned case studies clearly illustrate that appropriate LC–MS characterization methods of critical reagents are important to maintain robust LBA methods efficiently in support of biologic programs. In addition, LC–MS can provide critical comparative information between new and previously qualified lots of the reagent, and, in the case of conjugated reagents, estimate label incorporation efficiency and reproducibility. These characterization data can be shared with the manufacturer of critical reagents to aid in monitoring and implementing optimal production systems. Taking steps to elucidate and minimize changes in physicochemical attributes or impurities during critical reagent production can help ensure consistency across batch to batch, which leads to reliable LBA performance. Furthermore, analysis of protein modifications with LC–MS may also provide insights into critical reagent stability (freeze/thaw, temperature, storage, etc.) and help define "best practices" for the implementation of a process for maintaining and storing critical reagents. The proactive, front loading characterization of critical reagents is a relatively small investment that will, in turn, reduce inefficient LBA troubleshooting activities.

Although demonstrating molecular comparability between critical reagents is beneficial, the effort should encompass what is deemed reasonable and suitable for the required outcome. Applicable questions to maintain such a "fit-for-purpose" approach might include the following:

1) Does a significant change exist?
2) Is this change considered "vital" → impact on LBA performance or reagent shelf life?
3) Does this change represent a trend or pattern?

For example, a change in the N-linked carbohydrate profile of a reagent mAb Fc domain may not require full characterization of the oligosaccharide; on the other hand, production of high levels of aglycosylated species or identification of an additional glycosylation profile located in the Fab may require further evaluation.

A thorough characterization of critical reagents can also provide the foundation of documenting the "chain of custody" of these reagents during its use in supporting regulated bioanalysis. Although currently there are limited regulatory requirements for controlling the generation and characterization process of critical reagents used in GLP studies, any critical reagents used to support GLP or clinical studies are under regulatory scrutiny (US Food and Drug Administration 2001, European Medicines Agency 2012). Documentation of analytical characterization, storage conditions, and use of these reagents during the drug development process is an important part of maintaining the integrity of the regulated studies. Records of analysis describing adequate biophysical and biochemical characterization should be issued by end users in collaboration with manufacturers (O'Hara and Theobald 2013, O'Hara et al. 2012). These documents become critical during the investigation of given bioanalytical methods used in regulated bioanalysis to demonstrate the high-quality control of the process.

Through our own internal experience, we have gained an understanding of the importance of monitoring critical reagent profiles and the impact on assay performance. Although the functional response of the LBA remains the principal factor in determining the continued use of the critical reagent, we believe that consideration should also be given to the analytical characteristics. The monitoring and management of the production, storage, and qualification of critical reagents, through implementation of comprehensive functional and biophysical/biochemical characterization at appropriate points in the reagent production process, can provide valuable insights into the quality of these reagents. Such analytical characterization of critical reagents during LBA troubleshooting can often expedite the effort and aid in long-term solutions to the LBA problems at hand, notwithstanding potential savings in resources and minimization of time loss. Lastly, the greatest benefit attributed to analytical characterization of LBA critical reagents, particularly in a regulated environment, may be centered on the confidence in the reported bioanalytical results over the drug program's life cycle. With the growing industry initiative dedicated to effective project management, the efficient use of time and effort, and number of resources needed for the characterization of critical reagents, which may include LC–MS, becomes fully justified when it contributes to higher quality bioanalytical data and the successful delivery of biotherapeutic program milestones.

Acknowledgment

We would like to acknowledge Gopi Shankar, David Uhlinger, and Sanjeev Bhardwaj for critical review of the chapter, and Yuxin Dong for technical assistance.

References

Acchione M, Kwon H, Jochheim CM, et al. (2012). "Impact of linker and conjugation chemistry on antigen binding, Fc receptor binding and thermal stability of model antibody-drug conjugates." *mAbs* **4**, 362–372.

Banks DD, Hambly DM, Scavezze JL, et al. (2009). "The effect of sucrose hydrolysis on the stability of protein therapeutics during accelerated formulation studies." *J Pharm Sci* **98**, 4501–4510.

Beck A, Bussat MC, Zorn N, et al. (2005). "Characterization by liquid chromatography combined with mass spectrometry of monoclonal anti-Igf-1 receptor antibodies produced in CHO and NS0 cells." *J Chromatogr B: Anal Technol Biomed Life Sci* **819**, 203–218.

Beck A, Wagner-Rousset E, Ayoub D, et al. (2013). "Characterization of therapeutic antibodies and related products." *Anal Chem* **85**, 715–736.

Bertolotti-Ciarlet A, Wang W, Lownes R, et al. (2009). "Impact of methionine oxidation on the binding of human IgG1 to Fc Rn and Fc gamma receptors." *Mol Immunol* **46**, 1878–1882.

Bondarenko PV, Second TP, Zabrouskov V, et al. (2009). "Mass measurement and top-down HPLC/MS analysis of intact monoclonal antibodies on a hybrid linear quadrupole ion trap-Orbitrap mass spectrometer." *J Am Soc Mass Spectrom* **20**, 1415–1424.

Chelius D, Jing K, Lueras A, et al. (2006). "Formation of pyroglutamic acid from N-terminal glutamic acid in immunoglobulin gamma antibodies." *Anal Chem* **78**, 2370–2376.

Chevreux G, Tilly N and Bihoreau N (2011). "Fast analysis of recombinant monoclonal antibodies using IdeS proteolytic digestion and electrospray mass spectrometry." *Anal Biochem* **415**, 212–214.

Chirino AJ and Mire-Sluis A (2004). "Characterizing biological products and assessing comparability following manufacturing changes." *Nat Biotechnol* **22**, 1383–1391.

Cordoba AJ, Shyong BJ, Breen D, et al. (2005). "Non-enzymatic hinge region fragmentation of antibodies in solution." *J Chromatogr B: Anal Technol Biomed Life Sci* **818**, 115–121.

Damen CW, Chen W, Chakraborty AB, et al. (2009). "Electrospray ionization quadrupole ion-mobility time-of-flight mass spectrometry as a tool to distinguish the lot-to-lot heterogeneity in N-glycosylation profile of the therapeutic monoclonal antibody trastuzumab." *J Am Soc Mass Spectrom* **20**, 2021–2033.

DeForge LE, Loyet KM, Delarosa D, et al. (2010). "Evaluation of heterophilic antibody blocking agents in reducing false positive interference in immunoassays for IL-17AA, IL-17FF, and IL-17AF." *J Immunol Meth* **362**, 70–81.

DeSilva B, Smith W, Weiner R, et al. (2003). "Recommendations for the bioanalytical method validation of ligand-binding assays to support pharmacokinetic assessments of macromolecules." *Pharm Res* **20**: 1885–1900.

European Medicines Agency (2012). "Guideline on bioanalytical method validation." http://www.ema.

europa.eu/docs/en_GB/document_library/Scientific_guideline/2011/08/WC500109686.pdf

Evaluate Pharma (2013). "Return to growth for pharmaceutical sector: surge in drug approvals, R&D productivity and investor confidence to drive Expansion Through 2018." Evaluate Ltd.

Gaza-Bulseco G, Faldu S, Hurkmans K, et al. (2008). "Effect of methionine oxidation of a recombinant monoclonal antibody on the binding affinity to protein A and protein G." *J Chromatogr B: Anal Technol Biomed Life Sci* **870**, 55–62.

Gaza-Bulseco G and Liu H (2008). "Fragmentation of a recombinant monoclonal antibody at various pH." *Pharm Res* **25**, 1881–1890.

Geist BJ, Egan AC, Yang TY, et al. (2013). "Characterization of critical reagents in ligand-binding assays: enabling robust bioanalytical methods and lifecycle management." *Bioanalysis* **5**, 227–244.

Goetze AM, Liu YD, Arroll T, et al. (2012). "Rates and impact of human antibody glycation *in vivo*." *Glycobiology* **22**, 221–234.

Hall MP, Gegg C, Walker K, et al. (2010). "Ligand-binding mass spectrometry to study biotransformation of fusion protein drugs and guide immunoassay development: strategic approach and application to peptibodies targeting the thrombopoietin receptor." *AAPS J* **12**, 576–585.

Hambly DM, Banks DD, Scavezze JL, et al. (2009). "Detection and quantitation of IgG 1 hinge aspartate isomerization: A rapid degradation in stressed stability studies." *Anal Chem* **81**, 7454–7459.

Houde D, Peng Y, Berkowitz SA, et al. (2010). "Post-translational modifications differentially affect IgG1 conformation and receptor binding." *Mol Cell Proteomics* **9**, 1716–1728.

Hutterer KM, Hong RW, Lull J, et al. (2013). "Monoclonal antibody disulfide reduction during manufacturing: untangling process effects from product effects." *mAbs* **5**, 608–613.

ICH (1997). "Validation of analytical procedures: text and methodology." International Conference on Harmonisation of Technical Requirements for Registration of Pharmaceuticals for Human Use, Q2(R1).

Kaltashov IA, Bobst CE, Abzalimov RR, et al. (2012). "Advances and challenges in analytical characterization of biotechnology products: mass spectrometry-based approaches to study properties and behavior of protein therapeutics." *Biotechnol Adv* **30**, 210–222.

King LE, Farley E, Imazato M, et al. (2014). "Ligand binding assay critical reagents and their stability: recommendations and best practices from the Global Bioanalysis Consortium Harmonization Team." *AAPS J* **16**, 504–515.

Kling J (2014). "Fresh from the biotech pipeline – 2013." *Nat Biotechnol* **32**: 121–124.

Kozlowski S and Swann P (2006). "Current and future issues in the manufacturing and development of monoclonal antibodies." *Adv Drug Deliv Rev* **58**, 707–722.

Lee JW and Salimi-Moosavi H (2012). "Bioanalysis of target biomarker and PK/PD relevancy during the development of biotherapeutics." *Bioanalysis* **4**, 2513–2523.

Li F, Fast D and Michael S (2011). "Absolute quantitation of protein therapeutics in biological matrices by enzymatic digestion and LC-MS." *Bioanalysis* **3**, 2459–2480.

Li H, Ortiz R, Tran L, et al. (2012). "General LC-MS/MS method approach to quantify therapeutic monoclonal antibodies using a common whole antibody internal standard with application to preclinical studies." *Anal Chem* **84**, 1267–1273.

Liu H and May K (2012). "Disulfide bond structures of IgG molecules: Structural variations, chemical modifications and possible impacts to stability and biological function." *mAbs* **4**, 17–23.

Lowenthal MS, Gasca-Aragon H, Schiel JE, et al. (2011). "A quantitative LC-MS/MS method for comparative analysis of capture-antibody affinity toward protein antigens." *J Chromatogr B: Anal Technol Biomed Life Sci* **879**, 2726–2732.

Luo Q, Joubert MK, Stevenson R, et al. (2011). Chemical modifications in therapeutic protein aggregates generated under different stress conditions. *J Biol Chem* **286**, 25134–25144.

Manning MC, Chou DK, Murphy BM, et al. (2010). "Stability of protein pharmaceuticals: an update." *Pharm Res* **27**, 544–575.

Nicholson R, Lowes S, Caturla MC, et al. (2012). "Conference report: 6th GCC focus on LBA: critical reagents, positive controls and reference standards; specificity for endogenous compounds; biomarkers; biosimilars." *Bioanalysis* **4**, 2335–2342.

O'Hara DM and Theobald V (2013). "Life cycle management of critical ligand-binding reagents." *Bioanalysis* **5**, 2679–2696.

O'Hara DM, Theobald V, Egan AC, et al. (2012). "Ligand binding assays in the 21st century laboratory: recommendations for characterization and supply of critical reagents." *AAPS J* **14**, 316–328.

Paul R, Graff-Meyer A, Stahlberg H, et al. (2012). "Structure and function of purified monoclonal antibody dimers induced by different stress conditions." *Pharm Res* **29**, 2047–2059.

Rehder DS, Chelius D, McAuley A, et al. (2008). "Isomerization of a single aspartyl residue of anti-epidermal growth factor receptor immunoglobulin gamma2 antibody highlights the role avidity plays in antibody activity." *Biochemistry* **47**, 2518–2530.

Robinson NE (2002). "Protein deamidation." *Proc Natl Acad Sci USA* **99**, 5283–5288.

Roskos LK, Schneider A, Vainshtein I, et al. (2011). "PK-PD modeling of protein drugs: implications in assay development."*Bioanalysis* **3**, 659–675.

Rup B and O'Hara D (2007). "Critical ligand binding reagent preparation/selection: when specificity depends on reagents." *AAPS J* **9**, E148–E155.

Sahin E, Grillo AO, Perkins MD, et al. (2010). "Comparative effects of pH and ionic strength on protein-protein interactions, unfolding, and aggregation for IgG1 antibodies." *J Pharm Sci* **99**, 4830–4848.

Shankar G, Devanarayan V, Amaravadi L, et al. (2008). "Recommendations for the validation of immunoassays used for detection of host antibodies against biotechnology products." *J Pharm Biomed Anal* **48**, 1267–1281.

Shankar G, Shores E, Wagner C, et al. (2006). "Scientific and regulatory considerations on the immunogenicity of biologics." *Trends Biotechnol* **24**, 274–280.

Sinha S, Zhang L, Duan S, et al. (2009). "Effect of protein structure on deamidation rate in the Fc fragment of an IgG1 monoclonal antibody." *Protein Sci: A Public Protein Soc* **18**, 1573–1584.

Siuzdak G (1996), *Mass Spectrometry for Biotechnology Academic Press*, Burlington, MA.

Sola RJ and Griebenow K (2009). "Effects of glycosylation on the stability of protein pharmaceuticals." *J Pharm Sci* **98**, 1223–1245.

Srebalus Barnes CA and Lim A (2007). "Applications of mass spectrometry for the structural characterization of recombinant protein pharmaceuticals." *Mass Spectrom Rev* **26**, 370–388.

Staack RF, Stracke JO, Stubenrauch K, et al. (2011). "Quality requirements for critical assay reagents used in bioanalysis of therapeutic proteins: what bioanalysts should know about their reagents." *Bioanalysis* **3**, 523–534.

Staub A, Guillarme D, Schappler J, et al. (2011). "Intact protein analysis in the biopharmaceutical field." *J Pharm Biomed Anal* **55**, 810–822.

Stubenrauch K, Wessels U and Lenz H (2009). "Evaluation of an immunoassay for human-specific quantitation of therapeutic antibodies in serum samples from non-human primates." *J Pharm Biomed Anal* **49**, 1003–1008.

Tatarewicz S, Miller JM , Swanson SJ, et al. (2010). "Rheumatoid factor interference in immunogenicity assays for human monoclonal antibody therapeutics." *J Immunol Meth* **357**, 10–16.

Thway TM, Macaraeg CR, Calamba D, et al. (2010). "Bioanalytical method requirements and statistical considerations in incurred sample reanalysis for macromolecules." *Bioanalysis* **2**, 1587–1596.

Tipton JD, Tran JC, Catherman AD, et al. (2011). "Analysis of intact protein isoforms by mass spectrometry." *J Biol Chem* **286**, 25451–25458.

US FDA Guidance for Industry (1999). "Q6B specifications: test procedures and acceptance criteria for biotechnological/biological products." http://www.fda.gov/downloads/drugs/guidancecomplianceregulatoryinformation/guidances/ucm073488.pdf.

US FDA Guidance for Industry (2005). "Q5E comparability of biotechnological/biological products subject to changes in their manufacturing process." http://www.fda.gov/downloads/Drugs/GuidanceComplianceRegulatoryInformation/Guidances/ucm073476.pdf.

US Food and Drug Administration (2001). "Bioanalytical method validation guidance for industry." http://www.fda.gov/downloads/Drugs/Guidance/ucm070107.pdf

Viswanathan CT, Bansal S, Booth B, et al. (2007). "Quantitative bioanalytical methods validation and implementation: best practices for chromatographic and ligand binding assays." *Pharm Res* **24**: 1962–1973.

Vlasak J, Bussat MC, Wang S, et al. (2009). "Identification and characterization of asparagine deamidation in the light chain CDR1 of a humanized IgG1 antibody." *Anal Biochem* **392**, 145–154.

Vlasak J and Ionescu R (2008). "Heterogeneity of monoclonal antibodies revealed by charge-sensitive methods." *Curr Pharm Biotechnol* **9**, 468–481.

Wang W, Nema S and Teagarden D (2010). "Protein aggregation—pathways and influencing factors." *Int J Pharm* **390**, 89–99.

Wei Z, Feng J, Lin HY, et al. (2007). "Identification of a single tryptophan residue as critical for binding activity in a humanized monoclonal antibody against respiratory syncytial virus." *Anal Chem* **79**, 2797–2805.

Xie H, Chakraborty A, Ahn J, et al. (2010). "Rapid comparison of a candidate biosimilar to an innovator monoclonal antibody with advanced liquid chromatography and mass spectrometry technologies." *mAbs* **2**, 379–394.

Z. Zhang, H. Pan, X. Chen, Mass spectrometry for structural characterization of therapeutic antibodies. *Mass Spectrom Rev* **28**, 147–176 (2009).

Zheng K, Bantog C and Bayer R (2011). "The impact of glycosylation on monoclonal antibody conformation and stability." *mAbs* **3**, 568–576.

Zheng S, McIntosh T and Wang W (2015). "Utility of free and total target measurements as target engagement and efficacy biomarkers in biotherapeutic development – opportunities and challenges." *J Clin Pharmacol* **55** S3, S75–S84.

11

Application of Liquid Chromatography-High Resolution Mass Spectrometry in the Quantification of Intact Proteins in Biological Fluids

Stanley (Weihua) Zhang[1]*, Jonathan Crowther*[1]*, and Wenying Jian*[2]

[1] *Ortho Clinical Diagnostics, Raritan, NJ, USA*
[2] *Janssen Research and Development, Johnson & Johnson Spring House, PA, USA*

11.1 Introduction

The demand for quantification of peptides and proteins has rapidly increased in recent years due to their proven promise as therapeutic agents and the need for reliable peptide and protein biomarkers as tools for the drug research and development. Currently, immunoassays such as enzyme-linked immunosorbent assays (ELISA) are the preferred bioanalytical methods for quantification of peptides and proteins that support pharmacokinetic and biomarker studies. ELISA assays are highly sensitive and easy to automate but suffer from issues such as cross-reactivity, prolonged development time, and susceptibility to matrix effect.

The interest in liquid chromatography–mass spectrometry (LC–MS)-based techniques is growing in the field of large molecule quantification due to the technique's unique advantages over conventional immunoassays, such as ELISA. A few of the most important advantages of LC–MS assays are high specificity, ability to detect posttranslational modification (PTM), and a potentially short development time. First, LC–MS has exceptionally high specificity. This is a clear advantage over ELISA assay, which is often subject to cross-reactivity or interference. While the presence of other proteins resembling the target protein could interfere with an ELISA assay, an LC–MS method can typically differentiate these interferences from the target analyte. Second, LC–MS can be used to detect protein isoforms, degradation products, and PTMs. In addition, the newer generations of mass spectrometers and software have expanded the capability of postacquisition data mining (PADM), which facilitates the analysis and characterization of the overall properties of a target sample. Third, LC–MS assays may take less time to develop. A typical LC–MS-based method, when immunoaffinity enrichment of target analyte is not required, may take just days to weeks to develop. In contrast, the resource- and time-consuming process of antibody production and selection contributes to a prolonged assay development time for immunoassays. For example, it may take up to several months to develop a sandwich ELISA assay, which requires two specific antibodies, for toxicological or clinical studies. Therefore, LC–MS-based methods are very attractive for discovery-stage activities before substantial time and resources are committed to raise antibodies.

LC–MS assays must overcome some major challenges in order to be routinely used in bioanalytical work for large molecule quantitation. First, detection sensitivity of large molecules on an LC–MS platform is usually lower than that of an immunoassay. The detection limit of a typical LC–MS method for a protein target is usually in the low μg/mL range if affinity enrichment is not applied, whereas the lower limit of quantification for a typical ELISA assay can be in the ng/mL range or lower. Second, the throughput for an LC–MS-based large molecule quantification workflow is much lower than that of an immunoassay. The current LC–MS approach for protein analysis often requires multiple steps of analyte enrichment, sample cleanup, or fractionation, many of which are time-consuming and labor-intensive. It may take several days to process one batch of samples for LC–MS analysis, whereas only hours are needed for an ELISA assay. The data acquisition for LC–MS also involves long hours of injections and runs, yet it only takes a couple of minutes to read an ELISA plate. Nevertheless, it is believed that these roadblocks will be overcome, assisted by the evolving LC–MS technologies and emerging novel sample preparation methods. These advances will lead to the routine application of the LC–MS platform to the bioanalytical workflow for large molecules in pharmaceutical research and development.

There are two common approaches for LC–MS-based quantification of proteins: (i) Bottom-up approach, which

Protein Analysis using Mass Spectrometry: Accelerating Protein Biotherapeutics from Lab to Patient, First Edition.
Edited by Mike S. Lee and Qin C. Ji.

involves the enzymatic cleavage of a protein into small peptides followed by the LC–MS/MS analysis of one or more of the proteolytic peptides, which are called signature peptides, as surrogates. (ii) Top-down approach, which refers to analysis of an intact target molecule. This approach is often limited by the size of the analyte and is most suitable for proteins smaller than 30 kDa (Behnken et al. 2014).

The bottom-up approach has been evolved from proteomics research. Many proteomics methods are based on enzymatic digestion of proteins into small peptides that are readily analyzed by mass spectrometers. Over the years, sensitive and selective bottom-up methods capable of identifying thousands of proteins within a single sample have been developed (Graumann et al. 2008). The bottom-up approach, however, has limitations since only specific peptides of a protein are identified, rather than the whole protein itself. Thus, critical information such as PTMs, sequence variants and isoforms, and truncation and degradation products may be missed. The top-down method, on the other hand, overcomes many of these shortcomings. Since no proteolytic digestion is involved, the top-down approach analyzes intact proteins. Therefore, PTMs such as phosphorylation and acetylation, sequence variants such as mutants, amino acid polymorphisms, and isoforms, as well as truncation or degradation products may be detected. A particular advantage of the top-down approach with full-scan MS acquisition mode is its data-mining capacity after LC–MS data is acquired. Full-scan mode captures all information of a sample, and such data can be reprocessed postacquisition based on different requirements.

To quantify an intact protein from biological fluids by LC–MS, either multiple reaction monitoring (MRM), using triple quadrupole MS, or full scan, using high-resolution mass spectrometry (HRMS), can be employed. MRM analysis involves monitoring the parent ions of intact proteins at different charge states and their product ions generated in collision-induced dissociation (Ji et al. 2003, 2005). This MRM procedure requires pre-knowledge of the expected mass/charge (m/z) values and fragmentation pattern for assay setup, as well as efficient fragmentation for sensitive detection, which is often absent for large proteins. Therefore, full-scan HRMS has increasingly become the method of choice for top-down intact protein analysis. In this case, a mass spectrometer with sufficient resolving power is used to analyze a protein target without any fragmentation. Figure 11.1 represents hypothetical spectrums of three small compounds with the same nominal mass (133) but slightly different exact masses. For the purpose of simplicity, assume these compounds are eluted from the LC at the same time, a

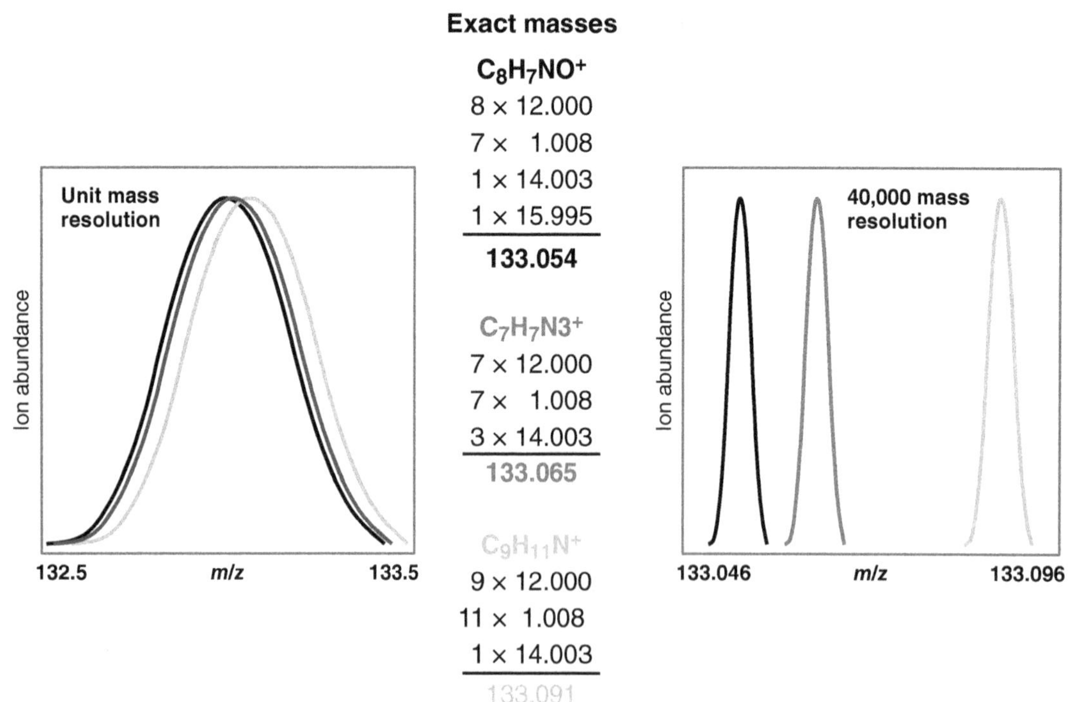

Figure 11.1 Effect of resolution on separation of three compounds with close molecular weights. (*Source*: Adapted from United States Environmental Protection Agency. Ion Composition Elucidation (ICE). National Environmental Research Laboratory, Environmental Sciences Division, Las Vegas, NV (www.epa.gov/esd/chemistry/ice/faq.htm).)

mass spectrometer with unit mass resolution (capable of separating two analytes differing by one mass unit) will not separate these compounds, whereas a mass spectrometer with 40,000 mass resolution can easily resolve them. The situation is similar to the case of intact protein analysis. Granted, different proteins may be chromatographically separated by the high-performance liquid chromatography (HPLC) front end that is coupled to a mass spectrometer. However, in biological fluids there may be hundreds of thousands or more of different protein species (Anderson and Anderson 2002) and chances are that some proteins with close *m/z* values will be eluted together under a given HPLC condition. Therefore, HRMS is a preferred platform for intact protein analysis in biologic fluids.

The requirement of high resolution also means that the suitable protein targets for the LC-HRMS-based quantification are those with a molecular weight up to 30 kDa, based on current mass spectrometry technologies available at most laboratories. For proteins larger than 30 kDa, isotopic resolution is not achievable and it will be difficult to separate signals of the target analytes from those of matrix components. The suitable strategy for quantification of these proteins is LC–MS/MS-based bottom-up approach, which is out of the scope of this chapter but has been discussed elsewhere (Zhang and Jian 2014).

In this chapter, we review the recent advances and trends in the application of LC-HRMS in intact protein analysis in biological fluids, along with an introduction to the basic workflow, analytical approaches, and common issues. The primary focus of this review is the targeted, absolute quantification of therapeutic biologics and biomarkers, excluding proteomics work, which is often conducted in a nontargeted, relative quantification fashion. Note that the generic term "protein" is used throughout this chapter and refers to both peptides and proteins larger than 5 kDa or so. The discussion of peptides smaller than 5 kDa, where conventional LC–MS-based methods for small compounds are applied, is beyond the scope of this chapter.

11.2 Workflows for Quantification of Proteins Using Full-Scan LC-HRMS

Due to the highly diverse properties of proteins and variations in different sample matrices, there is no universal workflow suitable for all LC-HRMS assays. Sample preparation can range from a single step of solid-phase extraction (SPE) to multiple steps that may involve depletion of high-abundant proteins or affinity enrichment (Lu et al. 2009) and depends on the size of a target analyte, sample composition, and sensitivity requirement. A variety of liquid chromatography modes can be selected to achieve optimal separation of a protein target from matrix components. For detection, traditional and novel HRMS have been explored for full scan analysis of intact protein. In this chapter, recent advances in major techniques of sample preparation, liquid chromatography, and mass spectrometer detection are reviewed for their application in bioanalysis of intact proteins.

11.2.1 Sample Preparation

11.2.1.1 Solid-Phase Extraction (SPE)

For proteins smaller than 10–15 kDa, SPE is the preferred sample preparation method to separate the protein from biological matrices. The relatively small sizes make these proteins ideal candidates for HRMS analysis. Besides providing surface chemistries, SPE takes advantage of the size-exclusion principle by excluding large proteins that cannot enter the pores on the SPE sorbent. For example, a pore size of 40–80 Å can roughly eliminate proteins larger than 20 kDa. Very often, ion exchange SPE can provide more selective cleanup as well as an orthogonal mode of separation when coupled with reversed-phase liquid chromatography (RPLC). In our laboratory, reversed-phase SPE is used to extract different glycosylation forms of apolipoprotein C3 (ApoC3) proteins, which have molecular weights of roughly 8.8–9.7 kDa, from human plasma (Jian et al. 2013).

11.2.1.2 Affinity Enrichment

For most proteins larger than 10–15 kDa, there are very limited options to separate them from biological matrices. Specific or nonspecific affinity enrichment strategies may be applied to enrich a protein analyte. Examples of such enrichment include capture of a target protein by specific antibodies (if they are available), affinity enrichment of antibody drugs containing the Fc region of IgG by protein A and/or protein G, and capture of PEGylated proteins by anti-PEG antibody (Xu et al. 2010, Liu et al. 2011). Affinity enrichment can be performed on-line by using a column loaded with a capture agent, such as an antibody, and coupled to a mass spectrometer (Dufield and Radabaugh 2012).

11.2.1.3 Depletion of High-Abundant Proteins

If there is no affinity enrichment method available for a target protein, then a sample can be cleaned by depletion of high-abundant proteins in the sample matrix. Human plasma, as an example, has a total concentration of endogenous proteins as high as 60–80 mg/mL. Complexity of the human plasma proteome, in which the dynamic range of protein abundance covers 11 orders of magnitude, presents a great challenge to the quantification of selected proteins (Anderson and Anderson 2002). Therefore, enrichment procedures are often needed for reducing interference and improving sensitivity for LC–MS analysis of a target protein, especially those circulating at low concentrations. One such procedure is the depletion of high-abundant plasma proteins. It is estimated that 99% of the plasma's total protein mass is due to the top 20 most abundant protein species. Human serum albumin, present at 35–50 mg/mL, is the most abundant protein. Depletion kits using chemical affinity or immunoaffinity for removal of serum albumin, immunoglobulin, and other high-abundant proteins have been shown to reduce protein content by up to 85% (Echan et al. 2005). Polaskova et al. (2010) studied six commercial products for high-abundance protein removal and showed that while performance varied, these products in general improved protein detection on 2D gel. Current commercially available technologies allow up to 20 high-abundance proteins to be removed and should greatly improve the sensitivity of LC–MS detection.

11.2.1.4 Solution Fractionation

Another potential strategy for reduction of sample complexity is solution fractionation, which is often designed to achieve orthogonal separation in conjunction with the subsequent LC–MS analysis. A complex sample is separated into multiple fractions based on various characteristics of target proteins, such as charge, isoelectric point (pI), hydrophobicity, molecular weight, or a combination of two or more of them. Fractionation based on charges of proteins is one of the most popular fractionation methods and can be achieved using either anion (Roth et al. 2008) or cation exchange chromatography (Ning et al. 2008). Fractionation based on pI may be achieved through chromatofocusing (Chong et al. 2001) or isoelectric focusing (Lubman et al. 2002). Reversed-phase (RP) chromatography has been used as a fractionation method, and the resulting fractions were subsequently analyzed by orthogonal hydrophilic interaction (Pesavento et al. 2006) or even another RP-HPLC coupled with MS detection (Ning et al. 2008). Fractionation based on molecular weight does not separate large proteins from small ones straightforwardly, since the majority of proteins normally form multiprotein complexes rather than exist as isolated molecules in a biological sample. Nevertheless, size exclusion chromatography has been used as a fractionation method for separation of protein complexes (Olinares et al. 2010, Kristensen et al. 2012, Kirkwood et al. 2013).

It is important to note that solution fractionation has been used extensively in the proteomics field, where a small number of samples are normally studied and the increased number of LC–MS runs resulted from sample fractionation can be reasonably handled. However, this approach has not been reported in bioanalysis studies, where a larger number of samples are often analyzed. When solution fractionation is considered as a sample preparation method in protein quantification, the number of LC–MS runs may be reduced by analyzing only the target-containing fractions. Further desalting of fractions is often needed to make them amenable for LC–MS analysis, for example, when ion exchange is used for fractionation.

11.2.1.5 Protein Precipitation for PEGylated Proteins

Protein precipitation is a common sample preparation method for small compounds. During the procedure, acetonitrile, methanol, or other water miscible organic solvents are added to the samples; proteins will precipitate out while small compounds remain in the solvent. PEGylated proteins, however, represent exceptions for this method. Attachment of polyethylene glycol (PEG) to peptides or proteins has been employed as a strategy to extend their *in vivo* circulatory half-life as well as to improve their chemical and physical stability, solubility, and potentially to reduce immunogenicity (Veronese and Mero 2008). Due to the hydrophilic PEG moiety, a PEGylated protein can remain in the extract supernatant under certain protein precipitation conditions. Wu et al. (2011) reported a protein precipitation method in which PEGylated proteins (protein drugs of 11–12 kDa coupled to 40 kDa PEG) were extracted using acidic isopropanol. Xu et al. (2010) used protein precipitation to extract a peptide of 38 amino acids attached to a 40 kDa branched PEG from human plasma.

11.2.2 LC-HRMS

11.2.2.1 HPLC

Protein separation has mostly been achieved by RP-HPLC. Two-dimensional LC using orthogonal separation mechanism such as ion-exchange chromatography (IEC)-RPLC or RPLC-HILIC has also been reported to be utilized to fractionate and clean up samples, thus improving the sensitivity of the detection (Julka et al. 2011).

Particle sizes and dimensions of columns suitable for protein separation are generally the same as those used for small molecule applications. However, pore sizes of conventional LC packing materials, normally between 60 and 150 Å, are often too small for proteins since they cannot get access to the surface chemistries within these small pores. Instead, packing materials with larger pore sizes (300–1000 Å) have been used for protein separation (Wagner et al. 2012). For analysis of antibody–drug conjugate (ADC), a pore size as high as 4000 Å has been reported (Xu et al. 2011). For reversed-phase separation, packing materials with shorter hydrocarbon chains (e.g., C3, C4, and C8) may be chosen over commonly used C18 due to increased hydrophobicity of proteins. The experiences gained from proteomics field have pushed the application of capillary and nanocolumns with inner diameter (I.D.) as narrow as 75 μm in protein quantification. The low flow rate (a typical flow rate for a 75 μm I.D. nanocolumn is 200–300 nL/min) of these columns greatly improves ionization efficiency of ESI (Sowell et al. 2004, You et al. 2011, Wu et al. 2014). A challenge for capillary and nano HPLC is the need for sophisticated instrumentation to give accurate and reproducible delivery of flows at the low μL to nL/min range. Clogging of columns or instruments also happens at nano flow rate when biological fluids such as plasmas and sera are analyzed. HPLC manufacturers have been struggling over these challenges but slowly achieving acceptable performance.

The HPLC mobile phase is crucial for LC-HRMS analysis because it affects not only chromatographic separation but also ionization of target protein analytes (i.e., ionization efficiency, different charge states, or adduct ions). Ion-pairing agents such as trifluoroacetic acid (TFA) are commonly used as mobile phase additives for protein chromatographic separations due to their ability to promote retention of proteins on reversed-phase HPLC by ion pairing with these proteins, thereby increasing their hydrophobicity (Cai and Li 1999). In addition, they improve peak shape by suppressing undesirable interactions between protein and the stationary phase (Sharma et al. 2007). The trade-off for using TFA is that TFA is known to suppress ESI signal intensity due to its ability to form gas-phase ion pairs with positively charged ions. Addition of acetic acid (0.5%), propionic acid (1%), or formic acid (0.2–0.3%) to the TFA-containing mobile phase can effectively reduce ion suppression (Chong et al. 2001, Shou and Weng 2005, Jian et al. 2013). In some cases, formic acid has been used to replace TFA altogether (You et al. 2011). In a study of wheat glutenin subunits, Lagrain et al. (2013) described a study where formic acid replaced TFA in the mobile phases, and the overall peak intensities as well as the number of peaks increased. However, the chromatographic separation of high molecular weight glutenin subunits, their main targets, was worse.

11.2.2.2 Full-Scan HRMS Data Acquisition and Analysis

A mass spectrometer with high resolution and accuracy is necessary for protein analyte detection. Specificity of the analysis is dependent on the resolving power to separate the isotope peaks of an analyte from the noise, as well as on the high mass accuracy. In the last several years, LC-HRMS technologies have been greatly accelerated due to the rapid development of advanced and novel HRMS and the need to quantify biologics and biomarkers in pharmaceutical and biotechnology industry. Great improvements have been made toward the performance of quadrupole time-of-flight (Q-ToF) type of mass spectrometers, including higher resolution and better mass accuracy. These improvements, along with the intrinsic fast scan rate and high mass range in full-scan mode, have made Q-ToF an ideal choice for LC-HRMS analysis. Another popular option is Orbitrap, a novel high-resolution mass analyzer developed recently. With expanded mass range and faster scan rate, the newer generation of Orbitrap has quickly established itself as an important tool for top-down protein analysis (Zubarev and Makarov 2013).

A typical HRMS spectrum of a protein contains a charged envelope that is made of multiple peaks of different charge states, and each charge state has its own multiple isotopic peaks (Figure 11.2b and c). Such spectra are considerably more complex than those of singly charged small compounds. Attention should be given to the optimization of mobile phase components and ionization conditions so that the optimal charge states and peak quality can be reproducibly obtained. Upon evaluation of the raw data, a data analysis strategy needs to be developed regarding which charge states should be selected, how many isotopic peaks should be integrated, and the width of the extraction window to be used. One or two isotopic peaks of high quality from the dominant charge state(s) may be chosen if the peak intensities are high, whereas as many peaks as possible should be included to improve the signal intensity and reproducibility of the quantification if the peak intensities are weak. However, the trade-off for the latter approach is that the noise level is also often increased. Results from different approaches may be compared so that the one that gives the most selective and reproducible quantification performance is chosen. Finally, the nontargeted full-scan approach allows the possibility of PADM, which may reveal other potential analytes, such as biotransformation products, different PTM species, or even different proteins. PADM can be conducted on an as-needed basis. For example, in a clinical biomarker study, PADM of the patient data may help to reveal important information relevant to the disease status or treatment outcome, as well as to identify other potential biomarkers.

11.3 Internal Standard Strategy

A well-designed strategy that features the use of an internal standard is critical for successful development of an LC-HRMS-based protein quantification assay. Variations may be introduced in any step in a workflow during sample processing, liquid chromatography, and ionization, but a good internal standard should track and compensate for such variations as much as possible.

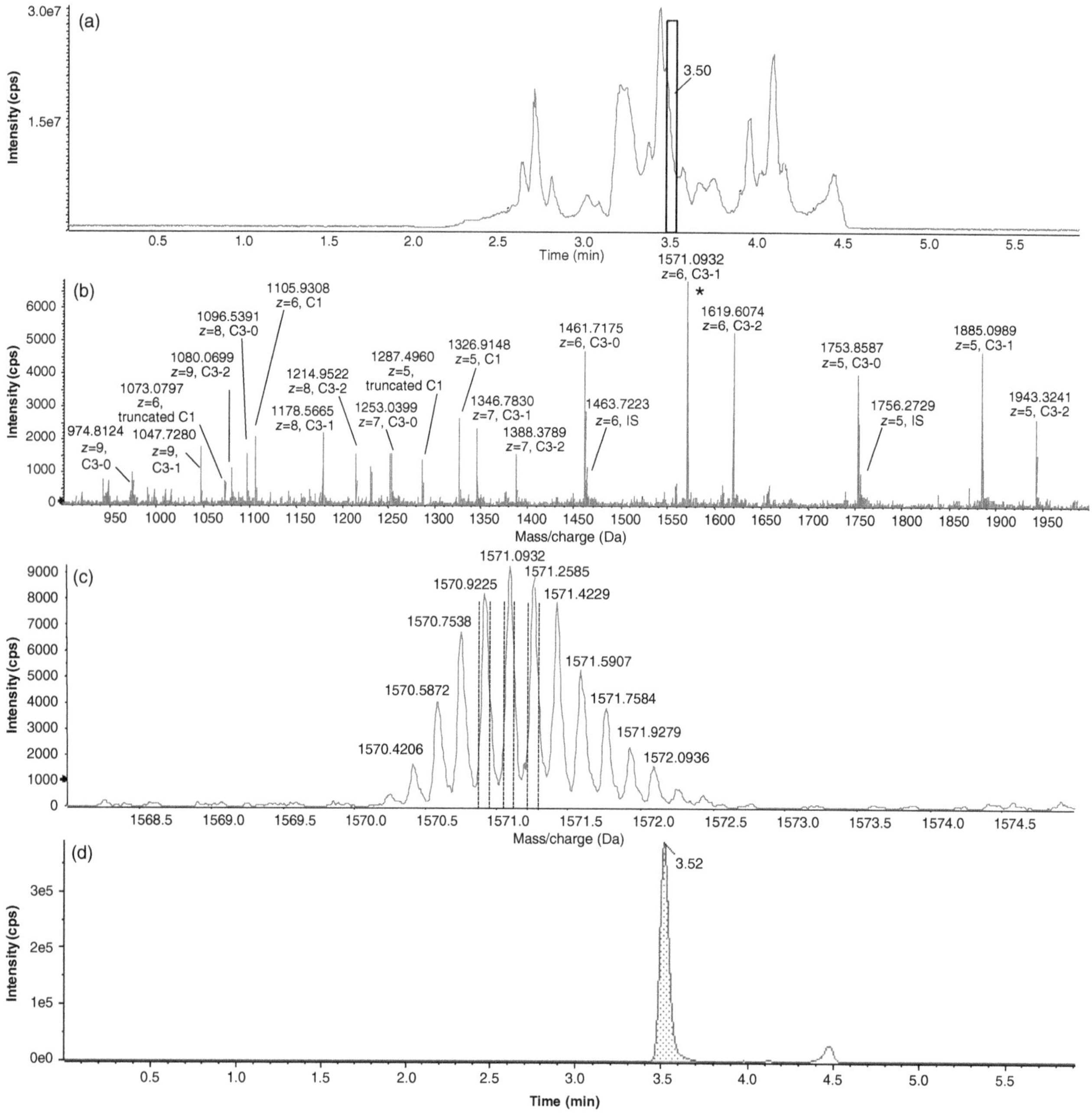

Figure 11.2 LC-ToF MS analysis of ApoC3 in human plasma. (a) TIC of full ToF MS scan. (b) Mass spectrum at retention time of 3.5 min (as selected in a). (c) Enlarged spectrum of ApoC3-1 at charge state 6 (as indicated by * in b). The extraction window for peak integration was shown by the shade. (d) Chromatographic peak of ApoC3-1 by integrating three most abundant isotopic peaks at each of charge states 5 and 6. (*Source*: Jian et al. (2013). Reproduced with permission of American Chemical Society.)

11.3.1 Stable Isotope Labeled Protein

The optimal internal standard for a protein analyte is the same protein labeled with stable isotopes. Such internal standards have identical physicochemical properties and behaviors to those of the target analyte, therefore, should precisely emulate the analyte. Internal standards are usually added at the start of sample processing, or even during sample collection, and thus, better control throughout the steps of the quantification workflow is provided and improvement of the accuracy and precision of the bioanalytical assay is achieved. There are different ways to produce a stable-isotope-labeled protein internal standard. For a small protein, the stable-isotope-labeled version can be chemically synthesized (Jian et al. 2013). Protein internal standards may also be made by *in vitro* protein synthesis in the presence of stable labeled amino acids in a cell-free system (Brun et al. 2007). Alternatively, standards can be produced by stable isotope labeling with amino acids in cell culture (SILAC) (Ong et al. 2002). Heudi et al. (2008) obtained their stable labeled antibody from antibody producing cells (SP2/0 Ag 14.0 cells) grown in medium supplied with stable labeled threonine. To facilitate the quantification of a large number of different antibody drug candidates in early discovery, Li et al. (2012) proposed a generic stable-isotope-labeled antibody as a common internal standard for different antibody quantification assays.

11.3.2 Protein Analog

Even though they are considered the best internal standards, stable-isotope-labeled proteins are costly and take relatively long time to make. In the absence of a stable-isotope-labeled protein, a protein analog may be used as an internal standard. A protein analog internal standard should resemble the target protein as much as possible to better track the behavior of the target protein. For example, horse myoglobin has been used as internal standard for human myoglobin (Mayr et al. 2006), and chicken lysozyme has been used as internal standard in a human lysozyme analysis (Ruan et al. 2011). Also, a protein analog can be made by introducing a point mutation into the amino acid sequence of the protein analyte using molecular biology techniques (Liu et al. 2013b).

11.4 Calibration and Quality Control (QC) Sample Strategy

A calibration curve is necessary for absolute quantification of proteins by LC-HRMS. A calibration curve should be prepared in the blank matrix (i.e., the target biological sample devoid of the analytes) whenever possible. If a matrix has endogenous analyte, then a surrogate matrix with similar property may be used. For example, a calibration curve may be prepared in monkey plasma if study samples are human plasma (Ruan et al. 2011). A similar approach was used by Cao et al. (2010) to quantify carbonyl reductases CBR1 and CBR3 in human liver against calibration curves made in pooled rat liver cytosol.

The same principle applies to quality control (QC) samples. QC samples should be prepared in a blank matrix if possible or in a surrogate matrix when a true blank matrix is not available. It is important to evaluate a method using QC samples prepared at least at one concentration level using authentic matrix to demonstrate the absence of potential bias introduced by using surrogate matrix (Jian et al. 2012).

Calibration standards and QC samples may exhibit different sample extraction recovery from incurred samples due to different matrix contents, and thus, lead to biased quantification results. Liu et al. used standard-addition approach to verify the ruggedness of their assay for incurred samples. After the initial measurement of an incurred sample, the sample was mixed with a standard sample containing a known amount of analyte in one-to-one ratio. The resulting sample was measured again, and the result was compared to the theoretical concentration (original + added) to reveal the accuracy (Liu et al. 2013b).

For assay validation, the similar parameters as those for LC–MS/MS methods of small molecules, such as accuracy, precision, sensitivity, selectivity, and stability, should be evaluated. To address the unique aspects of protein quantification, procedures such as high-abundant protein depletion, affinity enrichment, and solution fractionation should be evaluated for their efficiency, depending on the specific workflow.

11.5 Common Issues in Quantification of Proteins Using LC-HRMS

Compared to small compound applications, quantification of proteins by LC–MS, including LC-HRMS, has similar challenges, such as stability, adsorption, and solubility. Protein quantification also has its unique problems, such as specific protein binding and PTMs. Because of the nature of proteins, methods to address these challenges and issues may be quite different from those of small compounds.

11.5.1 Stability

Stability of proteins is a major concern in protein bioanalysis. Many different factors affect protein stability. Proteins, especially at low concentration, are susceptible to proteolysis caused by naturally occurring protease in all organisms. To address this issue, protease inhibitors may be added in one or more steps of sample preparation process (Olivieri et al. 2001, Rai et al. 2005, Olinares et al. 2010). Protease inhibitors are commercially available in individual or cocktail format and are easy to use. Protein samples should be handled on ice or at 4 °C since most proteases function optimally at room temperature to 37 °C. Antibacterial agents such as sodium azide may be added at the final concentration of about 0.02–0.05% (w/v) if a protein solution is stored at 4 °C for a prolonged time.

Some amino acids are prone to oxidation, including methionine and cysteine and, to a lesser extent, tryptophan and histidine. Prolonged exposure to atmospheric oxygen should be minimized. Since cysteine tends to form intra- and intermolecule disulfide bonds, a common practice is to reduce disulfide bonds completely by dithiothreitol (DTT) or 2-mercaptoethanol treatment, followed by alkylation with iodoacetamide or iodoacetic acid. Such reduction and alkylation steps can eliminate the variability of disulfide formation (Herbert et al. 2001, Kirsch et al. 2007).

Another important factor for stability is pH. Fragmentation occurring at the C-terminus of an Asp residue is one of the most frequent degradation pathways of monoclonal antibodies (mAbs) under mildly acidic conditions (Vlasak and Ionescu 2011). Deamidation of asparagine and glutamine also happens at acidic conditions (Robinson and Robinson 2004). Very low or high pH will denature a protein, and a pH too close to the pI of a protein may cause precipitation. Therefore, protein samples should be processed and stored in a buffered solution at near-neutral pH as much as possible.

Protein aggregation and solubility is a potential issue especially when dealing with antibody drugs and other biologics (den Engelsman et al. 2011). Proteins at high concentration tend to aggregate, so a simple dilution may help. A salt containing buffer, such as PBS, should be used to make protein solution, but too high or too low salt concentration may lead to protein aggregation and precipitation. High concentration of organic solvent will precipitate protein.

Finally, protein samples should be handled gently. For example, repetitive freeze/thaw, vigorous vortexing, and foaming can cause damage to proteins and should be avoided during sample preparation or storage.

11.5.2 Adsorption

The amphiphatic nature of proteins makes them readily adsorb to most surfaces, which can lead to inaccurate quantification results (Rabe et al. 2011). There are many different ways to prevent protein adsorption, but each method has to be tested empirically. Adsorption is more severe when a protein is at a very low concentration in a matrix-free aqueous solution, so it is good practice to make high-concentration protein stocks and spike them directly into plasma or serum samples. Adding displacement agents such as a protein-rich solution (e.g., bovine serum albumin) may limit adsorption (Lassen and Malmsten 1996). Using organic-aqueous solution is another option, but it should be evaluated case by case for different proteins (Staub et al. 2010). It is common practice to use specially treated tubes and plates that are commercially available for low protein binding, but attention should still be paid to each individual product. For example, Bratcher and Gaggar (2013) showed that even though less adsorption was observed on siliconized or prelubricated "low-binding" polypropylene tubes for some proteins, more adsorption was found for Surfactant Protein D (SP-D), their target protein, compared to untreated polypropylene tubes.

11.5.3 Specific Protein Binding

A potential issue for the LC–MS-based protein quantification assay is the specific protein binding in test samples. Specific protein binding can be illustrated by variations encountered in quantification of growth hormone (GH) when different immunoassays were used. Growth hormone-binding protein (GHBP) has a high affinity to GH, and up to 50% of GH forms complex with GHBP in human serum and plasma samples. As a result, GH antibodies might not get access to all the epitopes in an immunoassay due to steric hindrance, which led to underestimation of GH concentrations (Bidlingmaier 2008).

With the rapid advancement of biologics in pharmaceutical industry, one situation that deserves attention is the formation of antidrug antibodies (ADAs). Biologics have been shown to be immunogenic in animals and human patients (Warnke et al. 2012). The resulting ADAs pose challenges in quantification of these biologics. On the one hand, a protein drug has to be separated from its ADAs in the sample preparation process in order to avoid any misrepresentation of the total drug concentration in a specimen. For example, 8 M guanidine hydrochloride has been used to denature monkey plasma samples to disassociate a small protein drug from its ADAs before SPE (Ji et al. 2007). On the other hand, ADAs themselves may be the targets for quantification when the immunogenic potential of a protein drug is assessed during the development phase. Current methodologies for measuring ADAs, such as ELISA, surface plasmon resonance (SPR, e.g., Biacore), and cell-based assay are often subject to interference from high circulating concentration of the protein drug. In an LC–MS assay, Neubert et al. took advantage of the affinity between a PEGylated human growth hormone analog (hGHA) and its ADAs. They first captured the hGHA-ADAs complex with Protein G magnetic beads and then quantified the hGHA, whose concentration was proportional to that of ADAs (Neubert et al. 2008).

11.5.4 Posttranslational Modifications (PTMs)

PTMs are covalent processing events that change the properties of a protein by addition of a modifying group to one or more amino acids, by proteolytic cleavage, or, sometimes, by removal of a functional group from an amino acid. Observed in both prokaryotes and eukaryotes, PTMs are more widespread in eukaryotes due to the presence of organelles such as endoplasmic reticulum (ER) and Golgi apparatus (Huq and Wei 2007). Some common types of PTM are glycosylation, oxidation, phosphorylation, ubiquitination, nitrosylation, methylation, acetylation, lipidation, formation of disulfide bond, and truncation.

Since they may potentially occur to many proteins, PTMs have to be considered when a quantification workflow is designed. For example, N-linked glycosylation may be removed from a protein by treatment of PNGase F enzyme when the total concentration of the protein is quantified. On the other hand, proteins with PTMs of particular interest may be the targets for quantification. The quantification of a protein with particular PTMs has usually been accomplished by LC–MS/MS-based bottom-up approach, in which a protein sample is first digested by a protease, and the peptide that contains the PTM is quantified by LC–MS/MS (Liu et al. 2013a). A big challenge for such approaches is that the PTM-containing peptides may not be amenable to LC–MS/MS analysis. They are often too hydrophilic to be retained on an HPLC column, or too large to fit the detection range of a mass analyzer. The ionization efficiency for many of these peptides is low compared to the unmodified ones. The emerging full-scan LC-HRMS top-down approach provides another option to look at PTMs by analyzing intact proteins, and thus, this approach may avoid the above issues associated with PTM-containing peptides. Using this approach, Thakur and colleagues identified over 60 different glycoforms from the α subunit of a recombinant human chorionic gonadotropin protein (Thakur et al. 2009).

Moreover, LC-HRMS can readily provide the relative quantification information of PTMs since both modified and unmodified proteins may be analyzed in the same run. A good example is the determination of drug-to-antibody ratios (DAR) of ADC, which is an antibody conjugated with a chemical drug and thus can be considered as a product of "artificial" PTM (Xu et al. 2011). Measurement of the relative quantification of each DAR species using HRMS can provide critical mechanistic insights into understanding the stability and bioactivity of ADC *in vivo*. In our laboratory, we performed relative quantification of different ApoC3 protein glycoforms in human plasma samples, which is discussed in detail in the following section (Jian et al. 2013).

11.6 Examples of LC-HRMS-Based Intact Protein Quantification

Early experiments of LC-HRMS-based intact protein quantification were mostly conducted on Fourier transform mass spectrometers (FTMS). Padley et al. demonstrated the quantification of lysozyme, a 14 kDa protein using an electrospray source followed by a linear ion trap coupled to an FTMS. No internal standard was included and the linearity covered only 1.5 orders of magnitude in this early study (Padley et al. 1997). Gordon and Muddiman quantified cyclosporin A (CsA), a small cyclic peptide immunosuppressant using a similar setup of an FTMS instrument. Even though the target analyte was small (~1.2 kDa) and only pure standards in neat solutions were analyzed, this study had the basic elements of the full-scan LC-HRMS quantification approach. A calibration curve using an analog protein (cyclosporin G, or CsG) as the internal standard was established (Gordon and Muddiman 1999). Using a similar approach, Gordon's group also developed a method to measure the concentrations of equine heart cytochrome c. Bovine heart cytochrome c was used as the internal standard in this study. Each one of the charge states of 7^+–14^+ was isotopically resolved and the top four most abundant isotopic peaks from the dominating charge states of 7^+ and 8^+ were averaged for quantification (Gordon et al. 1999).

Ruan et al. developed an assay for the quantification of lysozyme in human plasma using the recently developed HRMS LTQ Orbitrap. The calibration curve was generated by spiking human lysozyme in monkey plasma. Chicken lysozyme was used as the internal standard. The plasma samples were successfully processed by SPE. This assay achieved sufficient sensitivity for the quantification of endogenous lysozyme in human plasma samples (Ruan et al. 2011). Liu and colleagues reported the measurement of a recombinant human mAb in monkey serum by a Q-ToF instrument. The target antibody was first enriched by Protein A affinity capture and then treated by limited Lys-C digestion. The resulting 47 kDa Fab fragment was subjected to LC-Q-ToF MS full-scan analysis. An isotopically labeled, otherwise identical antibody was used as the internal standard throughout the sample preparation and analysis to ensure accurate quantification. The actual quantification data were based on the intensities of Fab peaks in the deconvoluted mass spectra. Even though isotopic resolution could not be achieved for the 47 kDa fragment using current MS technology, the concentrations calculated based on deconvoluted peak areas were in good agreement with those obtained from an ELISA assay when the levels of the human mAb in multiple monkey serum samples were quantified (Liu et al. 2011). Lately, Gucinski and Boyne (2012) developed and validated quantification assays for two insulin variants and the human GH using an LTQ Orbitrap.

In our laboratory, we have developed an LC-HRMS-based method for quantification of various glycoisoforms of intact ApoC3 in human plasma using a Q-ToF mass spectrometer. ApoC3 protein exists mainly in three glycoisoforms: ApoC3-0, ApoC3-1, and ApoC3-2. Bound to threonine in position 74 of the protein sequence, the O-linked sugar moiety consists of one residue of galactose, one residue of *N*-acetyl-galactosamine, and one and two residues of *N*-acetylneuraminic acid (NeuNAc, known as sialic acid) for ApoC3-1 and ApoC3-2, respectively, while ApoC3-0 has no sugar chain at all (Figure 11.3). Changes in the levels of these ApoC3 glycoisoforms have been observed in patient samples of several diseases, including obesity, kidney diseases, liver diseases, and sepsis. Therefore, study of ApoC3 glycosylation may provide important information that can be used for diagnosis, prognosis, and evaluation of therapeutic responses of these diseases.

Since currently available antibodies could not differentiate these ApoC3 glycoisoforms, immunoassays were not suitable for quantification analysis. The sizes of these molecules (8.8–9.7 kDa), however, made them good targets for LC-HRMS analysis. Human plasma samples were first processed with reversed-phase SPE. The extractions were then analyzed by LC-HRMS in full-scan mode using a Q-ToF mass spectrometer. For each ApoC3 glycoisoform, the three most abundant isotopic peaks at two charge states (5 and 6) were extracted using a window of 50 mDa and integrated into a chromatographic peak (Figure 11.2), and the peak area ratios of ApoC3-1/ApoC3-0 and ApoC3-2/ApoC3-0 were then calculated. These ratios were evaluated as potential diabetes biomarkers in a preliminary study using plasma samples collected from normal, prediabetic, and diabetic subjects. In addition, PADM was conducted to identify additional proteins of interest in these samples, including ApoC1 and its truncated form shown in Figure 11.2(b) (Jian et al. 2013).

11.7 Conclusion and Future Perspectives

Even though LC-HRMS-based protein quantification has shown promise as an emerging platform for large molecule bioanalysis, this analytical platform is still in

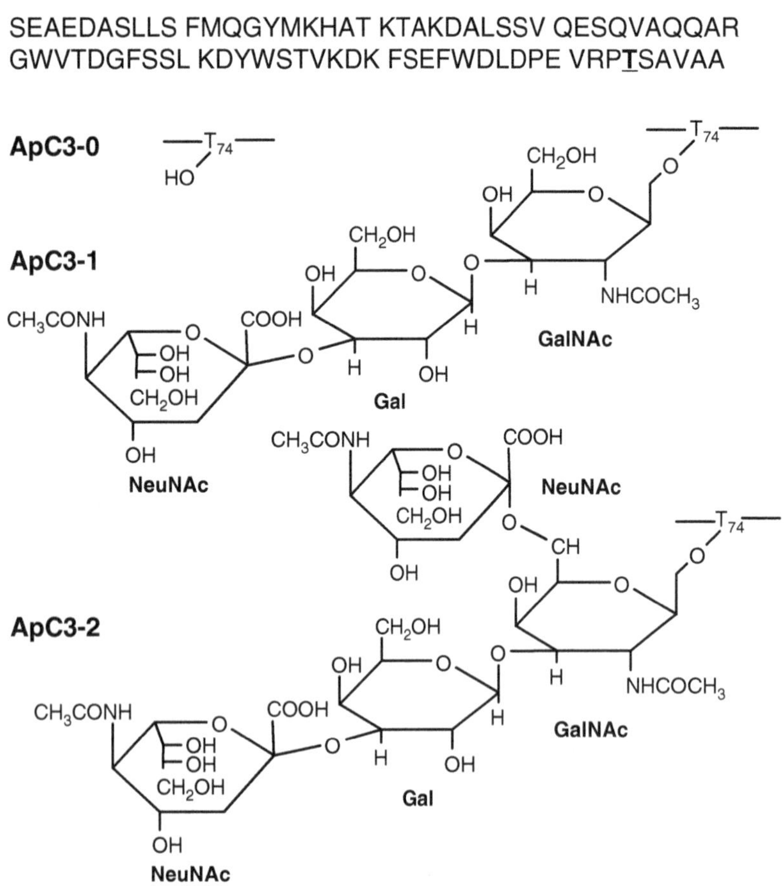

Figure 11.3 Amino acid sequence of ApoC3 and the structures of different glycol-isoforms. Gal, galactose; GalNAc, *N*-acetyl-galactose; NeuAc, *N*-acetyl-neuraminic acid. ***T*** (Thr) indicates the position for glycosylation. (*Source*: Jian et al. (2013). Reproduced with permission of American Chemical Society.)

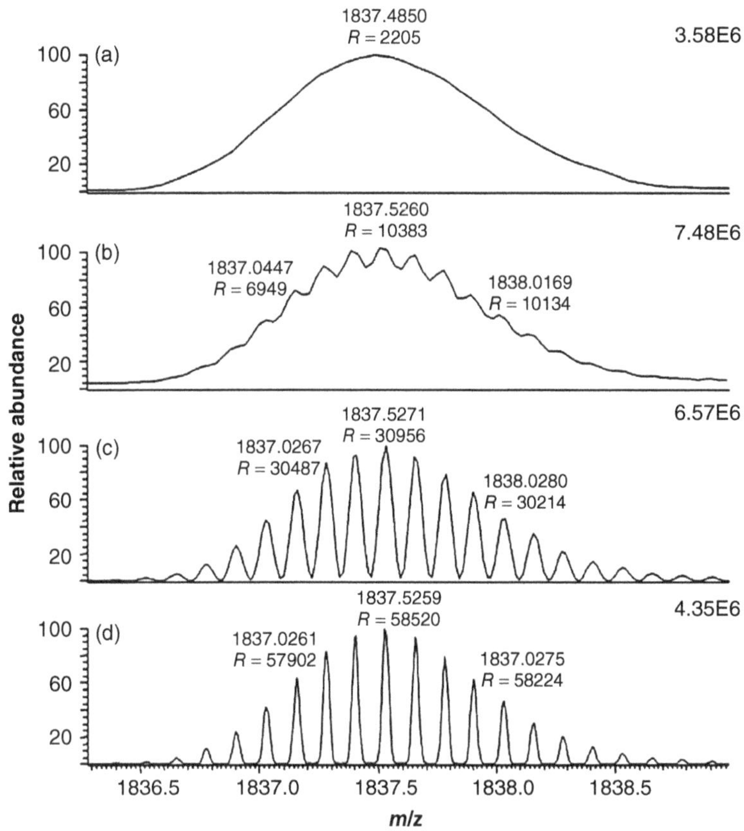

Figure 11.4 Effect of resolving power on separation of isotopic peaks of octuply charged lysozymes. (a) Resolving power of 2 k, (b) resolving power of 10 k, (c) resolving power of 30 k, and (d) resolving power of 60 k. (*Source*: Ruan et al. (2011). Reproduced with permission of American Chemical Society.)

the early stages of development and there are significant challenges associated with the technique. One of the issues for the workflow is that the current technology is more suitable for proteins smaller than 30 kDa. For quantification of bigger proteins, a mass spectrometer with higher resolving power is preferred. The higher the resolving power of a mass spectrometer, the better its selectivity and, therefore, the higher sensitivity, since a narrower extracted ion chromatogram (EIC) window can be applied to remove the potential interference. Figures 11.4 and 11.5 illustrate the effect of resolving power on the separation of isotopic peaks and the effect of EIC window on selectivity and sensitivity, respectively. Based on a model proposed by Ruan and colleagues, 600,000 practical resolving power is necessary to quantify an intact antibody of about 150 kDa (Ruan et al. 2011). Mass spectrometers that cover higher mass range are also in demand because some proteins may not carry sufficient charges to fit their *m/z* values into the narrow detection range of current mass spectrometers. Besides the requirement for high-end mass spectrometers, versatile software that can handle multiple charge states and isotopic peaks resulted from intact protein analysis is also needed. Currently, in our laboratory, the chosen charge states and isotopic peaks were summed manually to obtain final quantification results. Software packages that can perform automatic summing, quantitative deconvolution, and other quantification-friendly features are expected to be developed in the near future.

Another major bottleneck for LC-HRMS-based workflows is their relative low throughput in sample preparation. Automation of major sample preparation steps is necessary to expand the application of LC-HRMS. Sample preparation is a universal problem for almost all LC–MS-based protein bioanalysis technologies and methods. The automation of some sample preparation procedures for small compounds, such as SPE, is well established and can be transferred to protein bioanalysis fairly easily. Automation of immunoenrichment has been attempted and may be adapted for protein quantification workflows (Berna and Ackermann 2009).

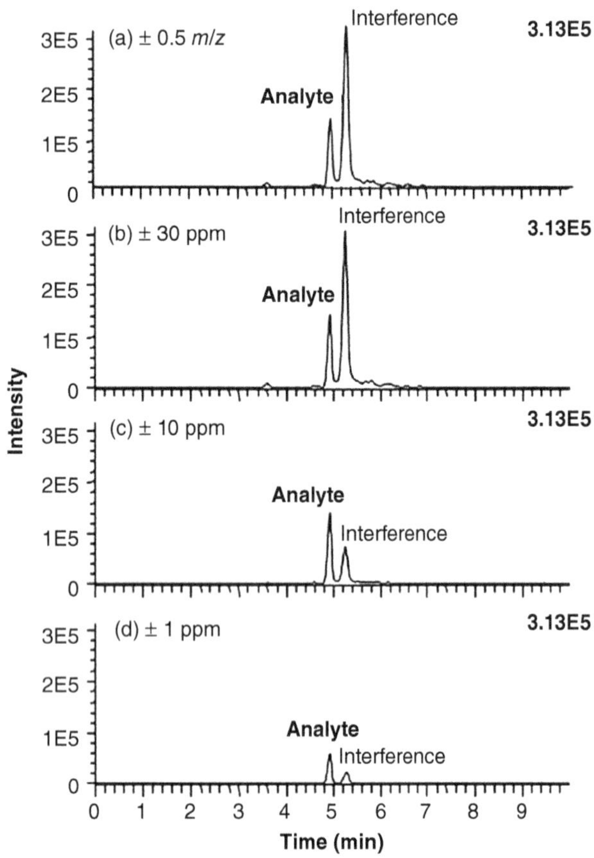

Figure 11.5 Effect of EIC window on assay selectivity and sensitivity. (a) EIC window, ±0.5 *m/z*; (b) EIC window, ±30 ppm; (c) EIC window, ±10 ppm; and (d) EIC window, ±1 ppm. (*Source*: Ruan et al. (2011). Reproduced with permission of American Chemical Society.)

A number of commercial suppliers have also made considerable efforts in the automation of proprietary sample preparation methods, including the automation of Dynobeads procedure by Life Technologies and MSIA technology by Thermo Fisher Scientific. These individual technologies will need to be extensively tested, standardized, and ultimately, linked together to improve the overall throughput of the LC–MS-based protein quantification workflows.

In summary, we believe that quantification of proteins with the LC-HRMS-based top-down approach has its unique advantages and is complementary to the LC–MS/MS-based bottom-up approach as well as conventional protein bioanalysis assays, such as ELISA. Technical hurdles still exist for LC-HRMS, but technology advancement should address many current issues and bottlenecks to expand its suitable applications. Growing needs from pharmaceutical and diagnostics industries as well as biomedical research will continue to push the LC-HRMS-based methodology as one of the important approaches for protein bioanalysis.

Acknowledgment

We thank Drs. Anna Luczak and Steve Alexander, and Nathan Connolly for critical reading and helpful discussions of this review.

References

Anderson NL and Anderson NG (2002). "The human plasma proteome: history, character, and diagnostic prospects." *Mol Cell Proteomics* **1**(11): 845–867.

Behnken HN, Ruthenbeck A, Schulz JM, et al. (2014). "Glycan analysis of prostate specific antigen (PSA) directly from the intact glycoprotein by HR-ESI/TOF-MS." *J Proteome Res* **13**(2): 997–1001.

Berna M and Ackermann B (2009). "Increased throughput for low-abundance protein biomarker verification by liquid chromatography/tandem mass spectrometry." *Anal Chem* **81**(10): 3950–3956.

Bidlingmaier M (2008). "Problems with GH assays and strategies toward standardization." *Eur J Endocrinol* **159** (Suppl 1): S41–44.

Bratcher PE and Gaggar A (2013). "Characterization and prevention of the adsorption of surfactant protein D to polypropylene." *PLoS One* **8**(9): e73467.

Brun V, Dupuis A, Adrait A, et al. (2007). "Isotope-labeled protein standards: toward absolute quantitative proteomics." *Mol Cell Proteomics* **6**(12): 2139–2149.

Cai B and Li J (1999). "Evaluation of trifluoroacetic acid as an ion-pair reagent in the separation of small ionizable molecules by reversed-phase liquid chromatography." *Anal Chim Acta* **399**(3): 249–258.

Cao J, Gonzalez-Covarrubias V, Straubinger RM, et al. (2010). "A rapid, reproducible, on-the-fly orthogonal array optimization method for targeted protein quantification by LC/MS and its application for accurate and sensitive quantification of carbonyl reductases in human liver." *Anal Chem* **82**(7): 2680–2689.

Chong BE, Yan F, Lubman DM, et al. (2001). "Chromatofocusing nonporous reversed-phase high-performance liquid chromatography/electrospray ionization time-of-flight mass spectrometry of proteins from human breast cancer whole cell lysates: a novel two-dimensional liquid chromatography/mass spectrometry method." *Rapid Commun Mass Spectrom* **15**(4): 291–296.

den Engelsman J, Garidel P, Smulders R, et al. (2011). "Strategies for the assessment of protein aggregates in pharmaceutical biotech product development." *Pharm Res* **28**(4): 920–933.

Dufield DR and Radabaugh MR (2012). "Online immunoaffinity LC/MS/MS. A general method to increase sensitivity and specificity: How do you do it and what do you need?" *Methods* **56**(2): 236–245.

Echan LA, Tang HY, Ali-Khan N, et al. (2005). "Depletion of multiple high-abundance proteins improves protein profiling capacities of human serum and plasma." *Proteomics* **5**(13): 3292–3303.

Gordon EF, Mansoori BA, Carroll CF, et al. (1999). "Hydropathic influences on the quantification of equine heart cytochrome c using relative ion abundance measurements by electrospray ionization Fourier transform ion cyclotron resonance mass spectrometry." *J Mass Spectrom* **34**(10): 1055–1062.

Gordon EF and Muddiman DC (1999). "Quantification of singly charged biomolecules by electrospray ionization Fourier transform ion cyclotron resonance mass spectrometry utilizing an internal standard." *Rapid Commun Mass Spectrom* **13**(3): 164–171.

Graumann J, Hubner NC, Kim JB, et al. (2008). "Stable isotope labeling by amino acids in cell culture (SILAC) and proteome quantitation of mouse embryonic stem cells to a depth of 5,111 proteins." *Mol Cell Proteomics* **7**(4): 672–683.

Gucinski AC and Boyne MT, 2nd (2012). "Evaluation of intact mass spectrometry for the quantitative analysis of protein therapeutics." *Anal Chem* **84**(18): 8045–8051.

Herbert B, Galvani M, Hamdan M, et al. (2001). "Reduction and alkylation of proteins in preparation of two-dimensional map analysis: why, when, and how?" *Electrophoresis* **22**(10): 2046–2057.

Heudi O, Barteau S, Zimmer D, et al. (2008). "Towards absolute quantification of therapeutic monoclonal antibody in serum by LC-MS/MS using isotope-labeled antibody standard and protein cleavage isotope dilution mass spectrometry." *Anal Chem* **80**(11): 4200–4207.

Huq MDM and Wei L-N (2007). Protein posttranslational modification: a potential target in pharmaceutical development, in *Handbook of Pharmaceutical Biotechnology*, S. C. Gad (ed.). Hoboken, NJ, USA, John Wiley & Sons, Inc.

Ji QC, Rodila R and El-Shourbagy TA (2007). "A sample preparation process for LC-MS/MS analysis of total protein drug concentrations in monkey plasma samples with antibody." *J Chromatogr B Anal Technol Biomed Life Sci* **847**(2): 133–141.

Ji QC, Rodila R, Gage EM, et al. (2003). "A strategy of plasma protein quantitation by selective reaction monitoring of an intact protein." *Anal Chem* **75**(24): 7008–7014.

Ji QC, Rodila R, Morgan SJ, et al. (2005). "Investigation of the immunogenicity of a protein drug using equilibrium dialysis and liquid chromatography tandem mass spectrometry detection." *Anal Chem* **77**(17): 5529–5533.

Jian W, Edom RW, Wang D, et al. (2013). "Relative quantitation of glycoisoforms of intact apolipoprotein C3 in human plasma by liquid chromatography-high-resolution mass spectrometry." *Anal Chem* **85**(5): 2867–2874.

Jian W, Edom RW and Weng N (2012). "Important considerations for quantitation of small-molecule biomarkers using LC-MS." *Bioanalysis* **4**(20): 2431–2434.

Julka S, Folkenroth J and Young SA (2011). "Two dimensional liquid chromatography-ultraviolet/mass spectrometric (2DLC-UV/MS) analyses for quantitation of intact proteins in complex biological matrices." *J Chromatogr B Analyt Technol Biomed Life Sci* **879**(22): 2057–2063.

Kirkwood KJ, Ahmad Y, Larance M, et al. (2013). "Characterization of native protein complexes and protein isoform variation using size-fractionation-based quantitative proteomics." *Mol Cell Proteomics* **12**(12): 3851–3873.

Kirsch S, Widart J, Louette J, et al. (2007). "Development of an absolute quantification method targeting growth hormone biomarkers using liquid chromatography coupled to isotope dilution mass spectrometry." *J Chromatogr A* **1153**(1–2): 300–306.

Kristensen AR, Gsponer J and Foster LJ (2012). "A high-throughput approach for measuring temporal changes in the interactome." *Nat Methods* **9**(9): 907–909.

Lagrain B, Brunnbauer M, Rombouts I, et al. (2013). "Identification of intact high molecular weight glutenin subunits from the wheat proteome using combined liquid chromatography-electrospray ionization mass spectrometry." *PLoS One* **8**(3): e58682.

Lassen B and Malmsten M (1996). "Structure of protein layers during competitive adsorption." *J Colloid Interf Sci* **180**(2): 339–349.

Li H, Ortiz R, Tran L, et al. (2012). "General LC-MS/MS method approach to quantify therapeutic monoclonal antibodies using a common whole antibody internal standard with application to preclinical studies." *Anal Chem* **84**(3): 1267–1273.

Liu G, Ji QC, Dodge R, et al. (2013a). "Liquid chromatography coupled with tandem mass spectrometry for the bioanalysis of proteins in drug development: practical considerations in assay development and validation." *J Chromatogr A* **1284**: 155–162.

Liu H, Manuilov AV, Chumsae C, et al. (2011). "Quantitation of a recombinant monoclonal antibody in monkey serum by liquid chromatography-mass spectrometry." *Anal Biochem* **414**(1): 147–153.

Liu X, Jin Z, O'Brien R, et al. (2013b). "Constrained selected reaction monitoring: quantification of selected post-translational modifications and protein isoforms." *Methods* **61**(3): 304–312.

Lu Q, Zheng X, McIntosh T, et al. (2009). "Development of different analysis platforms with LC-MS for pharmacokinetic studies of protein drugs." *Anal Chem* **81**(21): 8715–8723.

Lubman DM, Kachman MT, Wang H, et al. (2002). "Two-dimensional liquid separations-mass mapping of proteins from human cancer cell lysates." *J Chromatogr B Anal Technol Biomed Life Sci* **782**(1–2): 183–196.

Mayr BM, Kohlbacher O, Reinert K, et al. (2006). "Absolute myoglobin quantitation in serum by combining two-dimensional liquid chromatography-electrospray ionization mass spectrometry and novel data analysis algorithms." *J Proteome Res* **5**(2): 414–421.

Neubert H, Grace C, Rumpel K, et al. (2008). "Assessing immunogenicity in the presence of excess protein therapeutic using immunoprecipitation and quantitative mass spectrometry." *Anal Chem* **80**(18): 6907–6914.

Ning ZB, Li QR, Dai J, et al. (2008). "Fractionation of complex protein mixture by virtual three-dimensional liquid chromatography based on combined pH and salt steps." *J Proteome Res* **7**(10): 4525–4537.

Olinares PD, Ponnala L and van Wijk KJ (2010). "Megadalton complexes in the chloroplast stroma of *Arabidopsis thaliana* characterized by size exclusion chromatography, mass spectrometry, and hierarchical clustering." *Mol Cell Proteomics* **9**(7): 1594–1615.

Olivieri E, Herbert B and Righetti PG (2001). "The effect of protease inhibitors on the two-dimensional electrophoresis pattern of red blood cell membranes." *Electrophoresis* **22**(3): 560–565.

Ong SE, Blagoev B, Kratchmarova I, et al. (2002). "Stable isotope labeling by amino acids in cell culture, SILAC, as a simple and accurate approach to expression proteomics." *Mol Cell Proteomics* **1**(5): 376–386.

Padley HR, Bashir S and Wood TD (1997). "Quantification of biomolecules by external electrospray ionization Fourier transform mass spectrometry." *Anal Chem* **69**(15): 2914–2918.

Pesavento JJ, Mizzen CA and Kelleher NL (2006). "Quantitative analysis of modified proteins and their positional isomers by tandem mass spectrometry: human histone H4." *Anal Chem* **78**(13): 4271–4280.

Polaskova V, Kapur A, Khan A, et al. (2010). "High-abundance protein depletion: comparison of methods for human plasma biomarker discovery." *Electrophoresis* **31**(3): 471–482.

Rabe M, Verdes D and Seeger S (2011). "Understanding protein adsorption phenomena at solid surfaces." *Adv Colloid Interface Sci* **162**(1–2): 87–106.

Rai AJ, Gelfand CA, Haywood BC, et al. (2005). "HUPO Plasma Proteome Project specimen collection and handling: towards the standardization of parameters for plasma proteome samples." *Proteomics* **5**(13): 3262–3277.

Robinson NE and Robinson AB (2004). *Molecular Clocks: Deamidation of Asparaginyl and Glutaminyl Residues in Peptides and Proteins*. Cave Junction, OR, USA, Althouse Press.

Roth MJ, Parks BA, Ferguson JT, et al. (2008). "Proteotyping:' population proteomics of human leukocytes using top down mass spectrometry." *Anal Chem* **80**(8): 2857–2866.

Ruan Q, Ji QC, Arnold ME, et al. (2011). "Strategy and its implications of protein bioanalysis utilizing high-resolution mass spectrometric detection of intact protein." *Anal Chem* **83**(23): 8937–8944.

Sharma S, Simpson DC, Tolic N, et al. (2007). "Proteomic profiling of intact proteins using WAX-RPLC 2-D separations and FTICR mass spectrometry." *J Proteome Res* **6**(2): 602–610.

Shou WZ and Weng N (2005). "Simple means to alleviate sensitivity loss by trifluoroacetic acid (TFA) mobile phases in the hydrophilic interaction chromatography-electrospray tandem mass spectrometric (HILIC-ESI/MS/MS) bioanalysis of basic compounds." *J Chromatogr B Analyt Technol Biomed Life Sci* **825**(2): 186–192.

Sowell RA, Koeniger SL, Valentine SJ, et al. (2004). "Nanoflow LC/IMS-MS and LC/IMS-CID/MS of protein mixtures." *J Am Soc Mass Spectrom* **15**(9): 1341–1353.

Staub A, Comte S, Rudaz S, et al. (2010). "Use of organic solvent to prevent protein adsorption in CE-MS experiments." *Electrophoresis* **31**(19): 3326–3333.

Thakur D, Rejtar T, Karger BL, et al. (2009). "Profiling the glycoforms of the intact alpha subunit of recombinant human chorionic gonadotropin by high-resolution capillary electrophoresis-mass spectrometry." *Anal Chem* **81**(21): 8900–8907.

Veronese FM and Mero A (2008). "The impact of PEGylation on biological therapies." *BioDrugs* **22**(5): 315–329.

Vlasak J and Ionescu R (2011). "Fragmentation of monoclonal antibodies." *MAbs* **3**(3): 253–263.

Wagner BM, Schuster SA, Boyes BE, et al. (2012). "Superficially porous silica particles with wide pores for biomacromolecular separations." *J Chromatogr A* **1264**: 22–30.

Warnke C, Hermanrud C, Lundkvist M, et al. (2012). "Anti-drug antibodies." *Drugs Therapy Stud* **2**(1): 56–63.

Wu ST, Ouyang Z, Olah TV, et al. (2011). "A strategy for liquid chromatography/tandem mass spectrometry based quantitation of pegylated protein drugs in plasma using plasma protein precipitation with water-miscible organic solvents and subsequent trypsin digestion to generate surrogate peptides for detection." *Rapid Commun Mass Spectrom* **25**(2): 281–290.

Wu Z, Wei B, Zhang X, et al. (2014). "Efficient separations of intact proteins using slip-flow with nano-liquid chromatography-mass spectrometry." *Anal Chem* **86**(3): 1592–1598.

Xu K, Liu L, Saad OM, et al. (2011). "Characterization of intact antibody-drug conjugates from plasma/serum *in vivo* by affinity capture capillary liquid chromatography-mass spectrometry." *Anal Biochem* **412**(1): 56–66.

Xu Y, Mehl JT, Bakhtiar R, et al. (2010). "Immunoaffinity purification using anti-PEG antibody followed by two-dimensional liquid chromatography/tandem mass spectrometry for the quantification of a PEGylated therapeutic peptide in human plasma." *Anal Chem* **82**(16): 6877–6886.

You J, Wang L, Saji M, et al. (2011). "High-sensitivity TFA-free LC-MS for profiling histones." *Proteomics* **11**(16): 3326–3334.

Zhang SW and Jian W (2014). "Recent advances in absolute quantification of peptides and proteins using LC-MS " *Reviews in Analytical Chemistry* **33**(1): 31–47.

Zubarev RA and Makarov A (2013). "Orbitrap mass spectrometry." *Anal Chem* **85**(11): 5288–5296.

12

LC–MS/MS Bioanalytical Method Development Strategy for Therapeutic Monoclonal Antibodies in Preclinical Studies

Hongyan Li, Timothy Heath, and Christopher A. James

Amgen Inc., Thousand Oaks, CA, USA

12.1 Introduction: LC-MS/MS Bioanalysis of Therapeutic Monoclonal Antibodies

Quantitative analysis of therapeutic monoclonal antibodies (mAbs) in biological matrices has become increasingly important with the growing number of novel therapeutic antibody-based therapeutics. Novel therapeutic mAbs are registered or in development for treatment of a variety of human diseases including cancer, immunological disorders, infectious diseases, osteoporosis, respiratory disorders, muscular dystrophy, Alzheimer's disease, and pain (Chames et al. 2009, Beck et al. 2010, Weiner et al. 2010, Reichert 2011). Ligand binding assays (LBAs), typically enzyme-linked immunosorbent assay (ELISA), have been widely used to provide quantitative bioanalytical data for pharmacokinetic (PK), pharmacodynamic (PD), and toxicokinetic (TK) studies for mAb lead candidate selection and development. However, LBAs have some limitations and challenges, notably, method development is often time-consuming and costly particularly as specific unique critical reagents are required. These assays can be susceptible to interference from the matrix proteins, and the resources to develop and validate multiple LBAs for antibody drug candidates in different species can be substantial (Savoie et al. 2010).

Liquid chromatography–mass spectrometry (LC–MS/MS) is emerging as a promising alternative and complementary assay platform for quantitative analysis of therapeutic antibodies (Dubois et al. 2008, Hagman et al. 2008, Heudi et al. 2008, Duan et al. 2012a, 2012b, Furlong et al. 2012, Li et al. 2012, Furlong et al. 2013, Jiang et al. 2013, Li et al. 2013, Xu et al. 2014, 2015, Jenkins et al. 2015). LC–MS/MS has been widely used for small molecule and peptide bioanalysis, particularly with SRM (selected reaction monitoring) detection performed on triple quadrupole mass spectrometers (QqQ). In the SRM mode, the analyte molecular ions with a specific *m/z* (mass-to-charge ratio) are selected as the precursor ions by the first quadrupole (Q1) filter, and then fragmented in the second quadrupole (Q2) by collision-induced dissociation (CID). Subsequently, the third quadrupole (Q3) is set to allow product ions with a specific *m/z* to reach the detector. A signal is registered only when the predefined product ions arise from the predefined precursor ions. Since the SRM detection is based on the unique ion transitions (precursor ion *m/z* → product ion *m/z*) for each analyte, it is thus highly specific and capable of analyte measurement in complex biological matrices (e.g., plasma). In addition, SRM operates at high ion transmission efficiency and high duty cycle; as a result, SRM-based LC–MS/MS methods exhibit superior sensitivity and wide dynamic range for quantification.

Therapeutic antibodies have a much higher molecular weight (around 150 kDa) and greater structural complexity than small molecules and peptides, and direct LC–MS/MS quantitation is impractical for multiple reasons. Notably, electrospray ionization (ESI) of an intact antibody would generate numerous molecular ions with a widespread charge distribution, and most of these ions would be outside the *m/z* range of a typical triple quadrupole mass spectrometer. Therefore, the current practice for antibody quantitation by LC–MS/MS is to focus on analysis of characteristic peptides (surrogate peptides) after enzymatic digestion. Proteolytic peptides with a unique sequence and appropriate physicochemical properties can act as surrogates to represent the antibody molecule for quantitation. Among many available proteases, trypsin is generally preferred since tryptic peptides often have the appropriate size (~5–30 amino acids) for LC separation and good ionization efficiency for MS detection. Tryptic peptides also produce predictable MS/MS fragment ions, mostly b-ions and y-ions, and they can be utilized to simplify SRM selection and optimization during method development. The well-established LC–MS/MS methodology and the

Protein Analysis using Mass Spectrometry: Accelerating Protein Biotherapeutics from Lab to Patient, First Edition.
Edited by Mike S. Lee and Qin C. Ji.

advantages in sensitivity, specificity, linear dynamic range, accuracy, and precision for small molecule and peptide bioanalysis have generally been found suitable for therapeutic mAb bioanalysis with some adaption of approaches to meet the unique challenges associated with antibody molecules. For example, therapeutic mAb may circulate as the free mAb or bound to antidrug antibodies (ADAs), soluble targets, or other binding proteins in serum (White and Bonilla 2012). LC–MS/MS methods of small molecules and peptides generally measure total concentrations. However, for LC–MS/MS of mAb selective enrichment approaches in particular can allow a specific form of the mAb to be measured, for example, free mAb, mAb bound to a soluble target or ADAs, or the total mAb (sum of free and bound) (Fernandez Ocana et al. 2012). Appropriate assay design and development is thus important in order to understand exactly what is being measured, to provide appropriate PK interpretation, and for meaningful correlation if data is compared with LBAs (Fernandez Ocana et al. 2012, Jiang et al. 2013, Li et al. 2013). In addition, therapeutic mAbs often are heterogeneous in nature due to glycosylation, or other posttranslational modifications. mAb quantitation can be performed targeting surrogate peptides selected where modifications to the mAb are unlikely to occur, or using surrogate peptides selected to reflect specific modifications that may provide insights into how the modifications affect the overall pharmacokinetic properties of the antibody (Yin et al. 2013).

A typical LC–MS/MS assay workflow for mAb quantification includes biological sample preparation, enzymatic digestion, and surrogate peptide LC–MS/MS analysis for quantitation. Increasing number of publications reflect substantial improvements being made in each of these analytical steps to improve assay sensitivity, selectivity, robustness, and throughput for therapeutic antibody bioanalysis. This chapter highlights some recent method development strategies and provide case studies to represent utility of this approach in support of preclinical therapeutic mAb discovery research.

12.2 Highlights of Recent Method Development Strategies

12.2.1 Strategy for Surrogate Peptide Selection and Optimization

Surrogate peptides serve as the accurate quantitative representation of the analyte antibody for measurement. As a result, the identification, evaluation, and selection of surrogate peptides are critical for method development. The strategy for the identification and selection of optimal surrogate peptides varies between different laboratories. Various *in silico* tools, including MRMPilot, Skyline (MacLean et al. 2010), Pinpoint, and PeptideSieve (Mallick et al. 2007), are available and useful to aid in surrogate peptide selection. However, these tools should be used with caution since some potential surrogate peptide candidates can be inaccurately filtered out. An experimental approach is the most reliable and effective way to find the optimal surrogate peptides for most antibodies. Figure 12.1 summarizes our strategy and workflow for surrogate peptide selection and optimization. Briefly, it involves (i) *in vitro* target antibody digestion and peptide identification; (ii) peptide specificity, LC–MS/MS response, and stability assessment in the target biomatrices; and (iii) *in vivo* assay performance assessment in a pilot study using multiple surrogate peptides prior to the final surrogate peptide selection.

After initial *in vitro* antibody digestion, tryptic peptides are identified from the digest by a data-dependent LC–MS/MS experiment performed on a Q-TOF or LTQ mass spectrometer. The experimental MS/MS data of each identified peptide are then utilized to establish suitable SRM ion transitions. Specific peptides are avoided due to potential analytical challenges. For example, the N-terminal peptide of an antibody containing glutamine or glutamic acid is excluded due to spontaneous cyclization to pyroglutamic acid. The C-terminus peptide of the heavy chain is not considered due to potential for partial removal of lysine by carboxypeptidases during cell culture production. In addition, peptides that contain methionine are excluded due to significant propensity for methionine oxidation. The SRMs established from MS/MS data for each peptide are verified and optimized. All peptide candidates are then thoroughly assessed for specificity, interferences, and stability in the target biomatrices. In addition to *in silico* sequence similarity search with the basic local alignment search tool (BLAST), peptide sequence uniqueness is verified by the absence of interference from the blank matrix digest and should be evaluated separately for each biomatrix. Isobaric SRM interferences may also present and may be identified and managed through LC optimization. Digests of biomatrix spiked with antibody analyte are used to assess peptide LC retention behavior, MS responses, and the optimal LC parameters for high-throughput bioanalysis. It is not unusual that peptide candidates with high sensitivity and selectivity can be unstable, and thus, unsuitable for antibody quantification (Duan et al. 2012a, 2012b). Autosampler stability up to 48 h at a designated temperature is generally adequate for peptide stability assessment. Peptide LC–MS/MS response is also affected by digestion kinetics, which is discussed in detail in Section 12.2.3. The stable peptide candidates are then assessed for their performance using a pilot *in vivo* study, and this test provides the opportunity to evaluate assay reliability using multiple peptide candidates. Peptides with liabilities that are evident *in vivo* can then be

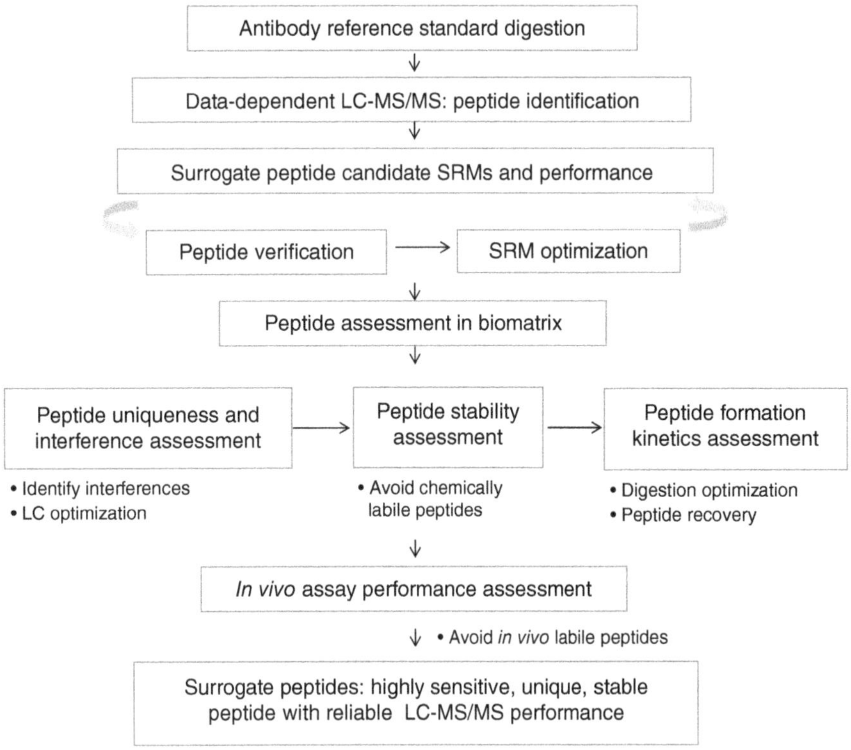

Figure 12.1 Workflow for the selection of appropriate signature peptides.

filtered out. Multiple surrogate peptides from different antibody domains can also be useful for confirmatory tests to assess the quantitation accuracy and the integrity of the antibody drug in the study samples (Dubois et al. 2008, Jiang et al. 2013). For the final analytical method, the peptide with the optimal performance in terms of sensitivity, specificity, stability, accuracy, and precision is selected as the surrogate peptide for antibody bioanalysis.

Therapeutic mAbs are now often fully human IgG monoclonal antibodies (e.g., IgG1, IgG2, and IgG4); Figure 12.2 illustrates a human IgG1 structure as an example. Human IgG1 consists of two identical heavy chains (γ) and two identical light chains (either κ or λ). Each chain has domains that are held together by disulfide bonds. The light chain has one variable domain and one constant domain. The heavy chain has one variable domain (VH) and three constant domains (CH1, CH2, and CH3). The amino acid sequences are the same for each IgG subclass in the constant domains but different in variable domains. Each variable domain has three short, separated, highly distinct sequences, called complementarity determining regions (CDRs): CDR1, CDR2, and CDR3. CDRs form unique antigen binding sites, which play an important role in antigen recognition.

Regardless of each domain's biological functions, the unique molecular structure of the antibody enables two distinct assay approaches for antibody bioanalysis. The first one is a highly specific assay approach using the surrogate peptides from the CDRs in the variable domains

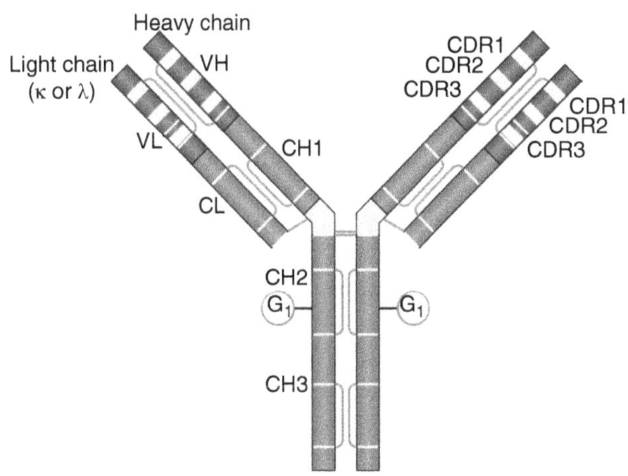

Figure 12.2 Schematic structure of the human IgG1 antibody. V indicates variable domains; C indicates constant domains; VL and CL are domains of the light chain; VH, CH1, CH2, and CH3 are domains of the heavy chain; CDR1, CDR2, and CDR3 are complementary determining regions in variable domains of light and heavy chains; G1 indicates the glycan and lines between light chain and heavy chain indicate – S—S – disulfide bond.

(unique surrogate peptide). Such assays are specific to the individual antibody and are appropriate in support of both preclinical and clinical studies (Dubois et al. 2008, Fernandez Ocana et al. 2012, Li et al. 2013, Xu et al. 2014, 2015) that include studies with multiple coadministered mAbs and surrogate mAb preclinical studies in the same species. It should be noted that due to the limited number of CDR peptides for each antibody, generation and selection of surrogate peptides that contain CDR peptides is therefore typically more challenging, and sometimes alternative enzymatic digestions (other than trypsin) should be considered. The second approach is a more generic assay using surrogate peptides from the constant domains of an antibody ("universal surrogate peptide") (Furlong et al. 2012, Li et al. 2012, Furlong et al. 2013, Zhang et al. 2014). In general, this is only appropriate for bioanalytical support when humanized therapeutic antibodies are administered to preclinical species where the endogenous IgGs from the animal species are different from humans. Due to the identical amino acid sequence in the constant domains of mAbs of the same IgG subclass, it is possible to develop a single LC–MS/MS method applicable to all mAb candidates of the same subclass (e.g., IgG1) for use in preclinical studies, which is similar to the generic ligand binding assays using anti-human Fc reagents for both capture and detection. This approach can be very valuable as it reduces time and resources required for method development for preclinical evaluation of therapeutic mAbs.

Table 12.1 lists a complete library of identified surrogate peptide candidates in the constant domains of human IgG1 monoclonal antibody including the sequence, location, ion transitions, and suitability as surrogate peptide for quantitation. Figure 12.3 contains representative chromatograms of the identified universal surrogate peptide candidates of human IgG1 antibody from Table 12.1, which further demonstrates the retention behavior and LC–MS/MS response under optimized conditions. The relative intensity of each peptide is dependent on LC–MS/MS conditions and other method details. Peptides, such as EPQ, TPE, and THT, are not suitable for quantitation due to poor chromatographic characteristics and low SRM responses as shown in Figure 12.3, while VVS and TVA have demonstrated robust analytical performance for human IgG1 quantitation (Furlong et al. 2012, Furlong et al. 2013, Jiang et al. 2013, Nouri-Nigjeh et al. 2014, Zhang et al. 2014). Assays using other universal surrogate peptides on the list can be expected. Similar method libraries can be found in the literature (Lesur et al. 2010, Duan et al. 2012a, 2012b, Zhang et al. 2014) and potentially could also be established for IgG2 and IgG4 antibodies.

12.2.2 Sample Preparation

Because of the complex nature of biological samples, efficient sample preparation to remove unwanted biomatrix components and to selectively extract/enrich the analytes of interest is an essential part of the bioanalytical workflow for sensitive and selective LC–MS/MS assays. This is especially important for antibody bioanalysis because endogenous immunoglobulins share many physicochemical characteristics with the therapeutic antibodies. In addition, during digestion many peptides can also be generated from other abundant endogenous proteins such as serum albumin, which may potentially interfere with the detection of the surrogate peptide from the targeted antibody analyte. The strategies for sample preparation of antibody LC–MS/MS bioanalysis are divided into two main categories: immunoaffinity-based sample preparation and nonimmunoaffinity-based sample preparation. In general, sample cleanup requirements are more challenging for antibody quantitation than for small-molecule LC–MS/MS bioanalysis.

12.2.2.1 Immunoaffinity-Based Sample Preparation

Immunoaffinity-based sample preparation or immunoaffinity capture (IC) involves selective purification using a specific antibody or binding protein against the target antibody. Immunoaffinity purification makes downstream digestion more efficient and produces a less complex digest. The obvious benefit is the potential enhancement in assay detection limits. As a result, lower limit of quantitation (LLOQ) at nanogram per milliliter level of mAbs becomes feasible for antibody bioanalysis by LC–MS/MS. In addition, immunoaffinity capture reagents with various selective mechanisms provide potential opportunities to measure free or bound drug forms when ADAs, soluble targets, or other binding proteins might interact with the antibody drug in serum.

Figure 12.4 illustrates various immunoaffinity capture mechanisms, which can be utilized for mAb measurements. For free or active antibody measurement, the capture reagent can either be the target (Dubois et al. 2008) or an anti-idiotype antibody (Fernandez Ocana et al. 2012, Xu et al. 2014, 2015), which bind to the same epitope of the analyte antibody that binds the target. For total antibody measurement (free and bound mAb), the capture reagent can be an antibody or protein that binds to a different part of the analyte antibody, which does not compete for target binding. The total antibody measurement allows more generic immunoaffinity capture approaches; for example, therapeutic mAbs can be captured by IgG binding proteins such as protein A or G (Fernandez Ocana et al. 2012, Furlong et al. 2013). However, highly abundant endogenous immunoglobulins can affect the capture of the therapeutic antibody due to cross-reactivity with protein A or G and thus affect assay sensitivity and performance. Another more useful generic approach for therapeutic mAbs utilizes an antibody reagent against a specific region of the heavy chain such as human Fc (Li et al. 2012, Li et al. 2013) or a specific type of the human light chain either κ or λ, which therefore allows a more specific differentiation between therapeutic mAbs and endogenous IgGs.

Table 12.1 Universal surrogate peptide candidates in the constant domains of human IgG1 monoclonal antibody with amino acid sequence, location, ion transition, and analytical suitability for quantitation.

Abbreviation	Peptide	Domain	Precursor	Product	Suitability
TVA	TVAAPSVFIFPPSDEQLK	Hu IgG1 LC CL	973.5171++	913.5+	Yes
SGT	SGTASVVCLLNNFYPR	Hu IgG1 LC CL	580.9628+++	435.2+	Yes
DST	DSTYSLSSTLTLSK	Hu IgG1 LC CL	751.8829++	448.3+	No
VYA	VYACEVTHQGLSSPVTK	Hu IgG1 LC CL	606.9733+++	444.3+	No
GPS	GPSVFPLAPSSK	Hu IgG1 HC CH1	593.8270++	699.4+	Yes
STS	STSGGTAALGCLVK	Hu IgG1 HC CH1	632.8319++	519.3+	Yes
THT	THTCPPCPAPELLGGPSVFLFPPKPK	Hu IgG1 HC CH2	910.8097+++	566.4+	No
TPE	TPEVTCVVVDVSHEDPEVK	Hu IgG1 HC CH2	694.6735+++	328.2+	No
FNW	FNWYVDGVEVHNAK	Hu IgG1 HC CH2	559.9388+++	697.4+	No
VVS	VVSVLTVLHQDWLNGK	Hu IgG1 HC CH2	603.3403+++	805.44++	Yes
ALP	ALPAPIEK	Hu IgG1 HC CH2	419.7553++	327.7+	Yes
EPQ	EPQVYTLPPSR	Hu IgG1 HC CH3	643.8406++	456.3+	No
NQV	NQVSLTCLVK	Hu IgG1 HC CH3	552.8077++	763.4+	Yes
GFY	GFYPSDIAVEWESNGQPENNYK	Hu IgG1 HC CH3	848.7153+++	764.4+	No
TTP	TTPPVLDSDGSFFLYSK	Hu IgG1 HC CH3	937.4645++	836.4++	Yes

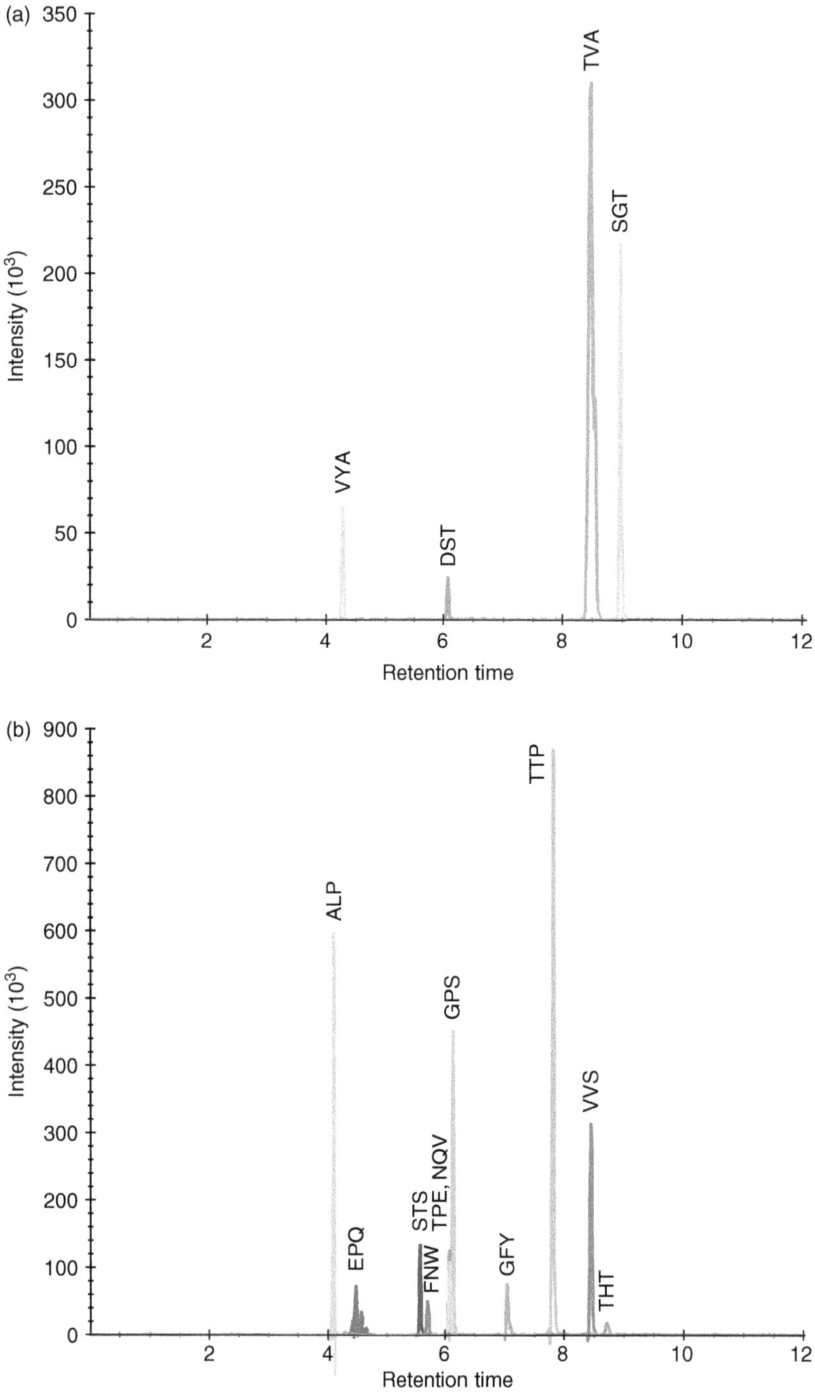

Figure 12.3 Representative chromatograms of identified universal surrogate peptide candidates of human IgG1. (a) Surrogate peptide candidates from the light chain constant domain of human IgG1. (b) Surrogate peptide candidates from the heavy chain constant domain of human IgG1.

Improved throughput can be achieved by the use of immunoaffinity capture in a 96-well format in combination with liquid handling robotic systems. Streptavidin-coupled magnetic beads, affinity pipette tips, and ELISA plates serve as the main platform to perform affinity capture. Among them, streptavidin-coupled magnetic beads have been used with great success (Dubois et al. 2008, Li et al. 2012, Li et al. 2013, Xu et al. 2014, 2015) due to good binding reproducibility and the ability to readily adjust binding capacity by increasing or decreasing the beads volume used for a particular assay. The bead has a high surface area with a monolayer of recombinant

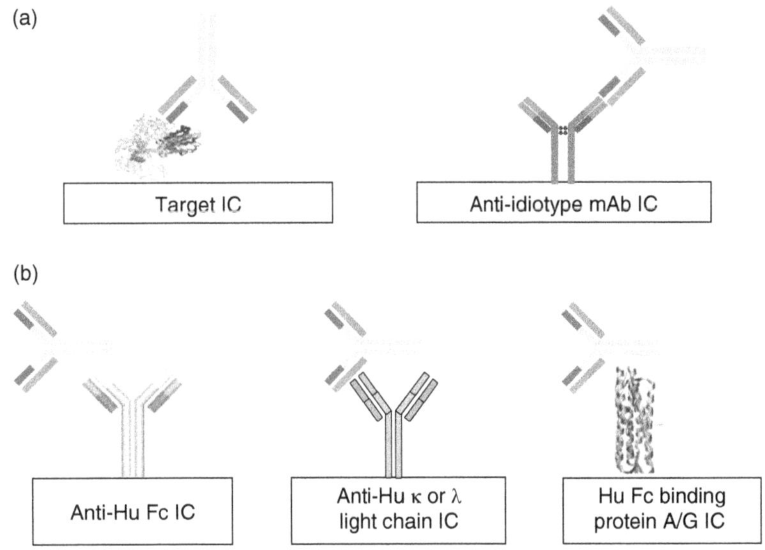

Figure 12.4 Selective immunoaffinity capture approaches for antibody LC–MS/MS quantitation. (a) Immunoaffinity capture approaches for free or active antibody measurement. (b) Immunoaffinity capture approaches for total antibody measurement.

streptavidin covalently attached to the surface. The streptavidin allows high affinity binding of biotinylated capture antibody ($K_d = 10^{-15}$), and thus, the beads provide high capacity capture of the therapeutic antibody. The beads also have low sedimentation rate and a high iron content, which allow rapid magnetic separation and downstream handling. The binding capacity used for biological sample capture defines the assay calibration linear dynamic range and should be assessed carefully. Figure 12.5 demonstrates the effect of binding capacity on the concentration/response of a human IgG2 antibody in 25 μL rat plasma; here a biotinylated antihuman Fc mouse antibody was used as the capture antibody. When the affinity binding capacity is in great excess compared to the antibody concentration in the plasma, the concentration/response for the antibody are both linear when using 25 and 50 μL magnetic beads. When affinity binding capacity becomes a limiting factor as the antibody concentration in the plasma increases, the concentration/response becomes nonlinear. As shown in Figure 12.5, the assay linear dynamic range increases with the binding capacity in a linear manner, the use of 25 μL beads is appropriate for quantitation in the linear calibration range of 50–12,500 ng/mL, while the use of 50 μL beads extends the upper linear range to 25,000 ng/mL in the same 25 μL rat plasma.

12.2.2.2 Nonimmunoaffinity-Based Sample Preparation

When total antibody measurement will meet study requirements and sensitivity (LLOQ) is not a critical

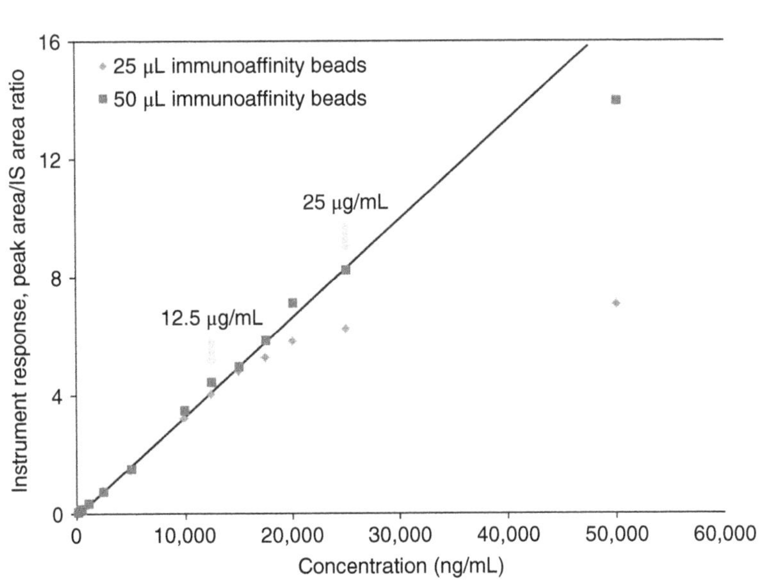

Figure 12.5 Effect of binding capacity on concentration/response of a human IgG2 antibody in 25 μL rat plasma. Diamond: 25 μL beads. Square: 50 μL beads.

limiting factor, simple and cost-effective nonimmuno-affinity-based approaches, such as protein precipitation and selective serum albumin removal, can be considered.

Protein precipitation is commonly used for removing matrix proteins prior to the quantitation of small molecules, and can also be used for the quantification of therapeutic antibodies. The antibody drug is coprecipitated with the matrix proteins and recovered in the pellet. Organic solvents, such as methanol or acetonitrile, have been used for protein precipitation, followed by pellet enzymatic digestion (Heudi et al. 2008, Duan et al. 2012a, 2012b, Furlong et al. 2012, Yuan et al. 2012, Furlong et al. 2013, Jiang et al. 2013, Zhang et al. 2014). Lipids, salts, soluble peptides/proteins are removed in the supernatant. This approach is simple and straightforward. However, protein precipitation/pellet digestion has several undesirable effects mainly due to the fact that the endogenous matrix proteins are codigested, forming a large number of abundant tryptic peptides. Since the physicochemical properties of the matrix peptides are often rather similar to those of the surrogate peptides, this can result in a lack of assay sensitivity (LLOQ at ~μg/mL level) (Jiang et al. 2013) and high levels of interferences, which may demand more rigorous LC separation, for example, 2D HPLC separation.

Albumin is the most abundant protein in plasma or serum at the level of approximately 30–50 mg/mL (Cohn et al. 1947); therefore, selective removal of serum albumin will clear more than 50% of the total protein content and result in a much cleaner matrix compared to direct protein precipitation using organic solvents. Selective removal of serum albumin has been achieved previously by immunodepletion using albumin binding protein or antibody reagents (Hagman et al. 2008, Lu et al. 2009). However, high variability and significant sample losses are typically related to this approach, which make it likely unsuitable for antibody quantitative analysis. Recently, selective removal of serum albumin using a chemical approach has emerged as an interesting alternative due to its simplicity and effectiveness. One example is the use of isopropanol containing 1% trichloroacetic acid (TCA) (Liu et al. 2014). Serum albumin and TCA form a stable complex, which has good solubility in isopropanol. By adding a 10-fold volume excess of isopropanol containing 1% TCA to plasma and discarding the supernatant, serum albumin can be selectively removed with an efficiency of up to 95% while immunoglobulins are retained in the pellet with a recovery close to 100% for a therapeutic mAb. As a result, significant improvement in sensitivity and selectivity are achieved compared to pellet digestion after protein precipitation by organic solvents. This approach is also compatible with automation for high-throughput antibody bioanalysis. Additional chemical approaches for selective removal of serum albumin or selective precipitation of IgGs can be anticipated.

12.2.3 Accelerated Trypsin Digestion

An efficient trypsin digestion is important for high-throughput antibody bioanalysis (Arsene et al. 2008, Lesur et al. 2010). This requires reliable and reproducible generation of surrogate peptides with adequate yield in a minimal amount of time. The enzymatic digestion of the antibody is often preceded by predigestion steps such as denaturation, reduction, and alkylation. The predigestion procedures help to unfold the antibody tertiary structure and open the disulfide bonds in each domain, thereby making the antibody more accessible to the digestion enzyme. For many assays, there is scope to streamline these predigestion procedures. For Cys-free surrogate peptides (i.e., VVSVLTVLHQDWLNGK, ALPAPIEK), a single-step predigestion procedure can be adopted by combining the denaturation and reduction without cysteine alkylation. Cysteine alkylation is traditionally performed to further block the reactive free thiols for cysteines; however, our recent work indicated cysteine alkylation is not absolutely necessary as long as there is residual reducing agent TCEP (tris(2-carboxyethyl)phosphine) in the final digest for LC–MS/MS analysis.

After predigestion treatment, conventional trypsin digestion methods often involve up to 12–16 h of incubation, which makes it the major rate-limiting step for high-throughput bioanalysis. Alternative methods have been introduced to accelerate enzymatic digestions in proteomic researches including heating, ultrasonic energy, high pressure, infrared energy, microwave energy, alternating electric fields, and immobilized trypsin microreactors. Among them microwave irradiation has been successfully utilized to speed up protein digestion for protein identification and characterization (Sandoval et al. 2007, Lesur et al. 2010, Reddy et al. 2010). In a review paper on microwave-assisted proteomics, Lill et al. (2007) described the kinetics in the microwave-assisted incubation as different from temperature-assisted digestion, in that proteolysis was greatly enhanced by microwave radiation and particularly with tightly folded proteins. Lesur et al. (2010) reported three different formation kinetics of peptides after enzymatic digestion of a human monoclonal antibody: (i) fast forming peptides reaching a plateau, which are preferred for quantitative analysis; (ii) slow forming peptides that never reach a maximum, which may be considered as long as the digestion is consistent and detection limit is adequate for the analysis; and (iii) rapidly forming intermediate peptides that further digest over time, which are unlikely to be useful for quantitation. An example of a surrogate peptide generation in 15 min by microwave-assisted trypsin digestion showed equivalent efficiency to a traditional overnight digestion at 37 °C. The formation kinetics of individual tryptic peptide depends on multiple factors including the peptide sequence, digestion temperature, microwave irradiation power, enzyme to substrate ratio, and predigestion treatment of the antibody.

However, to date, only limited information has been published about applications of microwave-assisted digestion for quantitative bioanalysis, particularly related to reproducibility when considering microwave-assisted trypsin digestion for high-throughput antibody bioanalysis in 96-well plate format. Our laboratory has established evaluation and optimization strategies in terms of hardware, formation kinetics of surrogate peptides, and digestion reliability for optimal assay performance. A microwave digestion system from Hudson Surface Technology was used and customized for 96-deep well plate digestion. This system provides uniform microwave output and precise temperature control. The formation kinetics of surrogate peptides from the targeted antibody is performed by monitoring the surrogate peptide SRM response over a time course under specified conditions as shown in Figure 12.6, using surrogate peptide VVS (Human IgG1 mAb) as an example. In this work, 200 µL of trypsin digestion solutions with concentrations from 0.1 to 1.75 mg/mL were used to perform the microwave-assisted digestion (400 W at 30 °C). The SRM response demonstrated that the trypsin concentration or the enzyme to substrate ratio needed for digestion of the target antibody is critical. The enzyme amount should be sufficient to perform the digestion, but not too high that results in trypsin autolysis and additional nonspecific digestion of the targeted antibody. Figure 12.6 shows three representative formation kinetics of VVS peptide with trypsin concentrations at 0.1, 0.25, and 1 mg/mL. 0.25 mg/mL trypsin appeared to be optimal with the highest overall SRM response, and the SRM response reached a plateau with 5 min of microwave irradiation in the 30-min kinetic assessment. It appears that 0.1 mg/mL trypsin was insufficient to perform the digestion since the

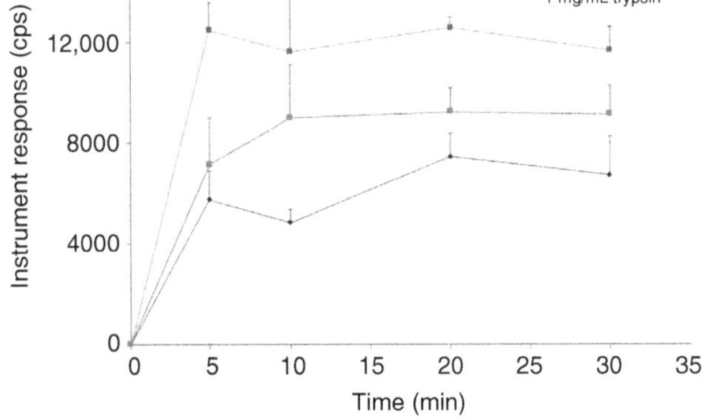

Figure 12.6 Formation kinetics of surrogate peptide VVSVLTVLHQDWLNGK using microwave-assisted digestion (400 W at 30 °C) with trypsin concentrations at 0.1, 0.25, and 1 mg/mL.

plateaued SRM response was much lower than that from digestion using 0.25 mg/mL trypsin while trypsin concentration at 1 mg/mL and above (data not shown) caused an evident drop in peptide SRM response.

For antibodies in biological matrices, it becomes even more complicated in that the antibody analyte may react differently with trypsin in a different matrix environment. Less effective digestions have been observed when model proteins were digested in complex mixtures compared to when digested separately (Hustoft et al. 2011), which may be caused by cleavage site competition when multiple proteins are digested together. Therefore, we found it important to assess the amount of trypsin to be used for an optimal digestion of the antibody in the biological sample.

The digestion consistency was tested by monitoring the surrogate peptide SRM response across the 96-well plate with the same amount of antibody in each well. Percentage of CV (%CV) was calculated for each row and each column as well as for the entire plate. All were within 10% clearly indicating the digestion homogeneity across the 96-well plate with 10-min microwave accelerated trypsin digestion. Further method qualification using calibration standards and quality controls (QCs) demonstrated reliable assay performance in terms of precision and accuracy.

In conclusion, microwave-assisted digestion enables accelerated trypsin digestion in a 96-well plate format for high-throughput mAb bioanalysis. Careful case-by-case assessment of the digestion kinetics and reliability is highly recommended to provide insights into obtaining optimal assay performance for method development and optimization (Lesur et al. 2010, Hustoft et al. 2011). In addition, the streamlined predigestion treatment and accelerated trypsin digestion not only enables high-throughput capability for antibody bioanalysis but also minimizes assay variation due to a highly efficient analytical workflow, and potentially can enable a simplified internal standard strategy as described in the following section.

12.2.4 Internal Standard Selection

The use of internal standards is a near universal practice in quantitative LC–MS/MS bioanalysis to ensure assay accuracy, precision, and reproducibility. For small-molecule bioanalysis use of stable isotope-labeled internal standards is strongly preferred as they have essentially identical physicochemical properties to the analyte. However, production of stable isotope-labeled internal standards (SIL-ISs) becomes more complicated with increasing size and complexity of the analyte molecule. Stable isotope-labeled internal standard strategies for antibody quantification can be mainly categorized into three groups: the SIL-peptide, cleavable flanking SIL-peptide, and SIL-mAb.

Analog or non-SIL internal standard can potentially be used for mAb quantitation (Szapacs et al. 2010, Li et al. 2013) but will not be discussed here as SIL internal standards are strongly preferred for all types of LC–MS/MS analysis.

12.2.4.1 SIL-Peptide IS

This internal standard is based on the selected antibody surrogate peptide while substituting one or more amino acids labeled with stable isotopes ($^{13}C/^{15}N$) to yield sufficient mass difference for distinctive SRM detection (Dubois et al. 2008, Li et al. 2012, Liu et al. 2014, Xu et al. 2014, 2015). SIL-peptide internal standards are relatively easy to synthesize or purchase from vendors for fast method development. A SIL-peptide would be added after antibody digestion in the analytical workflow and expected to only correct for the assay variability occurring in postdigestion sample processing, LC separation, and variability in the SRM detection due to sample matrix. The applicability of this type of IS may in particular be limited for assays with complicated extraction and digestion workflows.

12.2.4.2 Cleavable Flanking SIL-Peptide IS

The correction of variations in digestion step can be improved by using a SIL-peptide internal standard with additional cleavable sequences on either the N-terminus or C-terminus or on both (Li et al. 2012, Jiang et al. 2013). The cleavable flanking SIL-peptide internal standard would be added in the predigestion step after antibody extraction in the analytical workflow. This internal standard strategy combines relatively easy and affordable synthesis with the ability to compensate for variability that occurs in the digestion step in addition to the role of SIL-peptide internal standard during LC–MS/MS analysis and postdigestion sample preparation. The applicability of this approach is limited as it cannot correct for the variability that occurs in the antibody extraction step, and digestion of flanking peptide might be different from that of the intact antibody.

12.2.4.3 SIL-mAb IS

Potentially, the ideal internal standard for antibody bioanalysis would be the SIL-mAb, which has near identical behavior to the analyte antibody in the entire analytical workflow (Heudi et al. 2008, Li et al. 2012, Zhang et al. 2014). SIL-mAb can be produced by incorporation of isotope-labeled amino acids in the target mAb through labeling of whole cells in culture medium containing the SIL-amino acids. SIL-mAb internal standard would be added in the first step of the analytical workflow and can potentially correct for the variability that occurs in all the steps of analytical process; therefore, SIL-mAb would be the IS of choice when the assay requires extensive sample preparation and complicated enzymatic digestion. However, SIL-mAb internal standards are not easily available and require substantial time, expertise, and resources to generate (from protein expression to purification).

In summary, the internal standard strategy generally relies on available resources and most importantly on the assay design and workflow. Although SIL-mAb internal standards are potentially the ideal option, the resources needed to generate these internal standards are significant and likely not available for most assays. However, with increasingly better assay design and streamlined workflows, cleavable flanking SIL-peptide internal standards appear to work extremely well and offer the best compromise of improved method performance with materials that can be obtained relatively easily. Cleavable flanking SIL-peptide internal standards have also been successfully used for full method validations to meet the stringent assay acceptance criteria (Jiang et al. 2013).

12.3 Case Studies of Preclinical Applications of LC–MS/MS for Monoclonal Antibody Bioanalysis

12.3.1 Case Study #1

Fully validated LC–MS/MS assay for the simultaneous quantitation of coadministered therapeutic antibodies in cynomolgus monkey serum (Jiang et al. 2013).

mAb analytes	Human IgG4 therapeutic antibodies: mAb-A and mAb-B (molecular weight ~ 150 kDa) in cynomolgus monkey serum			
Surrogate peptides	Surrogate peptides from the CDRs to distinguish them from each other and endogenous peptides			
	Analyte	Surrogate peptide	Q1 (m/z)	Q3 (m/z)
	mAb-A	GLEWXXXXXXR	662.5 (2+)	725.5 (y6)
	mAb-B	ASGIXXXXXXMHWVR	825.5 (2+)	597.3 (y4)
Internal standards	Cleavable flanking SIL-peptide IS:SIL-f-GLEW, SIL-f-ASGI containing stable isotopically labeled amino acids ([$^{13}C_6$,^{15}N]leucine or [$^{13}C_5$,^{15}N]valine) and flanking amino acids on the N-terminal and C-terminal			
Sample preparation for mAb extraction	Protein precipitation (for pellet digestion)			

12.3.1.1 Key Analytical Method Features

An LC–MS/MS assay was developed and fully validated for the simultaneous quantitation of two coadministered human monoclonal antibodies (IgG4), mAb-A and mAb-B, in monkey serum. After methanol denaturation and precipitation, the retained serum protein pellets were reduced, alkylated (IAM), and digested with trypsin. The unique CDR surrogate peptides for each mAb were simultaneously quantified by LC–MS/MS in the multiple reaction-monitoring mode. Cleavable stable isotopically labeled peptides were used as internal standards. The LC–MS/MS assay had an LLOQ at 5 μg/mL for mAb-A and at 25 μg/mL for mAb-B. The intra- and interassay precision (%CV) was within 10.0% and 8.1%, respectively, and the accuracy (%Dev) was within ±5.4% for all the peptides. Other validation parameters, including sensitivity, selectivity, dilution linearity, processing recovery and matrix effect, autosampler carryover, run size, stability, and data reproducibility, were all evaluated according to current regulatory guidelines and internal SOPs for small molecule bioanalysis. The acceptance criteria of ±15% for calibration standards and QCs (±20% at the LLOQ) were applied. In addition, two confirmatory peptides from different IgG4 domains were also monitored to confirm the quantitation accuracy and the integrity of the biotherapeutic drugs in the study samples.

Cross-validation between the LC–MS/MS assay and the LBA were performed for mAb-A. The ELISA assay for mAb-A employed an antidrug-specific antibody for capture and antihuman IgG4 antibody for detection. The QC results (≤7.8% difference at different concentration levels) demonstrated that the LC–MS/MS data were comparable to the LBA data. Data correlation of the toxicokinetic study samples from the coadministration group indicated that the LC–MS/MS data were highly correlated with the LBA data. The good agreement between LC–MS/MS and LBA data may reflect that there were no significant ADAs or soluble target present in the serum, which otherwise might lead to apparent differences or the two difference assay platforms.

Protein precipitation/pellet digestion is simple and straightforward; however, sensitivity will likely be limited due to ion suppression and interferences caused by the large number of tryptic peptides from the endogenous matrix proteins. In this analysis, the 5 μg/mL LLOQ for mAb-A and 25 μg/mL LLOQ for mAb-B from 25 μL serum samples were adequate for the intended toxicokinetic studies. For preclinical applications requiring much lower LLOQs, more selective sample enrichment or cleanup would likely be required, as is demonstrated in the next case study, which uses immunoaffinity capture for simultaneous measurement of four IgG2 mAbs in a rat PK study.

12.3.2 Case Study #2

Simultaneous analysis of multiple monoclonal antibody biotherapeutics by LC–MS/MS method in rat plasma following cassette-dosing (Li et al. 2013).

mAb analytes	Human IgG2 therapeutic antibodies: αDA, αDB, αK, and mAC (molecular weight ~ 150 kDa) in rat plasma			
Surrogate peptides	Surrogate peptides from the CDRs to distinguish them from each other and endogenous peptides			
	Analyte	Signature peptide	Q1 (*m*/*z*)	Q3 (*m*/*z*)
	αDA	LLIYAASSLQSGVPSR	554.65 (3+)	602.33 (y6)
	αDB	LIYAASSLQSGVPLR	525.63 (3+)	628.38 (y6)
	αK	LIYAASSLQSGVPSR	774.92 (2+)	730.38 (y7)
	mAC	LLIYDASTR	526.29 (2+)	825.41 (y7)
Internal standards	SIL-mAb IS: αDA uniformly labeled with [$^{13}C_6$]-Leu during recombinant synthesis in cell culture. The isotopic purity was ~95%			
Sample preparation for mAb extraction	Immunoaffinity capture: magnetic streptavidin beads coated with biotinylated antihuman Fc (b-Ab35) (for on-bead digestion)			

12.3.2.1 Key Analytical Method Features

An LC–MS/MS assay was developed for simultaneous quantitation of four cassette-dosed human monoclonal antibodies (IgG2), αDA, αDB, αK, and mAC, in rat plasma. After immunoaffinity capture using streptavidin-coupled magnetic beads coated with biotinylated antihuman Fc antibody, the antibodies on the beads were denatured, reduced, and alkylated (IAM). The samples were then digested with trypsin overnight at ambient temperature. The digests were then desalted and enriched using solid-phase extraction for LC–MS/MS analysis. A unique surrogate peptide for each IgG2 mAb was selected at a similar CDR location and SIL-LLIYAASSLQSGVPSR peptide from the SIL-αDA IS was used as the internal standard for all four surrogate peptides.

The high-affinity immunocapture-enabled analyte enrichment provided a much cleaner sample to improve detection limits. The on-bead digestion greatly simplified the analytical workflow, circumvented the elution and vacuum dry-down steps used in previous methods (Li et al. 2012), providing more than an eightfold increase in sensitivity. As a result, an LLOQ of 0.5 μg/mL was achieved for αDB and αK, while 0.1 μg/mL LLOQ was reached for αDA and mAC in 25 μL rat plasma samples. Assay qualification in terms of QC accuracy and precision were obtained from six replicates at each QC concentration for each mAb. Good assay precision (%CV) and accuracy (%Bias) were observed as shown in Table 12.2, especially for αDA given that SIL-mAb IS was deployed. The use of the analog peptide SIL-IS situated at the similar CDR location provided acceptable IS correction for mAC, αDB, and αK antibodies. The method was applied to samples from discrete- and cassette-dosed

Table 12.2 Intraday accuracy and precision of the mixed QCs (N=6) of 4 IgG2 mAbs with SIL-αDA antibody internal standard.

	αDA		αDB		αK		mAC	
	%CV	%Bias	%CV	%Bias	%CV	%Bias	%CV	%Bias
QC 0.3	4.08	4.89	NA	NA	NA	NA	8.09	6.50
QC 1.5	5.10	5.78	9.55	0.89	17.7	2.78	7.93	7.44
QC 6	6.59	0.06	6.43	0.94	6.66	5.94	9.03	0.08
QC 12	2.40	−8.19	5.23	−7.78	10.4	1.53	11.2	−5.94

Source: Li et al. 2012. Reproduced with permission of Springer.

rat PK studies and compared to the results obtained from an ELISA assay, which employed the same anti-huFc clone as the capturing agent as the LC–MS/MS method, and another anti-huFc clone used for detection.

For each of the four discrete dosed rat PK studies, the concentrations of the four mAbs analyzed by LC–MS/MS and ELISA correlated well. For the cassette-dosed samples, the multiplexed LC–MS/MS method efficiently provided concentration data using their own unique CDR surrogate peptide for each mAb. On the other hand, the nonspecific ELISA could only quantify the total mAbs. To compare the LC–MS/MS results with those of the ELISA, the sum of the individual mAb concentrations by LC–MS/MS was calculated for the total mAbs. Figure 12.7 shows the comparative PK profiles of each cassette dosed rat from LC–MS/MS and ELISA methods. Good agreement was observed for the majority of samples, apart from late time points after 840 or 336 h of rats #1 and #3 (ADA positive), which may be a result of interference of ADA on the ELISA assay. The overall agreement demonstrated the potential applicability and robustness of LC–MS/MS methodology.

12.4 Conclusion and Future Perspectives

LC–MS/MS has become a viable alternative and complementary assay approach for therapeutic antibody bioanalysis in support of pharmacokinetic, pharmacodynamic, and toxicokinetic studies, for which bioanalysis has been historically performed using ligand binding assays. The advantages of LC–MS/MS include (i) fast method development, with assays less dependent on critical reagents as required in LBAs; (ii) enhanced assay specificity, assay established in one species are readily applicable to other preclinical species with minor modification; and (iii) capability for simultaneous quantification of multiple mAbs. While there are an increasing number of successful implementations of LC–MS/MS assays for mAbs being reported, technical challenges still need to be addressed, notably associated with assay sensitivity, more efficient and robust analytical workflows, and comparison and alignment with LBAs. These challenges certainly need further refinement of processes for antibody extraction/enrichment, accelerated enzymatic digestion, and chromatographic separation.

LC–MS/MS for antibody bioanalysis is now considered feasible for supporting regulated studies, and recent publications have demonstrated utility for clinical tests of therapeutic mAbs and proteins (Dubois et al. 2008, Fernandez Ocana et al. 2012, Jiang et al. 2013, Gong et al. 2014). However, there are potential regulatory challenges due to the long history with LBA for supporting PK of biologic drugs, and as current regulatory guidelines for bioanalysis have been developed in consideration of LC–MS/MS for small-molecule drug entities, or LBA for large molecule drugs. Utilization of LC–MS/MS for mAb and other large molecule drugs may require some modification of guidances used for regulated large molecule bioanalysis (Jenkins et al. 2015).

Besides mAb therapeutics, there has been an increased interest in novel multidomain modalities, such as bispecific antibodies. A bispecific antibody drug combines the specificity of two targets in a single entity with favorable efficacy and safety profiles obtained from individual single-target drug development. Instead of implementing multiple separate LBAs, a multiplexed surrogate peptide LC–MS/MS method can be utilized to support these hybrid molecule constructs, such as for the quantification of different functional regions of the bispecific mAbs. With appropriate selection of surrogate peptides, it can also provide insights into the potential *in vivo* biotransformation of such hybrid molecules.

Therapeutic antibodies can induce immune responses leading to the development of ADAs, which may potentially lead to reduced drug efficacy or toxicity.

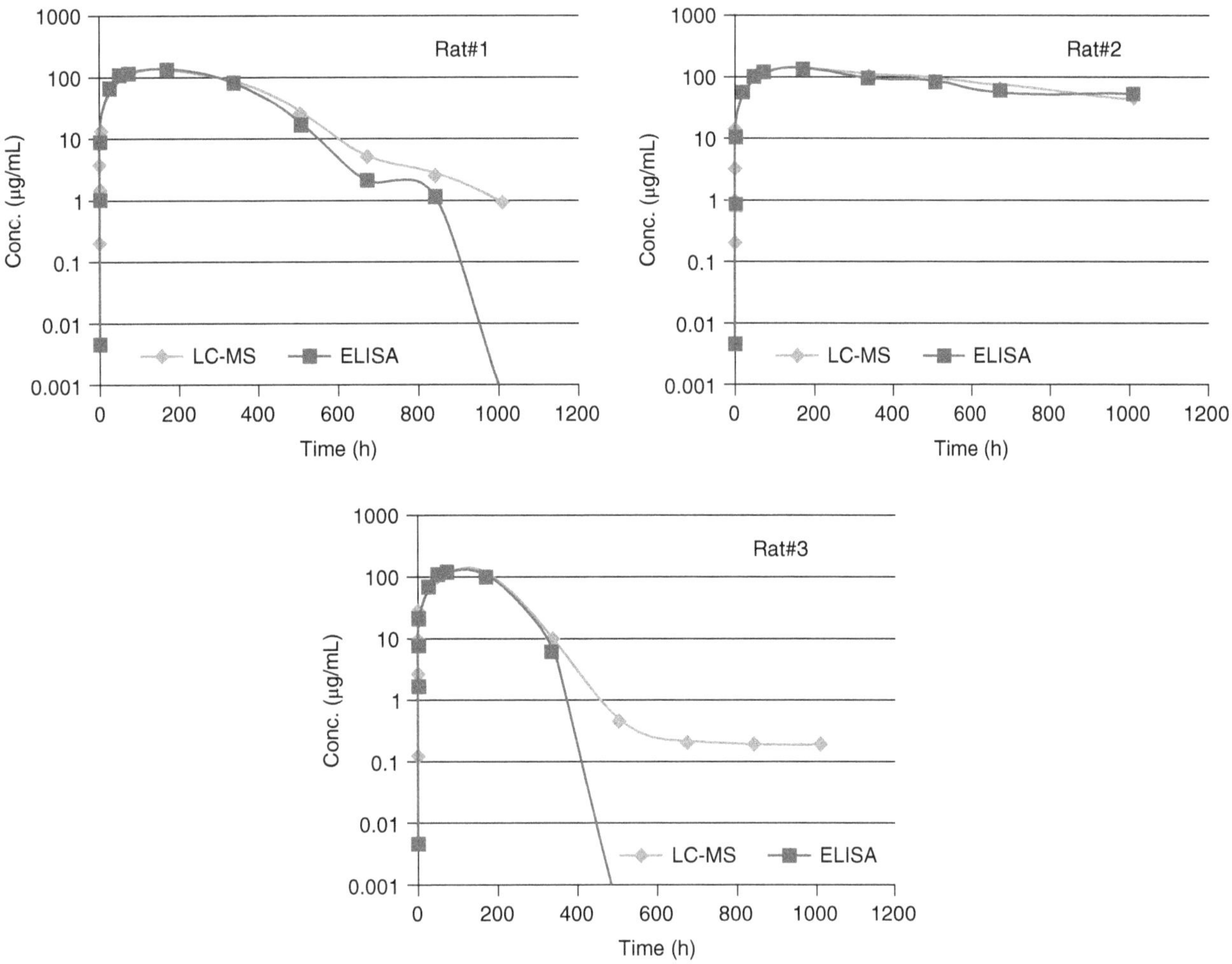

Figure 12.7 Comparison of PK profiles of total mAbs measured by LC–MS/MS and ELISA. The LC–MS/MS concentrations were the sum of the mAb concentrations. *Source:* Li et al. 2012. Reproduced with permission of Springer.

Immunogenicity assessment is thus important to develop safe and efficacious biotherapeutics during the drug development phase and for postmarketing surveillance. Ligand binding assays, cell-based assays, and Biacore assays are routinely used for ADA assessment but are subject to interference from high circulating drug concentrations. Approaches, such as acid dissociation, can improve the ability to detect ADAs in samples containing excess drug (Patton et al. 2005, Bourdage et al. 2007, Xu et al. 2014, 2015). However, concerns remain regarding acid-mediated loss of ADA activity. Alternative methodologies to improve the assay drug tolerance are desirable. For example, an immunoprecipitation of ADA–drug complex followed by quantitative LC/MS of the bound drug inferred the presence of total ADA and demonstrated its feasibility to monitor ADA responses to therapeutic proteins in the presence of high circulating concentrations of the therapeutic protein (Neubert et al. 2008). Since mass spectrometric detection is based on targeted protein sequences, it has the potential for development of drug tolerable ADA assays. Further LC–MS/MS assay strategies for indirect and direct ADA assessment can be expected.

In summary, the rapidly growing number of novel therapeutic antibodies for treatment of a variety of human diseases presents significant challenges and opportunities to LC–MS/MS bioanalysis of these proteins in biological matrices. The selection of a particular LC–MS/MS method approach depends on many factors, including the study measurement requirements, the analyte choices, required limit of quantitation, matrix types, and availability of reagents and internal standards. Approaches can range from simple protein precipitation sample preparation to various sophisticated immunoaffinity enrichment strategies and from generic (universal surrogate peptides) to analyte specific (unique CDR surrogate peptides) LC–MS/MS assays. Assay throughput may be improved using accelerated digestion strategies depending upon surrogate peptide formation kinetics. It is anticipated that LC–MS/MS will play a major role in the bioanalysis of mAbs, mAb derivative therapeutics, and related ADA assessment in support of biotherapeutic discovery and development in the future.

References

Arsene CG, Ohlendorf R, Burkitt W, et al. (2008). "Protein quantification by isotope dilution mass spectrometry of proteolytic fragments: cleavage rate and accuracy". *Anal Chem* **80**(11): 4154–4160.

Beck A, Wurch T, Bailly C and Corvaia N (2010). "Strategies and challenges for the next generation of therapeutic antibodies". *Nat Rev Immunol* **10**(5): 345–352.

Bourdage JS, Cook CA, Farrington DL, Chain JS and Konrad RJ (2007). "An Affinity Capture Elution (ACE) assay for detection of anti-drug antibody to monoclonal antibody therapeutics in the presence of high levels of drug." *J Immunol Methods* **327**(1-2): 10–17.

Chames P, Van Regenmortel M, Weiss E and Baty D (2009). "Therapeutic antibodies: successes, limitations and hopes for the future." *Br J Pharmacol* **157**(2): 220–233.

Cohn EJ, Hughes WL, Jr. and Weare JH (1947). "Preparation and properties of serum and plasma proteins; crystallization of serum albumins from ethanol water mixtures." *J Am Chem Soc* **69**(7): 1753–1761.

Duan X, Abuqayyas L, Dai L, Balthasar JP and Qu J (2012a). "High-throughput method development for sensitive, accurate, and reproducible quantification of therapeutic monoclonal antibodies in tissues using orthogonal array optimization and nano liquid chromatography/selected reaction monitoring mass spectrometry." *Anal Chem* **84**(10): 4373–4382.

Duan X, Dai L, Chen SC, Balthasar JP and Qu J (2012b). "Nano-scale liquid chromatography/mass spectrometry and on-the-fly orthogonal array optimization for quantification of therapeutic monoclonal antibodies and the application in preclinical analysis." *J Chromatogr A* **1251**: 63–73.

Dubois M, Fenaille F, Clement G, et al. (2008). "Immunopurification and mass spectrometric quantification of the active form of a chimeric therapeutic antibody in human serum." *Anal Chem* **80**(5): 1737–1745.

Fernandez Ocana M, James IT, Kabir M, et al. (2012). "Clinical pharmacokinetic assessment of an anti-MAdCAM monoclonal antibody therapeutic by LC-MS/MS." *Anal Chem* **84**(14): 5959–5967.

Furlong MT, Ouyang Z, Wu S, et al. (2012). "A universal surrogate peptide to enable LC-MS/MS bioanalysis of a diversity of human monoclonal antibody and human Fc-fusion protein drug candidates in pre-clinical animal studies." *Biomed Chromatogr* **26**(8): 1024–1032.

Furlong MT, Zhao S, Mylott W, et al. (2013). "Dual universal peptide approach to bioanalysis of human monoclonal antibody protein drug candidates in animal studies." *Bioanalysis* **5**(11): 1363–1376.

Gong C, Zeng J, Akinsanya B, et al. (2014). "Development and validation of an LC-MS/MS assay for the quantitation of a PEGylated anti-CD28 domain antibody in human serum: overcoming interference from antidrug antibodies and soluble target." *Bioanalysis* **6**(18): 2371–2383.

Hagman C, Ricke D, Ewert S, et al. (2008). "Absolute quantification of monoclonal antibodies in biofluids by liquid chromatography-tandem mass spectrometry." *Anal Chem* **80**(4): 1290–1296.

Heudi O, Barteau S, Zimmer D, et al. (2008). "Towards absolute quantification of therapeutic monoclonal antibody in serum by LC-MS/MS using isotope-labeled antibody standard and protein cleavage isotope dilution mass spectrometry." *Anal Chem* **80**(11): 4200–4207.

Hustoft HK, Reubsaet L, Greibrokk T, Lundanes E and Malerod H (2011). "Critical assessment of accelerating trypsination methods." *J Pharm Biomed Anal* **56**(5): 1069–1078.

Jenkins R, Duggan JX, Aubry AF, et al. (2015). "Recommendations for validation of LC-MS/MS bioanalytical methods for protein biotherapeutics." *AAPS J* **17**(1): 1–16.

Jiang H, Zeng J, Titsch C, et al. (2013). "Fully validated LC-MS/MS assay for the simultaneous quantitation of coadministered therapeutic antibodies in cynomolgus monkey serum." *Anal Chem* **85**(20): 9859–9867.

Lesur A, Varesio E and Hopfgartner G (2010). "Accelerated tryptic digestion for the analysis of biopharmaceutical monoclonal antibodies in plasma by liquid chromatography with tandem mass spectrometric detection." *J Chromatogr A* **1217**(1): 57–64.

Li H, Ortiz R, Tran L, et al. (2012). "General LC-MS/MS method approach to quantify therapeutic monoclonal antibodies using a common whole antibody internal standard with application to preclinical studies." *Anal Chem* **84**(3): 1267–1273.

Li H, Ortiz R, Tran LT, et al. (2013). "Simultaneous analysis of multiple monoclonal antibody biotherapeutics by LC-MS/MS method in rat plasma following cassette-dosing." *AAPS J* **15**(2): 337–346.

Lill JR, Ingle ES, Liu PS, Pham V and Sandoval WN (2007). "Microwave-assisted proteomics." *Mass Spectrom Rev* **26**(5): 657–671.

Liu G, Zhao Y, Angeles A, et al. (2014). "A novel and cost effective method of removing excess albumin from plasma/serum samples and its impacts on LC-MS/MS bioanalysis of therapeutic proteins." *Anal Chem* **86**(16): 8336–8343.

Lu Q, Zheng X, McIntosh T, et al. (2009). "Development of different analysis platforms with LC-MS for pharmacokinetic studies of protein drugs." *Anal Chem* **81**(21): 8715–8723.

MacLean B, Tomazela DM, Shulman N, et al. (2010). "Skyline: an open source document editor for creating and analyzing targeted proteomics experiments." *Bioinformatics* **26**(7): 966–968.

Mallick P, Schirle M, Chen SS, et al. (2007). "Computational prediction of proteotypic peptides for quantitative proteomics." *Nat Biotechnol* **25**(1): 125-131.

Neubert H, Grace C, Rumpel K and James I (2008). "Assessing immunogenicity in the presence of excess protein therapeutic using immunoprecipitation and quantitative mass spectrometry." *Anal Chem* **80**(18): 6907–6914.

Nouri-Nigjeh E, Zhang M, Ji T, et al. (2014). "Effects of calibration approaches on the accuracy for LC-MS targeted quantification of therapeutic protein." *Anal Chem* **86**(7): 3575–3584.

Patton A, Mullenix MC, Swanson SJ and Koren E (2005). "An acid dissociation bridging ELISA for detection of antibodies directed against therapeutic proteins in the presence of antigen." *J Immunol Methods* **304**(1-2): 189–195.

Reddy PM, Hsu WY, Hu JF and Ho YP (2010). "Digestion completeness of microwave-assisted and conventional trypsin-catalyzed reactions." *J Am Soc Mass Spectrom* **21**(3): 421–424.

Reichert JM (2011). "Antibody-based therapeutics to watch in 2011." *mAbs* **3**(1): 76–99.

Sandoval WN, Pham V, Ingle ES, Liu PS and Lill JR (2007). "Applications of microwave-assisted proteomics in biotechnology." *Comb Chem High Throughput Screen* **10**(9): 751–765.

Savoie N, Garofolo F, van Amsterdam P, et al. (2010). "2010 white paper on recent issues in regulated bioanalysis & global harmonization of bioanalytical guidance." *Bioanalysis* **2**(12): 1945–1960.

Szapacs MF, Urbanski JJ, Kehler JR, et al. (2010). "Absolute quantification of a therapeutic domain antibody using ultra-performance liquid chromatography–mass spectrometry and immunoassay." *Bioanalysis* **2**(9): 1597–1608.

Weiner LM, Surana R and Wang S (2010). "Monoclonal antibodies: versatile platforms for cancer immunotherapy." *Nat Rev Immunol* **10**(5): 317–327.

White JT and Bonilla LE (2012). "Free and total biotherapeutic evaluation in chromatographic assays: interference from targets and immunogenicity." *Bioanalysis* **4**(19): 2401–2411.

Xu K, Liu L, Maia M, et al. (2014). "A multiplexed hybrid LC-MS/MS pharmacokinetic assay to measure two co-administered monoclonal antibodies in a clinical study." *Bioanalysis* **6**(13): 1781–1794.

Xu W, Jiang H, Titsch C, et al. (2015). "Development and characterization of a pre-treatment procedure to eliminate human monoclonal antibody therapeutic drug and matrix interference in cell-based functional neutralizing antibody assays." *J Immunol Methods* **416** 94–104.

Yin S, Pastuskovas CV, Khawli LA and Stults JT (2013). "Characterization of therapeutic monoclonal antibodies reveals differences between *in vitro* and *in vivo* time-course studies." *Pharm Res* **30**(1): 167–178.

Yuan L, Arnold ME, Aubry AF and Ji QC (2012). "Simple and efficient digestion of a monoclonal antibody in serum using pellet digestion: comparison with traditional digestion methods in LC-MS/MS bioanalysis." *Bioanalysis* **4**(24): 2887–2896.

Zhang Q, Spellman DS, Song Y, et al. (2014). "Generic automated method for liquid chromatography-multiple reaction monitoring mass spectrometry based monoclonal antibody quantitation for preclinical pharmacokinetic studies." *Anal Chem* **86**(17): 8776–8784.

13

Generic Peptide Strategies for LC–MS/MS Bioanalysis of Human Monoclonal Antibody Drugs and Drug Candidates

Michael T. Furlong

PPD Bioanalytical Labs, Middleton, WI, USA

13.1 Introduction

Human monoclonal antibody (mAb)-based therapeutic agents occupy an increasingly important role in the treatment of a variety of human diseases (Strohl 2009, Nelson et al. 2010, Beck and Reichert 2011, 2014, Reichert 2012, 2014). The rapidly growing number of human mAbs entering drug development represents a significant challenge to bioanalytical laboratories engaged in the quantification of these drug candidates in biological matrices.

In the last several years, LC–MS/MS has emerged as a promising assay platform for therapeutic protein bioanalysis in plasma (Ezan et al. 2009, Li et al. 2011, van de Merbel et al. 2012, Bischoff et al. 2013, Hopfgartner et al. 2013, van den Broek et al. 2013, van de Merbel 2015). Compared with traditional ligand-binding-assay (LBA)-based protein quantification, LC–MS-based assays may offer wider dynamic ranges, better precision and accuracy, improved specificity, lower cost, shorter development lead times, and comparable throughput (Ezan and Bitsch 2009, van de Merbel et al. 2012, Bults et al. 2015). In addition to their primary application as quantitative tools, ligand binding and LC–MS/MS and LBA platforms can be deployed in an integrated manner during drug development to identify and mitigate quantitative challenges such as antidrug antibody (ADA) assay interferences (Wang et al. 2012, Furlong et al. 2014, Zhang et al. 2014a, b).

Bioanalysis of human therapeutic mAbs by LC–MS/MS is typically based upon quantification of "signature" surrogate peptides whose amino acid sequences are unique to the mAb analyte of interest (Yang et al. 2007, Dubois et al. 2008, Hagman et al. 2008, Heudi et al. 2008, Ji et al. 2009, Lesur et al. 2010, Fernandez Ocana et al. 2012, Jiang et al. 2013, Mekhssian et al. 2014, Xu et al. 2014). Signature peptides are found in the variable regions of human therapeutic mAb light and heavy chains (Figure 13.1). A significant shortcoming of the signature peptide approach is that a new LC–MS/MS assay must be developed for each new human therapeutic mAb. Method development can be especially time-, labor-, and cost-intensive during the earliest stages of drug development wherein multiple structural variants of an mAb drug candidate may need to be evaluated in various animal species to enable prioritization for further development. To overcome this burden, a single generic research grade LC–MS/MS assay capable of quantifying most or all human mAb drug candidates in the plasma/serum of all commonly used animal species would be of great value to bioanalytical laboratories. This chapter describes the development and successful deployment of generic surrogate peptide-based LC–MS/MS assays as useful alternatives to the traditional signature surrogate peptide-based assay approach to human therapeutic mAb bioanalysis. Throughout the chapter, emphasis is placed on challenges, tools, and solutions that apply not only to generic peptide LC–MS/MS assays but also to any LC–MS/MS bioanalytical assay that involves trypsin digestion of a protein analyte followed by analysis of a surrogate peptide.

13.2 A Universal Peptide LC–MS/MS Assay for Bioanalysis of a Diversity of Human Monoclonal Antibodies and Fc Fusion Proteins in Animal Studies[1]

13.2.1 Identification of a Candidate Universal Surrogate Peptide to Enable Quantification of Human mAb and Fc Fusion Protein Drug Candidates

Our first step in the development of a universal peptide LC–MS/MS assay was the identification of a universal

1 Throughout this chapter, the author uses the term "universal peptide" to describe peptides that he and his collaborators directly investigated. The alternate term "generic peptide" is applied to peptides/studies from other investigators.

Protein Analysis using Mass Spectrometry: Accelerating Protein Biotherapeutics from Lab to Patient, First Edition.
Edited by Mike S. Lee and Qin C. Ji.

Figure 13.1 Structures of a traditional human therapeutic monoclonal antibody and a human Fc-fusion hybrid protein.

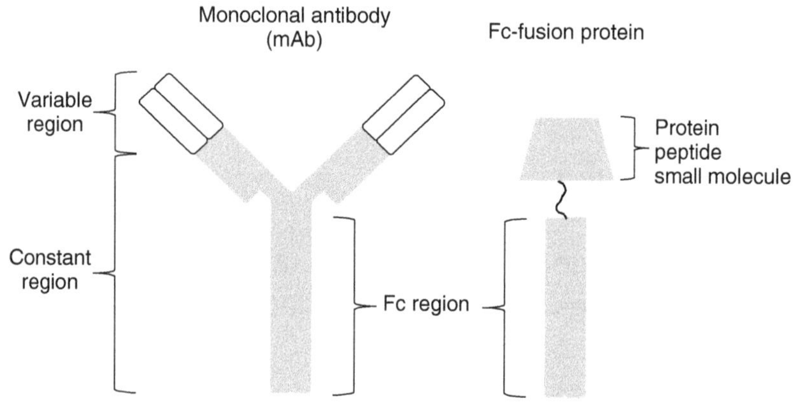

surrogate peptide (Furlong et al. 2012). An ideal universal surrogate peptide sequence would have four general characteristics:

1) *Universality:* The peptide sequence would have to be present in the majority of human mAb and Fc-fusion protein drug candidates (Figure 13.1), thus enabling its general applicability across many development programs involving a variety of both of these structural classes.
2) The peptide sequence would be reliably produced from trypsin digestion of human mAb and Fc-fusion protein analytes.
3) The sequence would possess favorable LC–MS/MS characteristics such as good chromatographic peak shape, adequate chromatographic retention, and efficient ionization.
4) *Selectivity:* The sequence would not be found in the protein sequences of any plasma/serum protein(s) present in any animal species typically used in nonclinical drug development studies, thus ensuring quantification that is free of plasma protein-derived interferences.

Using a combination of *in silico* and experimental approaches, we identified a single candidate universal peptide that met all of the aforementioned criteria – VVSVLTVLHQDWLNGK (Peptide 1; Tables 13.1 and 13.2). Peptide 1 is located in the Fc region of human IgG heavy chain subclasses 1 and 4, which are frequently used for human mAb and Fc-fusion protein drug candidates (Beck et al. 2008, 2010, Strohl 2009, Nelson et al. 2010, Beck and Reichert 2011, 2014, Myler et al. 2011, Reichert 2012).

13.2.2 Application of an Exploratory Universal (Peptide 1) LC–MS/MS Assay to a Monkey Pharmacokinetic Study

Upon identification of Peptide 1 as the leading universal peptide candidate, we immediately realized that we had by coincidence recently used this peptide during

Table 13.1 Representative human mAb heavy chain generic peptides.

Peptide	Human heavy chain subclass[a]
Peptide 1 VVSVLTVLHQDWLNGK[c]	IgG1, IgG4
Peptide 3 NQVSLTCLVK[b]	IgG1, IgG2, IgG4
Peptide 4 GPSVFPLAPSSK	IgG1
Peptide 5 YGPPCPPCPAPEFLGGPSVFLFPPKPK[b]	IgG4 (hinge-region-stabilized)
Peptide 6 GPSVFPLAPCSR[b]	IgG2, IgG4
Peptide 7 FNWYVDGVEVHNAK[c]	IgG1
Peptide 8 TTPPVLDSDGSFFLYSK[c]	IgG1
Peptide 9 GFYPSDIAVEWESNGQPENNYK[c]	IgG1, IgG2, IgG4

a) IgG3 not included because this subclass is rarely used.
b) Peptide quantified as the alkylated cysteine thiol derivative.
c) Dual universal peptide (heavy chain) candidate evaluated by Furlong et al. 2013.

Table 13.2 Representative human mAb kappa light chain candidates.

Peptide
Peptide 2 TVAAPSVFIFPPSDEQLK[b]
Peptide 10 SGTASVVCLLNNFYPR[a, b]
Peptide 11 DSTYSLSSTLTLSK[b]

a) Peptide quantified as the alkylated cysteine thiol derivative.
b) Dual universal peptide (light chain) candidate evaluated by Furlong et al. 2013.

LC–MS/MS bioanalysis of a human Fc-fusion protein in a non-GLP monkey pharmacokinetic (PK) study (Ouyang et al. 2012). In that study, Peptide 1 had been deployed only as a secondary, confirmatory surrogate peptide for Fc fusion protein quantification; the primary peptide used for quantification was a traditional signature peptide. Thus, we were afforded an opportunity to assess the potential utility of an exploratory Peptide-1-based universal LC–MS/MS assay.

As shown in a representative plasma concentration versus time profile Figure 13.2, quantitative data derived from Peptide 1 were in excellent agreement with data derived from a traditional signature surrogate peptide located in a different region of the protein. For each study sample, the percentage difference between the two observed concentrations was computed using the following calculation: Percent difference = (surrogate peptide conc. − confirmatory peptide conc.)/(mean of the two values) × 100 (Fast et al. 2009). In 39 of the 41 study samples quantified, the percentage difference between the observed concentrations was <15%. The overall mean percentage difference between the two peptides, calculated as the mean of all of the individual study sample percentage differences, was 0.83%. Human Fc fusion protein concentrations in the study samples were also quantified using an ELISA. PK profiles obtained using concentrations determined by LC–MS/MS and ELISA were nearly superimposable. Using the percentage difference calculation described above, the mean percentage difference between the ELISA and the LC–MS/MS data was 1.4%. Taken together, these data indicated that Peptide 1 could be used in a single, universal LC–MS/MS assay for the quantification of a variety of human mAb and Fc-fusion protein drug candidates in discovery-stage animal studies. Since our original reports on the Peptide-1-based universal assay (Furlong et al. 2012, Ouyang et al. 2012), numerous bioanalytical studies using this peptide have been published (Yuan et al. 2012, 2013, Furlong et al. 2013, Jiang et al. 2013, 2014, Liu et al. 2014, Gong et al. 2015, Xu et al. 2015, Zhao et al. 2015, Lanshoeft et al. 2016, Zhu et al. 2016).

It is interesting to note that Dubois et al. (2008) and Jiang et al. (2014) monitored Peptide 1, not for mAb quantification, but rather to assess nonspecific binding of *endogenous* human IgGs to magnetic beads during the immunocapture step or their sample preparation procedures (Xu et al. 2015).

13.2.3 Potential Applicability of a Peptide 1 Variant to Bioanalysis of Human IgG2-Based mAbs and Fc Fusion Proteins

As discussed earlier, Peptide 1 is found in the Fc region of human IgG1 and IgG4 subclasses, which enables its use in many human mAb and Fc-fusion protein drug development programs. However, certain human mAbs in development are structurally based upon the human IgG2 subclass (Strohl 2009, Nelson et al. 2010, Reichert 2012). The amino acid sequence of Peptide 1 is not found in the human antibody IgG2 subclass, and, therefore, an LC–MS/MS assay based upon this sequence would not be capable of providing nonclinical bioanalytical support for IgG2-based human mAb and Fc-fusion protein drug development programs. Fortunately, the Fc region of the human IgG2 molecule contains a very closely related sequence to Peptide 1 (VVSVLTV*V*HQDWLNGK), which differs from the sequence of Peptide 1 by a single

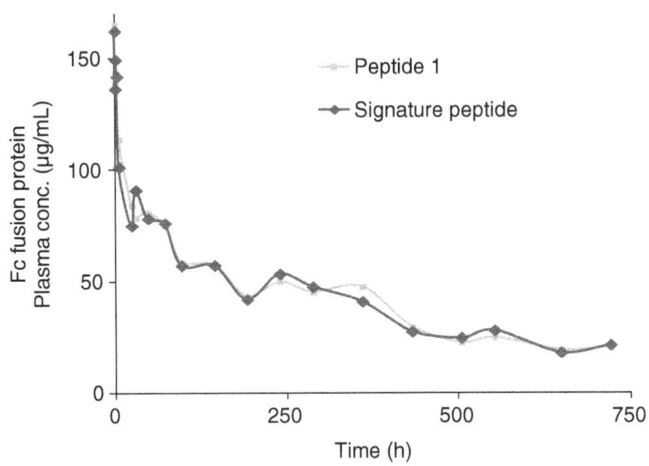

Figure 13.2 Plasma concentration versus time profile from a representative monkey after intravenous administration of a human Fc fusion protein drug candidate. Universal Peptide 1 and a traditional signature peptide were used for LC–MS/MS quantification. (*Source*: Ouyang et al. 2012. Reproduced with permission of Future Science Ltd.)

leucine-to-valine substitution at the eighth amino acid position. Given the relatively minor structural difference between Peptide 1 and the corresponding IgG2-derived tryptic peptide, it seemed likely that a very similar universal LC–MS/MS assay could be successfully developed based on the IgG2 version of Peptide 1 to support bioanalysis of human IgG2-based mAb and Fc-fusion proteins in animal studies. Furthermore, the IgG2 peptide sequence was not found in published protein sequences for commonly used animal species, and thus plasma protein-derived interferences would not be anticipated. To evaluate this possibility, we spiked a human IgG2 mAb into cynomolgus monkey plasma, digested the sample with trypsin, and analyzed the digest using the LC–MS/MS assay conditions employed in the human IgG1/IgG4 Peptide 1 universal LC–MS/MS assay (Furlong et al. 2012, Ouyang et al. 2012). Encouragingly, the IgG2 version of Peptide 1 behaved similarly to Peptide 1, exhibiting good peak shape and peak area response. In addition, no significant coeluting plasma interfering peaks were observed in the IgG2 Peptide 1 selected reaction monitoring (SRM) transition (Furlong et al. 2012). These data indicate that a single LC–MS/MS assay based upon the human IgG2 version of Peptide 1 could be developed to support animal studies involving all human IgG2-based mAb and Fc-fusion protein candidates in all commonly used animal species.

13.2.4 Impact of Peptide 1 Asparagine Deamidation on Human mAb Quantification Can Be Mitigated

The asparagine residue of Peptide 1 can potentially undergo deamidation to aspartic acid and isoaspartic acid during sample preparation (Figure 13.3) (Geiger and Clarke 1987, Patel and Borchardt 1990, Yan et al. 2009, Yang and Zubarev 2010). The deamidation products can potentially generate spurious responses in the SRM transition of Peptide 1 itself (Furlong et al. 2012). During our studies, we occasionally observed a minor peak eluting shortly after the Peptide 1 peak in the Peptide 1 analyte and internal standard chromatograms of calibration curve, QC and study samples (Figure 13.4). In order to determine if this minor peak was due to asparagine deamidation, authentic reference standards of Peptide 1, along with the corresponding aspartic acid and isoaspartic acid peptides, were spiked individually into blank monkey serum and extracted, and their respective retention times were determined by LC–MS/MS. Based on the observed retention times of the extracted peptide reference standards, presumptive aspartic- and isoaspartic-acid peaks were identified in the Peptide 1 chromatograms of the calibration curve/QC samples (Figure 13.4). The aspartic acid peptide peak was partially separated from the Peptide 1 peak, whereas the isoaspartic acid peptide was not separated from Peptide 1. Due to the poor chromatographic separation of these three peaks, the nonseparated Peptide 1/isoaspartic peptide peak and the partially separated aspartic acid peak, when present, were typically cointegrated prior to quantification.

Several lines of evidence indicate that asparagine deamidation of Peptide 1 will not significantly impact reliable quantification. First, study samples are treated exactly the same as calibration curve and QC samples during sample processing, and therefore any sample processing-related Peptide 1 deamidation and resultant

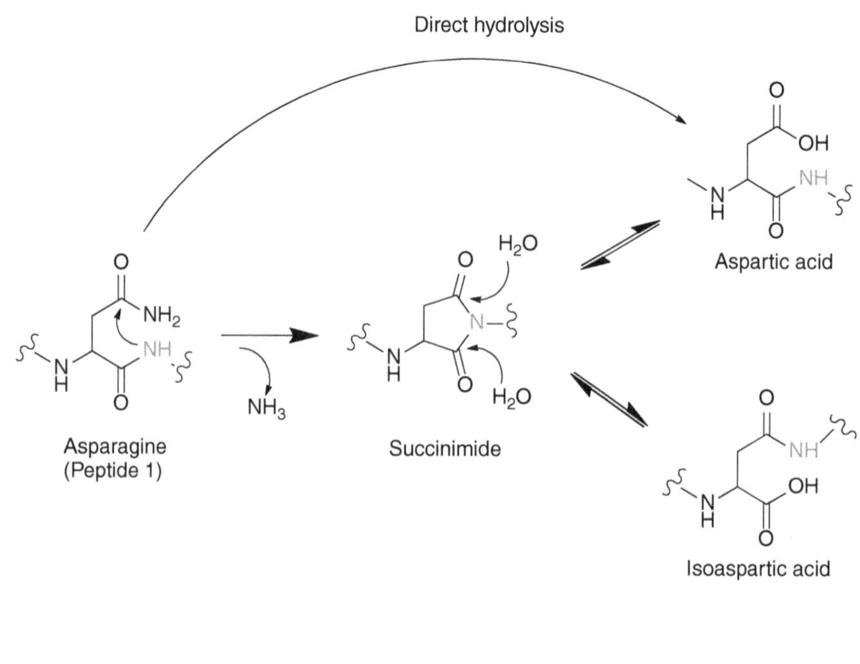

Figure 13.3 Mechanisms for deamidation of asparagine to aspartic acid and isoaspartic acid. (*Source*: Furlong et al. 2012. Reproduced with Permission of John Wiley & Sons.)

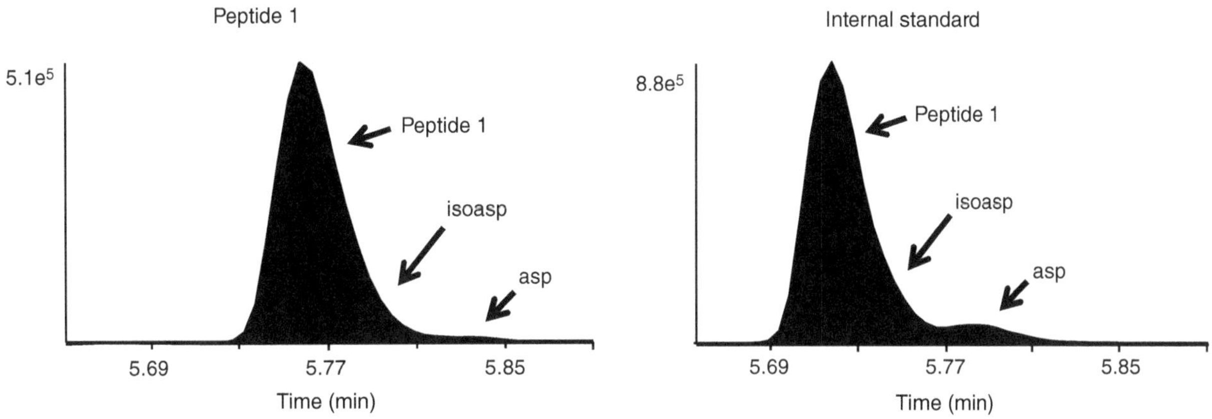

Figure 13.4 Representative Peptide 1 LC–MS/MS chromatograms. The samples were analyzed after digestion and extraction of monkey plasma spiked with an IgG1 human mAb drug candidate (Furlong et al. 2013). The retention times of Peptide 1 and the corresponding aspartic acid and isoaspartic acid-containing peptides are indicated by arrows. (*Source*: Furlong et al. 2013. Reproduced with permission of Future Science Ltd.)

peak area response change that may occur in the study samples would occur at the same rate, and to the same extent, in the calibration curve and QC samples, thus ensuring accurate quantification of the study samples (Jemal and Xia 2000). Second, a stable-isotope-labeled peptide or stable-isotope-labeled protein internal standard is used in various iterations of this assay prior to trypsin digestion (Ouyang et al. 2012, Furlong et al. 2013). Hence, any Peptide 1 deamidation that may occur during sample processing is predicted to be compensated for by the concurrent deamidation of the corresponding asparagine amino acid in the internal standard. Third, excellent accuracy and precision data were obtained using Peptide 1 as the surrogate peptide in the assay performance evaluation runs for three distinct human mAb analytes (Furlong et al. 2013). Finally, assay performance evaluation run and PK study sample concentration data obtained using Peptide 1 have consistently been in good agreement with data obtained using light chain surrogate peptides, even in cases wherein minor amounts of chromatographically unresolved deamidation peaks are present (Ouyang et al. 2012, Furlong et al. 2013).

Deamidation can be minimized by using a short digestion time of less than 1 h, a digestion buffer pH of not higher than 8, and a temperature of approximately 37 °C (Ren et al. 2009). In our hands, the extent of Peptide 1 deamidation was dependent on the sample preparation technique. No deamidation was observed when we used a "pellet digestion"-based sample preparation procedure (Ouyang et al. 2012 and unpublished data). However, a minor degree of deamidation was observed when we employed Protein-A-mediated analyte enrichment coupled to more stringent denaturation conditions as part of the sample preparation procedure (Figure 13.4) (Furlong et al. 2013).

13.3 An Improved "Dual" Universal Peptide LC–MS/MS Assay for Bioanalysis of Human mAb Drug Candidates in Animal Studies

Although the Peptide-1-based universal LC–MS/MS assay described above can provide sufficiently reliable data to support animal studies during early drug development, it may be tenuous to rely upon only one surrogate peptide to provide reliable concentration data for large proteins such as mAbs throughout the entire preclinical/clinical development cycle. As depicted in Figure 13.5, an LC–MS/MS assay based upon a single peptide may not be able to distinguish between structurally intact versus degraded forms of mAb-based analytes (Furlong et al. 2012). This concern holds true not only for single peptide *universal* LC–MS/MS assays but also for any surrogate peptide LC–MS/MS assay that relies on only one peptide for quantification. We reasoned that incorporation of a second universal surrogate peptide into the universal peptide assay would offer increased insights into the structural integrity of the protein analyte as well as overall increased confidence in the study sample concentration data (Figure 13.6) (Dubois et al. 2008, Furlong et al. 2012, 2013, 2014, Ouyang et al. 2012, Jiang et al. 2013, Liu et al. 2014, Mekhssian et al. 2014, Nouri-Nigjeh et al. 2014).

13.3.1 Identification and Evaluation of "Dual" Universal Peptide LC–MS/MS Assay Candidates

For quantification of mAb proteins, the light chain was a logical location to identify a second peptide for the proposed dual peptide assay (Figure 13.1). In addition, we reasoned that if the light chain peptide were chosen from

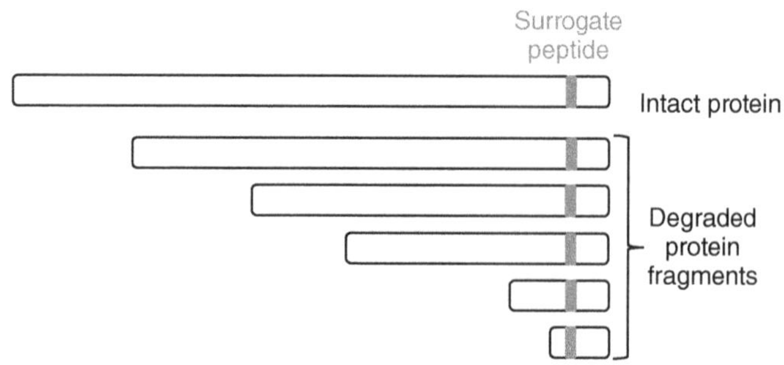

Figure 13.5 LC–MS/MS assays based on only one surrogate peptide may not be able to differentiate between an intact protein analyte and proteolytic fragments thereof.

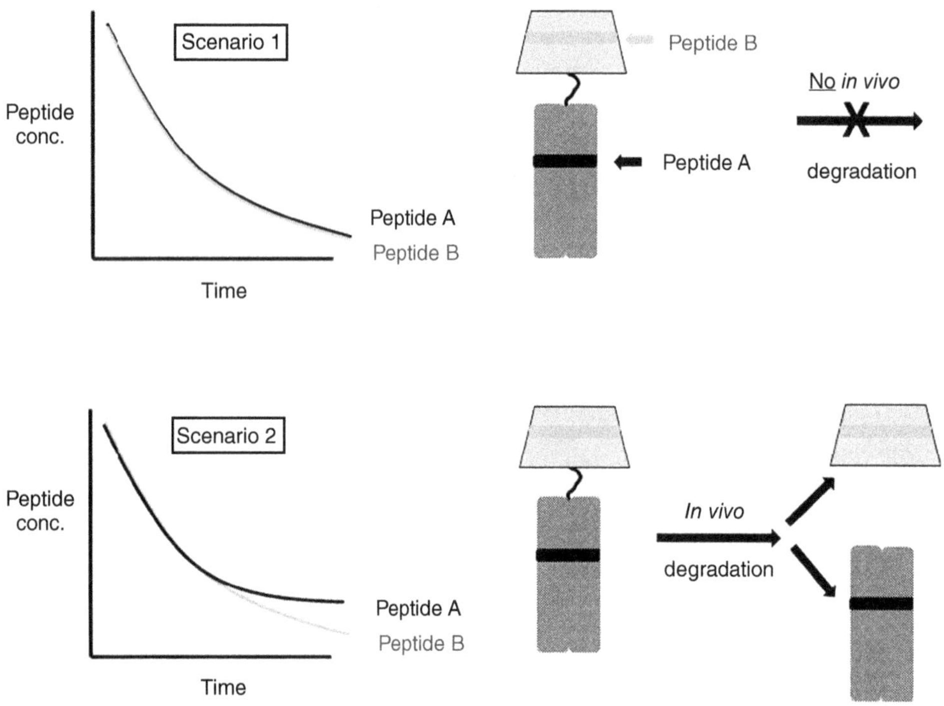

Figure 13.6 Hypothetical pharmacokinetic plasma sample concentration versus time profiles for an Fc fusion protein obtained by simultaneous quantification of surrogate Peptide A and Peptide B. In Scenario 1, the concordant profiles derived from both peptides suggest that the protein analyte undergoes little or no *in vivo* degradation over the time course of the pharmacokinetic study. In Scenario 2, the two profiles begin to diverge over the time course of the study, suggesting that (1) the protein is undergoing *in vivo* degradation over time and (2) the protein fragment that contains Peptide B is being cleared more rapidly from the bloodstream than the fragment that contains Peptide A. (*Source*: Furlong et al. 2012. Reproduced with Permission of John Wiley & Sons.)

the constant region, as was the case for the heavy chain peptide in the original universal assay, the resulting dual peptide assay would maintain its universality; that is, it could be used to support a diversity of human therapeutic mAb candidates in animal studies.

We identified and evaluated three universal light chain peptide candidates located in the constant region of the kappa class of human light chains. The kappa class was chosen due to its frequent use in human therapeutic mAb candidates (Myler et al. 2011, Reichert 2012). Candidate human light chain universal peptides were evaluated based upon the same criteria that were used to identify heavy chain universal Peptide 1 (Section 13.2.1). Concurrently with our evaluation of the human light chain peptide candidates, we opted to evaluate additional human heavy chain universal peptide candidates as potential alternative/backup peptides to Peptide 1. Additional heavy chain candidates had to be present in at least the human IgG1 subclass, which would enable their application to many human mAb drug development programs (Beck et al. 2010, Nelson et al. 2010, Myler et al. 2011, Reichert 2012). The amino acid sequences of the three human kappa light chain candidates (Peptides 2, 10, and 11) and four human heavy chain peptide candidates (Peptides 1, 7, 8, and 9) chosen for our dual universal peptide studies are shown in Tables 13.1 and 13.2.

13.3.2 Quantitative Evaluation and Comparison of Light and Heavy Chain Dual Universal Peptide Candidates

A single exploratory LC–MS/MS assay was developed to simultaneously quantify all of the light and heavy chain universal peptide candidates described above. Three human monoclonal antibodies – two containing a human IgG1 heavy chain and one containing human IgG4 – were each evaluated in three separate assay performance evaluation runs. The light chains of all three mAbs contained the kappa constant region.

The goals of the assay performance evaluation runs were to (i) evaluate the performance of the individual peptide candidates *vis à vis* calibration curve ranges, QC accuracy/precision data and assay sensitivity; (ii) identify the most promising light and heavy chain candidate peptides based upon their individual performance; and (iii) combine the most promising light and heavy chain candidates into a dual universal peptide LC–MS/MS assay capable of quantifying variety of mAb analytes in plasma/serum samples from animal PK/toxicokinetic studies.

Peptide 2 (TVAAPSVFIFPPSDEQLK) was the best overall performing kappa light chain constant region candidate in the three assay performance evaluation runs, with the lowest lower limits of quantitation (LLOQ) (50 ng/mL) and the largest calibration curve range (2000-fold). Peptide 2 has the additional advantage of not containing cysteine, and thus, similar to Peptide 1, can potentially be quantified without the need for the reduction and alkylation steps that are obligatory for quantification of cysteine-containing peptides. Peptide 1, previously identified in our laboratory as the original universal peptide (Furlong et al. 2012, Ouyang et al. 2012), was the best overall performing heavy chain peptide in the assay performance evaluation runs, with the lowest LLOQ (50 ng/mL) and the second largest calibration curve range (720-fold). Together, these assay performance evaluation results indicated that a dual universal peptide assay based upon Peptide 1 and Peptide 2 merited additional exploration. The availability of several backup peptide options (Tables 13.1 and 13.2) will be useful in the event that unexpected matrix interferences and/or sensitivity challenges are encountered during study samples analysis with a Peptide 1-/Peptide-2-based dual universal peptide assay.

13.3.3 Assessing the Level of Quantitative Agreement Between Peptide 1 and Peptide 2 in Assay Performance Evaluation Runs

Although Peptides 1 and 2 performed well individually in the assay performance evaluation runs, it was important to compare the quantitative data derived from both peptides. Excellent agreement between the two data sets derived from these peptides is a prerequisite for their successful deployment in a single universal assay.

To compare the assay performance evaluation run data sets, an approach based on incurred samples reanalysis (ISR) was used (Viswanathan et al. 2007, Fast et al. 2009). ISR is routinely implemented by the bioanalytical community to assess the reliability/reproducibility of study samples concentration data. A subset of study samples is reanalyzed and the repeat analysis data are quantitatively compared with the original data using the following calculation:

$$\text{Percent difference} = \frac{\left(\begin{array}{c}\text{repeat analysis concentration} - \\ \text{original analysis concentration}\end{array}\right)}{(\text{mean of the two values}) \times 100.}$$

For LC–MS/MS analysis, a study sample's concentration data set is typically considered reliable if two thirds of the percent difference values are within 20% (Fast et al. 2009). Given that ISR compares two concentration values for the same sample, this approach was deemed appropriate for assessment of the level of agreement between mAb concentration values for a given sample that are derived from two distinct surrogate peptides.

The percent differences between the observed concentrations from Peptides 1 and 2 were computed for all individual QC replicates in the three assay performance evaluation runs. Peptide 1- and Peptide 2-derived QC data sets were in excellent agreement for all three mAb analytes. One hundred percent of the evaluated QC replicates had percent difference values of less than 20% for two of the three mAbs evaluated; for the third mAb, 85.4% of QC replicates agreed within 20%. In addition, the overall mean percent difference between the two peptide data sets, calculated as the mean of all of the individual QC sample percent differences, ranged from −0.824% to −2.68%. The excellent agreement between the Peptide 1 and Peptide 2 QC data sets indicated that these two peptides could be combined into an exploratory dual universal peptide LC–MS/MS assay (Furlong et al. 2013).

During development and deployment of the exploratory dual universal peptide assay, we found that we could potentially achieve lower LLOQs with Peptide 1 compared to Peptide 2 (Furlong et al. 2013). Thus, the overall LLOQ for the dual universal peptide assay is essentially limited to that achievable for Peptide 2. In addition, Peptide 1 had a lower ULOQ and lower dynamic range compared to Peptide 2, which may be due to detector saturation of Peptide 1. Thus, the Peptide 1/Peptide 2 dual universal peptide assay could potentially be improved by additional assay refinement to align the sensitivities and dynamic ranges of these two universal peptides.

13.3.4 Deployment of the Exploratory Dual Universal Peptide Assay in Support of a Monkey Pharmacokinetic Study

The exploratory dual universal peptide assay was used to quantify serum samples from a monkey PK study. In this study, cynomolgus monkeys were dosed subcutaneously with a human IgG1 heavy chain/human kappa light chain mAb and serum samples were collected at several time points postadministration for LC–MS/MS quantification. The mAb analyte concentration values in the monkey study samples determined using Peptide 1 were in very good agreement with those obtained using Peptide 2 (Figure 13.7). Ninety-four percent of the 114 plasma samples analyzed in this study had percent difference values <20%; the overall mean percent difference (Peptide 1 vs Peptide 2) was −4.0% These results imparted confidence in the study samples concentration data as well as the *in vivo* structural integrity of the mAb analyte. Subsequent to our evaluation of Peptide 2 as a human kappa light chain universal peptide (Furlong et al. 2013), numerous bioanalytical studies of this peptide have been reported (Jiang et al. 2013, Nouri-Nigjeh et al. 2014, Remily-Wood et al. 2014, Zhang et al. 2014a, b, An et al. 2015, Gong et al. 2015, Xu et al. 2015).

Although our studies of the dual universal peptide assay have to date been limited to human IgG- and IgG4-based mAbs, the dual peptide approach could in principle be applied to human IgG2-based mAb bioanalysis. As described in Section 13.2.3, we have shown that the IgG2 variant of Peptide 1 exhibits favorable chromatographic peak shape and mass spectrometric response characteristics (Furlong et al. 2012). Thus, for human IgG2-based mAbs, it is likely that this Peptide 1 variant could be used as the heavy chain peptide component of a dual universal peptide assay.

13.3.5 Considerations for Calibration Curve/QC Replicate Acceptance Criteria When a Dual Peptide Assay Is Employed

When a dual peptide assay is used for the bioanalysis of mAb analytes, or any protein analyte for that matter, the use and interpretation of the two sets of concentration data (one data set from each surrogate peptide) must be established *a priori*. This requirement is the case regardless of whether the two surrogate peptides employed are generic or signature peptides.

We have proposed that each calibration curve or QC replicate in an analytical run can be accepted only if *both* of the concentration values pass the individual surrogate peptide acceptance criteria (Furlong et al. 2013). In addition, we recommended that the ISR-based percent difference calculation be applied to each calibration curve and QC replicate; the replicate would only be acceptable if the computed percent difference value was within 20%. These criteria ensure that calibration curve and QC replicates would be accepted only if the observed concentrations derived from each surrogate peptide are in good agreement, not only with the theoretical values but with each other as well. These criteria could be applied to calibration curve/QC replicates in study sample analytical runs as well as assay validation or assay performance evaluation runs. Mekhssian and colleagues applied the ISR-based criteria to calibration curve and QC sample data obtained during validation of a dual signature peptide LC–MS/MS assay for quantifying a human mAb in human plasma (Mekhssian et al. 2014).

13.3.6 Interpreting and Reporting Study Sample Concentration Data Generated with a Dual Peptide Assay

For study sample quantification, we have proposed that the ISR-based percent difference calculation could potentially be applied to each study sample (Furlong et al. 2013). If the computed percent difference value is within 20%, then the *mean* of the two values would be accepted and reported. If the 20% threshold is not met, then two scenarios are possible:

1) A concentration value for that sample would not be reported.
2) The mean concentration value would be reported for that sample with an accompanying note indicating that the 20% threshold was not met.

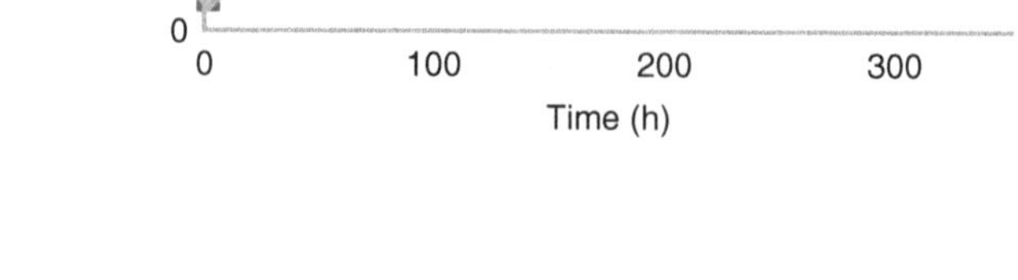

Figure 13.7 Human mAb serum concentration versus time profile from a representative monkey after a single subcutaneous dose. The dual universal peptide LC–MS/MS assay was used for quantification of the study samples. (*Source*: Furlong et al. 2012. Reproduced with Permission of John Wiley & Sons.)

If a significant proportion of the study samples do not meet the 20% threshold, then an investigation should be carried out to determine the reasons for the divergent results. If the divergence of concentrations is more frequent and more pronounced in later PK time points, then this result may indicate that (i) the protein is undergoing *in vivo* degradation over time; and (ii) the fragment that contains one of the surrogate peptides has a distinct PK profile compared to the fragment that contains the other surrogate peptide (Furlong et al. 2012, Bowen et al. 2015, Pearson and Rock 2015) (Figure 13.6).

Jiang and colleagues have reported an alternative approach to interpreting dual sets of human mAb study sample concentration data (Jiang et al. 2013). Instead of employing the above-proposed ISR-based approach to evaluate study sample data reliability and *in vivo* mAb structural integrity, they computed the *ratio* of the observed concentrations of a second "confirmatory" peptide to the observed concentrations of the primary quantitative peptide.

Nouri-Nigjeh et al. 2014 used Peptides 2 and 4 as surrogate peptides to quantify a human mAb in rat PK studies. To assess data reliability, data from the two peptides were compared in the context of individual sample concentration data as well as with PK parameters.

Regardless of the approach chosen for evaluation and comparison of dual peptide data sets, the stringency of the criteria could be relaxed or tightened in a "fit-for-purpose" manner for calibration curve/QC samples acceptance and study samples concentration reporting, depending on the stage of protein drug development, needs of a particular study, reliability/robustness of the LC–MS/MS assay, and so on. (Furlong et al. 2013).

13.3.7 Related Studies: Generic LC–MS/MS Assays for Human mAb Bioanalysis in Animal Studies

Independent of our universal peptide assay studies described in the previous section, Hongyan Li and his colleagues at Amgen explored a "general" approach to human mAb bioanalysis in animal studies (Li et al. 2012). They identified and investigated several promising common peptide sequences from human antibody heavy and light chain constant regions, none of which was Peptide 1. Eight different human IgG1 or IgG2 mAbs were subjected to single precision and accuracy runs using these common surrogate peptides for quantification. In all cases, percent bias and percent CV values met the typical LBA criteria. The general LC–MS/MS assay was applied to the quantification of two distinct human mAbs (one IgG1 and IgG2) in rat PK studies. A single common peptide – heavy chain constant region peptide NQVSLTCLVK (Peptide 3; Tables 13.1 and 13.2) – was used to quantify both mAbs. Concentration data derived using the general LC–MS/MS assay were in excellent agreement with data determined using a LBA.

Peptide 3 is of particular interest and value due to its presence in all three of the human IgG subclasses commonly used in therapeutic mAb drug candidates (IgG1, IgG2, and IgG4). In contrast, Peptide 1, as described above, is only present in human IgG1 and IgG4 subclasses. One limitation of Peptide 3 compared to Peptide 1 is the former peptide cannot be used for cynomolgus monkey studies due to its presence in the amino acid sequence of endogenous cynomolgus monkey antibodies. In addition, due to the presence of a cysteine residue Peptide 3, reduction and alkylation steps must be incorporated into the sample preparation procedure prior to LC–MS/MS quantification. By contrast, the absence of cysteine residues in Peptide 1 enables its use as a surrogate peptide with or without reduction/alkylation steps (Ouyang et al. 2012, Yuan et al. 2012, 2013, Furlong et al. 2013, Jiang et al. 2013).

Zhang et al reported their generic LC–MS/MS approach to the bioanalysis of human IgG1-based mAb drug candidates nonclinical studies (Zhang et al. 2014a, b). One of the human IgG1 heavy chain constant region peptides evaluated – GPSVFPLAPSSK (Peptide 4; Tables 13.1 and 13.2) – was shown to exhibit good linearity, precision, accuracy, and sensitivity. Peptide 4 was used to quantify a human IgG1 mAb in a cynomolgus PK study (Zhang et al. 2014a, b). Peptide 4 has since been used by other investigators (Ladwig et al. 2014, Law et al. 2014, Nouri-Nigjeh et al. 2014, Remily-Wood et al. 2014, An et al. 2015, Lanshoeft et al. 2016). Li's team also investigated an IgG2-/IgG4-specific variant of Peptide 4 (Peptide 6 – GPSVFPLAPCSR; Tables 13.1 and 13.2) as a potential common peptide for bioanalysis of human mAbs of these two subclasses in animal studies.

Nouri-Nigjeh et al. 2014 reported the quantification of a human mAb in a rat PK study using a dual generic peptide LC–MS/MS assay in which kappa light chain constant region Peptide 2 was coquantified with the IgG1-specific heavy chain constant region Peptide 4 (Tables 13.1 and 13.2). Excellent quantitative agreement (within ±15%) was noted between the mAb concentration data sets derived from each peptide. In principle, this novel dual peptide combination (Peptide 2 + Peptide 4) could be used to support bioanalysis of a diversity of human IgG1 heavy chain/kappa light chain-based mAbs in commonly used animal species.

Human mAb heavy chain peptides 7, 8, and 9 also showed promise in our exploratory dual peptide assay studies (FNWYVDGVEVHNAK, TTPPVLDSDGSFFLYSK, and GFYPSDIAVEWESNGQPENNYK, respectively; Tables 13.1 and 13.2) (Furlong et al. 2013). Peptide 9 is of particular interest due to its presence in the human heavy chain constant region amino acid sequences of all three commonly used subclasses (IgG1, IgG2, and IgG4). All three of these peptides have been further utilized for antibody bioanalysis by other investigators (Hong et al. 2013, Zhang et al. 2014a, b, Lebert et al. 2015, Lanshoeft et al. 2016). Similarly, Human mAb kappa light chain peptides 10 and 11 (SGTASVVCLLNNFYPR and DSTYSLSSTLTLSK, respectively; Tables 13.1 and 13.2) (Furlong et al. 2013) have been further investigated by Li et al. 2012 and Zhang et al. 2014a, b.

Lebert et al. 2015 evaluated human IgG1-, IgG2-, and IgG4-specific constant region peptides as tools to potentially enable simultaneous and selective quantification of multiple mAb isotypes in rat PK samples derived from coadministration studies.

In addition to their utility for mAb drug candidate bioanalysis in animal studies, human heavy and light chain constant region antibodies have also been used to quantify *endogenous* concentrations of immunoglobulins in clinical samples (Hong et al. 2013, Ladwig et al. 2014, Remily-Wood et al. 2014).

13.4 Extending the Universal Peptide Assay Concept to Human mAb Bioanalysis in Human Studies

The heretofore described universal LC–MS/MS assay approach to mAb bioanalysis is limited to animal studies and is not applicable to human studies (Figure 13.8a). This limitation is related to assay specificity, that is, all heavy and light chain constant region peptide sequences are present in both human therapeutic mAbs and endogenous human antibodies present in human matrix samples. Based upon these assay selectivity considerations, signature surrogate peptides from the variable regions of the human mAb light and/or heavy chains must be used for LC–MS/MS quantification of human mAb analytes in human samples (Figure 13.8b).

13.4.1 Potential Expansion of the Universal LC–MS/MS Assay Concept into Human Studies

Human IgG4 subclass antibodies are distinct from other subclasses in that they exhibit the inherit capacity to exchange half-molecules (Figure 13.9) (King et al. 1992). This phenomenon, commonly called "Fab-arm exchange," occurs *in vitro* (King et al. 1992, Deng et al. 2004) and *in vivo* (Labrijn et al. 2009, Stubenrauch et al. 2009). The exchange can occur between two human IgG4-based therapeutic mAb molecules, two endogenous human IgG4 antibody molecules, or between human IgG4-based therapeutic mAb molecules and endogenous human IgG4 antibody molecules upon administration of the therapeutic mAb to human subjects. Fab-arm interchange is potentially undesirable from a drug-stability and an efficacy perspective. Angal et al. (1993) and others (Labrijn et al. 2009, Stubenrauch et al. 2009) have shown that incorporation of a single amino substitution – serine

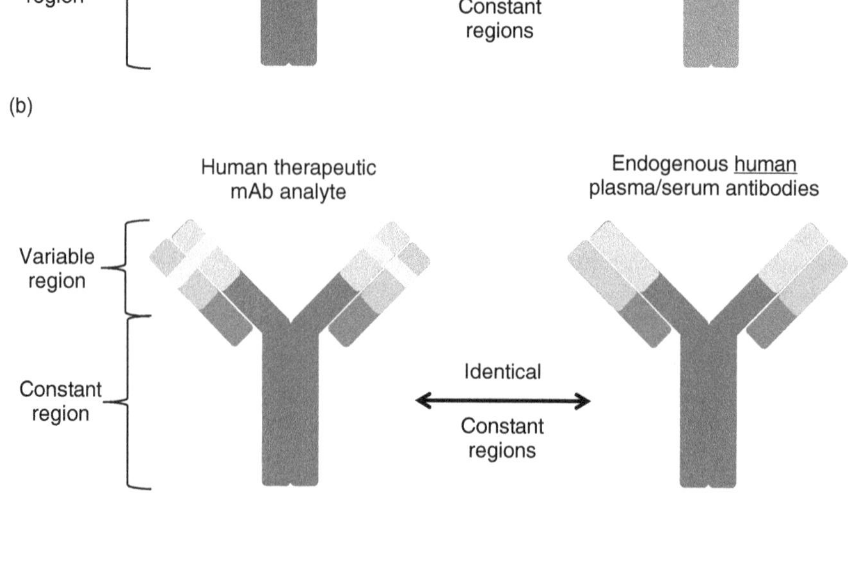

Figure 13.8 Surrogate peptide strategies for bioanalysis of human therapeutic mAbs are based in part on the species of the matrix sample. (a) For LC–MS/MS quantification of human therapeutic mAb analytes in *animal* samples, universal surrogate peptides, located in the constant regions of the human mAb light and heavy chains, may be used as alternatives to signature surrogate peptides from the variable regions. The locations of universal surrogate peptides in human mAb analytes are depicted as green bars. (b) For LC–MS/MS quantification of human mAb analytes in *human* samples, signature surrogate peptides, located in the variable regions of the human mAb light and heavy chains, must be used. The locations of the signature surrogate peptides in human mAb analytes are depicted as yellow bars. (*Source*: Furlong et al. 2014. Reproduced with permission of Future Science Ltd.) (Refer online version for the color representation of this figure)

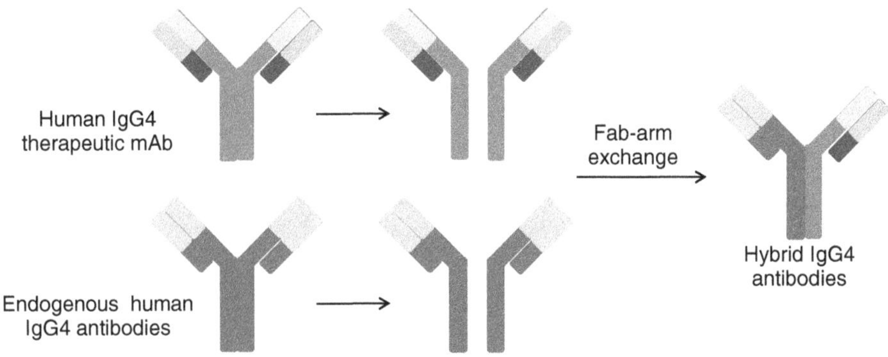

Figure 13.9 IgG4-based human therapeutic mAbs and endogenous IgG4 human antibodies undergo "Fab-arm exchange" *in vitro* and *in vivo*. Shown above is an exchange between an IgG4-based therapeutic mAb and an endogenous IgG4 human antibody; this exchange occurs in the bloodstream after administration of human IgG4-based mAbs to human subjects. (*Source*: Furlong et al. 2014. Reproduced with permission of Future Science Ltd.)

to proline at position 241 (Kabat numbering (Kabat et al. 1987)) – into the hinge region of human IgG4 mAbs substantially reduces Fab-arm exchange.

This structure-stabilizing substitution has subsequently been incorporated into many human IgG4-based therapeutic mAb drugs and drug candidates (International Immunogenetics Information System website n.d., Labrijn et al. 2009). The S241P amino acid substitution creates a predicted tryptic peptide (Peptide 5) with an amino acid sequence and mass that is distinct from the corresponding sequence in the endogenous human IgG4 antibodies present in human plasma/serum samples (Figure 13.10). We reasoned that a single LC–MS/MS assay based upon Peptide 5 should in theory enable quantification of all hinge-region-stabilized (S241P-containing) human IgG4 mAbs, free from interference from the large excess of endogenous IgG4 antibodies that would be present in human matrix samples, thus rendering the assay "universal" for this class of human therapeutic mAb drugs and drug candidates (Furlong et al. 2014).

13.4.2 Development and Evaluation of an Exploratory Universal IgG4 Clinical LC–MS/MS Assay

We developed an exploratory universal IgG4 LC–MS/MS assay based upon Peptide 5 (Tables 13.1 and 13.2) in human serum using a model hinge-region-stabilized human IgG4 therapeutic mAb drug candidate (Furlong et al. 2014). As predicted, Peptide 5 was detected following digestion of IgG4 mAb-spiked human plasma without interference from the corresponding peptide found in endogenous IgG4 (Figure 13.11). The exploratory assay was subjected to three assay performance evaluation runs. Intra- and interrun precision and accuracy was very good in these runs; mean percent bias and percent CV values were <10% at all QC levels. Human IgG4 mAb matrix stability experiments were also performed using Peptide 5 as the surrogate peptide. Matrix stability was demonstrated for three freeze–thaw cycles, 26 h at room temperature and 29 days at −70 °C.

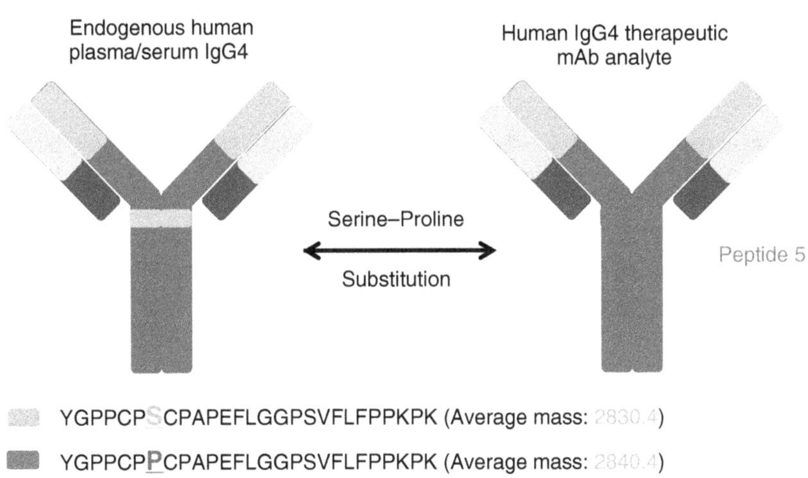

Figure 13.10 A single amino substitution (S241P) is incorporated into the hinge region of many human IgG4-based therapeutic mAbs to minimize Fab-arm exchange. Distinct predicted peptide sequences arising from trypsin digestion of S241P-stabilized human IgG4 therapeutic mAbs and endogenous human IgG4 antibodies are also shown. (*Source*: Furlong et al. 2014. Reproduced with permission of Future Science Ltd.)

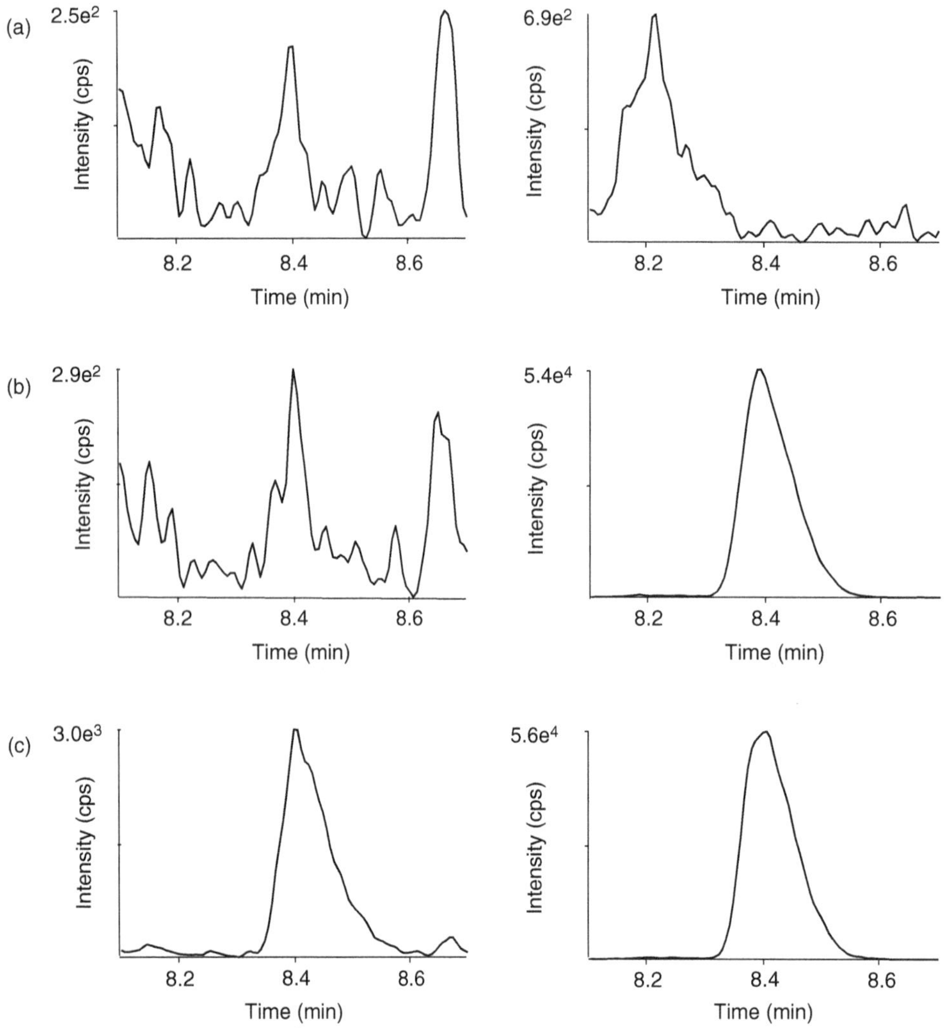

Figure 13.11 Representative Peptide 5 (Universal IgG4 peptide) LC–MS/MS chromatograms. (a) blank human serum, (b) blank human serum spiked with stable isotope-labeled protein internal standard, (c) serum spiked with a model hinge-region-stabilized human IgG4 therapeutic mAb drug candidate at the assay LLOQ (5 μg/mL) and stable-isotope-labeled protein internal standard. Left panels, analyte; right panels, internal standard. (*Source*: Furlong et al. 2014. Reproduced with permission of Future Science Ltd.)

To further evaluate the reliability of the universal IgG4 assay, a second confirmatory "signature" surrogate peptide (ASGITFSNSGMHWVR) located in the variable region of the heavy chain of the model human IgG4 mAb, was incorporated into the universal IgG4 LC–MS/MS assay. Very good quantitative agreement between the two peptides was observed in all three precision and accuracy runs. In 97% of the QCs evaluated over the three runs, the percent difference values for the two peptides were <20%. In addition, the overall mean percent difference values for all QC samples in each run ranged from 0.337% to 1.19%. Collectively, these data indicate that a Peptide-5-based LC–MS/MS assay could likely support bioanalysis of hinge-region-stabilized human IgG4 therapeutic mAb drug candidates in clinical samples. In addition to its potential utility as a quantitative peptide, Peptide 5 has also been used to assess nonspecific binding of a human stabilized IgG4 mAb drug candidate to magnetic beads used in a cell-based functional neutralizing antibody assay (Jiang et al. 2014, Xu et al. 2015).

The LLOQ for the exploratory universal IgG4 assay described herein was 5 μg/mL, which may be of insufficient sensitivity to support clinical studies in many cases. Preliminary experiments indicate the LLOQ of this assay could be lowered by approximately 10-fold if the serum tryptic digests quantified in these preliminary studies were subjected to solid-phase extraction prior to LC–MS/MS analysis (Furlong et al. 2014).

13.4.3 Evaluation of the Impact of Anti-mAb Antibodies on Exploratory Universal IgG4 LC–MS/MS Assay Performance

The administration of protein drugs to humans or animals can elicit the production of antidrug antibodies (ADA). ADA are typically polyclonal (heterogeneous) in nature, that is, they can bind to multiple unique regions of the protein drug (Figure 13.12). The binding of ADA to protein drugs in study samples can potentially interfere with accurate quantification of the total concentration of the protein drug, particularly if LBAs are deployed (Ezan and Bitsch 2009, White et al. 2011, Kelley et al. 2013). The susceptibility of LBA to ADA interference is due to their reliance upon binding interactions between the protein analyte and LBA reagents. ADA present in study samples can potentially prevent LBA reagents from binding to the protein analyte, thus impacting LBA reliability (Figure 13.12). In contrast, typical LC–MS analysis-enabling sample processing procedures can disrupt ADA–protein interactions, rendering LC–MS assays potentially less susceptible to ADA-mediated assay interference (Figure 13.12) (Ji et al. 2007, Heudi et al. 2008, Wang et al. 2012, Gong et al. 2014, Law et al. 2014).

In order to evaluate potential ADA impact on quantification of clinical samples using the exploratory universal IgG4 LC–MS/MS assay, blank human serum samples were cospiked with the model human IgG4 mAb and either (i) affinity-purified polyclonal antisera raised in monkeys against the model human IgG4 mAb or (ii) a mouse monoclonal antibody that binds to the target binding region of the model IgG4 mAb and interferes with its ability to bind to its drug target. As shown in Figure 13.13, the presence of up to 25-fold molar excess of ADA did not adversely impact the quantification of the model IgG4 mAb in the spiked samples. These results suggest that a fully validated version of this universal IgG4 LC–MS/MS assay should be able to support reliable quantification of hinge-region-stabilized human IgG4 therapeutic mAbs in clinical samples, even if ADA are present.

13.5 Internal Standard Options for Generic Peptide LC–MS/MS Assays

13.5.1 Stable Isotopically Labeled Peptide Internal Standards

Stable isotopically labeled (SIL) internal standards are preferred to structural analog internal standards for LC–MS/MS bioanalysis. This is because the highly similar chemical properties and chromatographic retention times of the analyte and stable isotope-labeled internal standard are thought to facilitate more reliable correction for lot-to-lot or subject-to-subject differences in extraction efficiency and/or suppression/enhancement by matrix components (Fu et al. 1998, Stokvis et al. 2005). SIL peptides, including Peptide 1, can be routinely synthesized on a sufficient scale for bioanalytical purposes, and thus are commonly used as analytical internal standards for bioanalysis of protein analytes, including mAbs (Barr et al. 1996, Barnidge et al. 2004a, b, Hagman et al. 2008, Bronsema et al. 2012, van den Broek et al. 2013, Voronin et al. 2014). We deployed a SIL peptide version of Peptide 1 as the internal standard in early drug development animal studies (Ouyang et al. 2012). Acceptable LC–MS/MS assay performance was achieved in these studies. Li et al. (2012) deployed a SIL peptide version of

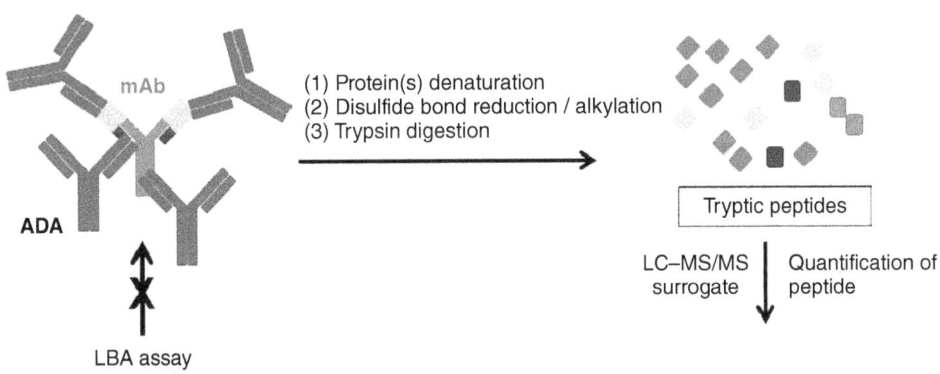

Figure 13.12 ADA binding to protein drugs (e.g., mAbs) *in vivo*: potential impact on bioanalysis. ADA can potentially prevent ligand binding assay (LBA) reagents from binding to the protein drug analyte, thus potentially impacting LBA assay quantification of the total concentration of the protein drug in the study sample. In contrast, typical LC–MS analysis-enabling sample processing procedures can disrupt ADA–protein drug interactions, rendering the LC–MS assays less susceptible to ADA-mediated assay interference. (*Source*: Furlong et al. 2014. Reproduced with permission of Future Science Ltd.)

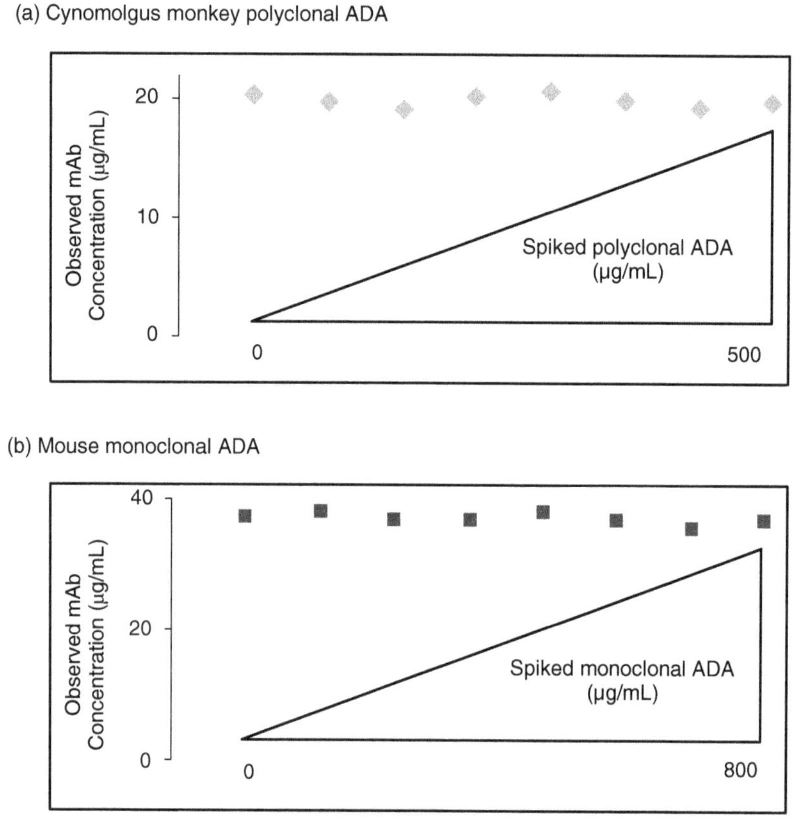

Figure 13.13 Impact assessment of antidrug antibodies (ADA) on the performance of the exploratory universal IgG4 LC–MS/MS assay. (a) Blank human serum was spiked with a model human IgG4 mAb (20 μg/mL) along with a series of concentrations (0–500 μg/mL) of affinity purified cynomolgus monkey polyclonal anti-IgG4 mAb antiserum. (b) Blank human serum was spiked with a model human IgG4 mAb (40 μg/mL) along with a series of concentrations (0–800 μg/mL) of a mouse monoclonal antibody that binds to the target binding region of the model human IgG4 mAb. Samples were extracted and quantified in triplicate by LC–MS/MS. Data plotted are mean concentrations. CV values for all replicate determinations were under 3%. (*Source*: Furlong et al. 2014. Reproduced with permission of Future Science Ltd.)

Peptide 3 as the internal standard, while Nouri-Nigjeh et al. (2014) had similar success with SIL versions of Peptides 2 and 4 in their version of a dual peptide assay.

13.5.2 Stable Isotopically Labeled Protein Internal Standards

While SIL peptide internal standards can normalize for sample-to-sample variability in surrogate peptide analyte extraction and matrix effects, they do not normalize for the protein extraction and trypsin digestion steps commonly employed during sample preparation. In contrast to SIL peptide internal standards, SIL protein internal standards are predicted to track the corresponding protein analytes through all sources of variability associated with bioanalytical processing, including protein extraction, trypsin digestion, peptide extraction, and matrix effects (Brun et al. 2007, Bronsema et al. 2012, 2013, van den Broek et al. 2013). SIL mAb proteins have been used as internal standards in mAb bioanalysis (Heudi et al. 2008, Ji et al. 2009, Lesur et al. 2010, Liu et al. 2011). However, they have typically been deployed in signature peptide-based assays.

Since signature peptides (located in the variable light or heavy chain regions) are unique to the mAb analyte of interest, the corresponding SIL mAb internal standard can only be used for quantification of that particular mAb. We reasoned that a SIL mAb could be used as a "universal" protein internal standard capable of supporting bioanalysis of a diversity of mAbs, provided that the surrogate peptides used for mAb quantification were universal peptides as opposed to signature peptides (Furlong et al. 2013).

To test this universal internal standard hypothesis, we biosynthetically labeled a single human mAb containing an IgG1 heavy chain constant region and a kappa light chain constant region with [$^{13}C_6$, ^{15}N]-L-Leucine and [$^{13}C_5$, ^{15}N]-L-Valine. The amino acid sequences of heavy chain universal Peptide 1 and kappa light chain universal Peptide 2 contain at least one valine or leucine (Tables 13.1 and 13.2), thus ensuring that a SIL peptide internal standard for both universal peptides would be generated from trypsin digestion of the universal SIL mAb protein. As described in Sections 13.3.2 through 13.3.4, the universal SIL mAb protein performed very well during quantification of three distinct human mAbs (two IgG1 mAbs and one IgG4 mAb) in both assay performance evaluation runs and a cynomolgus monkey PK study. In addition, no significant (i.e., ≤20%) internal standard-derived Peptide 1 or Peptide 2 analyte response was observed in the analyte SRM transitions. These results demonstrated the generality and reliability of a dual peptide assay that is universal not only for the surrogate peptide analytes but for the SIL protein internal standard as well (Furlong et al. 2013). In more recent experiments, we used a human IgG4-based SIL mAb protein internal standard as part of our exploratory experiments to extend the universal peptide concept from nonclinical to clinical bioanalysis (Furlong et al. 2014) (see Section 13.4).

In related internal standard studies, Li et al. 2012 deployed a SIL protein version of a single human IgG2-based mAb as the internal standard to enable bioanalysis of a human IgG1- and a human IgG2-based mAb drug candidate in separate rat PK studies. The ability of a single SIL mAb internal standard to perform across isotypes (IgG1 and IgG2) derived from the fact that the amino acid sequence of the surrogate peptide deployed for mAb quantification (Peptide 3) is found in both isotypes. Nouri-Nigjeh and colleagues reported the use of a SIL human IgG1-based mAb internal standard protein to quantify a human IgG1 heavy chain/kappa light chain mAb in rat PK studies using Peptides 2 and 4 as the surrogate peptides (Nouri-Nigjeh et al. 2014). Zhang et al. (2014a, b) evaluated a commercially available SIL mAb protein as an internal standard during bioanalysis of a human IgG1-based mAb drug candidate using their Peptide-4-based generic human IgG1 LC–MS/MS assay. The SIL human IgG1 mAb protein was uniformly labeled throughout its amino acid sequence with [^{13}C, ^{15}N] lysine and arginine. The commercially available SIL mAb protein internal standard performed well during a precision/accuracy run and subsequently during bioanalytical support of a monkey PK study.

13.5.3 "Flanked" Stable Isotopically Labeled Peptide Internal Standards

An intermediate form of SIL internal standard used for protein bioanalysis is the amino acid "flanked" SIL peptide (Barnidge et al. 2004a, b, Bronsema et al. 2012, 2013, Fernandez Ocana et al. 2012, Li et al. 2012, Jiang et al. 2013, Mekhssian et al. 2014, Nouri-Nigjeh et al. 2014). This class of peptide is similar to SIL peptides described earlier except that a few additional, unlabeled amino acids from the protein analyte sequence are added to the amino and carboxy side of the SIL peptide. Flanked SIL peptides have a potential advantage over simple SIL peptides in that the former may at least, in principle, be able to partially normalize sample-to-sample variability in trypsin digestion efficiency (Barnidge et al. 2004a, b, Bronsema et al. 2013). In addition, flanked SIL peptides are easier to obtain since they can usually be prepared by automated or semiautomatic synthesis, as opposed to the more complex, costly, and time-consuming biosynthetic labeling procedures required for SIL protein production. On the other hand, flanked SIL peptides are similar to simple SIL peptides in that they cannot normalize for sample-to-sample variability in protein extraction steps if such steps are employed in the sample preparation procedure.

Jiang et al. (2013) successfully deployed flanked SIL Peptides 1 and 2 as internal standards during quantification of two codosed human mAbs in a monkey PK study. In that study, Peptides 1 and 2 were quantified as confirmatory peptides rather than as primary quantitative peptides. Li et al. (2012) compared the internal standard performance of SIL Peptide 3 in the contexts of SIL peptide, flanked SIL peptide, and SIL mAb protein. In terms of precision and accuracy, the overall performance of the SIL mAb protein IS was superior to either the SIL peptide or the flanked SIL peptide. Nouri-Nigjeh et al. (2014) observed very good quality control sample precision and accuracy results during evaluation of SIL peptide, flanked SIL peptide and SIL mAb protein versions of Peptides 2 and 4; all three internal standard approaches were successfully used to quantify a human IgG1-kappa mAb in a rat PK study.

Regardless of the chosen generic internal standard strategy (SIL peptide, SIL protein, or flanked SIL peptide), the aforementioned studies suggest that only one or two such proteins/peptides would be required to have on hand as generic internal standards to enable bioanalysis of a wide variety of human mAb drug candidates in commonly used animal species.

13.6 Sample Preparation Strategies for Generic Peptide LC–MS/MS Assays

13.6.1 Direct Digestion, Pellet Digestion, and Solid-Phase Extraction

A variety of sample preparation strategies have been deployed in conjunction with the generic peptide assays described above. Yuan et al. (2012) evaluated direct trypsin digestion of plasma samples followed by LC–MS/MS analysis of a human mAb in cynomolgus monkey plasma using Peptide 1 as the surrogate peptide. We and others have used a pellet digestion-based approach to quantify human mAbs in animal studies using Peptide 1 and/or Peptide 2 as surrogate peptides (Furlong et al. 2012, Ouyang et al. 2012, Yuan et al. 2012, Jiang et al. 2013, Lanshoeft et al. 2016). In this approach, plasma/serum samples are treated with an organic solvent to precipitate the large protein fraction, which includes the mAb analytes. The protein pellet is subsequently suspended in buffer and digested with trypsin. The resulting peptide mix can be directly injected onto the LC–MS/MS system or optionally further processed by solid-phase extraction (SPE) prior to analysis. An and colleagues demonstrated that pretreatment of human mAb-containing plasma or tissue homogenate samples with sodium dodecyl sulfate prior to protein precipitation/pellet digestion improved the rate and yields of Peptides 2 and 4 formation (An et al. 2015). Liu et al. (2014) incorporated trichloroacetic acid into the protein precipitation step as a means to efficiently deplete albumin from the protein pellet. The investigators applied this sample preparation strategy to support human mAb bioanalysis in nonclinical studies using Peptide 1 as the confirmatory peptide (Liu et al. 2014, Zhao et al. 2015).

Peptides 1 and 2 do not contain cysteines and thus, in principle, do not require reduction and alkylation as part of the sample preparation procedure. However, these peptides have been successfully deployed using both reducing and nonreducing sample preparation strategies (Ouyang et al. 2012, Yuan et al. 2012, Jiang et al. 2013, Nouri-Nigjeh et al. 2014).

Jiang et al. (2013) found that reduction/alkylation of the resuspended protein pellet before trypsin digestion doubled the peak area response of Peptide 1 compared to pellet digestion without the reduction/alkylation steps. This result was presumably due to increased susceptibility of Peptide 1 to trypsin digestion as a result of improved mAb denaturation and unfolding under the reduction/alkylation conditions. To facilitate human mAb bioanalysis in a rat PK study using Peptides 2 and 4 as the surrogate peptides, Nouri-Nigjeh et al. (2014) successfully altered the sequence of the aforementioned protein precipitation–reduction/alkylation sample preparation strategy, reducing/alkylating the samples *prior to* protein precipitation and trypsin digestion.

13.6.2 Affinity Capture

Sample preparation strategies that rely upon affinity capture of the human mAb analyte prior to trypsin digestion and LC–MS/MS analysis offer the potential to generate very clean extracts. We developed an immobilized Protein-A-based sample preparation strategy to quantify human mAbs in animal studies using the Peptide 1/ Peptide 2 exploratory dual universal assay (Furlong et al. 2013). Protein A is a bacterial protein that binds with high affinity to the heavy chain within the Fc region of most human and animal immunoglobulins. During sample preparation, plasma/serum samples are mixed with magnetic bead immobilized Protein A in a 96-well plate format. Following bead capture and stringent washing, the bead-bound extracts will contain only the human mAb analyte along with an excess of endogenous animal immunoglobulins. All other plasma components, including abundant proteins such as serum albumin, do not bind to Protein A and are thus removed from the extract. The beads are then subjected to denaturation with RapiGest surfactant (Waters Corporation - Milford MA USA) along with cysteine reduction/alkylation. The denatured samples are then directly digested with trypsin, followed by acidification and analysis by LC–MS/MS. The coextracted endogenous animal antibodies are not predicted to interfere with the assay because their constant region sequences do not contain Peptide 1 and 2 sequences. However, co-eluting endogenous antibody-derived peptides can adversely impact assay sensitivity due to ion suppression. Isobaric coeluting endogenous antibody-derived peptides can also interfere with the assay (Furlong et al. 2013).

Li et al. (2012) developed a more selective sample preparation strategy based on immunocapture of human mAb analytes using an antibody reagent that binds to human heavy chain Fc regions. This antibody reagent is more selective than Protein A because it binds only to *human* heavy chain Fc regions; the human mAb analyte is captured from the animal plasma/serum during sample preparation, while the vast excess of animal antibodies present in the samples are not captured. The immunocaptured extracts, devoid of endogenous animal antibodies, are thus predicted to be cleaner than Protein-A captured extracts. This improvement in selectivity should result in reduced matrix-related ion suppression as well as reduced potential for matrix-derived coeluting assay interference peaks. Li's team used immunocapture sample preparation for bioanalysis of two distinct human mAbs (one IgG1 and IgG2) in rat PK studies using Peptide 3 as the generic peptide. Zhang et al. (2014a, b) were similarly successful using a human Fc-specific immunocapture reagent to quantify a human IgG1 mAb in a cynomolgus monkey study using Peptide 4 as the surrogate peptide.

13.6.3 Additional Sample Preparation Approaches for Generic Peptide LC–MS/MS Assays

Anderson et al. (2004) have developed a variant of immunocapture-based sample preparation: Stable Isotope Standards and Capture by Anti-Peptide Antibodies – SISCAPA. In this approach, the plasma/ serum samples are first digested with trypsin, followed by immunocapture using an antibody reagent that binds specifically to a protein analyte-derived tryptic peptide rather than the intact protein analyte. In principle, this technique could be incorporated into a generic peptide LC–MS/MS assay. Such a strategy would require the production of high-affinity monoclonal antibodies capable of selectively binding to human mAb-based generic peptide(s) of interest in the presence of all other analyte- and matrix protein-derived peptides that would be present in digested matrix samples.

The choice of sample preparation approach can increase the number of species in which generic LC–MS/MS assays can be used. For example, Zhang et al. (2014a, b) observed interfering matrix-derived peaks in several SRM transitions during development of an LC–MS/MS assay based upon Peptide 4 in cynomolgus monkey plasma (Tables 13.1 and 13.2). These interferences could potentially eliminate Peptide 4 as a generic peptide for quantifying any human IgG1 mAb in this important species. The investigators changed their sample preparation approach from pellet digestion (Ouyang et al. 2012) to human heavy chain Fc region immunocapture (Li et al. 2012). This change eliminated the monkey matrix-derived interference peaks and facilitated successful bioanalysis of a human IgG1 mAb in a monkey PK study. Importantly, this sample preparation improvement would in principle enable bioanalysis of all human IgG1 mAbs in cynomolgus monkey studies using Peptide 4 as the generic surrogate peptide.

Automation can enable increased throughput and efficiency in bioanalysis. This can be especially useful when implementing the complex multistep sample preparation strategies that are often required for mAb bioanalysis. Li et al. (2012) and Zhang et al. (2014a, b) both incorporated automation into several sample preparation steps, thus increasing the throughput and efficiency of their respective assays.

13.7 Limitations of Generic Peptide LC–MS/MS Assays

Most therapeutic mAbs interact with their pharmacological targets via the light and heavy chain variable regions of the mAbs (Figure 13.1). Disruptions to these variable regions (e.g., peptide bond hydrolysis, methionine oxidation) may disrupt the ability of the mAb to bind its target, which would in turn adversely impact the pharmacological potency of the mAb. Since generic LC–MS/MS mAb assays rely upon surrogate peptides located in the constant regions of the light and heavy chain regions, such assays would not be able to distinguish between pharmacologically active and inactive forms of the mAb being quantified (Figure 13.1). In contrast, "signature" peptide LC–MS/MS assays, which are based upon surrogate peptides found in the variable regions of the mAb light or heavy chains (Figure 13.1) (Yang et al. 2007, Dubois et al. 2008, Hagman et al. 2008, Heudi et al. 2008, Ji et al. 2009, Lesur et al. 2010, Fernandez Ocana et al. 2012, Jiang et al. 2013, Mekhssian et al. 2014, Xu et al. 2014), are more likely to be able to distinguish between pharmacologically active and inactive forms of the mAb.

Although the generic peptide LC–MS/MS assay approach to bioanalysis of mAbs is useful for bioanalytical support of animal studies involving a single human mAb analyte, this approach could not be used for studies that involve coadministration of more than one mAb, particularly if the coadministered mAbs in the study were of the same human heavy and light chain subclasses. The reason is all mAbs present in the study samples would give rise to the same generic peptides upon trypsin digestion, and, thus, the individual concentrations of each mAb in these samples could not be determined using generic peptides. The traditional signature peptide approach would need to be deployed for studies involving bioanalysis of coadministered human mAbs in animal studies. On the other hand, if the coadministered human mAbs were of distinct subclasses, then it is conceivable that a single generic peptide LC–MS/MS assay capable of quantifying subclass-specific generic peptides could be used (Lebert et al. 2015).

Jiang et al. (2013) used a creative approach to enable the use of generic Peptides 1 and 2 as confirmatory peptides during signature peptide-based quantification of two distinct human IgG4 heavy chain/kappa light chain mAbs following coadministration to cynomolgus monkeys. Signature peptides unique to each mAb, found in the heavy chain variable regions, were used for quantification. In order to evaluate the reliability of the signature peptide-based study sample concentration data, the concentration values derived from the two signature peptides in each study sample were summed and compared with the observed concentration values derived from Peptide 1 quantification. Similar comparisons were made between the summed signature peptide data and Peptide 2 data. In both comparisons, there was good agreement between the summed signature peptide-derived primary quantitative data and the generic peptide-derived confirmatory data, providing a higher level of confidence in the primary data.

As discussed in Section 13.4, the generic peptide assay approach is currently rather limited with respect to its application to bioanalysis of human mAbs in human studies. This is due to the fact that the constant regions of human therapeutic mAbs are identical to the constant regions of the endogenous antibodies found in human plasma/serum samples (Figure 13.8). This limitation can be overcome when quantifying hinge-region-stabilized human IgG4 mAbs, which contain a single amino acid substitution in the heavy chain constant region. This substitution results in a generic tryptic peptide that is distinguishable by LC–MS/MS from the corresponding tryptic peptide derived from endogenous IgG4 antibodies. Such selectivity enabled the development of an exploratory generic peptide LC–MS/MS assay that would likely be able in fully validated form to support bioanalysis of this class of human therapeutic mAb drugs and drug candidates in human studies (Furlong et al. 2014).

13.8 Conclusion

The large and rapidly growing number of human therapeutic mAb drug candidates in various stages of development presents a significant challenge to bioanalytical laboratories tasked with quantifying these proteins in biological matrices. The generic surrogate peptide LC–MS/MS assays described herein have the potential to enable bioanalysis of a diversity of human therapeutic mAb drug candidates in all commonly used animal species. This approach can potentially be extended to human studies if the human mAb analyte belongs to the hinge-stabilized IgG4 subclass.

Acknowledgments

The author wishes to acknowledge his collaborators on the universal peptide studies described in this chapter: Mohammed Jemal, Zheng Ouyang, William Mylott, Song Zhao, Rand Jenkins, Craig Titsch, Weifeng Xu, Hao Jiang, Jianing Zeng, Steven Wu, Bogdan Sleczka, James Tamura, Mian Gao, Vendana Hegde and Roger Demers.

References

An B, Zhang M, Johnson RW, et al. (2015). "Surfactant-aided precipitation/on-pellet-digestion (SOD) procedure provides robust and rapid sample preparation for reproducible, accurate and sensitive LC/MS quantification of therapeutic protein in plasma and tissues." *Anal Chem (Washington, DC, USA)* **87**(7): 4023–4029.

Anderson NL, Anderson NG, Haines LR, et al. (2004). "Mass spectrometric quantitation of peptides and proteins using stable isotope standards and capture by anti-peptide antibodies (SISCAPA)." *J Proteome Res* **3**(2): 235–244.

Angal S, King DJ, Bodmer MW, et al. (1993). "A single amino acid substitution abolishes the heterogeneity of chimeric mouse/human (IgG4) antibody." *Mol Immunol* **30**(1): 105–108.

Barnidge DR, Goodmanson MK, Klee GG, et al. (2004a). "Absolute quantification of the model biomarker prostate-specific antigen in serum by LC-MS/MS using protein cleavage and isotope dilution mass spectrometry." *J Proteome Res* **3**(3): 644–652.

Barnidge DR, Hall GD, Stocker JL, et al. (2004b). "Evaluation of a cleavable stable isotope labeled synthetic peptide for absolute protein quantification using LC-MS/MS." *J Proteome Res* **3**(3): 658–661.

Barr JR, Maggio VL, Patterson DG, Jr., et al. (1996). "Isotope dilution-mass spectrometric quantification of specific proteins: model application with apolipoprotein A-I." *Clin Chem (Washington, DC)* **42**(10): 1676–1682.

Beck A and Reichert J M (2011). "Therapeutic Fc-fusion proteins and peptides as successful alternatives to antibodies." *MAbs* **3**(5): 415–416.

Beck A and Reichert J M (2014). "Antibody–drug conjugates: present and future." *MAbs* **6**(1): 15–17.

Beck A, Wagner-Rousset E, Bussat M-C, et al. (2008). "Trends in glycosylation, glycoanalysis and glycoengineering of therapeutic antibodies and Fc-fusion proteins." *Curr Pharm Biotechnol* **9**(6): 482–501.

Beck A, Wurch T, Bailly C, et al. (2010). "Strategies and challenges for the next generation of therapeutic antibodies." *Nat Rev Immunol* **10**(5): 345–352.

Bischoff R, Bronsema KJ and van de Merbel NC (2013). "Analysis of biopharmaceutical proteins in biological matrices by LC-MS/MS I. Sample preparation." *TrAC, Trends Anal Chem* **48**: 41–51.

Bowen CL, Kehler J, Mencken T, et al. (2015). "Utilizing LC-MS/MS to provide adaptable clinical bioanalytical support for an extended half-life bioactive peptide fused to an albumin-binding domain antibody." *Anal Methods* **7**(1): 237–243.

Bronsema KJ, Bischoff R and van de Merbel NC (2012). "Internal standards in the quantitative determination of protein biopharmaceuticals using liquid chromatography coupled to mass spectrometry." *J Chromatogr, B Anal Technol Biomed Life Sci* **893–894**: 1–14.

Bronsema KJ, Bischoff R and van de Merbel NC (2013). "High-sensitivity LC-MS/MS quantification of peptides and proteins in complex biological samples: the impact of enzymatic digestion and internal standard selection on method performance." *Anal Chem (Washington, DC, USA)* **85**(20): 9528–9535.

Brun V, Dupuis A, Adrait A, et al. (2007). "Isotope-labeled protein standards: toward absolute quantitative proteomics." *Mol Cell Proteomics* **6**(12): 2139–2149.

Bults P, van de Merbel NC and Bischoff R (2015). "Quantification of biopharmaceuticals and biomarkers in complex biological matrices: a comparison of liquid chromatography coupled to tandem mass spectrometry and ligand binding assays." *Expert Rev Proteomics* **12**(4): 355–374.

Deng L, Wylie D, Tsao Yung S, et al. (2004). "Detection and quantification of the human IgG4 half-molecule, HL, from unpurified cell-culture supernatants." *Biotechnol Appl Biochem* **40**(Pt 3): 261–269.

Dubois M, Fenaille F, Clement G, et al. (2008). "Immunopurification and mass spectrometric quantification of the active form of a chimeric therapeutic antibody in human serum." *Anal Chem (Washington, DC, USA)* **80**(5): 1737–1745.

Ezan E and Bitsch F (2009). "Critical comparison of MS and immunoassays for the bioanalysis of therapeutic antibodies." *Bioanalysis* **1**(8): 1375–1388.

Ezan E, Dubois M and Becher F (2009). "Bioanalysis of recombinant proteins and antibodies by mass spectrometry." *Analyst (Cambridge, UK)* **134**(5): 825–834.

Fast DM, Kelley M, Viswanathan CT, et al. (2009). "Workshop report and follow-up – AAPS workshop on current topics in GLP bioanalysis: assay reproducibility for incurred samples—implications of Crystal City recommendations." *AAPS J* **11**(2): 238–241.

Fernandez Ocana M, James IT, Kabir M, et al. (2012). "Clinical pharmacokinetic assessment of an anti-MAdCAM monoclonal antibody therapeutic by LC-MS/MS." *Anal Chem (Washington, DC, USA)* **84**(14): 5959–5967.

Fu I, Woolf EJ and Matuszewski BK (1998). "Effect of the sample matrix on the determination of indinavir in human urine by HPLC with turbo ion spray tandem mass spectrometric detection." *J Pharm Biomed Anal* **18**(3): 347–357.

Furlong MT, Ouyang Z, Wu S, et al. (2012). "A universal surrogate peptide to enable LC-MS/MS bioanalysis of a diversity of human monoclonal antibody and human Fc-fusion protein drug candidates in pre-clinical animal studies." *Biomed Chromatogr* **26**(8): 1024–1032.

Furlong MT, Titsch C, Xu W, et al. (2014). "An exploratory universal LC-MS/MS assay for bioanalysis of hinge region-stabilized human IgG4 mAbs in clinical studies." *Bioanalysis* **6**(13): 1747–1758.

Furlong MT, Zhao S, Mylott W, et al. (2013). "Dual universal peptide approach to bioanalysis of human monoclonal antibody protein drug candidates in animal studies." *Bioanalysis* **5**(11): 1363–1376.

Geiger T and Clarke S (1987). "Deamidation, isomerization, and racemization at asparaginyl and aspartyl residues in peptides: succinimide-linked reactions that contribute to protein degradation." *J Biol Chem* **262**(2): 785–794.

Gong C, Zeng J, Akinsanya B, et al. (2014). "Development and validation of an LC-MS/MS assay for the quantitation of a PEGylated anti-CD28 domain antibody in human serum: overcoming interference from antidrug antibodies and soluble target." *Bioanalysis* **6**(18): 2371–2383.

Gong C, Zheng N, Zeng J, et al. (2015). "Post-pellet-digestion precipitation and solid phase extraction (SPE): a practical and efficient workflow to extract surrogate peptides for UHPLC-MS/MS bioanalysis of a therapeutic antibody in the low ng/mL range." *J Chromatogr A* **1424**: 27–36.

Hagman C, Ricke D, Ewert S, et al. (2008). "Absolute quantification of monoclonal antibodies in biofluids by liquid chromatography-tandem mass spectrometry." *Anal Chem (Washington, DC, USA)* **80**(4): 1290–1296.

Heudi O, Barteau S, Zimmer D, et al. (2008). "Towards absolute quantification of therapeutic monoclonal antibody in serum by LC-MS/MS using isotope-labeled antibody standard and protein cleavage isotope dilution mass spectrometry." *Anal Chem (Washington, DC, USA)* **80**(11): 4200–4207.

Hong Q, Lebrilla CB, Miyamoto S, et al. (2013). "Absolute quantitation of immunoglobulin G and its glycoforms using multiple reaction monitoring." *Anal Chem (Washington, DC, USA)* **85**(18): 8585–8593.

Hopfgartner G, Lesur A and Varesio E (2013). "Analysis of biopharmaceutical proteins in biological matrices by LC-MS/MS II. LC-MS/MS analysis." *TrAC, Trends Anal Chem* **48**: 52–61.

International Immunogenetics Information System website (n.d.) www.imgt.org.

Jemal M and Xia YQ (2000). "Bioanalytical method validation design for the simultaneous quantitation of analytes that may undergo interconversion during analysis." *J Pharm Biomed Anal* **22**(5): 813–827.

Ji C, Sadagopan N, Zhang Y, et al. (2009). "A universal strategy for development of a method for absolute quantification of therapeutic monoclonal antibodies in biological matrices using differential dimethyl labeling coupled with ultra performance liquid chromatography-tandem mass spectrometry." *Anal Chem (Washington, DC, USA)* **81**(22): 9321–9328.

Ji QC, Rodila R and El-Shourbagy TA (2007). "A sample preparation process for LC-MS/MS analysis of total protein drug concentrations in monkey plasma samples with antibody." *J Chromatogr B Anal Technol Biomed Life Sci* **847**(2): 133–141.

Jiang H, Xu W, Titsch CA, et al. (2014). "Innovative use of LC-MS/MS for simultaneous quantitation of neutralizing antibody, residual drug, and human immunoglobulin G in immunogenicity assay development." *Anal Chem (Washington, DC, USA)* **86**(5): 2673–2680.

Jiang H, Zeng J, Titsch C, et al. (2013). "Fully validated LC-MS/MS assay for the simultaneous quantitation of coadministered therapeutic antibodies in cynomolgus monkey serum." *Anal Chem (Washington, DC, USA)* **85**(20): 9859–9867.

Kabat E, Wu, TT, Reid-Miller, M, et al. (1987). *Sequences of Proteins of Imrnunological Interest.* United States Department of Health and Human Services, Washington, DC.

Kelley M, Ahene AB, Gorovits B, et al. (2013). "Theoretical considerations and practical approaches to address the effect of anti-drug antibody (ADA) on quantification of biotherapeutics in circulation." *AAPS J* **15**(3): 646–658.

King DJ, Adair JR, Angal S, et al. (1992). "Expression, purification and characterization of a mouse-human chimeric antibody and chimeric Fab' fragment." *Biochem J* **281**(2): 317–323.

Labrijn AF, Buijsse AO, van den Bremer ETJ, et al. (2009). "Therapeutic IgG4 antibodies engage in Fab-arm exchange with endogenous human IgG4 in vivo." *Nat Biotechnol* **27**(8): 767–771.

Ladwig PM, Barnidge DR, Snyder MR, et al. (2014). "Quantification of serum IgG subclasses by use of subclass-specific tryptic peptides and liquid chromatography-tandem mass spectrometry." *Clin Chem (Washington, DC, USA)* **60**(8): 1080–1088.

Lanshoeft C, Wolf T, Heudi O, et al. (2016). "The use of generic surrogate peptides for the quantitative analysis of human immunoglobulin G1 in pre-clinical species with high-resolution mass spectrometry." *Anal Bioanal Chem* **408**(6): 1687–1699.

Law WS, Genin J-C, Miess C, et al. (2014). "Use of generic LC-MS/MS assays to characterize atypical PK profile of a biotherapeutic monoclonal antibody." *Bioanalysis* **6**(23): 3225–3235.

Lebert D, Picard G, Beau-Larvor C, et al. (2015). "Absolute and multiplex quantification of antibodies in serum using PSAQ® standards and LC-MS/MS." *Bioanalysis* **7**(10): 1237–1251.

Lesur A, Varesio E and Hopfgartner G (2010). "Accelerated tryptic digestion for the analysis of biopharmaceutical monoclonal antibodies in plasma by liquid chromatography with tandem mass spectrometric detection." *J Chromatogr, A* **1217**(1): 57–64.

Li F, Fast D and Michael S (2011). "Absolute quantitation of protein therapeutics in biological matrices by enzymatic digestion and LC-MS." *Bioanalysis* **3**(21): 2459–2480.

Li H, Ortiz R, Tran L, et al. (2012). "General LC-MS/MS method approach to quantify therapeutic monoclonal antibodies using a common whole antibody internal standard with application to preclinical studies." *Anal Chem* **84**(3): 1267–1273.

Liu G, Zhao Y, Angeles A, et al. (2014). "A novel and cost effective method of removing excess albumin from plasma/serum samples and its impacts on LC-MS/MS bioanalysis of therapeutic proteins." *Anal Chem (Washington, DC, USA)* **86**(16): 8336–8343.

Liu H, Manuilov AV, Chumsae C, et al. (2011). "Quantitation of a recombinant monoclonal antibody in monkey serum by liquid chromatography-mass spectrometry." *Anal Biochem* **414**(1): 147–153.

Mekhssian K, Mess J-N and Garofolo F (2014). "Application of high-resolution MS in the quantification of a therapeutic monoclonal antibody in human plasma." *Bioanalysis* **6**(13): 1767–1779.

Myler HA, Given A, Kolz K, et al. (2011). "Biotherapeutic bioanalysis: a multi-indication case study review." *Bioanalysis* **3**(6): 623–643.

Nelson AL, Dhimolea E and Reichert JM (2010). "Development trends for human monoclonal antibody therapeutics." *Nat Rev Drug Discov* **9**(10): 767–774.

Nouri-Nigjeh E, Zhang M, Ji T, et al. (2014). "Effects of calibration approaches on the accuracy for LC-MS targeted quantification of therapeutic protein." *Anal Chem (Washington, DC, USA)* **86**(7): 3575–3584.

Ouyang Z, Furlong MT, Wu S, et al. (2012). "Pellet digestion: a simple and efficient sample preparation technique for LC-MS/MS quantification of large therapeutic proteins in plasma." *Bioanalysis* **4**(1): 17–28.

Patel K and Borchardt RT (1990). "Chemical pathways of peptide degradation. II. Kinetics of deamidation of an asparaginyl residue in a model hexapeptide." *Pharm Res* **7**(7): 703–711.

Pearson JT and Rock DA (2015). "Bioanalytical approaches to assess the proteolytic stability of therapeutic fusion proteins." *Bioanalysis* **7**(23): 3035–3051.

Reichert J M (2012). "Marketed therapeutic antibodies compendium." *MAbs* **4**(3): 413–415.

Reichert J M (2014). "Antibodies to watch in 2014: mid-year update." *MAbs* **6**(4): 799–802.

Remily-Wood ER, Benson K, Baz RC, et al. (2014). "Quantification of peptides from immunoglobulin constant and variable regions by LC-MRM MS for assessment of multiple myeloma patients." *Proteomics Clin Appl* **8**(9–10): 783–795.

Ren D, Pipes GD, Liu D, et al. (2009). "An improved trypsin digestion method minimizes digestion-induced modifications on proteins." *Anal Biochem* **392**(1): 12–21.

Stokvis E, Rosing H and Beijnen JH (2005). "Stable isotopically labeled internal standards in quantitative bioanalysis using liquid chromatography/mass spectrometry: necessity or not?" *Rapid Commun Mass Spectrom* **19**(3): 401–407.

Strohl WR (2009). "Therapeutic Monoclonal Antibodies: From Bench to Clinic. Zhiqiang An (Editor)"., ISBN: 978-0-470-11791-0, October 2009, pp. 3–50.

Stubenrauch K, Wessels U, Essig U, et al. (2009). "Evaluation of a generic immunoassay with drug tolerance to detect immune complexes in serum samples from cynomolgus monkeys after administration of human antibodies." *J Pharm Biomed Anal* **52**(2): 249–254.

van de Merbel NC (2015). "Advances in liquid chromatography tandem mass spectrometry (LC-MS-MS)-based quantitation of: biopharmaceuticals in biological samples." *LC-GC Eur* **28**(s10): 38–44.

van de Merbel NC, Bronsema KJ and Nemansky M (2012). "Protein quantification using LC-MS: can it make a difference?" *Bioanalysis* **4**(17): 2113–2116.

van den Broek I, Niessen WMA and van Dongen WD (2013). "Bioanalytical LC-MS/MS of protein-based biopharmaceuticals." *J Chromatogr B Anal Technol Biomed Life Sci* **929**: 161–179.

Viswanathan CT, Bansal S, Booth B, et al. (2007). "Quantitative bioanalytical methods validation and implementation: best practices for chromatographic and ligand binding assays." *Pharm Res* **24**(10): 1962–1973.

Voronin K, Allentoff AJ, Bonacorsi SJ, Jr., et al. (2014). "Synthesis of a stable isotopically labeled universal surrogate peptide for use as an internal standard in LC-MS/MS bioanalysis of human IgG and Fc-fusion protein drug candidates." *J Labelled Compd Radiopharm* **57**(9): 579–583.

Wang SJ, Wu ST, Gokemeijer J, et al. (2012). "Attribution of the discrepancy between ELISA and LC-MS/MS assay results of a PEGylated scaffold protein in post-dose monkey plasma samples due to the presence of anti-drug antibodies." *Anal Bioanal Chem* **402**(3): 1229–1239.

White JT, Golob M and Sailstad J (2011). "Understanding and mitigating impact of immunogenicity on pharmacokinetic assays." *Bioanalysis* **3**(16): 1799–1803.

Xu K, Liu L, Maia M, et al. (2014). "A multiplexed hybrid LC-MS/MS pharmacokinetic assay to measure two co-administered monoclonal antibodies in a clinical study." *Bioanalysis* **6**(13): 1781–1794.

Xu W, Jiang H, Titsch C, et al. (2015). "Development and characterization of a pre-treatment procedure to eliminate human monoclonal antibody therapeutic drug and matrix interference in cell-based functional neutralizing antibody assays." *J Immunol Methods* **416**: 94–104.

Yan B, Steen S, Hambly D, et al. (2009). "Succinimide formation at Asn 55 in the complementarity determining region of a recombinant monoclonal antibody IgG1 heavy chain." *J Pharm Sci* **98**(10): 3509–3521.

Yang H and Zubarev RA (2010). "Mass spectrometric analysis of asparagine deamidation and aspartate isomerization in polypeptides." *Electrophoresis* **31**(11): 1764–1772.

Yang Z, Hayes M, Fang X, et al. (2007). "LC-MS/MS approach for quantification of therapeutic proteins in plasma using a protein internal standard and 2D-solid-phase extraction cleanup." *Anal Chem (Washington, DC, USA)* **79**(24): 9294–9301.

Yuan L, Arnold ME, Aubry A-F, et al. (2012). "Simple and efficient digestion of a monoclonal antibody in serum using pellet digestion: comparison with traditional digestion methods in LC-MS/MS bioanalysis." *Bioanalysis* **4**(24): 2887–2896.

Yuan L, Aubry A-F, Arnold ME, et al. (2013). "Systematic investigation of orthogonal SPE sample preparation for the LC-MS/MS bioanalysis of a monoclonal antibody after pellet digestion." *Bioanalysis* **5**(19): 2379–2391.

Zhang Q, Spellman Daniel S, Song Y, et al. (2014a). "Generic automated method for liquid chromatography-multiple reaction monitoring mass spectrometry based monoclonal antibody quantitation for preclinical pharmacokinetic studies." *Anal Chem* **86**(17): 8776–8784.

Zhang YJ, Olah TV and Zeng J (2014b). "The integration of ligand binding and LC-MS-based assays into bioanalytical strategies for protein analysis." *Bioanalysis* **6**(13): 1827–1841.

Zhao Y, Liu G, Angeles A, et al. (2015). "Development and validation of a liquid chromatography tandem mass spectrometry assay for the quantitation of a protein therapeutic in cynomolgus monkey serum." *J Chromatogr B Anal Technol Biomed Life Sci* **988**: 81–87.

Zhu Y, D'Arienzo C, Lou Z, et al. (2016). "LC-MS/MS multiplexed assay for the quantitation of a therapeutic protein BMS-986089 and the target protein Myostatin." *Bioanalysis* **8**(3): 193–204.

14

Mass Spectrometry-Based Methodologies for Pharmacokinetic Characterization of Antibody Drug Conjugate Candidates During Drug Development

Yongjun Xue, Priya Sriraman, Matthew V. Myers, Xiaomin Wang, Jian Chen, Brian Melo, Martha Vallejo, Stephen E. Maxwell, and Sekhar Surapaneni

Celgene, Summit, NJ, USA

14.1 Introduction

Cytotoxic small molecules used as anticancer therapeutics lack target specificity, resulting in unintended toxicity to healthy tissues and poor therapeutic indices, while monoclonal antibody (mAb) drugs are target specific, but sometimes lack the desired efficacy (Stephan et al. 2011, Kaur et al. 2013, Beck and Reichert 2014). Antibody–drug conjugates (ADCs) offer a new platform for the synergistic use of a cytotoxic agent and an mAb connected via a chemical linker. The underlying principle for this new therapeutic modality is to combine the target selectivity of mAbs with the potency of a natural or synthetic cytotoxic agent, which is 100- to 1000-fold more potent than conventional chemotherapeutic agents but not suitable as a stand-alone anticancer therapy. ADCs are designed to bind to antigens that are overexpressed on the surface of the tumor target but minimally expressed on normal tissue. In principle, this new ADC therapy should be highly efficacious and safe because the mAb, as a perfect carrier for cytotoxic agents, can minimize unintended cytotoxin loss during systemic circulation. Key considerations for a successful ADC include target biology, mAb properties, linker chemistry, and payload characteristics. In 2000, the FDA (the US Food and Drug Administration) approved the first ADC drug, gemtuzumab ozogamicin (Mylotarg®), for the treatment of acute myeloid leukemia (AML), but it was withdrawn from the US market in 2010 due to lack of clinical benefit to patients. More recently, FDA and EMA (the European Medicine Agency) approved two more ADC drugs: brentuximab vedotin (Adcetris®) for hematological malignancies (Hodgkin's lymphoma, systemic anaplastic large cell lymphoma) and ado-trastuzumab emtansine [T-DM1] (Kadcyla®) for treatment of advanced breast cancer with overexpressed HER2 receptor. These two ADC drugs have largely overcome the shortcomings of gemtuzumab ozogamicin. Their success has led to a dramatic increase in the number of ADC candidates entering pharmaceutical development programs.

Similar to any other drug development candidate, a successful ADC requires extensive pharmacokinetic/pharmacodynamic (PK/PD) characterization throughout drug development to define its efficacy and safety. Because of ADC's inherent complexity, the assessment of its *in vivo* PK parameters is significantly more involved than that of either component: small molecule or therapeutic protein individually (Stephan et al. 2011, Kaur et al. 2013). Three discrete bioanalytical assays including total antibody, conjugated antibody or total conjugated drug, and drug catabolite are required to characterize an ADC drug in plasma or serum samples from preclinical safety evaluations as well as clinical safety and efficacy trials. Normally either conjugated antibody or total conjugated drug is measured but not both. In addition, methods are required to determine the drug-to-antibody ratio (DAR) in plasma or serum samples from at least one preclinical study and possibly a clinical trial. Data obtained from these methods enables accurate assessment of ADC *in vivo* stability, PK, safety, and efficacy. These methods include ligand binding assays (LBAs) for total antibody (and conjugated antibody), liquid chromatography–mass spectrometry (LC–MS/MS) analysis for total conjugated drug and catabolite quantification, and high-resolution MS for DAR distribution analysis. In this chapter, we briefly discuss ADC mechanism of action and describe in detail the analytical methodologies with an emphasis on MS-based methods.

14.2 Mechanism of Action

An ADC is composed of a small-molecule cytotoxin (payload), an mAb, and a chemical linker (Stephan et al. 2011, Kaur et al. 2013, Han and Zhao 2014, McCombs

Protein Analysis using Mass Spectrometry: Accelerating Protein Biotherapeutics from Lab to Patient, First Edition.
Edited by Mike S. Lee and Qin C. Ji.

and Owen 2015). The key characteristics of ADCs are summarized in Table 14.1. The antibody provides a way of targeting the toxin to specific cells, thus reducing off-target toxicity and extending the half-life of the small-molecule toxin in circulation. Typically, the payload is not active in its conjugated state, but upon release from the antibody produces a cytotoxic effect on the targeted cell. The mechanism of action of an ADC involves binding to its target antigen on a cell, internalization by receptor-mediated endocytosis, followed by trafficking to lysosomes and subsequent release of the toxin (Figure 14.1). Once released, the toxin (or catabolite) is free to exert a pharmacological effect in the targeted cells. Nonspecific uptake of ADCs by pinocytosis into

Table 14.1 ADC structure and characteristics.

Components	Characteristics	ADC structure
Antibody	• High binding affinity to target and readily internalized • Good PK (e.g., long plasma half-life, low clearance) • Maintains the same PK characteristics/target binding affinity after conjugation • Binds to an antigen preferentially expressed on target cells • Minimal nonspecific binding	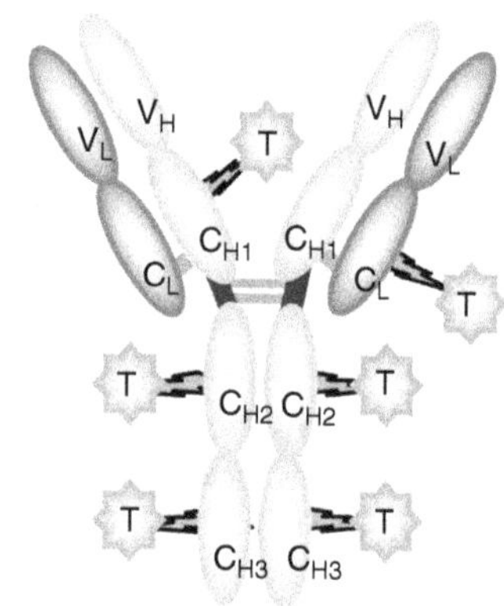
Cytotoxic agent	• Adequate stability in plasma • Nonimmunogenic • Nontoxic when conjugated to mAb • Highly potent once released from ADC	
Linker	• Stable in systemic circulation • Does not alter the mAb PK profile • Maintains toxin characteristics once reaching the target cell	

Picture showing structure of antibody-drug conjugate. C_{H1}–C_{H3} = constant regions of heavy chain, C_L = constant region of light chain, V_H = variable region of heavy chain, V_L = variable region of light chain, Tox = toxin/conjugated drug, Light black jagged line = linker, dark black lines = interchain disulfide bonds, and gray box = hinge region.

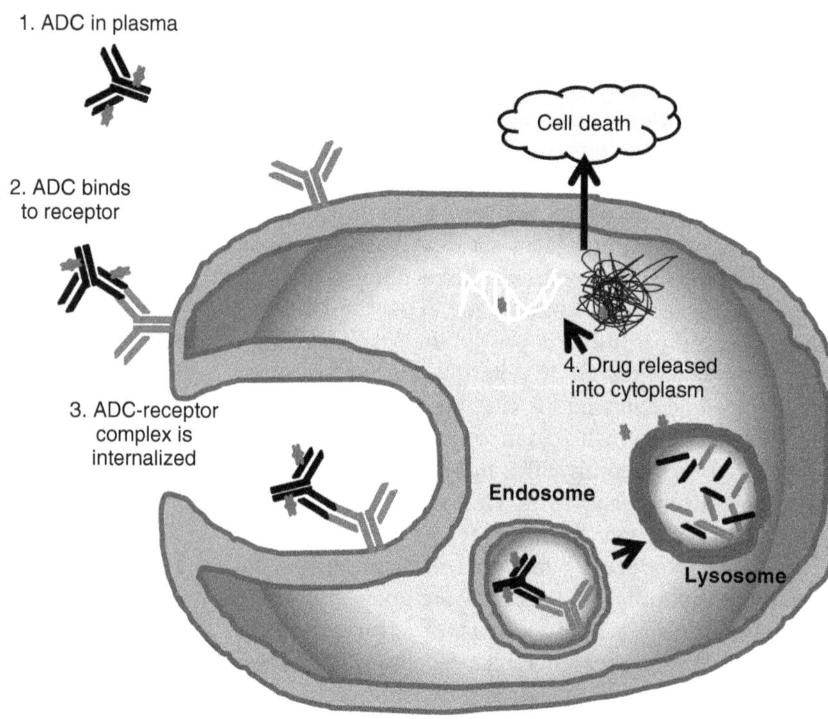

Figure 14.1 *ADC mechanism of action.* Schematic showing antibody-drug conjugate mechanism of action. 1. ADC in plasma travels through circulation as complete molecule (drug and antibody). 2. The ADC remains in circulation until it binds to the target receptor. 3. The ADC-target receptor complex is internalized via endocytosis forming and endosome. The endosome then fuses with a lysosome where the ADC linker is cleaved and the antibody is degraded or recycled. 4. The cleaved toxin (payload) is then released from the lysosome into the cell to disrupt the corresponding cellular processes resulting in cell death.

normal cells can also occur resulting in off-target cytotoxicity. If the toxin is membrane-permeable, diffusion from the target-expressing cells to neighboring cells can also cause cell death by the so-called "bystander effect." Toxin may also diffuse into the plasma and exhibit systemic effects, including toxicity, metabolism, and drug–drug interactions. If the target antigen is shed, high concentrations of shed antigen in the blood or even within the interstitial space of tumors can potentially limit effectiveness of ADCs especially for solid tumor indications.

14.2.1 Linker Chemistry

The conjugation of the toxin to the mAb is typically achieved via lysine residues, thiols, or engineered amino acids on the carrier antibody molecule and a chemical linker (Junutula et al. 2010, Stephan et al. 2011, Kaur et al. 2013), which can be classified as chemically or enzymatically cleavable or noncleavable. With the exception of conjugation at engineered cysteine or other specific amino acid residues, the conjugation reaction results in a heterogeneous mixture of ADC molecules with a range of DARs from 0 to *N* (the maximum number of drug molecules that can potentially be conjugated to the antibody) and multiple conjugation sites for toxin. The amount of the toxin attached to the antibody is generally reported as an average DAR. The release of the toxin in the cell, from the ADC, could be by deconjugation as with a cleavable linker, or by catabolism of the antibody as with a noncleavable linker. Linkers used in approved ADCs include the acid-labile hydrazone linker (Mylotarg®) (Sapra et al. 2011, Stephan et al. 2011), the enzymatically degradable linker para-aminobenzyl (PAB) group attached to a cathepsin-labile valine-citrulline dipeptide (Adcetris®), and the noncleavable alkyl and polymeric linkers such as *N*-maleimidomethylcyclo-hexane-1-carboxylate (Kadcyla®). Several ADCs with engineered proprietary linkers are in various stages of development. These new molecules are more homogeneous with DARs ranging from 0 to 2 (Beck and Reichert 2014). Understanding of the linker structure, conjugation, and cleavage chemistry is critical to the design of experimental approaches for ADME evaluation of the ADCs.

14.2.2 Toxins

Toxins that are a part of currently approved ADCs include tubulin inhibitors (auristatins, such as monomethyl auristatin E (MMAE) and monomethyl auristatin F (MMAF), and maytansine derivatives (DM), such as DM1 and DM4) and DNA-damaging agents (such as calicheamicins and duocarmycins) (Stephan et al. 2011, Adair et al. 2012, Kaur et al. 2013). Novel small-molecule toxins with high cytotoxic potential are being identified and evaluated in the context of next-generation ADCs. The toxin, once released, is subjected to metabolism, transport, and elimination pathways in a manner similar to other small molecules.

14.2.3 ADME

Each individual component of the biotherapeutic (mAb, linker, and toxin) contributes to the overall ADME properties of an ADC. To fully characterize the ADME of a novel ADC, the disposition and clearance of the mAb, the deconjugation and release of payload as well as the metabolism and excretion of the toxin need to be assessed. Evaluation of the antibody can include its target-mediated disposition, the dose required to saturate target-mediated clearance and circulating half-life. Evaluation of the linker's, toxin's, or catabolite's ADME properties tends to follow a path similar to that of traditional small molecule drugs and can include identification of active catabolites, metabolite profiling, plasma stability, and characterization of drug–drug interactions. Studies should investigate the pharmacokinetic, pharmacodynamic, and toxicological effects of the intact ADC, and their relationship between each of the above. Estimation of PK parameters in preclinical studies helps characterize the PK of an ADC, predict human PK when appropriate, and support the interpretation of efficacy and toxicity studies.

14.2.4 Unique Bioanalytical Challenges

To fully evaluate ADC PK in systemic circulation requires unique assays to characterize the intact ADC as well as the components that may be generated *in vivo*. Data from these assays should be taken together to gain an understanding of *in vivo* stability, the PK–PD relationship, and the exposure-toxicity profile of the ADC (Jenkins et al. 2015). From a technical standpoint, establishing these assays relies on combining experience from small molecule and biologics drug development, as well as on designing new methods specifically for the ADC species (e.g., DAR method) (Stephan et al. 2011, Gorovits et al. 2013, Kaur et al. 2013). Among these bioanalytical methods, some (total antibody, total conjugated drug, and catabolite methods) require full validation in the same matrix (either serum or plasma). Currently, there are no regulatory guidelines or best practices for ADC assay validations; therefore, the same criteria used for small molecule and biologics drugs are generally followed with additional provisions unique to ADCs (Dere et al. 2013). DAR methods are intended to profile DAR distribution in a preclinical study and are not required to be validated. In addition, due to catabolism, the ADC reference materials for calibration standard and quality control preparation may not be identical to the analytes *in vivo*. These issues present unique challenges for quantitative assay development. These four methodologies are discussed in detail in the following sections.

14.3 Mass Spectrometry Measurement for DAR Distribution of Circulating ADCs

ADCs are designed to deliver toxins to targeted tissue while minimizing side effects of these agents on normal body tissues. Thus, it is important that ADC molecules remain relatively stable during systemic circulation. Earlier ADC linkers were chemically labile and suffered from degradation, releasing the toxic payload into the blood, which resulted in toxicity. Recent advances in ADC linker chemistry (e.g., noncleavable or enzymatically cleavable linkers) have partially addressed such stability issues and helped enhance both efficacy and safety of ADC drug candidates (Junutula et al. 2010, Xu et al. 2011). Regardless of the conjugation mechanism, ADC drug deconjugation is still an unavoidable factor *in vivo* whether due to linker cleavage or some alternate degradation mechanism. This change in the ADC conjugation status will alter the efficacy and safety of an administered ADC, making it difficult to pinpoint the active species. In addition, an ADC drug starts with a heterogeneous mixture of conjugation products with different drug–antibody ratios. These factors highlight the need to monitor changes in the DAR distribution *in vivo*, as it can provide important information regarding structural stability and activity of ADC drugs (Xu et al. 2013).

While ligand binding assays have been used to characterize ADCs and can, if configured properly, effectively distinguish ADC from drug-free antibody, they cannot differentiate between different DAR species. High-resolution mass spectrometry, however, has proven to be a valuable tool for investigating the DAR distribution of ADCs by not only effectively differentiating between conjugated antibody and drug-free antibody but also detecting an array of DAR distributions generated by the manufacturing process, deconjugation, or degradation. Efforts to accurately characterize DAR distribution in ADC molecules commonly employ various types of mass spectrometry techniques from electrospray ionization (ESI) of partially digested analyte (e.g., light and heavy chains) to intact protein analysis. While the partial proteolytic digestion of ADCs can make DAR measurement simpler with smaller ADC components, this process can result in the loss of structural information pertinent to understanding the DAR and drug loss (Wagner-Rousset et al. 2014). With significant advances in high-resolution mass spectrometry in terms of sensitivity and resolution, measuring the DAR distribution of intact ADC has become practical and widely used. Therefore, it is the focus of the following discussions. This approach involves three key steps: immunocapture, deglycosylation, and high-resolution mass spectrometry analysis.

14.3.1 Immunocapture of ADCs from Plasma or Serum

While high-resolution mass spectrometry offers accuracy, precision, and specificity for small molecule measurements, it cannot achieve the desired sensitivity and selectivity needed for the DAR measurement without extensive sample cleanup and enrichment. The comparatively larger molecular weight of ADCs, which are less favorable for mass spectrometry, and the highly abundant and heterogeneous nature of protein content in plasma or serum samples add an additional layer of complexity to sample analysis. As a result, affinity capture approaches have become routine practice for sample preparation and enrichment of the target ADC during DAR measurement.

There are three common methods of capturing ADCs during DAR measurement: (i) target the Fc portion of the antibody with a polyclonal antihuman IgG (immunoglobulin G) Fc antibody (monkey absorbed), which selectively binds to the Fc region of human IgGs; (ii) capture the ADC using an immobilized form of target antigen, which will bind to the variable portions of the Fab region; or (iii) bind the conjugated drug using an anti-idiotype antibody specific to the ADC (Figure 14.2). While all of these methods have been utilized for DAR measurements, one must also consider the type of sample, where on the ADC the drug is conjugated and what reagents might be commercially available (Dubois et al. 2008, Dere et al. 2013) when choosing an option for sample preparation.

ADCs are normally humanized or fully human antibodies derived from the IgG class and possess conserved sequences that can be targeted using anti-IgG antibodies. Use of an anti-Fc enrichment method offers a generally applicable and robust capture method targeting the conserved Fc portion of an antibody and has been successfully implemented in studies used for immunocapture and quantification of the human IgG1 mAb, αPCSK9, in cynomolgus monkey serum with subsequent LC-ESI-MS of surrogate peptides (Stubenrauch et al. 2010). This same anti-Fc capture approach is an ideal option for the capture and measurement of all DAR species in preclinical studies, where the same capture procedure can be used for multiple structural variants of conjugated antibody or multiple species. It is also important to note that this method cannot be used for DAR measurement in human samples as it cannot differentiate ADCs from endogenous human IgG.

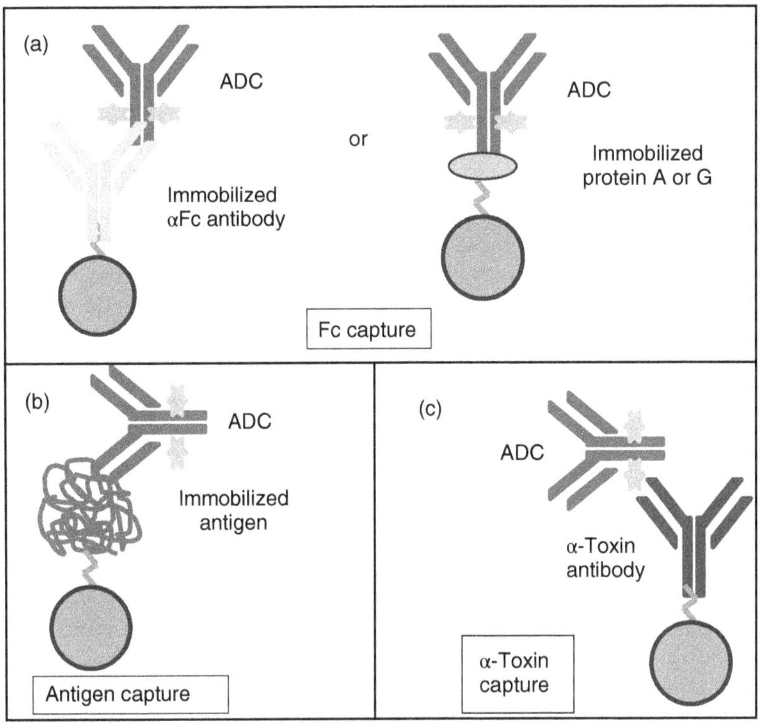

Figure 14.2 *Immunocapture methods.* Schematic of possible immunocapture methods for analyzing ADC molecules. (a) Pictures representing Fc capture where general αFc antibodies or protein A/G are conjugated to beads and used to pull down the ADC from complex samples. Pictures (b) and (c) depict more specific methods for capturing an ADC using an immobilized antigen specific for the ADC or an immobilized α-toxin antibody, respectively.

In addition to anti-Fc capture methods, some groups have demonstrated success using immobilized streptavidin beads and capture of the ADC via biotinylated ligand. One such method was used to capture biotinylated ligand (MUC16) bound to the Fab region of a cysteine engineered anti-MUC16 ADC, followed by LC–MS/MS to determine the DAR distribution (Xu et al. 2011). This methodology has since been expanded to investigate site-specific cysteine-labeled trastuzumab antibodies conjugated to MMAE as well as more heterogeneous ADC targets including those with lysine (Shen et al. 2012a, 2012b, Xu et al. 2013) and interchain disulfide cysteine linkages (Shen et al. 2012a, 2012b, Xu et al. 2013), and has been patented for the analysis of antibody drug conjugates (Kaur 2013a, 2013b). This approach can be used for DAR distribution measurement in both animals and humans; it will pull down all DAR forms (from 0 to N) of ADCs with exception of those where both ligand binding sites are occupied. The availability of commercial kits to biotinylate desired substrates, such as ligands or a portion of the antibody from the ADC itself, presents a myriad of options for possible enrichment strategies designed to target DAR or other ADC measurements.

Lastly, enrichment of an ADC species for DAR measurement can be performed using an antitoxin antibody to capture the drug on the ADC. While this method requires the generation of novel reagents against the conjugated drug, it is specific for the toxin portion of the ADC and avoids interference that may arise from endogenous antibodies. For example, an antitoxin antibody designed to capture MMAE conjugated to an antibody was used to measure the DAR values of the conjugated species (Sanderson et al. 2005). This approach can be used for DAR distribution measurement in both animals and humans; however, it will not pull down the DAR0 form and can exhibit differential binding affinities toward various DAR species.

Regardless of the enrichment method, it is evident that using a capture and elute strategy as part of sample preparation is required to accurately measure the DAR distribution present in *in vivo* samples. In addition, all three capture reagents discussed above may not have the same binding affinity to each DAR form due to steric hindrance from the linker and/or toxin; therefore, it is important to examine immunocapture with individual DARs to assess whether there is any bias for DARs in this step and then optimize the step for more quantitative or consistent pull-down of each DAR form.

14.3.2 Deglycosylation for Captured ADCs

For human IgG therapeutic antibodies produced by cell-based systems, there is one conserved N-linked glycosylation site located on the Fc region of each heavy chain at ASN-297 (Arnold et al. 2005). The combination of the glycans at each of the two glycosylation sites will generate considerable heterogeneity even within each batch of antibody production. For therapeutic mAbs, the characterization of this glycan structure can be very important because it plays a critical role in the efficacy of the protein drug (Abès and Teillaud 2010). But for measuring the DAR values of ADCs, the glycan structure complicates the MS spectrum of the intact protein and can reduce the accuracy of the calculation. Thus, a deglycosylation step is required after the immunocapture step. *N*-Glycanase (PNGase F) is the most commonly used and effective enzyme to remove N-linked oligosaccharides from glycosylated proteins (Tarentino and Plummer 1994) and deglycosylation kits or reagents can be obtained from various commercial sources. The general procedure involves incubation with the enzyme in either nondenaturing or denaturing conditions for 3h to overnight at 37°C (Prozyme, *Glycopro Enzymatic Deglycosylation Kit.* 2013). The deglycosylation reaction is considered close to quantitatively complete and the deglycosylated samples can be analyzed by high-resolution mass spectrometry without further enrichment or purification.

14.3.3 Mass Spectrometry Measurement for DAR Distribution of Circulating ADCs

Some early methods designed to characterize DAR profiles employed matrix-assisted laser desorption for ionization (MALDI) of the ADC sample (Siegel et al. 1991, Siegel et al. 1997, Quiles et al. 2010). MALDI-ToF-MS (matrix-assisted laser desorption ionization–time-of-flight mass spectrometry) has also been utilized to assess DAR of multiple conjugated species to PSMA (prostate-specific membrane antigen)-specific monoclonal antibodies (huJ591) conjugated to either a radio-labeled form of 1,4,7,10-tetraazacyclododecane-*N*,*N*0,*N*00,*N*000-tetraaceticacid (DOTA) or the cytotoxic drug maytansine (Lu et al. 2005). While these methods represent simple yet viable approaches to obtain DAR measurements, their analysis can be complicated due to overlapping spectra from multiple isotopic species leading to difficulties in differentiating between variants with minute differences (Wakankar et al. 2011).

LC-ESI-MS/MS methods currently dominate the field for detection and quantitation of DAR distribution for ADCs. Typically, sample analysis has been performed using a quadrupole time-of-flight mass spectrometer (Q-ToF) in positive ESI mode; however, more current ADC studies have employed an orbitrap mass spectrometer. Electrospray ionization produces a multiply charged envelope of ADC species. Subsequent software analysis uses deconvolution of the full mass spectrum to generate a profile of ADCs including varying forms of conjugated antibody with different drug loads. These types of analyses make possible the temporal determination of drug distribution and stability. Relative amounts of the individual conjugated ADC species can be determined by comparing the deconvoluted peak area for each DAR present. In order to compare peak areas, it is important to ensure each DAR ionizes with similar efficiency and there is no bias in the measurement. Q-ToF coupled with both reverse-phase (RP) and size-exclusion HPLC have been utilized to compare the DAR distribution of the recombinant human C242 antibody as well as its maytansinoid conjugates (Lazar et al. 2005). While multiple mass spectrometry platforms allow for DAR measurement, ADC drug conjugation methods may dictate a specific type of sample analysis for proper DAR measurement.

Methods for analysis of lysine-linked ADCs and engineered cysteine-linked residues are relatively straightforward employing typical RP-LC–MS (Q-ToF) and have been well studied (Lazar et al. 2005, Wakankar et al. 2011, Xu et al. 2011). However, ADCs utilizing reduced interchain cysteine residues for drug conjugation which, unlike lysine or engineered cysteine-linked ADCs, are held together by noncovalent interaction have been problematic for analysis by traditional RP-LC–MS (Basa 2013). DAR analysis of this particular type of ADC has relied on hydrophobic interaction chromatography (HIC) using buffer not especially suitable for MS analysis. More recent studies compared both native and denaturing approaches for average DAR measurements (Valliere-Douglass et al. 2012) and extended the standard use of MS to incorporate nano-ESI and native MS to achieve direct determination of the intact mass and to calculate the average DAR (Chen et al. 2013). A combination of HIC-UV and SEC-MS (size exclusion chromatography mass spectrometry) was also used to determine DARs and the mass of specific ADC drug loaded forms (Pan et al. 2014). In some cases, intact protein analysis alone is insufficient to understand the nature of DAR distribution and peptide mapping of an ADC is required. However, this process can result in the loss of structural information, which is pertinent to understanding the DAR and drug loss from the ADC. In a monkey study for Kadcyla®, biotinylated extracellular domain (ECD) of recombinant HER2 or anti-DM1 antibody was to capture T-DM1 from plasma, followed by on-bead deglycosylation with PNGase F (Prozyme®). Subsequently, all DAR analytes were eluted with 30% acetonitrile in water containing 1% formic acid. Reverse-phase, large-pore-size-capillary LC, Q-ToF was used to measure a gradual DAR shift to lower values over 28 days, providing evidence that standards/QCs may not represent the analytes measured at later PK time points (Dere et al. 2013, Kaur et al. 2013). The downward shift in DAR is the result of drug deconjugation and potential difference in clearance rates with different DAR species *in vivo* (Xu et al. 2013).

In summary, while quantitative total antibody, total conjugated drug, and small-molecule catabolite assays provide data necessary to characterize key PK/TK parameters required during ADC drug development, they might not be able to provide all the vital biotransformation information (e.g., DAR species, plasma stability) regarding ADCs in circulation. Although most DAR distribution analysis methods are not extremely sensitive (μg on column) and exploratory in nature, they are necessary for the characterization of the ADC. An accurate and consistent DAR measurement can not only be used to ensure the quantitative PK assays are properly developed for the determination of ADC species *in vivo* (Xu et al. 2013) but also to help optimize linker chemistry, provide a better understanding of drug stability issues, and troubleshoot other ADC-related assays. Utilizing current state-of-the-art high-resolution mass spectrometry to provide detailed DAR distribution data and to monitor *in vivo* DAR changes should be included as a vital part of ADC bioanalytical strategy at least in one preclinical species (usually primate) or even in humans.

14.4 Total Antibody Quantitation by Ligand Binding or LC–MS/MS

Total antibody, as described in the context of bioanalysis of ADCs, is defined as the antitarget antibody with a DAR of 0–*N*. It is important to maintain a distinction between free or semifree antibody, which can bind to target versus total antibody, which is the sum of free antibody and antibody, which cannot bind to target because its binding sites are obstructed with bound soluble target or antidrug antibodies (ADAs) (Ahene 2011, Lee et al. 2011). The ideal ADC drug is a completely conjugated antibody with a specific DAR. However, in practice, the dosed drug may contain a small percentage of naked or unconjugated antibody from the dosing material, which could be detected at early time points in the PK profile if present in sufficient quantities. In addition, linker cleavage could result in an increase in naked antibody concentration over time (Stephan et al. 2011, Gorovits et al. 2013). In the case of noncleavable linkers, where the payload is released by the degradation of the antibody, naked antibody returning to circulation should not be significant.

Although the therapeutically active molecule is the ADC, with the mAb serving the targeting role, the naked mAb itself may have therapeutic properties such as antibody-dependent cellular cytotoxicity (ADCC) or complement-dependent cytotoxicity (CDC) (Presta 2002). Thus, the naked antibody may retain some efficacy of the ADC. The naked antibody retains its ability to bind to target and can compete with the ADC for target binding. This may alter the PK properties and the efficacy of the ADC, as both now become a function of the ratio of unconjugated to conjugated antibody rather than just the concentration of the ADC. It is possible that over time, a buildup of naked antibody in circulation, due to the extended half-life of antibodies, could cause a shift in the proportion of naked and conjugated antibodies with a corresponding shift in efficacy. Therefore, it is important to measure concentrations of total antibody and conjugated antibody using the difference between these two measurements as inferred concentration of naked antibody (Kaur et al. 2013).

However, the decision to continue to measure both total antibody and ADC concentrations throughout development needs to be made based on the linker chemistry as well as early preclinical and clinical data. In the case of stable or noncleavable linkers, if early preclinical studies or the first clinical study demonstrates that the ratio of concentrations of total antibody to concentration of ADC does not change significantly over time (i.e., concentrations of naked antibody remain approximately constant), this may justify not measuring total antibody concentrations in subsequent studies. Currently, there are two assay types for total antibody measurement: LBA and LC–MS/MS (Stephan et al. 2011, Dere et al. 2013, Kaur et al. 2013, van den Broek et al. 2013).

14.4.1 Ligand Binding Assay

Total antibody measurement by LBA is typically accomplished by a sandwich ELISA method (Figure 14.3) (Stephan et al. 2011, Kaur et al. 2013). The target protein or an anti-idiotypic antibody may be used as capture reagent. In the case of ADCs that use a humanized or human monoclonal antibody, the detection reagent is often a readily available antihuman-Fc antibody. The same assay developed for the preclinical setting can be used in the clinical setting, following a full validation of the same assay in human matrix. A large body of literature, experience, and regulatory guidance documents exist in the development and validation of ligand binding assays for mAb therapeutics (FDA, 2001, Smolec et al. 2005, European Medicines Agency, 2011). The development and validation of the total antibody assay should follow these guidelines. For preclinical PK/TK studies, a more generic assay can be used, with polyclonal antihuman antibodies that bind to the nonvariable regions of the antibody as both the capture and detection reagents. These affinity-purified antibodies are commercially available and do not cross-react with endogenous IgGs in animal plasma or serum.

However, in the context of ADC therapeutics, a few additional tests may be important during method development to gain confidence that the assay is indeed measuring what it is intended to measure, that is, all forms of the antibody irrespective of DAR. It is important to demonstrate that the assay being used to measure total antibody is insensitive to the DAR (Kaur et al. 2013). This can be accomplished by preparing and testing quality

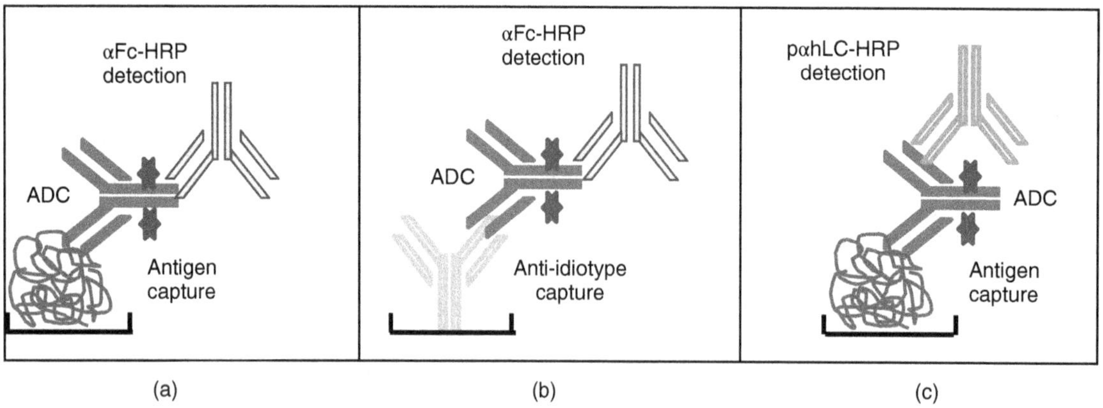

Figure 14.3 *Examples of sandwich ELISA Methods.* Panels (a)–(c) represent potential strategies for ELISA analysis of total ADC. (a) Antigen used as capture reagent for ADC and subsequent detection using αFc antibody. (b) Anti-idiotype antibody used to capture the ADC, which is then detected with αFc detection antibody. (c) Antigen used to capture reagent for ADC but subsequent detection using αLC antibody.

control samples that each has a different DAR or a different mean DAR along with quality control samples prepared with naked antibody and comparing them to reference standards containing intact ADC (similar to the dosing solution). Applying accuracy criteria to each of the quality controls tested will demonstrate that the assay is insensitive to the drug–antibody ratio. In the case of site-specific conjugation, it may be important to demonstrate that the site of conjugation does not interfere in the detection of the antibody in the assay (Gorovits et al. 2013). This can be achieved by comparing antibodies conjugated at one or more sites against the naked antibody in the assay. The primary challenge in performing these experiments is in the availability of the tool reagents that are necessary, such as ADC preparations with different DARs or mean DAR, anti-idiotype antibodies and target antigen.

For Kadcyla® that has a noncleavable thioether linker, an LBA was used to measure total trastuzumab in monkey and human serum with recombinant HER2 ECD as the capture and F(ab′)2 goat antihuman IgG Fc for detection. The validated total antibody assays were capable of quantifying all T-DM1 DARs, including conjugated T-DM1, as well as partially unconjugated and fully unconjugated T-DM1 that are immunoreactive to HER2 ECD (Dere et al. 2013, Kaur et al. 2013). A generic LBA for Kadcyla® has also reported (Xu et al. 2013).

14.4.2 LC–MS/MS Assay for Total Antibody Quantitation

Even though there are a growing number of publications on therapeutic protein (e.g., mAb) quantitation by LC–MS/MS (Yang et al. 2007, Dubois et al. 2008, Fernandez Ocana et al. 2012, Furlong et al. 2012, Li et al. 2012, Ouyang et al. 2012, Furlong et al. 2013, Jiang et al. 2013, Neubert et al. 2013, An et al. 2014), none of these publications specifically discussed the potential application of the technology to total antibody quantitation of ADC drugs. By this approach, a peptide generated by protease digestion is used as a surrogate for total antibody, and quantified by LC–MS/MS. There are three types of MS-based methods: (i) immunocapture LC–MS/MS – using immunocapture to enrich the total antibody; (ii) solvent-mediated coprecipitation LC–MS/MS – using protein precipitation to enrich the total antibody; and (iii) direct LC–MS/MS – no enrichment at all. These LC–MS/MS methods involve the following sample processing and analysis steps: predigestion enrichment (for the first two methods only), enzymatic digestion, postdigestion treatment, and LC–MS/MS analysis.

14.4.2.1 Predigestion Treatment

Highly abundant endogenous proteins in plasma or serum pose a challenge to LC–MS/MS analysis. The purpose of predigestion treatment is (i) enriching the ADC from a biological matrix to increase mass spectrometry signal of the total antibody for the desired detection limit; (ii) removing abundant endogenous components to reduce the background noise of mass spec measurements; and (iii) increasing digestion efficiency and reducing enzyme consumption. Currently, immunocapture (Dubois et al. 2008, Fernandez Ocana et al. 2012, Li et al. 2012, Neubert et al. 2013) and solvent-mediated coprecipitation (Furlong et al. 2012, Ouyang et al. 2012, Jiang et al. 2013) are the most commonly used methods of enrichment while immunodepletion or filtration enrichment is less frequently used and, therefore, are not discussed here.

14.4.2.1.1 Immunocapture

As discussed in Section 14.3.1, immunocapture selectively removes most abundant endogenous components (e.g., albumins, IgGs) by enriching for the specific ADC. Immunocapture for total antibody measurement can be less stringent than that for DAR measurement because chromatographic separation is much more powerful in small peptides than large intact ADCs. The most common immunocapture reagents used are (i) protein A or G, which binds to the heavy chain Fc region of many mammalian IgGs with various degrees of the binding affinity and is therefore less selective; (ii) polyclonal antihuman IgG Fc antibody (monkey absorbed) as discussed in the DAR measurement section, which is very selective to human IgGs; and (iii) anti-idiotype, which selectively binds to the variable region of the ADC, thus very specific to only unbound forms of the ADC (Figure 14.2). For preclinical PK/TK studies, protein A or G pulls down the ADC along with endogenous animal IgGs present in excess; while an antihuman IgG Fc would pull down only the ADC in animal samples. Anti-idiotype pulls down only unbound forms of the ADC with the targeted idiotype, which means it cannot be used for other ADCs. Therefore, antihuman IgG Fc can serve as a generic immunocapture reagent for any ADCs in animal studies. For clinical trials, anti-idiotype pulls down unbound forms of the given ADC, while protein A/G or antihuman IgG Fc pulls down the total antibody (unbound, bound) of the ADC along with excess endogenous human IgGs; the latter approach consumes a lot more of the pull-down reagents and also leads to a less sensitive assay due to background noise from the endogenous IgGs.

Similar to the discussions in the DAR measurement section, the capture reagents discussed above may not offer the same binding affinity to each DAR form; therefore, it is important to examine this step with individual DARs and screen multiple capture reagents for quantitative or consistent pull-down of each DAR form.

14.4.2.1.2 Solvent-Mediated Coprecipitation

Solvent-mediated coprecipitation uses an organic solvent (e.g., methanol) to separate the ADC (along with other insoluble serum proteins) from soluble serum components (e.g., peptides, phospholipids, and salts); the precipitate is then reconstituted for enzymatic digestion (Furlong et al. 2012, Ouyang et al. 2012, Jiang et al. 2013). This predigestion sample cleanup step is simple and cost-effective but is nonselective and most endogenous proteins will still be recovered with the ADC. Thus, it is more suitable for ADC concentrations close to μg/mL or above.

14.4.2.2 Enzymatic Digestion

Many proteases are available for therapeutic protein digestion, each having their own cleavage specificity and efficiency, such as trypsin, Arg-C, Asp-N, Glu-C, and Lys-C. Trypsin is the most widely reported and considered as the gold standard in therapeutic protein digestion (Dubois et al. 2008, Fernandez Ocana et al. 2012, Furlong et al. 2012, Li et al. 2012, Ouyang et al. 2012, Jiang et al. 2013, Neubert et al. 2013). It is highly specific in cleaving the peptide bonds C-terminal to the basic residues Lys and Arg, except when followed by Pro. With Lys and Arg being relatively abundant and well distributed throughout in ADCs, the tryptic peptides produced contain 7–20 amino acids in multiple charges and are well suited for LC–MS/MS analysis. A typical protocol for tryptic digestion involves reduction of disulfide bridges using dithiothreitol (DTT) and subsequent alkylation of the cysteines by iodoacetamide.

14.4.2.3 Postdigestion Treatment

The sample extracts from immunocapture (with anti-idiotype or antihuman Fc for animal samples) and enzymatic digestion should be very clean, and no postdigestion treatment is required. However, the sample extracts captured using protein A & G or antihuman Fc (in the case of human samples) still contain a large excess of peptides from endogenous human or animal IgGs, and the sample extracts from solvent-mediated coprecipitation or direct digestion are much dirtier than those from immunocapture. In these cases, further sample cleanup (e.g., mixed-mode solid-phase extraction (SPE) or 2D-SPE) is warranted to remove salt or other potentially interfering molecular species and achieve better detection limits (Yang et al. 2007, Rose 2014). In many cases, postdigestion sample cleanup is less effective because the surrogate peptide may not be easily isolated from numerous peptides with very similar amino acid sequences or composition, and as a result it is sparsely used.

14.4.2.4 LC–MS/MS Analysis

The first step of LC–MS/MS method development is to select a suitable surrogate peptide for the total antibody quantitation by *in silico* trypsin digestion and a protein database search. This surrogate peptide should have an amino acid sequence unique to the ADC and be absent in other endogenous proteins, especially IgGs. The peptide should be stable and sensitive using mass spectrometry, although multiple charge states further dilute its mass spectrometry response. High m/z precursor and daughter ions are normally monitored for low background noise and high assay specificity. For animal plasma or serum assays, a universal surrogate peptide for human IgGs (e.g., VVSVLTVLHQDWLNGK for human IgG1 and IgG4) is sufficient and widely reported for mAb determination (Furlong et al. 2012, Li et al. 2012, Furlong et al. 2013, Jiang et al. 2013), while for human plasma or serum assays, a more specific surrogate peptide from the variable region is required (Yang et al. 2007, Dubois et al. 2008, Fernandez Ocana et al. 2012, Neubert et al. 2013).

The sample extracts, especially those from coprecipitation/digestion or direct digestion, consist of many endogenous peptides with very similar amino acid compositions or identical SRM transitions as the surrogate peptide. As a result, more extensive chromatographic separation is required to resolve the surrogate peptide peak from isobaric peptide signals. Normally, an analytical column (e.g., 1 × 100 mm C18) is used with slow gradient and extended run time to achieve the desired separation. For better sensitivity, micro- or nano-LC can also be used (Dubois et al. 2008), although the method might be less rugged than those using traditional HPLC and analytical columns. Triple-quadrupole instruments (e.g., Sciex 6500) operating in SRM mode are ideally suited for the surrogate peptide analysis, and any advances in mass spectrometry sensitivity and selectivity can further push the LLOQ below the low ng/mL level.

In summary, method sensitivity is better with immunocapture LC–MS/MS than either solvent-mediated coprecipitation or direct LC–MS/MS. Depending on the sensitivity requirements and reagent availability, a typical workflow may involve one of these predigestion enrichments to facilitate the measurement of the total antibody concentrations in animal and human plasma or serum. For toxicology studies, coprecipitation LC–MS/MS or even direct LC–MS/MS should be sufficient to measure the μg/mL total antibody concentrations, while immunocapture LC–MS/MS using anti-Fc will allow for the low ng/mL measurement for other low-dose preclinical studies. As for clinical trials, the use of anti-idiotype will result in a more sensitive and specific assay for unbound forms of the ADC, while the use of antihuman Fc or protein A & G will be much less preferred because it also pulls down endogenous human IgGs.

14.4.3 Ligand Binding versus LC–MS/MS Assays

Ligand binding assays for the total antibody rely on the binding of the ADC or naked antibody to its target or to a reagent antibody. Such specific binding is a function of the protein sequence and tertiary structure, especially in the binding region. LBAs may be susceptible to matrix effects that result from cross-reactivity with other proteins, such as antidrug antibody, soluble ligand or other endogenous proteins. Pros and cons for ligand binding assays are summarized in Table 14.2.

LC–MS/MS analysis uses a surrogate peptide for the measurement of the total antibody and is less prone to matrix effect. However, the surrogate peptide does not carry the information for the whole ADC. Therefore, structural differences beyond the surrogate peptide (e.g., degradation products) will not be measured. Pros and cons for LC–MS/MS methods are summarized in Table 14.2.

For total antibody measurement, ligand binding and LC–MS/MS assays might produce different results due to intrinsic differences in the methodology or reagents used for LBAs compared to immunocapture. Thus, it is important to determine which type of methodology is more appropriate for the measurement of the total antibody in a development ADC program. No matter whether a ligand binding or LC–MS/MS assay is used, a full method validation is required in animal and human plasma or serum to ensure the adequate method performance for

Table 14.2 Ligand binding assays versus LC–MS/MS for the total antibody quantitation of ADC drugs.

Platform	Pros	Cons
ELISA	Excellent sensitivity (pg/mL) High throughput Free or total antibody measurement Inexpensive instrumentation Extensive validation experience	Dependent on critical reagents Matrix interference Multiplexing challenging Narrow curve range Different binding affinity to each DAR
Immunocapture and LC–MS/MS	Wide curve range Good sensitivity (ng/mL) Good assay specificity Short method development time Easy to multiplex	Expensive instrumentation Total antibody measurement (free with anti-idiotype) Different binding affinity to each DAR Limited validation experience Dependent on critical reagents for immunocapture
Direct LC–MS/MS (coprecipitation LC–MS/MS)	Wide curve range Good assay specificity No need for critical reagents Short method development time Easy to multiplex	Low sensitivity (μg/mL) Expensive instrumentation Total antibody measurement Limited validation experience

the total antibody measurement. A typical method validation includes, but is not limited to, accuracy, precision, selectivity, sensitivity, reproducibility, limit of detection, reagent stability, dilutional linearity, and matrix effect. The ±20/25% rule for traditional therapeutic proteins is followed. Furthermore, some components that are specific to the total antibody measurement of ADCs are evaluated during method validation (FDA, 2001, Jenkins et al. 2015).

ADCs, being biologic, may cause an immunogenic reaction in the host, which results in the production of ADAs. Such ADAs may neutralize the ADC and prevent its binding to the target. It is also possible that circulating target protein neutralizes the ADC. Such neutralized ADC may not be active. It is important to understand whether the total antibody method measures total or free form of the ADC.

14.5 Total Conjugated Drug Quantitation by Ligand Binding or LC–MS/MS

As discussed in the previous sections, ADCs are designed to be stable until reaching intended target. Conjugated drug is capable of exerting tumor targeted efficacy, while minimizing toxicity to cells not expressing target. Free drug, on the other hand, may exert nontumor-specific cytotoxicity. It can thus be argued that total conjugated drug exposure is more reflective of efficacy and target-mediated toxicity, while free drug exposure is more reflective of nontarget-mediated toxicity. Quantitation of the total conjugated drug is intended to measure the total cytotoxic drug load that is still linked to the antibody regardless of DAR species and to provide total targeted drug exposure. Thus, accurately measuring the total conjugated drug concentrations in serum or plasma is critical for assessing the overall PK/PD of an ADC. Currently, there are two assay formats: LBAs or LC–MS/MS (Stephan et al. 2011, Dere et al. 2013, Kaur et al. 2013). Both of these assay formats typically include one ligand binding step, where the antibody portion of the ADC is required to bind either to the target or to a capture antibody. The second step is the detection of the cytotoxic drug that is bound to this captured antibody part. This second step is achieved either using a labeled anticytotoxin antibody (in the case of LBAs) or by releasing the toxin *in vitro*, and measuring the released toxin using LC–MS/MS.

14.5.1 Ligand Binding Assays for ADC Quantitation

ADC, by definition, has the cytotoxic drug still linked to the mAb. A typical sandwich ELISA method, with one binding event directed toward the antibody and one binding event directed toward the cytotoxin, can be used for the measurement of ADC (Figure 14.3). Generation of an antibody with affinity and specificity to the cytotoxin is perhaps the biggest challenge in the development of this assay. For this purpose, the toxin is typically linked, ideally using the same linker chemistry as the ADC, to an immunogen/adjuvant such as KLH (Keyhole limpet hemocyanin), and this conjugate is then used to immunize animals to raise antibodies. Antibodies specific to the cytotoxin are affinity-purified from the resulting antisera. The ELISA assay can be designed as outlined in the following sections.

14.5.1.1 DAR-Sensitive Total Conjugated Drug Assay

The capture reagent is the recombinant target protein or an antibody directed toward the mAb portion of the ADC. The interaction between the mAb portion of the ADC and the target protein or anti-idiotype antibody has typically high affinity; hence, the capture step can be efficient and specific, leading to good assay sensitivity. It must be demonstrated at method development stage that this capture step is not affected by steric increasing DARs. The detection reagent is a labeled anticytotoxin antibody. The number of anticytotoxin antibodies that will bind to the captured ADC may depend on the number of cytotoxins, leading to a DAR-sensitive assay. However, the relationship of assay response to the number of cytotoxins may not always be linear and will have to be characterized during method development and validation. The efficiency of binding and the binding ratio of the anticytotoxin antibody to the toxin should be independent of overall DAR with no steric interference at high DARs. For example, a biotinylated anti-DM1 or anti-MMAF antibody was used to detect anti-CD22-MCC–DM1 or MC–MMAF ADCs, respectively, captured either by the human CD22-ECD or a goat antihuman IgG antibody, Fc or F(ab′)2 fragment specific (Stephan et al. 2008).

14.5.1.2 DAR-Insensitive Total Conjugated Antibody Assay

The capture reagent is the anticytotoxin antibody and the detection reagent is a labeled antibody directed toward the mAb portion of the ADC, or labeled target protein. The use of the anti-mAb/recombinant target for detection should render the assay insensitive to DAR. However, the affinity of the anticytotoxin antibody to the toxin determines the efficiency of the capture step. Lower affinity antibodies may not be efficient in capturing ADCs with low DARs (Xu et al. 2013), but ADCs with higher DARs may be more efficiently captured due to the increased avidity (multiple toxins interacting with the anticytotoxin antibody). DAR insensitivity should be tested during method development and demonstrated during method validation, using ADCs with specific DARs or ADCs enriched for specific DARs, covering the range of DARs expected in the patient samples. This assay measures conjugated antibody concentrations, and it can be argued that this assay does not provide a direct measurement of the efficacious entity (conjugated drug), as an antibody with one cytotoxin is expected to yield the same assay response as an antibody with multiple cytotoxins, and the efficacy from each of the two can be expected to be different. For monkey and human PK analysis of trastuzumab–maytansinoid conjugates, an anti-DM1 antibody was used as capture and the biotinylated HER2-ECD followed by streptavidin–HRP as detection; the conjugated antibody assay was not sensitive to drug payload with DARs ranging from 1.9 to 3.8. When reference material with an average DAR of 3.5 was used for standard/QC preparation, the assays accurately quantified T-DM1 with DARs ranging from 2.58 to 4.10 (Lewis Phillips et al. 2008, Dere et al. 2013, Kaur et al. 2013, Xu et al. 2013). However, for anti-STEAP1 ADC, the conjugated-antibody assay used an antitoxin mAb for capture and a biotinylated anti-idiotype mAb for detection significantly underquantified DAR1 by 89% and DAR6 by 36% (Kaur et al. 2013).

Each of the above LBA methods has limitations that may be difficult to overcome with assay design. Highly sensitive LC–MS/MS methods can serve as an alternative to ligand binding assays, especially when using an immunocapture-LC–MS/MS-based approach to help utilize the efficiencies of both the ligand binding step (at the antibody end) and quantitative detection of the released cytotoxin by LC–MS/MS, which will be discussed next.

14.5.2 LC–MS/MS for the Total Conjugated Drug Quantitation

The LC–MS/MS assay format provides direct measurement of the total conjugated drug and is sensitive to changes in drug load. Because of the high potency of cytotoxic drugs used for ADCs, even a small change in the drug payload could affect both safety and efficacy. Thus, the ability to detect small changes in drug payload is very important. However, this assay format cannot provide the DAR distribution information for the ADC and has not been widely reported yet. The total conjugated drug quantitation by LC–MS/MS involves these key steps: predigestion treatment (optional), enzymatic digestion, postdigestion treatment, and LC–MS/MS analysis (Buckwalter et al. 2004, Stephan et al. 2011, Dere et al. 2013, Kaur et al. 2013).

14.5.2.1 Predigestion Treatment

As previously discussed in the section Immunocapture, immunocapture methods have been used to enrich for targets of interest. The same pull-down reagents as those used in total antibody quantitation can be applied for enrichment of total conjugated drug (Immunocapture section and Figure 14.2). However, as mentioned, immunocapture methods should be chosen to appropriately meet the requirements of the studies being performed whether preclinical or clinical. In cases where the total conjugated drug is highly concentrated (e.g., μg/mL) and/or free of interference from catabolites with different SRM transitions, predigestion treatment can be eliminated (direct digestion).

14.5.2.2 Enzymatic or Chemical Digestion

Most ADCs currently undergoing preclinical, early clinical, or late stage clinical development contain linkers, which fall into two broad categories: cleavable or noncleavable. For ADCs with acid-labile cleavable linkers (hydrazone) (Bross et al. 2001, Hamann et al. 2002, Stephan et al. 2011, Gorovits et al. 2013, Kaur et al. 2013), a pH-dependent release mechanism can be used to release the drug (+linker) by adjusting plasma or serum pH from 7.3–7.5 to below 4.5 for the total conjugated drug analysis. For ADCs with disulfide cleavable linkers (Widdison et al. 2006, Chari 2008, Stephan et al. 2011, Gorovits et al. 2013, Kaur et al. 2013), common small-molecule redox reagents, such as dithiothreitol (DTT), can be used to cleave disulfide bonds releasing the drug (with thiol group), which is further stabilized by alkylating the thioether bond (e.g., iodoacetamide). For ADCs with peptide cleavable linkers (e.g., Val-Cit, Phe-Lys) (Doronina et al. 2003, Carter and Senter 2008, Gorovits et al. 2013, Kaur et al. 2013), lysosomal proteases (e.g., cathepsins and plasin) can be used to release the drug under more acidic conditions. For ADCs with β-glucuronide cleavable linkers (Jeffrey et al. 2006, Gianolio et al. 2012), enzymatic hydrolysis with β-glucuronidase can be used to release the drug from the antibody. In each case, the efficiency of the release step has to be characterized and linearity between the amount of conjugated drug present in the input ADC and amount of released drug has to be demonstrated during method validation using ADC of different DARs and covering the different sites of possible conjugation. The released cytotoxin should have adequate stability throughout sample preparation along with appropriate levels of MS sensitivity. In addition, the digestion should be close to 100% complete or at least consistent among samples and ADC concentrations (and even different DARs). For example, the total conjugated drug concentrations for anti-STEAP1 ADC via a protease-cleavable peptide linker was determined using immunocapture/enzymatic digestion/LC–MS/MS because the corresponding ELISA could not accurately measure DAR1 (Kaur et al. 2013). However, using enzymatic or chemical digestion for ADCs with noncleavable linkers can face two challenges: (i) no suitable enzymatic or chemical digestion available;

(ii) incomplete digestion to clip the conjugated drug from mAb can lead to multiple species of the conjugated drug (drug + linker or even amino acids), which would make quantitation impractical. For example, the total conjugated drug concentrations of Kadcyla® (with a noncleavable linker) were measured indirectly by ELISA (total conjugated antibody assay) not immunocapture/digestion-LC–MS/MS (Stephan et al. 2011, Dere et al. 2013, Kaur et al. 2013) possibly due to the same concerns.

14.5.2.3 Postdigestion Treatment

Similar to total antibody quantitation, preclinical serum or plasma samples with anti-idiotype or antihuman Fc pull-down followed by digestion should be very clean, and no postdigestion treatment is required. In contrast, human serum or plasma samples from protein A/G or antihuman Fc may require further sample cleanup to remove endogenous IgGs (or its digested peptides). The sample extracts from direct digestion are more complex than those from immunocapture; therefore, further sample cleanup/enrichment (e.g., SPE or liquid–liquid extraction (LLE)) may be necessary to remove endogenous components and achieve a desired detection limit. It should be noted that postdigestion cleanup or enrichment should be very effective, since physicochemical properties of the released cytotoxin from digestion are very different from those of endogenous components.

14.5.2.4 LC–MS/MS Analysis

The first step of LC–MS/MS method development is to scan the digestion product from the ADC reference material and run various MS scans to identify the released cytotoxin for LC–MS/MS analysis. This released cytotoxin should represent total drug payload of the ADC and may include linker. The sample extracts from direct digestion still contain peptides from many endogenous components, and more extensive chromatographic separation may be required to resolve the released cytotoxin from potential interference. However, much like a small molecule, it is normally more hydrophobic than endogenous components with the exception of phospholipids, and reverse-phase chromatographic columns should provide adequate separation. A stable-labeled ADC internal standard should be used to track the released cytotoxin from immunocapture, digestion, further cleanup to LC–MS/MS analysis. As an alternative, a stable-labeled released cytotoxin might work, but it will not track immunocapture and digestion. For better sensitivity, micro-LC can also be used, although the method might be less rugged than conventional chromatography. Typically, triple-quadrupole instruments operating in SRM mode are used for these methods and should be able to accommodate the required assay ranges. For example, a hybrid affinity-MS-based conjugated drug assay provided a direct measurement of plasma stability and robust clinical PK data over several doses for anti-STEAPI ADC (Kaur et al. 2013).

In summary, immunocapture LC–MS/MS provides better selectivity and sensitivity while a simpler direct LC–MS/MS approach can provide adequate sensitivity when drug levels are in the μg/mL range. The use of anti-idiotype for immunocapture provides an added level of specificity relative to antihuman Fc or protein A/G. Depending on the sensitivity requirements and reagent availability, a typical workflow may involve a predigestion enrichment to facilitate the measurement of the total conjugated drug concentration in animal and human serum or plasma. Typically, toxicology studies are conducted at dose levels where direct LC–MS/MS may be sufficient for the measurement of total conjugated drug. For low-dose preclinical studies, immunocapture LC–MS/MS using anti-Fc may be required. For clinical trials, the use of anti-idiotype will result in a more sensitive assay for the total conjugated drug (unbounded forms), while the use of antihuman Fc or protein A & G will result in a less sensitive assay for the total conjugated drug.

14.5.3 Ligand Binding versus LC–MS/MS

For ligand binding assays, it is challenging to generate specific antibodies to the toxin payload, and antibodies with low affinity or specificity can limit assay sensitivity and selectivity. In addition, variable affinity to individual DAR species can lead to differential recovery. While the relationship of assay response to the number of cytotoxins may not always be linear, it is hard to establish a simple relationship for DAR-sensitive total conjugated drug assay. Pros and cons for ligand binding assays are summarized in Table 14.3.

On the other hand, LC–MS/MS can provide a direct and more sensitive measurement of the total conjugated drug via the released cytotoxin with minimal interference from endogenous components resulting in an assay that is more accurate and precise relative to some LBA alternatives. In addition, the total conjugated drug assay provides information that is distinctly different from the total antibody assay and together these data provide a simple way to define a heterogeneous DAR mixture in terms of the total amount of the two key molecular components (antibody, toxin) of the ADC. Pros and cons for LC-MS/MS methods are summarized in Table 14.3.

Overall, LC–MS/MS is preferred to measure conjugated ADC PK, when technically feasible. Regardless of the methodology being used, ligand binding or LC–MS/MS, a full method validation is required in animal and human matrix to ensure the adequate method performance for total conjugated drug measurement, and the same validation acceptance criteria (the ±20–25% role)

Table 14.3 Ligand binding assays versus LC–MS/MS for the total conjugated drug quantitation of ADC drugs.

Platform	Pros	Cons
ELISA	Excellent sensitivity (pg/mL)	Low and variable binding to small molecule drug (e.g., DAR1), leading to differential recovery for individual DARs
	High throughput	Complex correlation between detection signal to drug load, lack drug load information
	Inexpensive instrumentation	Need to characterize the assay performance for individual DARs
	Extensive validation experience	
Immunocapture and LC–MS/MS	Good sensitivity (ng/mL)	No information on the amount of antibody conjugated to drug
	Direct drug load information	Expensive instrumentation
	Direct detection of small changes in drug load	Need to characterize with individual DARs for potential bias of immunocapture
	Good assay specificity	Limited validation experience
	Short method development time	
	Preferred choice	
Direct LC–MS/MS	Wide curve range	Low sensitivity (μg/mL)
	Good assay specificity	Expensive instrumentation
	No need for critical reagents	Total antibody measurement
	Short method development time	Limited validation experience

should apply. Furthermore, some components that are specific to the total conjugated drug measurement (e.g., potential interference from catabolite) of ADCs are evaluated during method validation (FDA, 2001, Jenkins et al. 2015).

As discussed in the total antibody section, ADA or circulating target protein may interfere with LBA assay or the immunocapture step of an LC–MS/MS assay, and it is important to evaluate their impacts on the method.

14.6 Catabolite Quantitation by LC–MS/MS

The mAb portion of an ADC is catabolized into amino acids and recycled into other proteins; the catabolic fate of the mAb portion is well understood with no safety concerns (Ezan et al. 2014, Hall 2014, Han and Zhao 2014). On the other hand, the extent of the catabolism (or deconjugation) for the payload is very important for the safety evaluation of an ADC due to off-target toxicity caused by circulating catabolites (Erickson and Lambert 2012, Prueksaritanont and Tang 2012). Less stable linkers or high DARs lead to increased release of catabolites. The catabolites formed include but not limited to free drug, free drug + linker and free drug + linker + amino acid. These catabolites can also be metabolized by cytochrome P450 enzymes and are subject to potential drug–drug interactions from P450 inhibitors or inducers. Investigation of ADC catabolism requires the use of both *in vitro* and *in vivo* systems. Appropriate *in vitro* studies including catabolism studies in target-expressing cell lines and plasma stability studies across species can elucidate disposition mechanisms, identify ADC catabolites, and establish the relevant preclinical species. *In vivo* studies are used to confirm the relevance of *in vitro* observations. For Kadcyla®, three catabolites MCC-DM1, Lys-MCC-DM1, and DM1 have been identified. Among them, DM1 may be metabolized by CYP2D6, CYP3A4, and CYP3A5 and is also a time-dependent inhibitor of CYP3A (Lin and Tibbitts 2012, Shen et al. 2012a, 2012b, Han and Zhao 2014).

Along with the measurements of total antibody and total conjugated drug in animal and human serum or plasma, the measurement of the major circulating catabolites is a key component for understanding PK/PD, especially for ADC safety evaluation. LBAs (e.g., competitive ELISA methods) were once used to detect circulating catabolites, but MS-based methods are now applied for this analysis because of improved dynamic range, sensitivity, and selectivity (Stephan et al. 2011, Kaur et al. 2013). Catabolite quantitation by LC–MS/MS involves two steps: sample preparation and LC–MS/MS analysis.

14.6.1 Sample Preparation

Sample preparation methods used for small molecules can be applied to catabolite sample cleanup. The simplest extraction method is protein precipitation (PPT). However, using PPT to extract the catabolites in serum or plasma samples may result in significant matrix effect due to interfering components remaining in the extracts. In addition, PPT may not offer the adequate sensitivity needed to effectively quantitate low level catabolites, which represent normally less than 1% of the molar concentration of the ADC. On the other hand, LLE and

SPE are far more effective methods yielding cleaner, more concentrated extracts when a lower LLOQ is required. It is also critical to evaluate catabolite release from the ADC during sample storage and processing, since even a minor release of catabolites can have a large impact on quantitation due to the low level of circulating catabolites.

14.6.2 LC–MS/MS Analysis

As with sample preparation, LC–MS/MS principles for small molecules also apply to catabolite analysis. Unlike surrogate peptides used for total antibody quantitation, catabolites normally exist in a single charge state, which makes tuning the mass spectrometer straightforward. With the catabolites being more hydrophobic than the majority of endogenous components, reverse-phase chromatography should provide adequate separation, but keep in mind that more extensive chromatographic separation may be required for more complex samples. A stable-labeled internal standard should be used to track the analytes from sample preparation to LC–MS/MS analysis using a triple-quadrupole instrument operating in SRM mode. The major circulating catabolite (DM1) of Kadcyla® was measured in animal and human plasma samples with validated PPT, LC–MS/MS methods, while two other minor catabolites (MCC-DM1, Lys-MCC-DM1) were measured using nonvalidated methods (Shen et al. 2012a, 2012b, Kaur et al. 2013).

Similar to total antibody or total conjugated drug quantitation, catabolite measurement should be conducted using validated methods in accordance with small-molecule validation acceptance criteria. Furthermore, some components specific to catabolite measurement of ADCs should be evaluated during method validation (FDA, 2001, Woolf et al. 2014, Jenkins et al. 2015). The major challenge for catabolite measurement is sensitivity. While a simple PPT, LC–MS/MS method might be sufficient to support high-dose toxicology studies, low-dose clinical trials may necessitate sample enrichment to quantitate low-abundance catabolites.

14.7 Preclinical and Clinical Pharmacokinetic Support

In comparison to preclinical and clinical development of a small molecule or therapeutic protein, the full pharmacokinetic characterization of an ADC is more complex as it consists of both an antibody and a cytotoxic drug. ADC PK is driven by the characteristics of the antibody due to an intrinsically long half-life, although it can also be influenced by linker chemistry (stability, polarity) and the payload. In addition, antibodies tend to have low clearance, low volumes of distribution and proteolysis-mediated catabolism. Extrinsic factors affecting antibody clearance and, therefore, the PK of ADCs are target-specific binding (sink), target turnover, neonatal Fc receptor (FCRn)-dependent recycling, and Fc effector functions. Total antibody quantitation can be used to characterize the antibody portion of the ADC drug and to confirm whether its key PK parameters are significantly altered by cytotoxic drug conjugation (Stephan et al. 2011, Lin and Tibbitts 2012, Kaur 2013a, 2013b, Kaur et al. 2013, Sapra et al. 2013). Direct comparison of total antibody and total conjugated drug data can further reveal ADC PK properties related to the linker chemistry and cytotoxic drug. Plasma or serum concentrations of the total antibody or total conjugated drug are expected to decline over time. A faster decline of the total conjugated drug indicates drug loss (deconjugation) from ADC and the formation of lower DAR species (or even DAR0) during systemic circulation. On the other hand, similar PK profiles among the DAR species along with low levels of catabolites can indicate the linker is relatively stable *in vivo*. The DAR measurement will help to further understand DAR distribution and its correlation to the mean DAR. A typical concentration–time profile for these three key components (total antibody, total conjugated drug, and catabolite) is depicted in Figure 14.4.

In preclinical studies, DAR distribution, total antibody, total conjugated drug, and catabolite assays are mainly

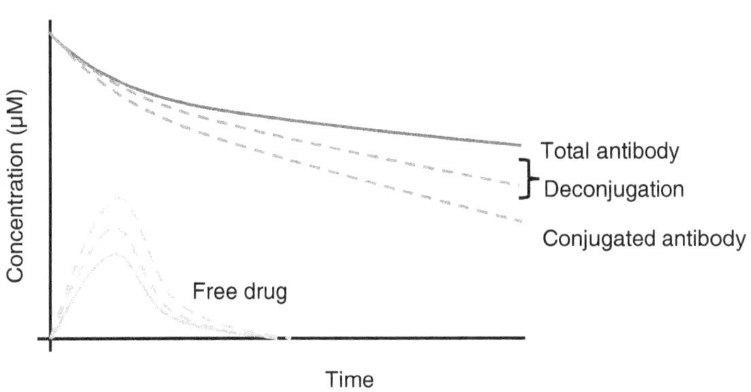

Figure 14.4 *Typical ADC concentration–time profiles.* Concentration versus time plots for three ADC components: Light black line for total antibody, dark black lines for total conjugated drug or conjugated antibody, and gray lines for catabolite.

used to define pharmacological activity, PK exposure relative to efficacy and safety, PK/PD and to project the starting dose for first-in-human studies (Junutula et al. 2010, Erickson et al. 2012). Cynomolgus monkeys are the most commonly used toxicology species because the human antibody is cross-reactive to the monkey target protein; clearance mechanisms, whether target- or non-target-mediated, are similar to those in humans. In clinical trials, these same measurements (DAR distribution in human optional) are conducted to establish the maximum tolerable dose in healthy subjects and patients and then used to characterize therapeutic window, PK variability, and the relationship of exposure to efficacy and safety. It should be noted that each of the above assays carries inherent method variability, which is likely to be compounded when the different assay data are integrated for quantitative PK analyses. Caution is advised in making direct comparison of ADC PK and PK/PD without careful consideration of the differences in analytical methods, assay formats, and materials. Despite these challenges, assessment and integration of ADC PK can be valuable not only in understanding a single ADC but also in evaluating multiple ADCs with different structural and pharmacologic characteristics; allowing improved design and development of these complex molecules.

For example, Kadcyla® is a humanized IgG1 anti-HER2 mAb covalently bound to DM1 via a stable thioether linker (Krop et al. 2010, Burris et al. 2011, Poon et al. 2013). Its total-trastuzumab concentration–time profile follows a pattern similar to that of conjugated trastuzumab in cynomolgus monkeys and humans. The difference of the total and conjugated trastuzumab concentrations indicates that T-DM1 gradually deconjugates *in vivo* to form low levels of DM1 (catabolite), consistent with DAR distribution data (Dere et al. 2013, Xu et al. 2013). Thus, the conjugated T-DM1 is considered the primary analyte for pharmacokinetic characterization with a volume of distribution of 3.3 L, clearance of 0.7 L/day, and a half-life of 4.5 days in humans. Adcetris® is a chimeric IgG1 anti-CD30 antibody with a protease cleavable linker to MMAE. This ADC drug has a favorable PK profile due to its relatively stable protease cleavable linker, which is deconjugated mainly in lysosomes (Sapra et al. 2013). The concentrations of the total antibody and the total conjugated MMAE are highly correlated; thus, the conjugated MMAE was considered the primary analyte for PK characterization with a volume of distribution of 8.2 L, clearance of 1.7 L/day, and a half-life of 4–6 days in human. Mylotarg® is a humanized IgG4 mAb against the CD33 antigen with an acid-labile cleavable linker (also disulfide bond) attaching a derivative of calicheamicin (*N*-acetyl-γ calicheamicin 1,2-dimethyl hydrazine dichloride). This ADC has a less favorable PK profile compared to Kadcyla® or Adcetris® due to its more labile linker (Sapra et al. 2013) with a volume of distribution of 10–18 L, clearance of 6.4 L/day, and a half-life of 3 days in humans.

14.8 Conclusion and Future Perspectives

By combining a cytotoxin's potency with an mAb's target specificity, ADCs represent a significant advance in cancer therapy as demonstrated by the recent regulatory approval of two compounds. Because of their structural complexity, ADCs represent a particular challenge for bioanalysis requiring four distinct bioanalytical methods to fully characterize the PK properties of an individual ADC and establish PK/PD as opposed to a single assay required for a small molecule or therapeutic protein. These include total antibody, total conjugated drug, and catabolite assays along with a DAR distribution assay. Because the reference standards may not truly represent the ADC species *in vivo*, additional method development and validation steps must be taken to accurately measure PK *in vivo*. In addition, some of these methods require merging multiple technologies (e.g., immunocapture, digestion, LC–MS/MS for total conjugated drug measurement) or deploying new technology – such as high-resolution mass spectrometry for DAR measurement – which has been proven to be a powerful tool in understanding the fate of ADCs.

Going forward, with more than 35 ADCs at various stages of drug development, more regulatory approvals will be expected within the next few years. New antibodies, cytotoxins, and linker chemistries are becoming available for the next generation of ADCs. In addition, as we increase our understanding of ADC mechanisms of action, it will lead to improved specificity and safety allowing expansion into new indications. With rapid advancement of mass spectrometry, new and more robust bioanalytical approaches will emerge to support ligand binding assays for improved tracking of ADC components.

References

Abès R and Teillaud J-L (2010). "Impact of glycosylation on effector functions of therapeutic IgG." *Pharmaceuticals* **3**(1): 146–157.

Adair JR, Howard PW, Hartley JA, et al. (2012). "Antibody-drug conjugates – a perfect synergy." *Expert Opin Biol Ther* **12**(9): 1191–1206.

Ahene AB (2011). "Application and interpretation of free and total drug measurements in the development of biologics." *Bioanalysis* **3**(11): 1287–1295.

An B, Zhang M and Qu J (2014). "Toward sensitive and accurate analysis of antibody biotherapeutics by liquid chromatography coupled with mass spectrometry." *Drug Metab Dispos* **42**(11): 1858–1866.

Arnold JN, Royle L, Dwek RA, et al. (2005). "Human immunoglobulin glycosylation and the lectin pathway of complement activation." *Adv Exp Med Biol* **564**: 27–43.

Basa L (2013). "Drug-to-antibody ratio (DAR) and drug load distribution by LC-ESI-MS." *Methods Mol Biol*: 285–293.

Beck A and Reichert JM (2014). "Antibody-drug conjugates: present and future." *MAbs* **6**(1): 15–17.

Bross PF, Beitz J, Chen G, et al. (2001). "Approval summary: gemtuzumab ozogamicin in relapsed acute myeloid leukemia." *Clin Cancer Res* **7**(6): 1490–1496.

Buckwalter M, Dowell JA, Korth-Bradley J, et al. (2004). "Pharmacokinetics of gemtuzumab ozogamicin as a single-agent treatment of pediatric patients with refractory or relapsed acute myeloid leukemia." *J Clin Pharmacol* **44**(8): 873–880.

Burris HA, 3rd, Rugo HS, Vukelja SJ, et al. (2011). "Phase II study of the antibody drug conjugate trastuzumab-DM1 for the treatment of human epidermal growth factor receptor 2 (HER2)-positive breast cancer after prior HER2-directed therapy." *J Clin Oncol* **29**(4): 398–405.

Carter PJ and Senter PD (2008). "Antibody-drug conjugates for cancer therapy." *Cancer J* **14**(3): 154–169.

Chari RV (2008). "Targeted cancer therapy: conferring specificity to cytotoxic drugs." *Acc Chem Res* **41**(1): 98–107.

Chen J, Yin S, Wu Y, et al. (2013). "Development of a native nanoelectrospray mass spectrometry method for determination of the drug-to-antibody ratio of antibody-drug conjugates." *Anal Chem* **85**(3): 1699–1704.

Dere R, Yi JH, Lei C, et al. (2013). "PK assays for antibody-drug conjugates: case study with ado-trastuzumab emtansine." *Bioanalysis* **5**(9): 1025–1040.

Doronina SO, Toki BE, Torgov MY, et al. (2003). "Development of potent monoclonal antibody auristatin conjugates for cancer therapy." *Nat Biotechnol* **21**(7): 778–784.

Dubois M, Fenaille F, Clement G, et al. (2008). "Immunopurification and mass spectrometric quantification of the active form of a chimeric therapeutic antibody in human serum." *Anal Chem* **80**(5): 1737–1745.

Erickson HK and Lambert JM (2012). "ADME of antibody-maytansinoid conjugates." *AAPS J* **14**(4): 799–805.

Erickson HK, Lewis Phillips GD, Leipold DD, et al. (2012). "The effect of different linkers on target cell catabolism and pharmacokinetics/pharmacodynamics of trastuzumab maytansinoid conjugates." *Mol Cancer Ther* **11**(5): 1133–1142.

European Medicines Agency (2011). Guideline on bioanalytical method validation In: *Committee for Medicinal Products for Human Use (EMEA/CHMP/EWP/192217/2009)* http://www.ema.europa.eu/docs/en_GB/document_library/Scientific_guideline/2011/08/WC500109686.pdf. Effective Feburary 2012.

Ezan E, Becher F and Fenaille F (2014). "Assessment of the metabolism of therapeutic proteins and antibodies." *Expert Opin Drug Metab Toxicol* **10**(8): 1079–1091.

FDA (2001). "Guidance for industry: bioanalytical method validation." *Fed Regist* **66**: 28526–28527.

Fernandez Ocana M, James IT, Kabir M, et al. (2012). "Clinical pharmacokinetic assessment of an anti-MAdCAM monoclonal antibody therapeutic by LC-MS/MS." *Anal Chem* **84**(14): 5959–5967.

Furlong MT, Ouyang Z, Wu S, et al. (2012). "A universal surrogate peptide to enable LC-MS/MS bioanalysis of a diversity of human monoclonal antibody and human Fc-fusion protein drug candidates in pre-clinical animal studies." *Biomed Chromatogr* **26**(8): 1024–1032.

Furlong MT, Zhao S, Mylott W, et al. (2013). "Dual universal peptide approach to bioanalysis of human monoclonal antibody protein drug candidates in animal studies." *Bioanalysis* **5**(11): 1363–1376.

Gianolio DA, Rouleau C, Bauta WE, et al. (2012). "Targeting HER2-positive cancer with dolastatin 15 derivatives conjugated to trastuzumab, novel antibody-drug conjugates." *Cancer Chemother Pharmacol* **70**(3): 439–449.

Gorovits B, Alley SC, Bilic S, et al. (2013). "Bioanalysis of antibody-drug conjugates: American Association of Pharmaceutical Scientists Antibody-Drug Conjugate Working Group position paper." *Bioanalysis* **5**(9): 997–1006.

Hall MP (2014). "Biotransformation and *in vivo* stability of protein biotherapeutics: impact on candidate selection and pharmacokinetic profiling." *Drug Metab Dispos* **42**(11): 1873–1880.

Hamann PR, Hinman LM, Hollander I, et al. (2002). "Gemtuzumab ozogamicin, a potent and selective anti-CD33 antibody-calicheamicin conjugate for treatment of acute myeloid leukemia." *Bioconjug Chem* **13**(1): 47–58.

Han TH and Zhao B (2014). "Absorption, distribution, metabolism, and excretion considerations for the development of antibody-drug conjugates." *Drug Metab Dispos* **42**(11): 1914–1920.

Jeffrey SC, Andreyka JB, Bernhardt SX, et al. (2006). "Development and properties of beta-glucuronide linkers for monoclonal antibody-drug conjugates." *Bioconjug Chem* **17**(3): 831–840.

Jenkins R, Duggan JX, Aubry AF, et al. (2015). "Recommendations for validation of LC-MS/MS bioanalytical methods for protein biotherapeutics." *AAPS J* **17**(1): 1–16.

Jiang H, Zeng J, Titsch C, et al. (2013). "Fully validated LC-MS/MS assay for the simultaneous quantitation of coadministered therapeutic antibodies in cynomolgus monkey serum." *Anal Chem* **85**(20): 9859–9867.

Junutula JR, Flagella KM, Graham RA, et al. (2010). "Engineered thio-trastuzumab-DM1 conjugate with an improved therapeutic index to target human epidermal growth factor receptor 2-positive breast cancer." *Clin Cancer Res* **16**(19): 4769–4778.

Kaur S (2013a). "Analysis of antibody drug conjugates by bead-based affinity capture and mass spectrometry."

Kaur S (2013b). "Bioanalysis special focus issue on antibody-drug conjugates." *Bioanalysis* **5**(9): 981–983.

Kaur S, Xu K, Saad OM, et al. (2013). "Bioanalytical assay strategies for the development of antibody-drug conjugate biotherapeutics." *Bioanalysis* **5**(2): 201–226.

Krop IE, Beeram M, Modi S, et al. (2010). "Phase I study of trastuzumab-DM1, an HER2 antibody-drug conjugate, given every 3 weeks to patients with HER2-positive metastatic breast cancer." *J Clin Oncol* **28**(16): 2698–2704.

Lazar AC, Wang L, Blattler WA, et al. (2005). "Analysis of the composition of immunoconjugates using size-exclusion chromatography coupled to mass spectrometry." *Rapid Commun Mass Spectrom* **19**(13): 1806–1814.

Lee JW, Kelley M, King LE, et al. (2011). "Bioanalytical approaches to quantify "total" and "free" therapeutic antibodies and their targets: technical challenges and PK/PD applications over the course of drug development." *Aaps j* **13**(1): 99–110.

Lewis Phillips GD, Li G, Dugger DL, et al. (2008). "Targeting HER2-positive breast cancer with trastuzumab-DM1, an antibody-cytotoxic drug conjugate." *Cancer Res* **68**(22): 9280–9290.

Li H, Ortiz R, Tran L, et al. (2012). "General LC-MS/MS method approach to quantify therapeutic monoclonal antibodies using a common whole antibody internal standard with application to preclinical studies." *Anal Chem* **84**(3): 1267–1273.

Lin K and Tibbitts J (2012). "Pharmacokinetic considerations for antibody drug conjugates." *Pharm Res* **29**(9): 2354–2366.

Lu SX, Takach EJ, Solomon M, et al. (2005). "Mass spectral analyses of labile DOTA-NHS and heterogeneity determination of DOTA or DM1 conjugated anti-PSMA antibody for prostate cancer therapy." *J Pharm Sci* **94**(4): 788–797.

McCombs JR and Owen SC (2015). "Antibody drug conjugates: design and selection of linker, payload and conjugation chemistry." *AAPS J* **17**(2): 339–351.

Neubert H, Muirhead D, Kabir M, et al. (2013). "Sequential protein and peptide immunoaffinity capture for mass spectrometry-based quantification of total human beta-nerve growth factor." *Anal Chem* **85**(3): 1719–1726.

Ouyang Z, Furlong MT, Wu S, et al. (2012). "Pellet digestion: a simple and efficient sample preparation technique for LC-MS/MS quantification of large therapeutic proteins in plasma." *Bioanalysis* **4**(1): 17–28.

Pan LY, Salas-Solano O and Valliere-Douglass JF (2014). "Conformation and dynamics of interchain cysteine-linked antibody-drug conjugates as revealed by hydrogen/deuterium exchange mass spectrometry." *Analytical Chemistry* **86**(5): 2657–2664.

Poon KA, Flagella K, Beyer J, et al. (2013). "Preclinical safety profile of trastuzumab emtansine (T-DM1): mechanism of action of its cytotoxic component retained with improved tolerability." *Toxicol Appl Pharmacol* **273**(2): 298–313.

Presta LG (2002). "Engineering antibodies for therapy." *Curr Pharm Biotechnol* **3**(3): 237–256.

Prueksaritanont T and Tang C (2012). "ADME of biologics-what have we learned from small molecules?" *AAPS J* **14**(3): 410–419.

Quiles S, Raisch KP, Sanford LL, et al. (2010). "Synthesis and preliminary biological evaluation of high-drug-load paclitaxel-antibody conjugates for tumor-targeted chemotherapy." *J Med Chem* **53**(2): 586–594.

Rose M, Tran L, Sharifi A, et al. (2014). "*Validation of LC-MS/MS Methods for the Determination of Large Molecules: Promises and Challenges*." AAPS NBC.

Sanderson RJ, Hering MA, James SF, et al. (2005). "*In vivo* drug-linker stability of an anti-CD30 dipeptide-linked auristatin immunoconjugate." *Clin Cancer Res* **11**(2 Pt 1): 843–852.

Sapra P, Betts A and Boni J (2013). "Preclinical and clinical pharmacokinetic/pharmacodynamic considerations for antibody-drug conjugates." *Expert Rev Clin Pharmacol* **6**(5): 541–555.

Sapra P, Hooper AT, O'Donnell CJ, et al. (2011). "Investigational antibody drug conjugates for solid tumors." *Expert Opin Investig Drugs* **20**(8): 1131–1149.

Shen BQ, Bumbaca D, Saad O, et al. (2012a). "Catabolic fate and pharmacokinetic characterization of trastuzumab emtansine (T-DM1): an emphasis on preclinical and clinical catabolism." *Curr Drug Metab* **13**(7): 901–910.

Shen BQ, Xu K, Liu L, et al. (2012b). "Conjugation site modulates the *in vivo* stability and therapeutic activity of antibody-drug conjugates." *Nat Biotechnol* **30**(2): 184–189.

Siegel MM, Hollander IJ, Hamann PR, et al. (1991). "Matrix-assisted UV-laser desorption/ionization mass spectrometric analysis of monoclonal antibodies for the determination of carbohydrate, conjugated chelator, and conjugated drug content." *Anal Chem* **63**(21): 2470–2481.

Siegel MM, Tabei K, Kunz A, et al. (1997). "Calicheamicin derivatives conjugated to monoclonal antibodies: determination of loading values and distributions by infrared and UV matrix-assisted laser desorption/ionization mass spectrometry and electrospray ionization mass spectrometry." *Anal Chem* **69**(14): 2716–2726.

Smolec J, DeSilva B, Smith W, et al. (2005). "Bioanalytical method validation for macromolecules in support of pharmacokinetic studies." *Pharm Res* **22**(9): 1425–1431.

Stephan JP, Chan P, Lee C, et al. (2008). "Anti-CD22-MCC-DM1 and MC-MMAF conjugates: impact of assay format on pharmacokinetic parameters determination." *Bioconjug Chem* **19**(8): 1673–1683.

Stephan JP, Kozak KR and Wong WL (2011). "Challenges in developing bioanalytical assays for characterization of antibody-drug conjugates." *Bioanalysis* **3**(6): 677–700.

Stubenrauch K, Wessels U, Essig U, et al. (2010). "Evaluation of a generic immunoassay with drug tolerance to detect immune complexes in serum samples from cynomolgus monkeys after administration of human antibodies." *J Pharm Biomed Anal* **52**(2): 249–254.

Tarentino AL and Plummer TH, Jr. (1994). "Enzymatic deglycosylation of asparagine-linked glycans: purification, properties, and specificity of oligosaccharide-cleaving enzymes from *Flavobacterium meningosepticum*." *Methods Enzymol* **230**: 44–57.

Valliere-Douglass JF, McFee WA and Salas-Solano O (2012). "Native intact mass determination of antibodies conjugated with monomethyl Auristatin E and F at interchain cysteine residues." *Anal Chem* **84**(6): 2843–2849.

van den Broek I, Niessen WM and van Dongen WD (2013). "Bioanalytical LC-MS/MS of protein-based biopharmaceuticals." *J Chromatogr B Analyt Technol Biomed Life Sci* **929**: 161–179.

Wagner-Rousset E, Janin-Bussat MC, Colas O, et al. (2014). "Antibody-drug conjugate model fast characterization by LC-MS following IdeS proteolytic digestion." *MAbs* **6**(1): 273–285.

Wakankar A, Chen Y, Gokarn Y, et al. (2011). "Analytical methods for physicochemical characterization of antibody drug conjugates." *MAbs* **3**(2): 161–172.

Widdison WC, Wilhelm SD, Cavanagh EE, et al. (2006). "Semisynthetic maytansine analogues for the targeted treatment of cancer." *J Med Chem* **49**(14): 4392–4408.

Woolf EJ, McDougall S, Fast DM, et al. (2014). "Small molecule specific run acceptance, specific assay operation, and chromatographic run quality assessment: recommendation for best practices and harmonization from the global bioanalysis consortium harmonization teams." *AAPS J* **16**(5): 885–893.

Xu K, Liu L, Dere R, et al. (2013). "Characterization of the drug-to-antibody ratio distribution for antibody-drug conjugates in plasma/serum." *Bioanalysis* **5**(9): 1057–1071.

Xu K, Liu L, Saad OM, et al. (2011). "Characterization of intact antibody-drug conjugates from plasma/serum *in vivo* by affinity capture capillary liquid chromatography-mass spectrometry." *Anal Biochem* **412**(1): 56–66.

Yang Z, Hayes M, Fang X, et al. (2007). "LC-MS/MS approach for quantification of therapeutic proteins in plasma using a protein internal standard and 2D-solid-phase extraction cleanup." *Anal Chem* **79**(24): 9294–9301.

15

Sample Preparation Strategies for LC–MS Bioanalysis of Proteins

Long Yuan and Qin C. Ji

Analytical & Bioanalytical Operations, Bristol-Myers Squibb, Princeton, NJ, 08543, USA

15.1 Introduction

Protein therapeutics, especially monoclonal antibody (mAb)-based therapeutics, represent the fastest growing sector in pharmaceutical industry over the past decade. Currently, more than 130 protein or peptide drugs (Leader et al. 2008) and more than 29 mAb drugs (Deng et al. 2012) have been approved by the US Food and Drug Administration (FDA) or European Medicines Agency (EMA), and many more (e.g., >500 mAbs (Deng et al. 2012)) are under development. The measurement of protein therapeutics in toxicokinetic and pharmacokinetic (PK) studies is critical in drug discovery and development. In addition, as more translational research integrates into drug discovery and development, the need to monitor endogenous proteins, such as protein biomarkers, the target proteins of drugs, enzymes, and transporters, has increased rapidly. All of them significantly increase the demands on fast, accurate, and reliable bioanalytical methods for the quantitation of proteins. Conventionally, ligand binding assays (LBAs) or immunoassays are used for the quantitation of proteins in biological matrices (Desilva et al. 2003). LBAs have the advantages of high throughput, low cost, and superior sensitivity, often can measure proteins as low as pg/mL level. However, LBAs require suitable capture and detection reagents. Developing the reagents takes time and resources, which may not be affordable in drug discovery and early development. In addition, LBAs sometimes lack specificity (e.g., cannot differentiate between protein variants) and may be affected by the presence of antidrug antibodies (ADA), antireagent antibodies, soluble targets (Ezan and Bitsch 2009, Hoofnagle and Wener 2009), or other nonspecific interfering components (Kelley and DeSilva 2007, DeSilva and Garofolo 2014) in the sample. LC–MS assays have unique advantages of high specificity, wide linear dynamic range, fast method development, less interference from antidrug antibodies, and the ability to quantify multiple proteins simultaneously. More importantly, LC–MS assays can significantly reduce the cycle time in drug discovery and development, since they usually have no or lesser requirements for capture and detection reagents. As a result, LC–MS assays have gained increasing attention and interest as alternative methods for quantitation of proteins in recent years (Yang et al. 2007, Dubois et al. 2008, Heudi et al. 2008, Yang et al. 2009, Wu et al. 2011, Duan et al. 2012a, 2012b, Fernández Ocaña et al. 2012, Li et al. 2012, Neubert et al. 2012, Jiang et al. 2013, Li et al. 2013, Liu et al. 2013, Palandra et al. 2013, Sleczka et al. 2014, Zhang et al. 2014a, 2014b, 2014c, An et al. 2015, Shen et al. 2015, Zhao et al. 2015).

There are three major strategies for the LC–MS bioanalysis of proteins (summarized in Figure 15.1). The most commonly used strategy for LC–MS bioanalysis of proteins is the surrogate peptide approach (Figure 15.1a). The protein sample is first digested to peptides. Then peptides with good sensitivity and specificity are selected as the surrogate analytes of the protein and quantified by LC–MS. For small proteins, direct mass spectrometric analysis of the intact protein can be used (Figure 15.1b). This approach is a "true" measurement of the whole target protein but is currently impractical for larger proteins (e.g., mAbs) because of the difficulties in chromatographic separation and mass spectrometric detection (Ji et al. 2003, Ruan et al. 2011). The third strategy, the middle-up or limited digestion approach, is a combination of the previous two approaches. A large protein can first undergo limited digestion to generate a few large fragments, which are then analyzed by LC–MS (Figure 15.1c).

One major challenge of LC–MS assays is their frequently poor sensitivity compared to LBAs, which often limit their wider application for protein bioanalysis. In biological matrices, there are huge amounts of endogenous proteins with very large dynamic concentration

Protein Analysis using Mass Spectrometry: Accelerating Protein Biotherapeutics from Lab to Patient, First Edition.
Edited by Mike S. Lee and Qin C. Ji.

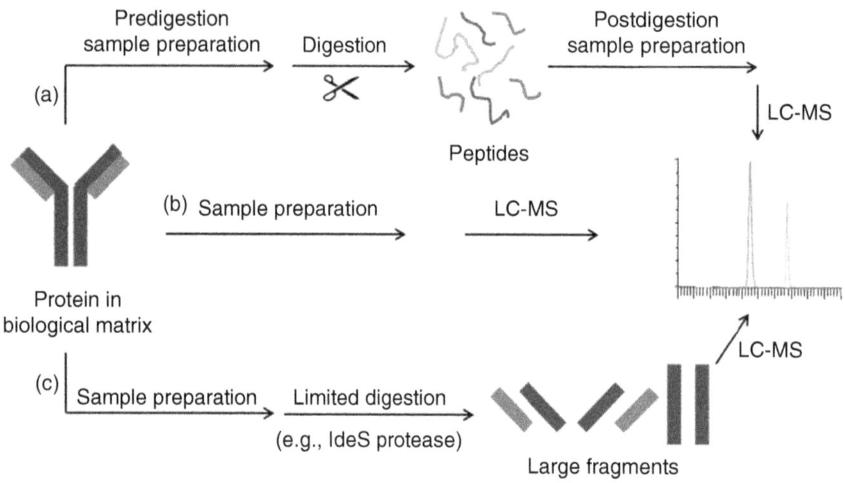

Figure 15.1 Scheme of major strategies for the bioanalysis of proteins: (a) the surrogate peptide approach, (b) the intact protein approach, and (c) the middle-up (limited digestion) approach.

ranges. For example, in human plasma, the dynamic concentration range of various proteins is >10 orders of magnitude (Anderson and Anderson 2002). The level of protein therapeutics in biological samples is usually low compared to that of the high-abundance endogenous proteins; the concentrations of protein biomarkers could be even lower. The analysis of low-concentration proteins of interest is inevitably affected by the high-abundance endogenous plasma/serum proteins, some of which have similar physicochemical properties. Without a selective protein purification prior to digestion, the endogenous proteins will also be digested along with the target protein and will generate vast amounts of background peptides. These background peptides are difficult to separate from the surrogate peptides and may cause severe ion suppression, high background noise, and potential interference with the MS detection (Yuan et al. 2013, Yuan et al. 2014). As a result, without a proper sample extraction procedure, the assay sensitivity is significantly reduced and usually can only achieve lower limit of quantitation (LLOQ) at the low μg/mL level. An appropriate sample preparation strategy that can selectively extract the target protein (before digestion) or surrogate peptide (after digestion) can significantly improve the sensitivity of LC–MS assays to a level similar to or even better than that of LBAs. We introduce various sample preparation strategies in improving assay sensitivity. The pros and cons of these strategies are briefly summarized in Table 15.1 and discussed in detail in this chapter.

Another issue of LC–MS assays is that in the absence of a selective sample preparation, they usually can only

Table 15.1 Comparison of different sample preparation strategies.

	Pros	Cons	Purification efficiency
PPT	Simple and fast operation Low cost Easy to automate	Limited sample purification	+
SPE	Various separation mechanisms to choose from Easy to automate	Requires extensive method development	++
Depletion	Most of the highly abundant proteins removed	Loss of target proteins by nonspecific binding High cost Low throughput	++
Immunocapture – generic	Commercially available reagents Generically applicable to different proteins containing a same structural scaffold	Lesser degree of purification than specific immunocapture High cost	++
Immunocapture – specific	Highly specific purification	Requires specific capture reagent High cost	+++

+, low; ++, moderate; +++, high.

measure total protein, while LBAs can measure either free protein (unbound active protein, the protein still has binding activity) or total protein using different capture reagents. This sometimes causes discrepancy in measured concentrations (and thus the calculated PK parameters) by LC–MS method and LBAs (Heudi et al. 2008, Wang et al. 2012). Measurement of free protein is usually a better choice when measurement of the functional molecules is needed (Zhang et al. 2014a, 2014b, 2014c). The progress of sample preparation strategies, especially the use of immunocapture, has made selective extraction of free protein from biological samples possible; therefore, LC–MS assays are able to measure free or active protein. In this chapter, we present various approaches to differentiate free, bound, and total protein, and to overcome interference from antidrug antibodies or soluble targets.

Efficient and reproducible protein digestion is critical for accurate and sensitive quantitation of protein using a surrogate peptide approach. The speed of digestion will affect throughput of the method; the completeness of digestion will affect sensitivity; the consistency of digestion will affect the accuracy and precision of the assay (Addona et al. 2009), especially when a stable-isotope-labeled protein internal standard is unavailable. We also discuss strategies to achieve fast, efficient, and reproducible digestion in support of quantitative analysis of proteins.

15.2 Sample Preparation Strategies to Improve Assay Sensitivity

15.2.1 Protein Precipitation

Protein precipitation (PPT) is a simple, fast, efficient, and cost-effective sample cleanup strategy. It is also easy to automate, making it a commonly used method in bioanalysis of small-molecule drugs. PPT uses water-miscible organic solvents (e.g., acetonitrile and methanol) to precipitate most of the proteins in plasma/serum samples and separate them from the supernatant by centrifugation. For small-molecule drugs, the target analytes remain in the supernatant, and thus, the supernatant is analyzed by LC–MS/MS. For proteins, depending on their solubility in the extraction solvent, the protein analytes could be retained in the supernatant or precipitated along with other background proteins in plasma/serum samples. Organic soluble proteins (e.g., PEGylated proteins (Wu et al. 2011, Dawes et al. 2013) or small proteins (Becher et al. 2006, Zhang et al. 2014a, 2014b, 2014c)) can be retained in the supernatant and separated from the precipitated endogenous proteins by centrifugation, making PPT a simple, efficient, and reliable approach for the sample cleanup of these proteins. Wu et al. 2011 compared different water-miscible organic solvents for PPT of PEGylated proteins. They found that acidified alcoholic solvents such as isopropanol usually had better extraction efficiency compared to acetonitrile. They also found that for PEGylated proteins with different physicochemical properties, screening of different extraction organic solvents is required to achieve optimized recovery. Using PPT, they successfully achieved simple, reproducible, and high-throughput LC–MS/MS quantitation of multiple PEGylated protein drug candidates in plasma. With appropriate optimization of acidified precipitation solvent and the solvent-to-serum ratio, almost 100% recovery from human serum, with minimum matrix effect, was achieved for the analysis of interferon-gamma-inducible protein-10 (IP-10), an 8.6 kDa protein (Zhang et al. 2014a, 2014b, 2014c). Using the optimized method, a highly sensitive LC–MS/MS method with an LLOQ of 31.62 pM was developed using only 50 μL of human serum, which was 100-fold better than a direct digestion method and the same as analyzing the neat sample (IP-10 in buffer). One advantage of PPT method is that it can eliminate interferences from ADAs, as shown in a validated LC–MS/MS assay for a PEGylated adnectin therapeutic protein in monkey plasma. In contrast, the ADA interference was observed in the LBA analysis (Dawes et al. 2013).

Large proteins (e.g., mAbs) mostly coprecipitate with endogenous proteins in serum/plasma and only very small amount of the analyte protein will remain in the supernatant. As a result, the supernatant after PPT is inappropriate for the analysis of large proteins. On the other hand, the pellet containing the precipitated proteins can be digested and used for the analysis. This approach can remove soluble proteins, salts, and a significant portion of the phospholipids in serum/plasma, therefore, provide some degree of sample cleanup (Ouyang et al. 2012, Yuan et al. 2012).

Recently, Liu et al. 2014 developed an innovative acid-assisted PPT method that can efficiently remove albumin, the most abundant endogenous protein in serum/plasma samples, while retaining the target proteins. For three therapeutic proteins tested, using isopropanol with 1.0% trichloroacetic acid as the precipitation solvent, 95% of the total albumin in human plasma samples was removed. At the same time, almost 100% for two out of the three proteins and 60–80% for the third were retained. Their simple and cost-effective approach resulted in cleaner samples, improved digestion consistency, and improved sensitivity (fivefold for the therapeutic protein tested).

15.2.2 Solid-Phase Extraction

Depending on the physicochemical properties of the target proteins, solid-phase extraction (SPE) may be applied before protein digestion to extract the intact protein or after digestion to extract the generated surrogate peptides. For small proteins (Ji et al. 2003, Ji et al. 2007, Ruan et al. 2011) and PEGylated proteins (Yang et al. 2009, Li et al. 2011), SPE is an effective approach to extract the intact target protein from biological matrices. For example, an Oasis HLB SPE plate was used to extract rK5, a protein drug candidate with a molecular weight of 10,464 Da, from plasma samples and achieved recovery of 72–85% (Ji et al. 2003). SPE sample preparation also greatly improved the assay ruggedness since it significantly removed background proteins, such as albumin, and enriched the rK5 analyte in the extracts (Ji et al. 2007).

For large proteins, SPE has been commonly applied for the extraction of the surrogate peptides of target proteins from digested samples (Yang et al. 2007, Heudi et al. 2008, Yuan et al. 2013, Bronsema et al. 2015a, 2015b). To achieve further purification of the peptides of interest, 2D SPE (reversed-phase (RP) followed by strong-cation exchange (SCX) SPE) has been applied to clean up the tryptic digests of plasma samples (Yang et al. 2007). The RP SPE was used to remove salts and highly hydrophobic components and SCX SPE to remove background peptides with basicity significantly different from the target peptides. As a result, the background and ion suppression were significantly reduced in LC–MS/MS analysis, the sensitivity was improved, and an LLOQ of 0.5 μg/mL was achieved for the analysis of a therapeutic mAb in serum.

An orthogonal SPE sample preparation can provide an additional dimension of separation to the chromatographic separation, and therefore, better sample cleanup is usually achieved compared to when an SPE with the same separation mechanism as the column is used. Sample cleanup with SCX SPE, RP SPE, and 2D SPE (RP SPE followed by SCX SPE) was systematically compared for the analysis of a test mAb in serum (Yuan et al. 2013). To evaluate the "cleanliness" of the extracted samples, the intensities of the LC–MS total ion chromatogram (TIC) scan over the range of m/z 200–1250 were compared. As shown in Figure 15.2, SCX SPE, which acted as an orthogonal sample preparation technique to the RP chromatographic separation, was found to be more efficient than RP SPE in removing the background peptides in serum tryptic digests. Most background peptides were removed using the orthogonal SCX SPE approach, and, as a result, the ion suppression and background noise were greatly reduced, and a much improved (fivefold compared to the sample without SPE) LLOQ of 0.20 μg/mL was achieved using only 25 μL serum samples and 5 μL injection volume. Although 2D SPE could further improve the "cleanliness" of the sample, the improvement was not significant compared to using SCX SPE alone. The decreased recovery of 2D SPE also minimized its ability to enhance the assay sensitivity.

15.2.3 Derivatization

Derivatization of peptides has been widely used in proteomics for qualitative work, such as enhancing protein identification and characterization (Foettinger et al. 2006, Xu et al. 2008, Suh et al. 2010), improving *de novo* sequencing of proteins (Madsen and Brodbelt 2009), and relative protein quantification (Sandra et al. 2008). Recently, we proposed and developed a novel selective peptide derivatization (SPD) strategy to improve

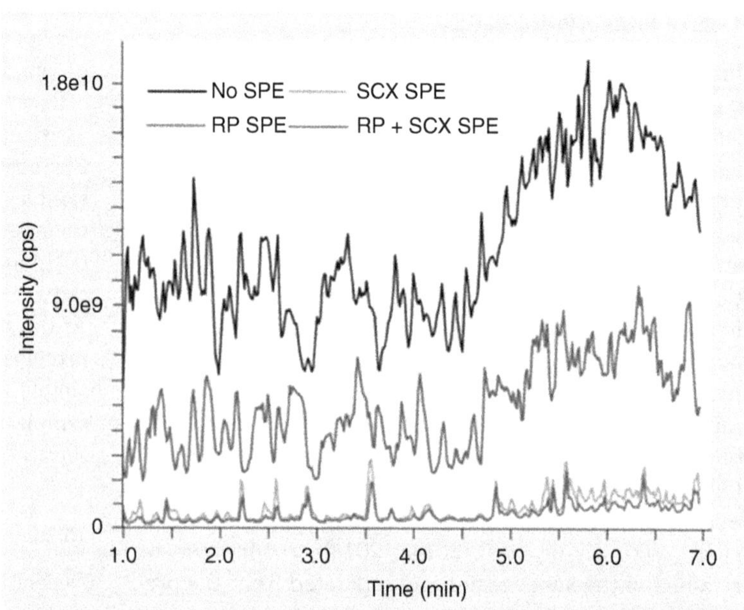

Figure 15.2 LC–MS scan (m/z 200–1250) total ion chromatograms of monkey serum tryptic digest samples processed with different cleanup methods. (*Source*: Yuan et al. 2013. Reproduced with permission of Future Medicine Ltd (Bioanalysis).)

sensitivity for the LC–MS/MS quantitative bioanalysis of proteins (Yuan et al. 2014). This strategy works by selectively derivatizing the surrogate peptide of the target protein while not derivatizing background peptides. SPD can enhance physicochemical differences between derivatized target peptides and underivatized peptides, therefore, improving their separation during extraction and chromatographic separation. In addition, SPD can also promote increased ionization efficiency, improved fragmentation pattern (e.g., generating more abundant product ions), and/or increased sample extraction recovery for target peptides. All of these effects result in significantly enhanced assay sensitivity. To evaluate if SPD can improve the assay sensitivity as expected, malondialdehyde (MDA), which can selectively derivatize arginine-containing peptides, was used to derivatize tryptic peptides (including the surrogate peptide of the target mAb). Digested monkey serum samples were treated with MDA derivatization, and their LC–MS/MS chromatograms were compared to nonderivatized samples. As shown in Figure 15.3, after MDA derivatization and SPE, the signal-to-noise ratio (S/N) of the sample increased more than fivefold compared to the nonderivatized/no SPE sample, and more than twofold compared to the nonderivatized/SPE sample. The SPD strategy was successfully applied in the development of a sensitive, accurate, and precise LC–MS/MS assay for the analysis of a test mAb in monkey serum. SPD provides a novel sample preparation strategy for improving sensitivity of LC–MS/MS bioanalytical assays for proteins, especially when a suitable immunocapture antibody is unavailable.

15.2.4 Depletion of High-Abundance Proteins

Depletion of high-abundance proteins (e.g., albumin and immunoglobulins) is widely used in proteomics to reduce sample complexity and improve detection of low-abundance proteins (Björhall et al. 2005, Zolotarjova et al. 2005, Whiteaker et al. 2007a, 2007b). The depletion can be done based on chemical affinity (e.g., cibacron blue dye) or immunoaffinity (e.g., antibodies). Cibacron blue dye has a high affinity for albumin. However, it also has affinity for nicotinamide adenine dinucleotide (NAD), flavin adenine dinucleotide (FAD), and adenosine triphosphate (ATP) binding sites of proteins, which often results in lower specificity with the removal of proteins of interest (Zolotarjova et al. 2005). Immunoaffinity depletion is usually based on a mixture of polyclonal antibodies, which can specifically target multiple epitopes on the proteins, and, therefore, can simultaneously remove multiple high-abundance proteins (Zolotarjova et al. 2005). Minimizing nonspecific loss of target proteins and improving the consistency and reproducibility of depletion are major challenges in immunodepletion. Another limitation is the low throughput of depletion due to the labor-intensive and time-consuming operation. Hagman et al. (2008) compared three dye-based and two antibody-based affinity kits for depletion of albumin. The antibody-based affinity kits generally had a higher specificity than dye-based affinity kits for albumin removal. Among the tested kits, the ProteoExtract albumin depletion kit showed both the best depletion ability and specificity, removing approximately 50% of the total serum protein. Using albumin depletion, they developed an accurate and reproducible LC–MS/MS method for the quantification of mAbs in

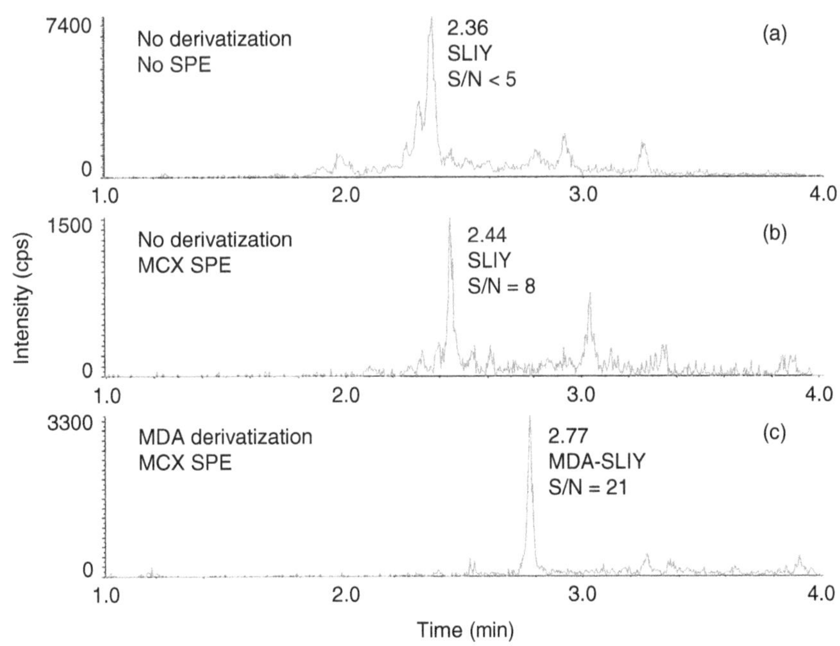

Figure 15.3 LC/MS/MS chromatograms of peptide SLIY and its MDA derivatized product, MDA-SLIY, in trypsin-digested monkey serum samples containing 1.00 µg/mL of BMS-986012. (*Source*: Yuan et al. 2014. Reproduced with permission of John Wiley & Sons.)

serum over the range of 2–1000 μg/mL. An assay using immunoaffinity depletion coupled with LC–SRM–MS was developed for the sensitive and accurate quantification of total and free prostate-specific antigen (PSA) in serum (Liu et al. 2012). After immunodepletion with a Seppro IgY14 LC10 column, 14 high-abundance proteins, which counted for approximately 95% of the total protein mass, were specifically removed from human plasma. Low ng/mL level detection for total (LLOQ 2.03 ng/mL) and free (LLOQ 0.86 ng/mL) PSA was consistently achieved in clinical serum samples. The results obtained using the developed immunodepletion assay showed a good correlation (R^2 ranging from 0.90 to 0.99) to those obtained using the conventional immunoassays.

15.2.5 Immunoaffinity Purification

In contrast to depletion of high-abundance proteins, the specific purification/enrichment of the target protein or the surrogate peptide is a more efficient way to increase the sensitivity. Immunoaffinity capture can efficiently and specifically extract low-concentration proteins or their corresponding surrogate peptides from the abundant endogenous proteins/peptides; thus, background peptides are removed and the sensitivity and selectivity for the LC–MS bioanalysis of proteins in biological matrices are greatly improved. Immunocapture has become the current method of choice for developing highly sensitive LC–MS method. In addition, it can selectively extract free or total protein by using different capture reagents (e.g., ligand, anti-idiotypic antibody, or noncompeting antibody; see Section 15.3 for more details). Immunocapture LC–MS methods combine the selectivity of immunoaffinity extraction with the sensitivity and specificity of LC–MS, therefore, achieving both high sensitivity/specificity and good selectivity. Immunocapture, with its unique advantages, has become a very useful and promising sample preparation technique and will definitely be more widely applied in LC–MS bioanalysis of proteins. Immunocapture can be done either offline using beads, tips, or plate format or online with a column-switching format. Generally, the more specific the immunocapture, the better the enrichment and purification of the target proteins/peptides that can be achieved, and as a result, the better sensitivity.

15.2.5.1 Immunocapture of a Specific Peptide

One commonly used immunocapture approach is to use an antipeptide antibody to specifically capture the surrogate peptides in digested samples (Anderson et al. 2004, Whiteaker et al. 2007a, 2007b, Kuhn et al. 2009, Whiteaker et al. 2010). The leading example is the stable-isotope standards with capture by antipeptide antibodies (SISCAPA) technique developed by Anderson et al. (2004). SISCAPA combines the advantages of immunocapture enrichment of target peptides and the use of stable-isotope-labeled internal standards, therefore, achieving highly sensitive and precise quantification of proteins in biological matrices. Immunocapture of target peptides is highly specific and efficient and usually able to achieve >100-fold enrichment, which could contribute sensitivity improvements of about two orders of magnitude (Anderson et al. 2004, Whiteaker et al. 2007a, 2007b). As shown in Figure 15.4, for the analysis of an endogenous serum protein, AAC, in human serum, the antipeptide antibody-enriched sample showed significantly higher peak intensity and much cleaner background compared to the sample without immunocapture treatment, resulting in an estimated enhancement of ion signal of 1453-fold (Whiteaker et al. 2007a, 2007b). In addition, since the samples are digested before immunocapture, there is no interference from ADAs, soluble targets, or other antitarget protein antibodies.

Simultaneous analysis of multiple proteins in a single assay can be achieved by multiplexed immunocapture using antipeptide antibodies targeting different peptides. For example, Kuhn et al. (2009) developed a multiplexed SISCAPA assay for simultaneous determination of two protein biomarkers, troponin I and interleukin-33, in plasma. The multiplexed assay showed excellent agreement with the separate assays using individual antipeptide antibody, and it also agreed well ($R = 0.89$) with an established commercial immunoassay for the analysis of patient samples. In another example, a multiplexed SISCAPA assay for simultaneous quantification of nine protein biomarkers in plasma was developed using an automated magnetic bead-based platform (Whiteaker et al. 2010). Their multiplexed SISCAPA process improved the sensitivity by 100- to 1000-fold and yielded detection limits at physiologically relevant ng/mL level with good precision (median CV 12.6%). The assay sensitivity can be even increased to the low pg/mL range of protein concentrations using enrichment of peptides from larger volumes of plasma (1 mL).

15.2.5.2 Immunocapture of a Specific Protein

Another commonly used immunocapture strategy works by using anti-idiotypic antibody or the receptor/ligand of the target protein to specifically capture the target protein. This strategy has been widely applied in the bioanalysis of various proteins, including protein therapeutics (Dubois et al. 2008, Xu et al. 2014) and protein/peptide biomarkers (Berna et al. 2006, Oe et al. 2006, Berna et al. 2007, Winther et al. 2009, Dufield et al. 2010, Ocaña and Neubert 2010, Rafalko et al. 2010, Torsetnes et al. 2014, Chen et al. 2015). Berna et al. (2007) developed and validated an immunocapture LC–MS/MS assay to measure myosin light chain 1 (Myl3), a 23-kDa protein biomarker

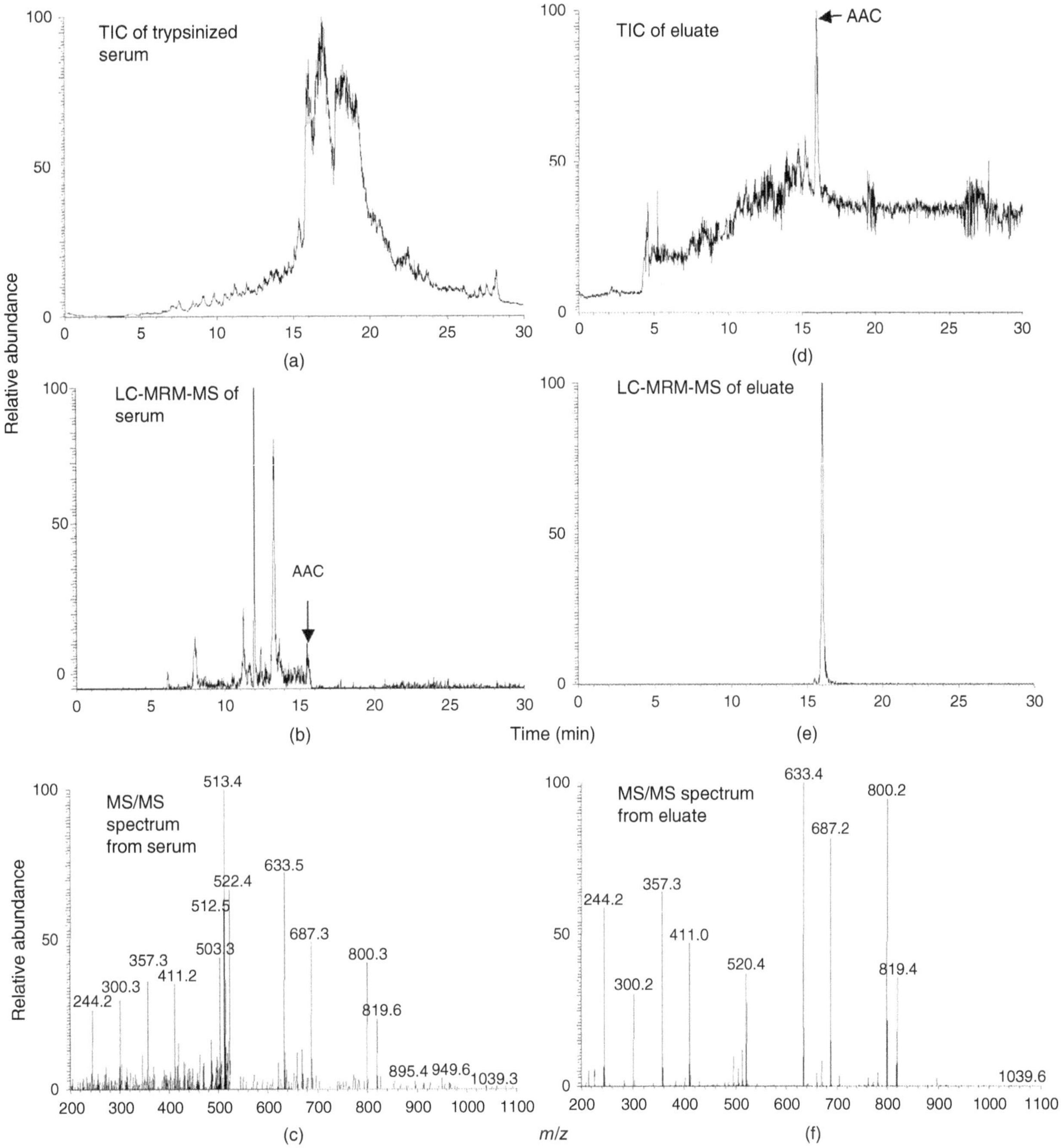

Figure 15.4 Enhancement of endogenous AAC ion signals following antibody enrichment assessed by LC–MS. (a) 30-min LC–MS analysis of 20 μL depleted and digested human serum. (b) 30-min LC–MRM–MS analysis of AAC from same serum digest. (c) MS-MS spectrum of AAC from serum digest. (d) 30-min LC–MS analysis of antibody-enriched AAC from 20 μL depleted and digested human serum. (e) 30-min LC–MRM–MS analysis of antibody captured AAC. (f) MS-MS spectrum of AAC from antibody capture. Equivalent amounts of peptide were injected for LC–MS analysis for each sample (before and after capture). (*Source*: Whiteaker et al. 2007a,b. Reproduced with permission of Elsevier.)

of cardiac necrosis, in rat serum. Myl3 protein was extracted from 100 μL of serum sample using an antirat Myl3 mouse monoclonal antibody. The assay was validated over the linear range of 0.0734–7.16 nM with interday accuracy and precision at 12.9% and 13.2%, respectively. Soluble epidermal growth factor receptor, the pharmacological target of cetuximab (an mAb drug), was used for specific immunocapture of cetuximab from human serum samples (Dubois et al. 2008). The immunocapture process significantly improved the sensitivity and achieved an LLOQ of 20 ng/mL, which was similar to that of ELISA methods.

Similar to the multiplexed SISCAPA assay, multiplexed immunocapture of proteins can be used for simultaneous analysis of multiple proteins (Torsetnes et al. 2014, Xu et al. 2014). For instance, a multiplexed method was developed for the simultaneous determination of pro-gastrin releasing peptide (ProGRP) and neuron-specific enolase (NSE) isovariants, the biomarkers for the small-cell lung cancer (Torsetnes et al. 2014). A combination of anti-NSE mAb and anti-ProGRP mAb-coated magnetic beads was used for the immunocapture purification of the ProGRP and NSE biomarkers. The multiplexing approach reduced the consumption of samples, as well as the sample preparation time. The performance of the multiplexed method was similar to that of the two previously validated methods for individual biomarker and showed good linearity, recovery, accuracy, and precision with an LLOQ of 24 pM (300 pg/mL) and 15 pM (700 pg/mL) for total ProGRP and γ-NSE, respectively.

Sequential immunocapture of target protein and its surrogate peptide can be applied to obtain highly enriched and purified samples; therefore, sensitivity significantly increased (Neubert et al. 2012, Palandra et al. 2013). Neubert et al. (2012) developed and validated a highly sensitive method for the quantification of human β nerve growth factor (NGF) in serum using this approach. They first used magnetic bead-based anti-NGF polyclonal antibody to extract NGF from serum. After tryptic digestion, to further purify the generated surrogate peptide, they applied an online peptide immunoaffinity enrichment using anti-NGF peptide antibody. In combination with sequential immunocapture and nanoflow LC–MS/MS, they achieved a very high assay sensitivity (LLOQ 7.03 pg/mL) with good accuracy and precision (<10% interassay relative error and <15% interassay CV).

15.2.5.3 Generic Immunocapture

One major limitation for the above protein-/peptide-specific immunocapture approaches is that, similar to LBAs (but to a lesser extent, since no detection reagent is needed), they require suitable capture reagents, which take time and resources to develop. A generic immunocapture approach uses reagents that can bind to a common region of proteins or a sequence of peptides from various proteins; thereby, one method can be used to extract different proteins. Commercially available immunocapture reagents (i.e., Protein A, Protein G, antispecies Fc antibody) have been used for the generic immunocapture purification of antibodies. For instance, Protein A or Protein G can effectively bind to the Fc region of immunoglobulin G (IgG); therefore, it can be used for immunocapture of antibody drugs containing an IgG Fc region (Lu et al. 2009, Fernández Ocaña et al. 2012). In addition to the relatively easy availability of the reagent, another advantage is that a single immunocapture method using the generic reagents can, individually or simultaneously, extract different proteins containing the same region of protein. A generic immunocapture method using an antihuman Fc antibody (anti-Fc Ab35) as the capturing reagent was developed for extraction of human mAbs in plasma/serum samples (Li et al. 2012). Anti-Fc Ab35 specifically binds to Fc region of human IgG, whereas does not bind to IgGs from other species. Consequently, various human mAbs can be purified using this generic immunocapture method. This method was applied to the individual analysis of eight different mAbs, as well as the simultaneous extraction of four different mAbs in rat plasma and achieved LLOQs of 100 ng/mL (Li et al. 2013). Sleczka et al. (2014) developed a generic immunocapture method using commercially available antihuman Fc antibody for the analysis of tissue samples. Using ustekinumab as the model protein, they achieved an LLOQ of 20 ng/mL in serum. The developed method was used for the quantification of human mAbs and Fc-fusion protein (containing a human IgG Fc region) in various types of mouse tissues.

The generic immunocapture strategy can also be used for other types of proteins that contain a common structural scaffold. An immunocapture method using commercially available anti-PEG antibodies, which can specifically capture the PEG portion of the PEGylated protein, was developed for the purification of MK-2662, a PEGylated therapeutic protein (Xu et al. 2010). The method is generally applicable to the purification of other PEGylated proteins/peptides. The validated assay using the optimized anti-PEG antibody immunocapture method achieved good precision (CV < 9.76%), accuracy (94.8–105.8%), and sensitivity (LLOQ 2 nM).

Generic immunocapture purification of multiple proteins can be achieved by capturing a peptide in the common structural scaffold of the proteins of interest. For example, peptide VVSVLTVLHQDWLNGK is a peptide universally existing in the Fc region of all human IgG1 and IgG4 (Furlong et al. 2012). An immunocapture method that can specifically capture this universal peptide could be used for the quantitative analysis of various human IgG1 and IgG4 Fc region-based protein therapeutics in animal species.

The generic immunocapture approach may not be applicable in all cases. For instance, the generic immunocapture method using an antihuman Fc antibody cannot be used to support human clinical studies due to the existence of large amounts of endogenous human IgGs in plasma or serum samples. In addition, the less-specific generic immunocapture usually results in a lower degree of sample purification and therefore, less improvement in sensitivity. Specific immunocapture using anti-CNTO736 antibody and generic immunocapture using Protein A were compared for the analysis of CNTO736, a 60 kDa protein drug, in serum (Lu et al. 2009). A sensitivity of 333 ng/mL was achieved using Protein A. In contrast, using anti-CNTO736 antibody, which specifically binds the idiotypic domain of CNTO736, highly efficient enrichment of CNTO736 was achieved, and the assay sensitivity was improved 100-fold to 3.3 ng/mL.

15.2.6 Online Sample Preparation

Offline sample preparation, such as SPE, usually involves multiple steps including sample transfer, dry down, and reconstitution. Any loss of analyte during these steps will result in lowered recovery and decreased sensitivity. Adsorption (nonspecific binding) to vessel walls (e.g., sample vials, pipette tips, and 96-well plates) is well known for peptides and proteins (van Midwoud et al. 2007); thus, overall recovery is further reduced. Online sample preparation can significantly reduce the sample loss that could happen during offline sample preparation; therefore, the extraction recovery increases and the assay sensitivity improves. In addition, online sample preparation is fully automated, which can significantly improve the efficiency and quality of the assay by avoiding labor-intensive manual operations and reducing human errors.

In the continuous efforts to further improve the sensitivity of LC–MS-based protein assays, low-flow LC (e.g., microflow or nanoflow LC) has drawn considerable attention since the ionization and ion-transfer efficiency of the analyte can increase significantly with the decrease in flow rate (Wilm and Mann 1996, Tang et al. 2004), which typically results in much improved sensitivity. However, due to the smaller size of the low-flow LC column, its loading capacity (the sample volume that can be directly injected into the column) is much smaller than that of a regular LC column. To fully utilize the ability of a low-flow LC in improving sensitivity, an online trapping device is often needed to allow the injection of a larger volume of sample. Online sample preparation is an ideal fit for low-flow LC, since it can combine the trapping (loading of the sample) and the purification/enrichment of the sample in a single operation.

The most commonly utilized online sample preparation approach uses column-switching with multiport valves. For a typical online sample preparation method, the sample is initially injected into a first column (loading column). After washing and chromatographic separation on the loading column, the multiport valve is switched and the LC fraction containing the analyte is transferred (back or forward flushed) to a second column (analytical column) for LC–MS analysis. The loading column could be an SPE column (online SPE) or an analytical column (online 2D-LC). For protein bioanalysis, due to the high complexity of the digested biological samples, the limited separation achieved by SPE columns (online SPE) may not provide enough sample cleanup and sensitivity improvement. Using an analytical column to achieve the first-dimensional separation, 2D-LC can yield a more complete resolution of the peptides, and it has been applied for the LC–MS bioanalysis of proteins (Xu et al. 2010, Shen et al. 2015). Orthogonal separation mechanisms that feature the use of ion exchange or hydrophilic interaction (HILIC) in the first dimension with reversed phase in the second dimension are often used in 2D-LC to achieve better separation.

Shen et al. (2015) developed an online 2D-LC–MS/MS method for the quantification of immunoglobulin A1 protease (IgAP), a therapeutic protein under development, in human serum. Orthogonal chromatographic separations were achieved using different pH mobile phases: a high-pH (pH 8) reversed-phase separation (Acquity UPLC BEH C18 column, 1.7 μm, 2.1 × 100 mm) was used in the first dimension and a low-pH (pH 3) reversed-phase separation (Acquity UPLC XBridge C8 column, 2.5 μm, 2.1 × 100 mm) was used in the second dimension. This approach obtained high resolving power in the first dimension and an easy transition from the first-dimension separation to the second-dimension separation. As shown in Figure 15.5, compared to the sample analyzed by the traditional 1D-LC, the sample analyzed by 2D-LC showed much better chromatographic separation, higher peak intensity, and significantly lower background. Overall, the online 2D-LC system obtained more than 40-fold improvement in sensitivity (achieved an LLOQ of 0.05 μg/mL).

For immunocapture, online sample preparation can also be achieved by column-switching using an immunoaffinity column in the first dimension. In the immunoaffinity column, usually polyclonal or monoclonal antibodies are immobilized on the surface of the column packing and used to selectively capture and enrich the peptides/proteins of interest. Online immunocapture has been used for the depletion of high-abundance proteins (Cellar et al. 2009) and the purification/enrichment of proteins (Hoos et al. 2006, Cingöz et al. 2010) or peptides (Berna et al. 2006, Li et al. 2007, Neubert et al. 2010). Online immunocapture can minimize the adsorption loss of the analyte from offline steps (transfer, dry down, and reconstitution) and increase the antibody capture efficiency, therefore achieving a significantly better sample purification/enrichment and higher sensitivity. This approach also significantly improves the assay automation by converting the time-consuming and labor-intensive offline immunocapture steps into a single online method. It can be even combined with

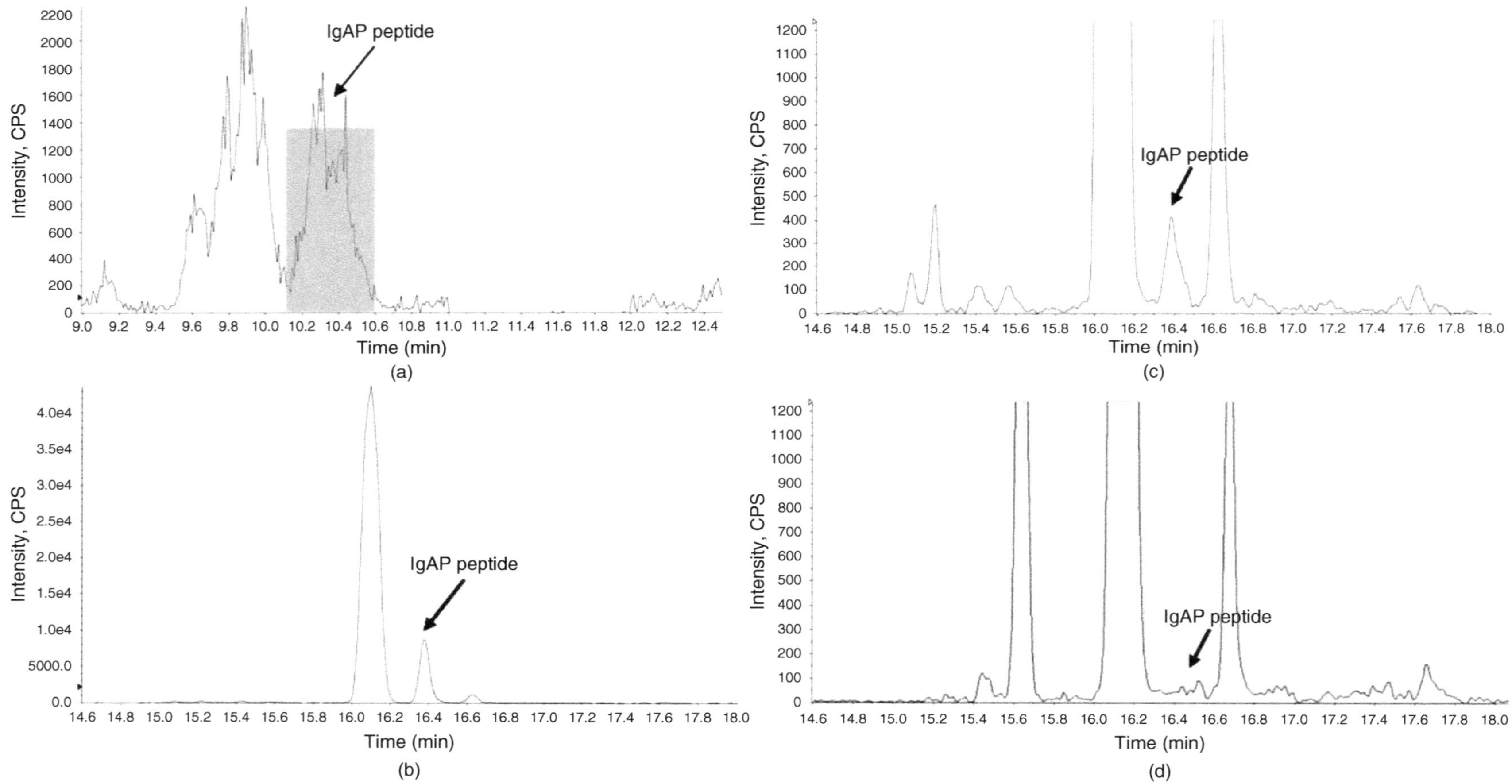

Figure 15.5 (a) MRM ion chromatogram of the spike-in surrogate peptide in serum tryptic peptides (equivalent to 2.0 μg/mL IgAP) separated by the first-dimension RPLC system; (b) MRM ion chromatogram of the surrogate peptide at the same concentration following separation by the 2D-LC system (the highlighted fraction in (a) was transferred to the second RP column); (c) MRM ion chromatogram of the spike-in surrogate peptide at 0.05 μg/mL following separation by the 2D-LC system; (d) surrogate peptide MRM ion chromatogram of the blank matrix following separation by the 2D-LC system. (*Source*: Shen et al. 2015. Reproduced with permission of American Chemical Society.)

online protein digestion techniques (using a column packed with immobilized enzymes) to achieve fully automated online sample preparation (Hoos et al. 2006, Cingöz et al. 2010).

15.3 Sample Preparation Strategies to Differentiate Free, Total, and ADA-Bound Proteins

The ability to measure free protein, or active protein with binding activity, is often required for a protein bioanalytical assay, since most therapeutic proteins function by binding to the target protein, which makes the measurement of the active proteins meaningful. A free (unbound) assay is designed to measure analyte, which still has a free binding site, and thus is able to bind the specific binding partner, whereas a total assay is designed to measure both free and bound analytes. For LBA, in a free assay, the ligand (target) or the anti-idiotypic antibody of the target protein is used as the capture reagent. Either reagent can specifically bind to the active binding site of the target protein; therefore, only the free or partially free (e.g., for an mAb, one binding site free and the other site bound) proteins in the sample are captured. In a total assay, an antibody that binds to a noncompeting region (regions not containing the active binding site) of the target protein is used as the capture reagent. Therefore, both the free and the bound proteins can be captured.

Similar immunocapture sample preparation strategies can be applied to LC–MS assay for the measurement of free and total protein (see Figure 15.6 for typical workflows). For example, two different immunocapture strategies were developed for the free and total LC–MS assay of an mAb therapeutic (PF-00547,659 a human IgG2 mAb), in human serum (Fernández Ocaña et al. 2012). The free assay used an anti-idiotypic mAb raised against the PF-00547,659 idiotype to selectively capture free or partially free PF-00547,659 in serum. The total assay used Protein G to generically extract all IgGs in human serum samples, including free and bound PF-00547,659. The use of an anti-idiotypic antibody in the free assay provided a higher degree of enrichment and selectivity for PF-00547,659, and achieved an LLOQ of 7.03 ng/mL. In contrast, Protein G used in the total assay is much less selective than the anti-idiotypic antibody, resulting in less purified samples and reducing the assay sensitivity by >100-fold to an LLOQ of 781 ng/mL.

Heinig et al. (2013) developed a sample preparation strategy using Protein A or G tips to differentiate between free and ADA-bound drug for an acylated peptide drug of MW 4.5 kDa. Plasma samples were first processed with Protein A or G. Thus, the ADA-bound drug was extracted by Protein A or G (through binding with the ADAs at the Fc region) and trapped in the tips, while the drug not bound to ADA, including free and other plasma protein bound drug, remained in the sample. The separated fractions were further processed and analyzed for ADA-bound or free drug. In addition, the total drug was measured by applying acid-assisted PPT (15% formic acid in acetonitrile) to break the drug-ADA bindings and release the drug. The sample was then purified by SPE and analyzed by LC–MS/MS. This differentiation strategy was also applied to other nonantibody proteins (Zhang et al. 2014a, 2014b, 2014c, Bronsema et al. 2015a, 2015b). For example, for the quantitation of total and free IP-10, a soluble target for the treatment of autoimmune diseases, in human serum, a Protein A filter spin

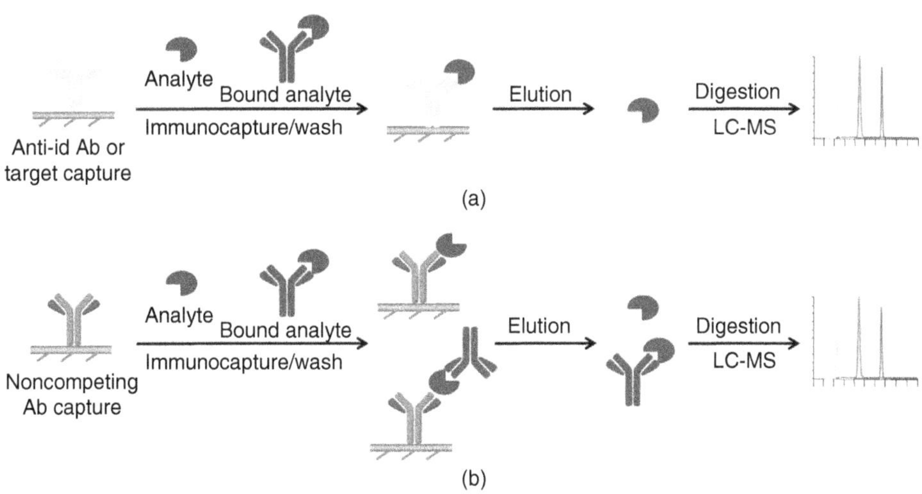

Figure 15.6 Immunocapture sample preparation strategies for LC–MS bioanalysis of free (a) and total (b) protein.

plate was used to remove antibody-bound IP-10, retaining free IP-10 in the antibody depleted serum, while an acid-assisted PPT method was used to extract the total IP-10 (Zhang et al. 2014a, 2014b, 2014c). Similarly, in an LC–MS/MS assay for the quantitation of ADA-bound and total recombinant human α-glucosidase in human plasma, Protein G was applied for the extraction of ADA-bound analyte and PPT was used for the extraction of the total analyte (Bronsema et al. 2015a, 2015b).

15.4 Sample Preparation Strategies to Overcome Interference from Antidrug Antibodies or Soluble Target

One common issue in LBA method development is the interference from ADAs or soluble targets in the samples, since they may interfere with the binding of the protein of interest to the capture or detection reagent and may result in underestimation of the analyte concentration (Hoofnagle and Wener 2009). For LC–MS assays, generally this type of interference is less of a concern, given that sample extraction procedures, such as PPT or SPE, can disrupt binding between target protein and ADAs or soluble targets, and the proteolytic digestion cleaves all the proteins in the sample. However, ADA-related interference has also been reported in LC–MS assays when using PPT or SPE method (Ji et al. 2007, Heinig and Wirz 2009, Heinig et al. 2013, Gong et al. 2014). This may be because the binding between the protein of interest and ADAs/soluble target is strong enough that it cannot be completely disrupted during sample extraction. Various strategies, including denaturation (Ji et al. 2007, Heinig and Wirz 2009, Shen et al. 2015), acid-assisted PPT (Heinig and Wirz 2009, Heinig et al. 2013), acid dissociation (Gong et al. 2014), and immunocapture (Kushnir et al. 2013), have been applied to overcome interferences from ADAs or soluble targets.

Ji et al. (2007) observed that the LC–MS/MS response of rK5, a protein drug candidate, significantly decreased with an increased amount of ADAs spiked into the monkey plasma sample, even though the wash steps in SPE involved the use of relatively strong acid (0.2% trifluoroacetic acid) and organic solvent (hexane). They evaluated various reagents, including guanidine (8 M), trifluoroacetic acid (0.5%), urea (8 M), saturated sodium chloride solution, and sodium dodecyl sulfate (SDS), for overcoming the ADA interference. The use of 8 M guanidine gave the best results and completely eliminated the interference, since high-concentration guanidine denatured the ADAs in the sample, disassociating the binding between rK5 and ADAs, and releasing the bound rK5 for SPE.

The presence of ADAs was found to cause low extraction recovery of taspoglutide, a human glucagon-like peptide-1 (GLP-1) analog: the recovery was only at 16–30% using SPE and 24–40% using PPT (Heinig and Wirz 2009). Acid-assisted PPT with different concentrations of formic acid (2% and 4%) in acetonitrile was assessed for eliminating the ADA interference. The recovery of taspoglutide was improved to 80–85% using 2% formic acid and 96–99% using 4% formic acid. The use of urea (6 M) or guanidine (6 M) prior to SPE was also evaluated. Urea did not improve the release of the drug from ADA complexes, while guanidine improved the recovery to 94–96%. These results demonstrated that using either formic acid or guanidine can efficiently dissociate the drug-ADA bindings and release the drug.

Acid dissociation can also be applied to overcome interferences from antidrug antibodies and soluble targets (Gong et al. 2014). The extraction recovery of a PEGylated domain antibody drug candidate using PPT was significantly lower in the presence of ADAs or CD28 protein, the soluble target. An acid dissociation pretreatment step using 10% formic acid in water or 1 M acetic acid in 40% acetonitrile in water before PPT resulted in full recovery of the drug, even in the presence of up to 4000 ng/mL of the ADA and 500 ng/mL of CD28 protein.

Kushnir et al. (2013) developed an LC–MS/MS assay for the measurement of thyroglobulin (Tg), a protein biomarker, in serum/plasma in the presence of endogenous anti-Tg autoantibodies (Tg-AAb). A rabbit polyclonal anti-Tg antibody was used to convert all free Tg in serum/plasma samples into antibody-bound form. The antibody-bound Tg along with the AAb-bound Tg was then precipitated with saturated ammonium sulfate. The precipitates containing total Tg were digested, the generated digests were further processed with antipeptide antibody, and the purified target surrogate peptide was analyzed by LC–MS/MS. The developed method achieved a good sensitivity at an LLOQ of 0.5 ng/mL (0.76 fmol/mL). The LC–MS/MS method showed good agreement with an available Beckman Access immunoassay for the quantification of Tg in samples free of Tg-AAb. In contrast, poor agreement was observed between the LC–MS/MS method and the immunoassay in Tg-AAb-positive samples, especially for samples with Tg at low concentrations, suggesting that the presence of Tg-AAb affected the immunoassay for the measurements of Tg but not the LC–MS/MS method.

15.5 Protein Digestion Strategies

Protein digestion is a critical sample preparation step to ensure the reproducible, accurate, and sensitive quantitation of proteins. The use of enzymes is the most commonly applied method for protein digestion. Various enzymes (trypsin, Lys-C, Asp-N, Glu-C, etc.) have their own characteristics specificity, efficiency, and digestion conditions, and can be applied for protein digestion in different situations (Switzar et al. 2013). Trypsin is the most commonly used due to its high specificity, good efficiency, and relatively low cost. Other enzymes can also be used, especially in specific cases when no suitable surrogate peptide can be identified in the tryptic digest due to the lack or overabundance of Lys and Arg in the protein sequence. To achieve more complete and consistent digestion, digestion conditions, including incubation buffer, time, temperature, and enzyme-to-protein ratio, need to be optimized. Various strategies, such as elevated digestion temperature and organic solvent, ultrasound, or microwave-assisted digestion, have been evaluated to accelerate the digestion process (Switzar et al. 2013).

Traditionally, a pretreatment, such as sequential denaturation, reduction, and alkylation, is applied prior to the actual enzymatic digestion. The pretreatment can help unfold the proteins and improve the accessibility of the digestion enzyme; therefore, the digestion efficiency and completeness is enhanced. However, the pretreatment is time-consuming and labor-intensive. The reagents used in the pretreatment (e.g., guanidine, urea, dithiothreitol (DTT), and iodoacetamide) may interfere with later digestion or LC–MS analysis. Ouyang et al. (2012) reported a simple, fast, and efficient "pellet digestion" methodology for digestion of large proteins. Organic solvents were used to precipitate plasma/serum sample. The protein pellet was separated by centrifugation and reconstituted into digestion buffer and digested with trypsin. Pellet digestion eliminated the time-consuming reduction and alkylation steps, and thus, the digestion throughput significantly improved. Pellet digestion was compared with traditional digestion-with-pretreatment or direct digestion methods using a test mAb (Yuan et al. 2012). Digestion efficiencies of these methods were evaluated based on the digestion yield of three surrogate peptides chosen from different regions of the antibody: peptide VVSV from the heavy chain Fc region, peptide SLIY from the light-chain complementary determining regions (CDRs), and peptide DIYY from the heavy-chain CDRs. Compared to direct digestion, pellet digestion achieved much better digestion efficiency for all of the three surrogate peptides, especially for those in the hard-to-digest regions of the protein (e.g., the digestion yield improved by ~30-fold for peptides SLIY and DIYY). More importantly, pellet digestion also provided similar or better digestion efficiency compared to the traditional digestion-with-pretreatment method for the test protein, including regions resistant to trypsin under direct digestion. Pellet digestion can also be used in combination with other sample pretreatment methods (e.g., reduction and alkylation) to further improve the digestion efficiency (Jiang et al. 2013, Mekhssian et al. 2014, Shen et al. 2015).

It should be noted that for quantitative bioanalysis, only one (or very few) surrogate peptide is required to be monitored. This is the fundamental difference compared to the qualitative work, which requires the detection of as many peptides as possible to obtain the most complete sequence coverage of the protein. Considering this difference, conventional pretreatment may not be necessary. It is worthwhile to evaluate pellet digestion or other digestion methods without pretreatment first to see if they can achieve a reliable and efficient digestion to produce the surrogate peptide.

Digestion using chemicals (e.g., acids, cyanogen bromide, 2-nitro-5-thiocyanobenzoate, and hydroxylamine), though less common, can be used as an alternative to enzymatic digestion (Switzar et al. 2013). For example, Fung et al. (2014) found that the aspartic acid-proline (Asp-Pro) amide bond is acid-labile and can be selectively cleaved in low-pH formic acid solution. They investigated various acid hydrolysis conditions, including the concentration of formic acid, and the temperature and duration of incubation time, for FGF21-AdPKE, a fusion protein. Using 2% formic acid at 90 °C for 2 h, selective hydrolysis of aspartyl–prolyl amide bonds was achieved with an efficiency of ~100%. This method was applied for the quantitative bioanalysis of FGF21-AdPKE in serum samples from a monkey toxicokinetic study.

Enzymes such as IdeS (immunoglobulin-degrading enzyme of *Streptococcus pyogenes*) (Chevreux et al. 2011, An et al. 2014, Fornelli et al. 2014), Lys-C (Gadgil et al. 2006) and papain (Adamczyk et al. 2000) have been applied to the limited proteolysis of IgG molecules at the hinge region to generate intact Fab and Fc fragments. These fragments can be further reduced to antibody domains (including light-chain, monomeric Fc (Fc/2), and Fd) of ~25 kDa, and then analyzed by LC–MS (the so-called middle-up approach). Compared to Lys-C and papain, IdeS has the advantages of high specificity, good yield, and simple digestion procedure (von Pawel-Rammingen et al. 2002, Vincents et al. 2004). IdeS also has a wider application as it can cleave all IgG subclasses at the hinge region, as well as Fc fusion proteins (Lynaugh et al. 2013, An et al. 2014).

15.6. Conclusion

LC–MS-based assays are rapidly gaining momentum and acceptance for quantitative bioanalysis of proteins, including protein therapeutics, protein biomarkers, and soluble targets. Progress made in sample preparation methods has played a critical role in this process. Combined with the continuous improvement in LC and

MS instrumentation, the sensitivity of LC–MS assays has been tremendously improved to be similar to or, in some cases, better sensitivity than LBAs. Among all sample preparation strategies described herein, immunocapture shows significant benefits and the potential for wider applications. The combination of immunocapture with LC–MS can greatly improve the assay sensitivity, as well as the specificity. By the use of different immunocapture reagents, it also enables LC–MS assays to differentiate the free, bound, and total proteins, therefore, to a great extent, broadening the applicability of LC–MS assays in protein bioanalysis.

Acknowledgment

The authors would like to thank Drs Mark E. Arnold and Anne-Françoise Aubry for their insightful comments and helpful review of this book chapter.

References

Adamczyk M, Gebler JC and Wu J (2000). "Papain digestion of different mouse IgG subclasses as studied by electrospray mass spectrometry." *J Immunol Methods* **237**(1–2): 95–104.

Addona TA, Abbatiello SE, Schilling B, et al. (2009). "Multi-site assessment of the precision and reproducibility of multiple reaction monitoring-based measurements of proteins in plasma." *Nat Biotechnol* **27**(7): 633–641.

An B, Zhang M, Johnson RW, et al. (2015). "Surfactant-aided precipitation/on-pellet-digestion (SOD) procedure provides robust and rapid sample preparation for reproducible, accurate and sensitive LC/MS quantification of therapeutic protein in plasma and tissues." *Anal Chem* **87**(7): 4023–4029.

An Y, Zhang Y, Mueller H-M, et al. (2014). "A new tool for monoclonal antibody analysis." *mAbs* **6**(4): 879–893.

Anderson NL and Anderson NG (2002). "The human plasma proteome: history, character, and diagnostic prospects." *Mol Cell Proteomics* **1**(11): 845–867.

Anderson NL, Anderson NG, Haines LR, et al. (2004). "Mass spectrometric quantitation of peptides and proteins using stable isotope standards and capture by anti-peptide antibodies (SISCAPA)." *J Proteome Res* **3**(2): 235–244.

Becher F, Pruvost A, Clement G, et al. (2006). "Quantification of small therapeutic proteins in plasma by liquid chromatography–tandem mass spectrometry: application to an elastase inhibitor EPI-hNE4." *Anal Chem* **78**(7): 2306–2313.

Berna M, Schmalz C, Duffin K, et al. (2006). "Online immunoaffinity liquid chromatography/tandem mass spectrometry determination of a type II collagen peptide biomarker in rat urine: investigation of the impact of collision-induced dissociation fluctuation on peptide quantitation." *Anal Biochem* **356**(2): 235–243.

Berna MJ, Zhen Y, Watson DE, et al. (2007). "Strategic use of immunoprecipitation and LC/MS/MS for trace-level protein quantification: myosin light chain 1, a biomarker of cardiac necrosis." *Anal Chem* **79**(11): 4199–4205.

Björhall K, Miliotis T and Davidsson P (2005). "Comparison of different depletion strategies for improved resolution in proteomic analysis of human serum samples." *Proteomics* **5**(1): 307–317.

Bronsema KJ, Bischoff R, Bouche M-P, et al. (2015a). "High-sensitivity quantitation of a Nanobody® in plasma by single-cartridge multidimensional SPE and ultra-performance LC–MS/MS." *Bioanalysis* **7**(1): 53–64.

Bronsema KJ, Bischoff R, Pijnappel WWMP, et al. (2015b). "Absolute quantification of the total and antidrug antibody-bound concentrations of recombinant human α-glucosidase in human plasma using protein G extraction and LC-MS/MS." *Anal Chem* **87**(8): 4394–4401.

Cellar NA, Karnoup AS, Albers DR, et al. (2009). "Immunodepletion of high abundance proteins coupled on-line with reversed-phase liquid chromatography: a two-dimensional LC sample enrichment and fractionation technique for mammalian proteomics." *J Chromatogr B* **877**(1–2): 79–85.

Chen Y-T, Tuan L-P, Chen H-W, et al. (2015). "Quantitative analysis of prostate specific antigen isoforms using immunoprecipitation and stable isotope labeling mass spectrometry." *Anal Chem* **87**(1): 545–553.

Chevreux G, Tilly N and Bihoreau N (2011). "Fast analysis of recombinant monoclonal antibodies using IdeS proteolytic digestion and electrospray mass spectrometry." *Anal Biochem* **415**(2): 212–214.

Cingöz A, Hugon-Chapuis F and Pichon V (2010). "Total on-line analysis of a target protein from plasma by immunoextraction, digestion and liquid chromatography–mass spectrometry." *J Chromatogr B* **878**(2): 213–221.

Dawes ML, Gu H, Wang J, et al. (2013). "Development of a validated liquid chromatography tandem mass spectrometry assay for a PEGylated adnectin in cynomolgus monkey plasma using protein precipitation and trypsin digestion." *J Chromatogr B* **934**: 1–7.

Deng R, Jin F, Prabhu S, et al. (2012). "Monoclonal antibodies: what are the pharmacokinetic and pharmacodynamic considerations for drug development?" *Expert Opin Drug Metab Toxicol* **8**(2): 141–160.

DeSilva B and Garofolo F (2014). "Matrix interference in ligand-binding assays: challenge or solution?" *Bioanalysis* **6**(8): 1029–1031.

Desilva B, Smith W, Weiner R, et al. (2003). "Recommendations for the bioanalytical method validation of ligand-binding assays to support pharmacokinetic assessments of macromolecules." *Pharm Res* **20**(11): 1885–1900.

Duan X, Abuqayyas L, Dai L, et al. (2012a). "High-throughput method development for sensitive, accurate, and reproducible quantification of therapeutic monoclonal antibodies in tissues using orthogonal array optimization and nano liquid chromatography/selected reaction monitoring mass spectrometry." *Anal Chem* **84**(10): 4373–4382.

Duan X, Dai L, Chen S-C, et al. (2012b). "Nano-scale liquid chromatography/mass spectrometry and on-the-fly orthogonal array optimization for quantification of therapeutic monoclonal antibodies and the application in preclinical analysis." *J Chromatogr A* **1251**: 63–73.

Dubois M, Fenaille F, Clement G, et al. (2008). "Immunopurification and mass spectrometric quantification of the active form of a chimeric therapeutic antibody in human serum." *Anal Chem* **80**(5): 1737–1745.

Dufield DR, Nemirovskiy OV, Jennings MG, et al. (2010). "An immunoaffinity liquid chromatography–tandem mass spectrometry assay for detection of endogenous aggrecan fragments in biological fluids: use as a biomarker for aggrecanase activity and cartilage degradation." *Anal Biochem* **406**(2): 113–123.

Ezan E and Bitsch F (2009). "Critical comparison of MS and immunoassays for the bioanalysis of therapeutic antibodies." *Bioanalysis* **1**(8): 1375–1388.

Fernández Ocaña M, James IT, Kabir M, et al. (2012). "Clinical pharmacokinetic assessment of an anti-MAdCAM monoclonal antibody therapeutic by LC-MS/MS." *Anal Chem* **84**(14): 5959–5967.

Foettinger A, Leitner A and Lindner W (2006). "Derivatisation of arginine residues with malondialdehyde for the analysis of peptides and protein digests by LC-ESI-MS/MS." *J Mass Spectrom* **41**(5): 623–632.

Fornelli L, Ayoub D, Aizikov K, et al. (2014). "Middle-down analysis of monoclonal antibodies with electron transfer dissociation Orbitrap Fourier transform mass spectrometry." *Anal Chem* **86**(6): 3005–3012.

Fung EN, Zambito F, Haulenbeek J, et al. (2014). "Targeting an acid labile aspartyl–prolyl amide bond as a viable alternative to trypsin digestion to generate a surrogate peptide for LC–MS/MS analysis." *Bioanalysis* **6**(22): 2985–2998.

Furlong MT, Ouyang Z, Wu S, et al. (2012). "A universal surrogate peptide to enable LC-MS/MS bioanalysis of a diversity of human monoclonal antibody and human Fc-fusion protein drug candidates in pre-clinical animal studies." *Biomed Chromatogr* **26**(8): 1024–1032.

Gadgil HS, Bondarenko PV, Pipes GD, et al. (2006). "Identification of cysteinylation of a free cysteine in the Fab region of a recombinant monoclonal IgG1 antibody using Lys-C limited proteolysis coupled with LC/MS analysis." *Anal Biochem* **355**(2): 165–174.

Gong C, Zeng J, Akinsanya B, et al. (2014). "Development and validation of an LC–MS/MS assay for the quantitation of a PEGylated anti-CD28 domain antibody in human serum: overcoming interference from antidrug antibodies and soluble target." *Bioanalysis* **6**(18): 2371–2383.

Hagman C, Ricke D, Ewert S, et al. (2008). "Absolute quantification of monoclonal antibodies in biofluids by liquid chromatography–tandem mass spectrometry." *Anal Chem* **80**(4): 1290–1296.

Heinig K and Wirz T (2009). "Determination of taspoglutide in human and animal plasma using liquid chromatography–tandem mass spectrometry with orthogonal column-switching." *Anal Chem* **81**(10): 3705–3713.

Heinig K, Wirz T, Schick E, et al. (2013). "Bioanalysis of therapeutic peptides: differentiating between total and anti-drug antibody bound drug using liquid chromatography–tandem mass spectrometry quantitation." *J Chromatogr A* **1316**: 69–77.

Heudi O, Barteau S, Zimmer D, et al. (2008). "Towards absolute quantification of therapeutic monoclonal antibody in serum by LC-MS/MS using isotope-labeled antibody standard and protein cleavage isotope dilution mass spectrometry." *Anal Chem* **80**(11): 4200–4207.

Hoofnagle AN and Wener MH (2009). "The fundamental flaws of immunoassays and potential solutions using tandem mass spectrometry." *J Immunol Methods* **347**(1–2): 3–11.

Hoos JS, Sudergat H, Hoelck J-P, et al. (2006). "Selective quantitative bioanalysis of proteins in biological fluids by on-line immunoaffinity chromatography–protein digestion–liquid chromatography–mass spectrometry." *J Chromatogr B* **830**(2): 262–269.

Ji QC, Rodila R and El-Shourbagy TA (2007). "A sample preparation process for LC–MS/MS analysis of total protein drug concentrations in monkey plasma samples with antibody." *J Chromatogr B* **847**(2): 133–141.

Ji QC, Rodila R, Gage EM, et al. (2003). "A strategy of plasma protein quantitation by selective reaction monitoring of an intact protein." *Anal Chem* **75**(24): 7008–7014.

Jiang H, Zeng J, Titsch C, et al. (2013). "Fully validated LC-MS/MS assay for the simultaneous quantitation of coadministered therapeutic antibodies in cynomolgus monkey serum." *Anal Chem* **85**(20): 9859–9867.

Kelley M and DeSilva B (2007). "Key elements of bioanalytical method validation for macromolecules." *AAPS J* **9**(2): E156–E163.

Kuhn E, Addona T, Keshishian H, et al. (2009). "Developing multiplexed assays for troponin I and interleukin-33 in plasma by peptide immunoaffinity enrichment and targeted mass spectrometry." *Clin Chem* **55**(6): 1108–1117.

Kushnir MM, Rockwood AL, Roberts WL, et al. (2013). "Measurement of thyroglobulin by liquid chromatography–tandem mass spectrometry in serum and plasma in the presence of antithyroglobulin autoantibodies." *Clin Chem* **59**(6): 982–990.

Leader B, Baca QJ and Golan DE (2008). "Protein therapeutics: a summary and pharmacological classification." *Nat Rev Drug Discov* 7(1): 21–39.

Li W, Nemirovskiy O, Fountain S, et al. (2007). "Clinical validation of an immunoaffinity LC–MS/MS assay for the quantification of a collagen type II neoepitope peptide: a biomarker of matrix metalloproteinase activity and osteoarthritis in human urine." *Anal Biochem* **369**(1): 41–53.

Li H, Ortiz R, Tran L, et al. (2012). "General LC-MS/MS method approach to quantify therapeutic monoclonal antibodies using a common whole antibody internal standard with application to preclinical studies." *Anal Chem* **84**(3): 1267–1273.

Li H, Ortiz R, Tran LB, et al. (2013). "Simultaneous analysis of multiple monoclonal antibody biotherapeutics by LC-MS/MS method in rat plasma following cassette-dosing." *AAPS J* **15**(2): 337–346.

Li H, Rose M, Holder J, et al. (2011). "Direct quantitative analysis of a 20 kDa PEGylated human calcitonin gene peptide antagonist in cynomolgus monkey serum using in-source CID and UPLC-MS/MS." *J Am Soc Mass Spectrom* **22**(9): 1660–1667.

Liu T, Hossain M, Schepmoes AA, et al. (2012). "Analysis of serum total and free PSA using immunoaffinity depletion coupled to SRM: correlation with clinical immunoassay tests." *J Proteome* **75**(15): 4747–4757.

Liu G, Ji QC, Dodge R, et al. (2013). "Liquid chromatography coupled with tandem mass spectrometry for the bioanalysis of proteins in drug development: practical considerations in assay development and validation." *J Chromatogr A* **1284**: 155–162.

Liu G, Zhao Y, Angeles A, et al. (2014). "A novel and cost effective method of removing excess albumin from plasma/serum samples and its impacts on LC-MS/MS bioanalysis of therapeutic proteins." *Anal Chem* **86**(16): 8336–8343.

Lu Q, Zheng X, McIntosh T, et al. (2009). "Development of different analysis platforms with LC–MS for pharmacokinetic studies of protein drugs." *Anal Chem* **81**(21): 8715–8723.

Lynaugh H, Li H and Gong B (2013). "Rapid Fc glycosylation analysis of Fc fusions with IdeS and liquid chromatography mass spectrometry." *mAbs* **5**(5): 641–645.

Madsen JA and Brodbelt JS (2009). "Simplifying fragmentation patterns of multiply charged peptides by N-terminal derivatization and electron transfer collision activated dissociation." *Anal Chem* **81**(9): 3645–3653.

Mekhssian K, Mess J-N and Garofolo F (2014). "Application of high-resolution MS in the quantification of a therapeutic monoclonal antibody in human plasma." *Bioanalysis* **6**(13): 1767–1779.

Neubert H, Gale J and Muirhead D (2010). "Online high-flow peptide immunoaffinity enrichment and nanoflow LC-MS/MS: assay development for total salivary pepsin/pepsinogen." *Clin Chem* **56**(9): 1413–1423.

Neubert H, Muirhead D, Kabir M, et al. (2012). "Sequential protein and peptide immunoaffinity capture for mass spectrometry-based quantification of total human β-nerve growth factor." *Anal Chem* **85**(3): 1719–1726.

Ocaña MF and Neubert H (2010). "An immunoaffinity liquid chromatography–tandem mass spectrometry assay for the quantitation of matrix metalloproteinase 9 in mouse serum." *Anal Biochem* **399**(2): 202–210.

Oe T, Ackermann BL, Inoue K, et al. (2006). "Quantitative analysis of amyloid β peptides in cerebrospinal fluid of Alzheimer's disease patients by immunoaffinity purification and stable isotope dilution liquid chromatography/negative electrospray ionization tandem mass spectrometry." *Rapid Commun Mass Spectrom* **20**(24): 3723–3735.

Ouyang Z, Furlong MT, Wu S, et al. (2012). "Pellet digestion: a simple and efficient sample preparation technique for LC–MS/MS quantification of large therapeutic proteins in plasma." *Bioanalysis* **4**(1): 17–28.

Palandra J, Finelli A, Zhu M, et al. (2013). "Highly specific and sensitive measurements of human and monkey interleukin 21 using sequential protein and tryptic peptide immunoaffinity LC-MS/MS." *Anal Chem* **85**(11): 5522–5529.

Rafalko A, Iliopoulos O, Fusaro VA, et al. (2010). "Immunoaffinity enrichment and liquid chromatography-selected reaction monitoring mass spectrometry for quantitation of carbonic anhydrase 12 in cultured renal carcinoma cells." *Anal Chem* **82**(21): 8998–9005.

Ruan Q, Ji QC, Arnold ME, et al. (2011). "Strategy and its implications of protein bioanalysis utilizing high-resolution mass spectrometric detection of intact protein." *Anal Chem* **83**(23): 8937–8944.

Sandra K, Moshir M, D'Hondt F, et al. (2008). "Highly efficient peptide separations in proteomics. Part 1.

Unidimensional high performance liquid chromatography." *J Chromatogr B Anal Technol Biomed Life Sci* **866**(1-2): 48–63.

Shen Y, Zhang G, Yang J, et al. (2015). "Online 2D-LC-MS/MS assay to quantify therapeutic protein in human serum in the presence of pre-existing antidrug antibodies." *Anal Chem* **87**(16): 8555–8563.

Sleczka BG, Mehl JT, Shuster DJ, et al. (2014). "Quantification of human mAbs in mouse tissues using generic affinity enrichment procedures and LC–MS detection." *Bioanalysis* **6**(13): 1795–1811.

Suh M-J, Clark DJ, Parmer PP, et al. (2010). "Using chemical derivatization and mass spectrometric analysis to characterize the post-translationally modified *Staphylococcus aureus* surface protein G." *Biochim Biophys Acta Proteins Proteomics* **1804**(6): 1394–1404.

Switzar L, Giera M and Niessen WMA (2013). "Protein digestion: an overview of the available techniques and recent developments." *J Proteome Res* **12**(3): 1067–1077.

Tang K, Page JS and Smith RD (2004). "Charge competition and the linear dynamic range of detection in electrospray ionization mass spectrometry." *J Am Soc Mass Spectrom* **15**(10): 1416–1423.

Torsetnes SB, Levernæs MS, Broughton MN, et al. (2014). "Multiplexing determination of small cell lung cancer biomarkers and their isovariants in serum by immunocapture LC-MS/MS." *Anal Chem* **86**(14): 6983–6992.

van Midwoud PM, Rieux L, Bischoff R, et al. (2007). "Improvement of recovery and repeatability in liquid chromatography–mass spectrometry analysis of peptides." *J Proteome Res* **6**(2): 781–791.

Vincents B, von Pawel-Rammingen U, Björck L, et al. (2004). "Enzymatic characterization of the streptococcal endopeptidase, IdeS, reveals that it is a cysteine protease with strict specificity for IgG cleavage due to exosite binding." *Biochemistry* **43**(49): 15540–15549.

von Pawel-Rammingen U, Johansson BP, and Björck L (2002). "IdeS, a novel streptococcal cysteine proteinase with unique specificity for immunoglobulin G." *EMBO J* **21**(7): 1607–1615.

Wang S, Wu S, Gokemeijer J, et al. (2012). "Attribution of the discrepancy between ELISA and LC-MS/MS assay results of a PEGylated scaffold protein in post-dose monkey plasma samples due to the presence of anti-drug antibodies." *Anal Bioanal Chem* **402**(3): 1229–1239.

Whiteaker JR, Zhang H, Eng JK, et al. (2007a). "Head-to-head comparison of serum fractionation techniques." *J Proteome Res* **6**(2): 828–836.

Whiteaker JR, Zhao L, Anderson L, et al. (2010). "An automated and multiplexed method for high throughput peptide immunoaffinity enrichment and multiple reaction monitoring mass spectrometry-based quantification of protein biomarkers." *Mol Cell Proteomics* **9**(1): 184–196.

Whiteaker JR, Zhao L, Zhang HY, et al. (2007b). "Antibody-based enrichment of peptides on magnetic beads for mass-spectrometry-based quantification of serum biomarkers." *Anal Biochem* **362**(1): 44–54.

Wilm M and Mann M (1996). "Analytical properties of the nanoelectrospray ion source." *Anal Chem* **68**(1): 1–8.

Winther B, Nordlund M, Paus E, et al.(2009). "Immuno-capture as ultimate sample cleanup in LC-MS/MS determination of the early stage biomarker ProGRP." *J Sep Sci* **32**(17): 2937–2943.

Wu ST, Ouyang Z, Olah TV, et al. (2011). "A strategy for liquid chromatography/tandem mass spectrometry based quantitation of pegylated protein drugs in plasma using plasma protein precipitation with water-miscible organic solvents and subsequent trypsin digestion to generate surrogate peptides for detection." *Rapid Commun Mass Spectrom* **25**(2): 281–290.

Xu K, Liu L, Maia M, et al. (2014). "A multiplexed hybrid LC–MS/MS pharmacokinetic assay to measure two co-administered monoclonal antibodies in a clinical study." *Bioanalysis* **6**(13): 1781–1794.

Xu Y, Mehl JT, Bakhtiar R, et al. (2010). "Immunoaffinity purification using anti-PEG antibody followed by Two-dimensional liquid chromatography/tandem mass spectrometry for the quantification of a PEGylated therapeutic peptide in human plasma." *Anal Chem* **82**(16): 6877–6886.

Xu Y, Zhang L, Lu H, et al. (2008). "Mass spectrometry analysis of phosphopeptides after peptide carboxy group derivatization." *Anal Chem* **80**(21): 8324–8328.

Yang Z, Hayes M, Fang X, et al. (2007). "LC-MS/MS approach for quantification of therapeutic proteins in plasma using a protein internal standard and 2D-solid-phase extraction cleanup." *Anal Chem* **79**(24): 9294–9301.

Yang Z, Ke J, Hayes M, , et al. (2009). "A sensitive and high-throughput LC-MS/MS method for the quantification of pegylated-interferon-Î±2a in human serum using monolithic C18 solid phase extraction for enrichment." *J Chromatogr B Anal Technol Biomed Life Sci* **877**(18-19): 1737–1742.

Yuan L, Arnold ME, Aubry A-F, et al. (2012). "Simple and efficient digestion of a monoclonal antibody in serum using pellet digestion: comparison with traditional digestion methods in LC–MS/MS bioanalysis." *Bioanalysis* **4**(24): 2887–2896.

Yuan L, Aubry A-F, Arnold ME, et al. (2013). "Systematic investigation of orthogonal SPE sample preparation for the LC–MS/MS bioanalysis of a monoclonal antibody after pellet digestion." *Bioanalysis* **5**(19): 2379–2391.

Yuan L, Mai A, Aubry A-F, et al. (2014). "Feasibility assessment of a novel selective peptide derivatization strategy for sensitivity enhancement for the liquid chromatography/tandem mass spectrometry bioanalysis of protein therapeutics in serum." *Rapid Commun Mass Spectrom* **28**(7): 705–712.

Zhang YJ, Olah TV and Zeng J (2014c). "The integration of ligand binding and LC-MS-based assays into bioanalytical strategies for protein analysis." *Bioanalysis* **6**(13): 1827–1841.

Zhang Q, Spellman DS, Song Y, et al. (2014b). "Generic automated method for liquid chromatography–multiple reaction monitoring mass spectrometry based monoclonal antibody quantitation for preclinical pharmacokinetic studies." *Anal Chem* **86**(17): 8776–8784.

Zhang H, Xiao Q, Xin B, et al. (2014a). "Development of a highly sensitive liquid chromatography/tandem mass spectrometry method to quantify total and free levels of a target protein, interferon-gamma-inducible protein-10, at picomolar levels in human serum." *Rapid Commun Mass Spectrom* **28**(13): 1535–1543.

Zhao Y, Liu G, Angeles A, et al. (2015). "Development and validation of a liquid chromatography tandem mass spectrometry assay for the quantitation of a protein therapeutic in cynomolgus monkey serum." *J Chromatogr B* **988**: 81–87.

Zolotarjova N, Martosella J, Nicol G, et al. (2005). "Differences among techniques for high-abundant protein depletion." *Proteomics* **5**(13): 3304–3313.

16

Characterization of Protein Therapeutics by Mass Spectrometry

Wei Wu, Hangtian Song, Thomas Slaney, Richard Ludwig, Li Tao, and Tapan Das

Molecular and Analytical Development, Biologics Development, Bristol-Myers Squibb Company, USA

16.1 Introduction

The landscape of the pharmaceutical industry has shifted from small molecules toward protein therapeutics in the past two decades. This transition has been driven mainly by the unique pharmaceutical properties of biomolecules – mainly high specificity and low toxicity – that allow protein therapeutics to move through the development pipeline more rapidly than small-molecule drug candidates. More recently, the revolutionary breakthrough in immuno-oncology to fight various types of cancers has further strengthened monoclonal antibodies as the predominant platform to deliver novel therapeutics for treating severe diseases such as cancer and providing unmet medical needs.

As successful as protein therapeutics are in the clinical studies and in the market place, a protein therapeutic candidate has to go through long and thorough discovery and development stages to be developed into a viable commercial product. During these processes, concurrent to drug product being produced to support clinical studies, a consistent and economical manufacturing process is being developed. A mature manufacturing process for protein therapeutics includes sophisticated cell culture and protein purification processes, optimized formulation and storage condition, and appropriate analytical methods for release testing and stability monitoring. The foundation for these three components and overall protein therapeutics development is an extensive knowledge of the proteins' molecular characteristics, degradation pathways, and structure–function relationships. This information is obtained through the execution of comprehensive characterization studies. Compared to the small-molecule drug candidates, the complex nature of biomolecules and the intricate manufacturing processes result in a significant increase in the extent of heterogeneous variants. These variants are generated due to error or incomplete processing at any point in the protein biosynthesis (i.e., sequence variants, different glycosylation forms, N and C-terminal variants, and disulfide related isoforms). In addition, degradation such as oxidation, deamidation, isomerization, and fragmentation can occur throughout the lifetime of therapeutic proteins. The characterization of protein therapeutics is in essence characterization of these isoforms. Unfortunately, most protein variants cannot be categorized individually, since it is often not feasible to isolate individual variants for characterization. In addition, conformational changes in which all covalent bonds are intact, such as denaturation and aggregation, can have more significant and detrimental impacts on properties of protein therapeutics . The assurance of efficacy and safety of protein therapeutics is achieved by well-controlled manufacturing processes that are capable of generating highly consistent products. Ultimately, the protein therapeutics and their associated manufacturing process are qualified by the satisfactory clinical trials.

This chapter focuses on the common molecular variants and degradation pathways of protein therapeutics generated under normal manufacturing and storage conditions. Omitted in this section are the size variants such as high-molecular-weight species that are generated by aggregation. Furthermore, impurities such as host cell proteins, host cell DNAs, endotoxins, microbials, and leachables, although critically important for protein therapeutics development, are also out of the scope of this chapter.

16.2 Variants Associated with Cysteine/Disulfide Bonds in Protein Therapeutics

Disulfide bond formation is an important posttranslational modification that helps to stabilize protein conformation (Betz 1993). However, nonnative disulfide bond

Protein Analysis using Mass Spectrometry: Accelerating Protein Biotherapeutics from Lab to Patient, First Edition.
Edited by Mike S. Lee and Qin C. Ji.

formation as a result of disulfide bond reshuffling can lead to aggregation and loss of biological activity (Hwa et al. 1999). Nonnative disulfide bonds have been introduced into proteins through molecular biology techniques in an attempt to increase protein stability, with the aid of molecular modeling and *in silico* calculation (Craig and Dombkowski 2013). Disulfide bond formation is closely associated with protein folding pathways (Mamathambika and Bardwell 2008). The kinetics and sequence of oxidative disulfide bond formation in fully reduced ribonuclease A, a model protein with four disulfide bonds in its native state, have been studied extensively (Shin et al. 2003, Arai et al. 2010). During initial stages of oxidative refolding, nonnative disulfide bonds may form, but are ultimately replaced by the correctly paired disulfide bonds, leading to conformational energy minimization (Carty et al. 2002). The mismatched disulfide bonds can be restored to their native forms *in vivo* by protein disulfide isomerase (PDI), an enzyme in the endoplasmic reticulum (ER) with four thioredoxin-like domains (Hatahet and Ruddock 2007).

The disulfide bond is about ~2 Å in length. A distance criterion can be utilized for disulfide bond prediction in proteins with their 3D structures available. For proteins without available structures, a distance criterion of 8 Å between α carbon positions of the two cysteines can be used for determination of a potential disulfide bond when the proteins are mapped onto a homologous structure (O'Connor and Yeates 2004). This criterion has an accuracy of ~80% (Mallick et al. 2002). The connectivity of cysteines can also be experimentally determined through enzymatic digestion of the cystinyl protein and identification of the disulfide-linked peptides by Edman sequencing or mass spectrometry. In cases when cysteines are located in close proximity and cannot be isolated into separate proteolytic fragments, strategies based on chemical modification of the sulfhydryls can be implemented to determine the disulfide structures, such as partial reduction/alkylation (Gray 1993) and cyanylation-induced cleavages (Wu and Watson 1998, Wu et al. 1998, Qi et al. 2001).

The presence of disulfide bonds in proteins reduces peptide sequence coverage by MS/MS, as the fragments resulting from cleavages in between the two cysteines are still held together by the disulfide bond. During top-down analysis of intact IgG1, sequence coverage ranged from 25% to 32% by different precursor ion selection mechanisms (Mao et al. 2013). The reason of such a low coverage is partly due to the presence of disulfide bonds. The lack of peptide bond fragmentation between disulfide bonded cysteines plus the mass shift of −2 Da from formation of a disulfide bond have been used as criteria for disulfide bond determination of a 200 kDa protein by electron capture dissociation (ECD) top-down mass spectrometry (Han et al. 2006). Top-down analysis of smaller proteins, such as lysozyme, showed that by selecting a low charge (+9) state precursor, CID fragmentation could yield fragments with concurrent bond cleavages at a peptide bond and up to three disulfide bonds. At least four pairs of products were observed as a result of cleavages along the C–S–S–C linkage (Chen et al. 2010).

A more conventional way to analyze disulfide bond is through analysis of proteolytic fragments from a protein, especially the disulfide-linked peptides. Proteolytic digestion in ^{18}O water yields a signature isotope pattern that can be detected by high-resolution mass spectrometry for the disulfide-linked peptides in a complex peptide mixture, such as a pepsin digest, because disulfide-bond-linked peptides have two C-termini as opposed to one for a normal peptide (Gorman et al. 2002). During MALDI analysis of disulfide-linked peptides, the disulfide bond dissociates as a result of prompt fragmentation or in-source fragmentation, resulting in detection of two peptides as if the disulfide that links the two peptides were reduced (Patterson and Katta 1994). Under negative-mode CID dissociation, disulfide bond is preferentially cleaved while most of the peptide backbone remains intact. The dissociation at the disulfide bonds generates a product with reduced cysteine containing a free sulfhydryl group, and other products with mass shifts of −34, −2, and +32 Da. This particular spacing in masses of the gas-phase dissociation products has been used as a signature feature for detection of the disulfide-containing peptides (Zhang and Kaltashov 2006).

16.2.1 Thiolation Isoforms

Protein S-thiolation isoforms are formed through covalent linkages of protein thiols to protein molecules. S-thiolation has been shown to be responsible for regulating cellular redox status and nitrogen oxide mediated signal transduction (Ward et al. 2000, Tao and English 2004). Thiolation isoform could change protein functions and complicate protein purification through introduced heterogeneity (Melchers et al. 2007). Thiolation isoforms including adducts of glutathione and coenzyme A intermediates has been identified through mass spectrometric analysis of recombinant proteins secreted from *Escherichia coli* cells (Liu et al. 2009).

16.2.2 Disulfide Isoforms

Disulfide isoforms differ from each other in the ways the cysteines are connected. For a protein with four cysteines to form two disulfide bonds, there are three different isoforms as shown in Figure 16.1. The number of possible isoforms increases faster than exponentially with the number of disulfide bonds, as shown in Table 16.1. For a

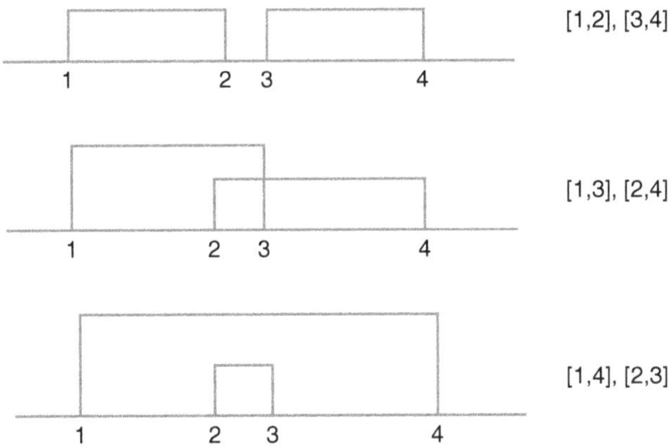

Figure 16.1 Three 2-disulfide isoforms.

Table 16.1 Number of disulfide isoforms and the number of disulfides.

Number of disulfides	Number of isoforms	Number of disulfide linkages
1	1	1
2	3	6
3	15	15
4	105	28
5	945	45
6	10,395	66
7	135,135	91
8	2,027,025	120

typical antibody with over 10 disulfide bonds, the number of theoretical disulfide isoforms is enormous. Most of the disulfide isoforms cannot be isolated due to their structural similarity.

The disulfide structure of a particular disulfide isoform can be denoted by its disulfide linkages, for example, the disulfide structure of the first disulfide isoform in Figure 16.1 can be denoted with [1,2], [3,4]. [*x*,*y*] represents the position of the cysteines that are involved in a disulfide bond. The number of disulfide linkages, on the other hand, is relatively small compared with the number of disulfide isoforms, as shown in Table 16.1. Software packages have been developed to assist assignment of disulfide linkages using MS/MS data (Huang et al. 2012a, Murad and Singh 2013).

The disulfide structures of immunoglobulin G (IgG) isotypes are similar due to the high sequence homology among them. The hinge region in IgG2 is special in that it has four cysteines, two of them adjacent to each other. Three different disulfide isoforms have been identified and their connectivity has been determined by mass spectrometry and N-terminal sequencing, as shown in Figure 16.2 (Martinez et al. 2008, Zhang et al. 2010). The difference in structure and conformation of the disulfide isoforms is large enough that they can be separated chromatographically (Dillon et al. 2008), electrophoretically (Guo et al. 2008), or by ion mobility mass spectrometry (Bagal et al. 2010). The high flexibility in the hinge region in IgG2 results in a slightly larger gas-phase collision cross section compared to an IgG1 (Bagal et al. 2010). This flexibility can also be probed by mass spectrometry analysis of the oxidized products in the hinge region by fast photochemical oxidation of proteins (FPOP) of different antibodies. Wild-type IgG2 showed a higher level of flexibility as evidenced by a higher level of oxidation products in the hinge region, compared to IgG1 and other IgG2 Cys-Ser mutants with reduced hinge flexibility (Jones et al. 2013). The difference in the heavy light interchain disulfide configurations in IgG1 and IgG4 is found responsible for the melting temperature of an IgG1 Fab domain to be up to 11 °C higher than

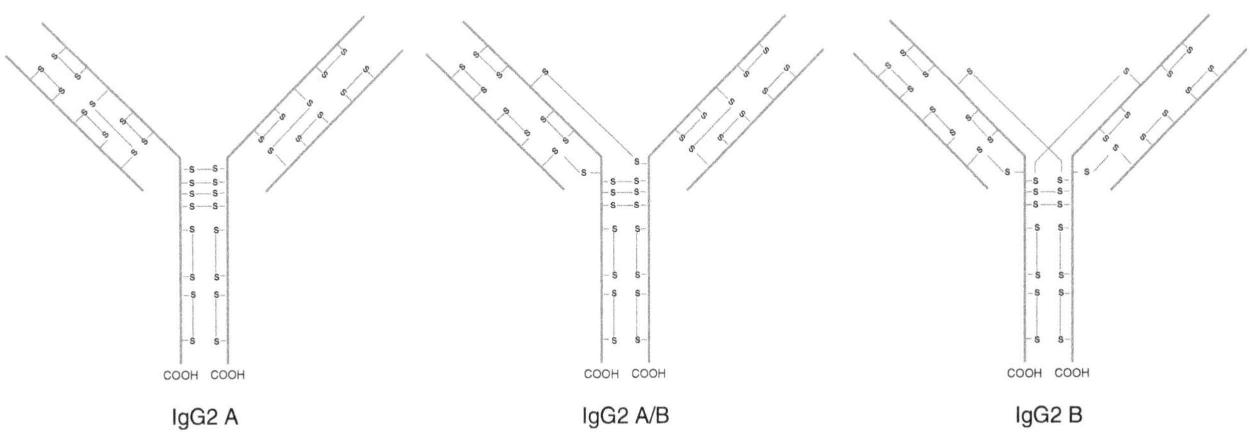

Figure 16.2 Three disulfide isoforms of IgG2.

the IgG4 Fab domain containing the same variable region (Heads et al. 2012).

16.2.3 Free Sulfhydryl

Usually, trace amounts of free thiols are present in antibody therapeutics and can be assayed by colorimetric assays using 5,5′-dithiobis-(2-nitrobenzoic acid) (DTNB). It was found under denaturing conditions, IgG1 and IgG2 contain 0.17–0.59 mol of free thiols per mole of antibody. An assay monitoring fluorescently labeled proteolytic fragments was used to map the majority of the free thiols to C_H1 domain (Huh et al. 2013).

Nonreduced capillary electrophoresis with sodium dodecyl sulfate (NR CE-SDS) is a useful tool to monitor dissociation of interchain disulfide bonds in antibodies. The relative areas associated with the prepeaks, containing light chain (L), heavy chain (H), and combinations of the two chains (HL, HH, HHL), and main peak (HHLL) can be used to monitor interchain disulfide reduction. It is reported that the susceptibility of products to reduction by thioredoxin and DTT depends on antibody class and light chain type in the order of IgG1λ > IgG1κ > IgG2λ > IgG2κ (Hutterer et al. 2013). One or two additional light chains have been observed to attach to an antibody (2H3L, 2H4L) through disulfide linkages (Lu et al. 2013).

During analysis of size variants in antibodies by nonreduced SDS-PAGE or CE-SDS, low-molecular-weight variants could be artificially generated during sample treatment. Alkylating agents have been used to mask free thiols and reduce method induced low-molecular-weight species during assay. *N*-ethyl maleimide was found to be a superior alkylating agent to iodoacetamide for this purpose (Zhu et al. 2013).

16.2.4 Thioether/Trisulfide Bond

The thioether modification on proteins (lanthionine) is the result of elimination of one of the sulfur atoms from a disulfide bond (—C—S—S—C—), producing a nonreducible thioether linkage (—C—S—C—). A variant similar in size as the low-molecular-weight species containing one heavy chain and one light chain was detected in the reducing capillary gel electrophoresis of an antibody (IgG1) and thioether bond in that species was confirmed by mass spectrometry (Tous et al. 2005). The same thioether bond between LC214-HC220 was found in an IgG1κ therapeutic antibody dosed in humans *in vivo* at a rate of about 0.1%/day in blood (Zhang et al. 2013a). Dehydroalanine through base-catalyzed β-elimination is proposed to be an intermediate for thioether formation as well as racemization of the cysteines at the antibody hinge region (Zhang and Flynn 2013).

Trisulfide bond was detected in antibody heavy/light and heavy/heavy interchain disulfide bonds (Gu et al. 2010). A possible pathway leading to the formation of this molecular variant is through a reaction between the native disulfide bonds and the H_2S gas generated during cell culture. Nonreduced peptide mapping with mass spectrometry detection can be used to detect and quantify the levels of trisulfide bond formation. It was found that the level of trisulfide bond correlates with the level of HHL in nonreduced CE-SDS (Aono et al. 2010). The level of trisulfide formation during cell culture has a strong positive correlation with cysteine levels in the feed media and can be controlled by lowering the cysteine level in the feed or switching to a cysteine-free feed during stationary phase of the culture (Kshirsagar et al. 2012). The trisulfide between the heavy and light chains in an IgG1was significantly reduced (from 13% to less than 1%) during purification of the IgG1 mAb via a cysteine wash step incorporated into Protein A affinity column chromatography. No disulfide scrambling or increase in free sulfhydryls was observed (Aono et al. 2010).

16.2.5 Disulfide Bond in Antibody Drug Conjugates

Conjugation of thioreactive small-molecule drugs to proteins through cysteine residues has been utilized to generate ADCs. The process involves a controlled partial reduction of interchain disulfide bonds, coupling of the drug to the nascent sulfhydryl groups and results in a heterogeneous mixture of ADCs that differ with respect to the site of conjugation, drug-to-antibody ratios (DARs), and the number of intact interchain disulfide bonds (Hamblett et al. 2004). A new generation of drug antibody linker was developed that takes advantage of the susceptibility of the interchain disulfide bonds to reduction. The maleimide-based linker would hold the interchain cysteines together through an S—C—C—S bond and deliver one drug per interchain disulfide bond (Figure 16.3, Schumacher et al. 2014).

To better control the DAR and to maintain native disulfide bonds, cysteine residues can be engineered into proteins as potential conjugation sites. Engineered free cysteine residues on the surface of a protein can form a dimer through disulfide bonds (Woo et al. 1991). It is also possible that introduced cysteines can form nonnative disulfide bonds through thiol disulfide bond exchange reaction with native disulfide bonds, resulting in disulfide bond scrambling and possibly protein inactivation (Wootton and Yoo 2003). The success of site-specific conjugation through introduction of cysteines requires a careful selection of proper sites that do not alter protein structure or function. The Phage ELISA-based selection of reactive thiols (PHESELECTOR) was developed by creating a phage library with reactive cysteine residues introduced into an antibody-Fab (trastuzumab-Fab 4D5)

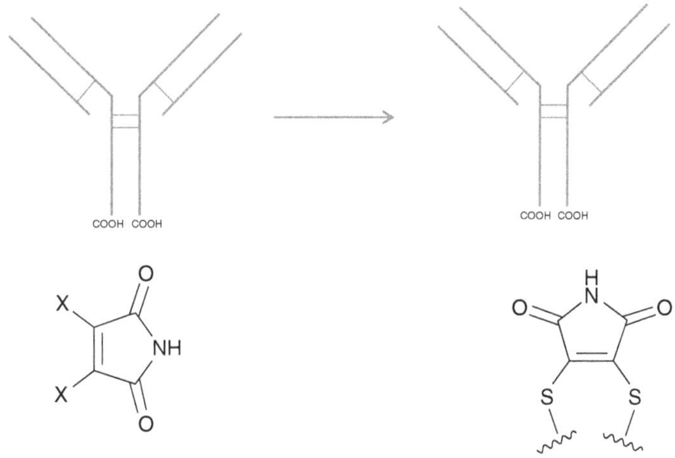
1, X = Br
2, X = PhS

Figure 16.3 Drug conjugation to antibody through a novel linker at antibody interchain disulfide positions.

at various sites and screening to identify the cysteine sites that do not negatively affect the antigen binding (Junutula et al. 2008).

Maleimide-based drug conjugation to antibody has been widely used due to compatibility with aqueous media and high levels of selectivity. The succinimide thioether product can undergo either hydrolysis or exchange with other thiols, through a retro-Michael reaction, to yield molecular variants *in vivo*, and thus alters the therapeutic efficacy and leads to undesired drug release and potential toxicity (Baldwin and Kiick 2011). Sulfone-based linkers to cysteine residues in antibody showed improved stability in human plasma at sites previously shown to be labile for maleimide conjugates (Cal et al. 2014, Patterson et al. 2014).

16.3 N–C-Terminal Variants

Pyroglutamic acid (pE) formation is a common modification at N-terminal residues of antibodies. The conversion from an N-terminal glutamine to pyroglutamic acid is spontaneous and near completion conversion (>95%) is usually observed for antibodies (Dick et al. 2007). An alternative, enzymatic approach of using glutaminyl-peptide cyclotransferase to convert the N-terminal glutamine into pyroglutamate was developed to drive the conversion to completion to simplify charge variant characterization (Xu et al. 2013). The conversion from glutamic acid to pyroglutamic acid at N-terminal residues occurs at a slower rate. The conversion rate from E to pyro E increases around 10-fold at the light chain N-terminus upon exposure to 6 M guanidine hydrochloride (from 0.00019 to 0.0022 day^{-1}), but the pE formation rate at heavy chain N-terminus remained relatively unchanged (k: 0.0025^{-1} day for native and 0.0027^{-1} day in 6 M guanidine HCl) (Liu et al. 2011). Besides pE formation at the N-terminal side, another type of N-terminal variant in antibodies results from incomplete processing of the leader sequence, a 20–30 amino acid peptide that binds to signal recognition particle, as a result of incomplete enzymatic cleavage from the signal peptidase (Kotia and Raghani 2010, Ambrogelly et al. 2012).

Additional N-terminal modification can arise from chemical modification with formulation components, such as citric acid (Chumsae et al. 2014), urea (Sun et al. 2014), or with feed media or metabolites in cell culture (Kim et al. 2001, Santora et al. 2006). Nonenzymatic cyclization of the first two amino acids to form a diketopiperazine at the N-terminus and subsequent cleavage has been reported in recombinant human growth hormone (rhGH), causing a truncated N-terminal variant (Battersby et al. 1994).

N-terminal methionine, which can be enzymatically removed by methionine aminopeptidase, serves as a prophylactic cap that prevents premature or inappropriate degradation through ubiquitin-dependent N-end rule pathway (Bradshaw et al. 1998). N-terminal methionine removal by methionine aminopeptidase is affected by the N-terminal penultimate residue, with a Q2A site directly mutagenesis facilitating the methionine removal (Humbard et al. 2009). The N-terminal penultimate residue also affects the level of N-(α)-acetylation and protein levels (Humbard et al. 2009).

The N-terminal peptides can be enriched and separated prior to mass spectrometric analysis through a two-step process termed combined fractional diagonal chromatography (COFRADIC). First, all the free amines, including the N-terminal amines, are modified by acetylation. The peptide mixture is then digested; all nascent N-terminal amines from the tryptic peptides are labeled with a highly hydrophobic reagent: 2,4,6-trinitrobenzenesulfonic acid

(TNBS). The acetylated, non-TNBS-labeled peptides from protein N-terminal side can be separated on a reversed-phase chromatography, from the TNBS-labeled tryptic peptides from the rest of the proteins (Gevaert et al. 2003). A similar approach, using Sulfo-NHS acetate (sulfosuccinimidyl acetate) to modify the N-terminal amines of proteins and NHS-agarose beads to enrich the modified peptides after tryptic digestion, has been used to enrich N-terminal peptides prior to MS analysis (Min et al. 2014).

All human IgGs contain a lysine residue at the C-terminus of the heavy chain. Lysine residue at the C-terminus of proteins is usually removed by cellular enzymes called carboxypeptidases. Typically, the removal is incomplete, resulting in two so-called lysine variants, with lysine on either one or both heavy chains. The level of residual C-terminal lysine on an antibody has a correlation with copper to zinc concentration ratio during cell culture from a Chinese hamster cell line (Luo et al. 2012). Variants with 0, 1, or 2 lysine had a similar higher-order structure as assayed by DSC (differential scanning calorimetry) and HDX (hydrogen/deuterium exchange) mass spectrometry (Tang et al. 2013). Lysine variants are not expected to affect efficacy or safety (Vlasak and Ionescu 2008). Similar processing occurs in natural human antibodies (Harris 1995). In addition, lysine removal was found to have no effect on CDC (Antes et al. 2007).

Common terminal variants in antibody drugs may not be considered as critical quality attributes (CQAs) compared to other posttranslational modifications such as glycosylation, due to their position at the end of the protein chain (Brorson and Jia 2014).

16.4 Glycation

Glycation is a nonenzymatic reaction of reducing sugar to proteins. Reducing sugars, such as glucose, can form an Amadori product with amine groups in proteins (N-terminus or lysine side chains) through a Schiff base intermediate. Spatial proximity of a potential glycated lysine to an acidic residue such as aspartic acid can help catalyze the Amadori rearrangement and promote the glycation product (Zhang et al. 2008). Further oxidation, dehydration, and cross-linking steps can then occur, such as through the formation of reactive dicarbonyl compounds (methylglyoxal) from glycolysis pathway. These latter compounds exhibit significantly enhanced reactivity for sites such as arginine and lysine residues on proteins, forming advanced glycation end products (AGEs).

Low levels of glycation can be detected during cell culture of antibody production and can be controlled by decreasing glucose concentration and modifying the feeding strategy (Yuk et al. 2011). The percentage of glycated species can be evaluated through boronate affinity chromatography, which separates the glycated species from the nonglycated proteins (Zhang et al. 2008).

Modifications of arginine residues by methylglyoxal lead to two adducts (dihydroxyimidazolidine and hydroimidazolone) with a mass shift of 72 and 54 Da, respectively. In addition, the modification by methylglyoxal causes the antibody to elute earlier in the weak cation exchange chromatogram (Chumsae et al. 2013).

Increased levels of high mannose structures has been linked to faster clearance of antibodies in blood, possibly through macrophage mannose receptor on mammalian cell surfaces (Goetze et al. 2011). Bovine serum albumin (BSA) had been used as a model system to study the effect of mannose on clearance. BSA derivatized by relatively high levels of 2-imino-2-methoxyethyl-1-thiomannoside (IMT-mannose) had been found to be cleared from blood faster in mice than the same protein with lower IMT-mannose levels (Opanasopit et al. 2001). To assess the effect of glycation of mannose on antibody clearance, samples with a low level of mannose glycation (1.4 mol of sugar/mol of mAb), and high level of mannose glycation (17.5 mol of sugar/mol of mAb), as well as a high level of IMT mannose modification (18.2 mol of sugar/mol of mAb) were prepared and subjected to mice clearance studies (Yang et al. 2015). The high and low mannose glycation samples showed similar clearance rates compared with the unmodified antibody, while the IMT mannose-modified antibody cleared much faster (13–17 times) than the unmodified antibody. It is hypothesized that the closed pyranose ring structure in the IMT-modified antibody, similar to that of the terminal mannose within N-linked glycan structures, is critical for binding to the mannose receptor, responsible for the antibody clearance (Yang et al. 2015).

Antibody glycated with glucose was found to aggregate at a faster rate than nonglycated antibody during stability evaluation (Banks et al. 2009). In another study, it was found that glycation by ribose on lysozyme does not induce structural changes. However, a decrease in pI and an increase in hydrophobicity contribute to the increased aggregation of lysozyme upon ribosylation (Adrover et al. 2014). The effect of glycation at particular lysines can be accessed through simulation and docking studies and a hindrance in enzymatic activities after glycation had been predicted in human paraoxonase (Saleem et al. 2012).

16.5 Oxidation

One of the most common chemical modifications observed on protein therapeutics is oxidation. Oxidation can occur at any point during manufacture and storage (Torosantucci et al. 2014), and stressors that may induce oxidation in protein therapeutics are numerous. These stressors include oxidizing agents (such as dissolved oxygen or peroxides), electroactive metals, light, and pH (Waterman et al. 2002, Manning et al. 2010). Since oxidation may affect drug efficacy and safety, proper process controls must be in place to minimize its occurrence (Wang et al. 2011, Gao et al. 2015, Torosantucci et al. 2014). Although any amino acid is a potential site of oxidation, it is most commonly observed on cysteine, histidine, methionine, phenylalanine, tryptophan, and tyrosine (Torosantucci et al. 2014).

16.5.1 Methionine Oxidation

One of the most labile targets to oxidation is methionine residues. Methionine oxidation is mostly pH-independent and able to proceed from a number of reactive oxygen species (ROS), including dissolved oxygen (Manning et al. 2010). Though the reaction rate may decrease at lower temperatures, solubility of dissolved oxygen also increases at lower temperatures, potentially increasing reactivity under refrigeration (Manning et al. 2010). Oxidation of methionine even occurs *in vivo* in endogenous proteins; thus enzymes, namely, the protein methionine sulfoxide reductases, exist that may reverse this oxidation (Oien and Moskovitz 2008).

Tertiary and quaternary structures are significant factors affecting the rate of methionine oxidation, as a buried residue is less likely to be oxidized than a solvent-exposed residue. As a result, oxidation is typically only observed at certain methionines on a protein. Certain methionines may have different physiological significance than others when oxidized. For example, a recent study of monoclonal antibody (mAb) methionine oxidation found that highly oxidized forced degradation products of an mAb exhibited a more severe loss of FcRn binding and shortened half-life than those degradation products occurring during prolonged storage (Wang et al. 2011). This was found to correlate with the different oxidation levels of two methionines on the Fc region of the antibody between the two test substances (Wang et al. 2011). To further study this effect, another study utilizing site-directed mutagenesis to compare the impact of oxidation at these two sites demonstrated that each methionine had a different impact on binding and aggregation of the mAb when oxidized (Gao et al. 2015). It is worth noting that while proteins may have well-protected methionines, methionines of smaller peptides are generally more susceptible to oxidation as they lack the higher-order structures present in proteins (Manning et al. 2010).

16.5.2 Metal-Catalyzed Oxidation (MCO)

MCO occurs when a redox active metal binds to a protein. Glycine, aspartate, histidine, and cysteine are known ligands of electroactive metals; however, only histidine and cysteine are the two that are susceptible to oxidation (Manning et al. 2010).

Histidine is known to have multiple oxidation products, although 2-oxo-histidine may be the most abundant (Manning et al. 2010). Histidine oxidation has been observed from copper(II) catalysis with ascorbic acid or hydrogen peroxide, but not from iron(II) or (III) when studied using parathyroid hormone as a model system (Ji et al. 2009). Copper(II) sulfate may be added to Chinese hamster ovary (CHO) cell cultures to improve disulfide bond formation, although this copper is not believed to be a major source of degradation (Chaderjian et al. 2005, Torosantucci et al. 2014). Consequences of 2-oxo-histidine formation may include aggregation, as has been observed in growth hormone and insulin (Torosantucci et al. 2014).

Addition of chelating agents such as ethylenediamine tetraacetic acid (EDTA) or diethylenetriamine pentaacetic acid (DTPA) has been shown to inhibit MCO, as well as to prevent conformational instability-related issues associated with bound metal ions (Zhou et al. 2010). While ascorbic acid may prevent some forms of oxidation, it is known to exacerbate MCO (Manning et al. 2010, Torosantucci et al. 2014).

16.5.3 Photooxidation

Photooxidation is a route of degradation in which photons are absorbed by photosensitive amino acids, exciting electrons and initiating oxidation. The amino acids susceptible to photooxidation include tryptophan, tyrosine, phenylalanine, and cysteine (Kerwin and Remmele 2007, Manning et al. 2010). Multiple oxidation products may exist for each amino acid, as described in-depth previously (Kerwin and Remmele 2007). Tryptophan oxidation has been associated with protein discoloration and loss of drug activity (Qi et al. 2009, Li et al. 2014). In addition, tryptophan that has been activated by UV light may be able to affect local amino acids, including cleavage of disulfide bonds, formation of a thioether linkage between cysteine and tryptophan, or backbone cleavage (Vanhooren et al. 2002, Kerwin and Remmele 2007, Manning et al. 2010). Tyrosine, after UV exposure, can generate a free radical, reacting with oxygen, cysteine, or other tyrosines, among other reactions (Kerwin and Remmele 2007). Phenylalanine will react to form tyrosine isomers but can also form additional products from free radical intermediates (Kerwin and Remmele 2007).

Photooxidation may also be exacerbated by excipients present within the formulated drug substance. Polysorbate has been known to contain reactive oxygen species, which can result in degradation of sample proteins (Wasylaschuk et al. 2007, Manning et al. 2010, Agarkhed et al. 2013). At higher concentrations, polysorbate was also found to increase the protein's susceptibility to photooxidation (Agarkhed et al. 2013).

16.5.4 Deamidation

Deamidation involves the loss of ammonia from the amide side chain of asparagine (as well as glutamine, to a lesser extent) (Liu et al. 2008, Manning et al. 2010). The mechanisms of deamidation have been explored in-depth elsewhere (Tonie Wright and Urry 1991, Catak et al. 2009). This degradation route is highly dependent on pH, as well as solvent accessibility of the asparagine. There are two routes for deamidation, dependent on pH (Tonie Wright and Urry 1991, Manning et al. 2010). Under acidic conditions (Ph < 4), the amide side chain of asparagine (or glutamine) is susceptible to direct hydrolysis, resulting in a loss of ammonia and formation of L-aspartic acid (or L-glutamate) (Tonie Wright and Urry 1991, Manning et al. 2010). At neutral or basic pH (>6), deamidation proceeds by the reaction of the amide with the backbone amine of the C-terminal adjacent amino acid, resulting in loss of an ammonium and formation of a cyclic succinimide intermediate (Tonie Wright and Urry 1991, Catak et al. 2009, Manning et al. 2010, Pace et al. 2013). This succinimide can then undergo nucleophilic attack from water to form aspartic or isoaspartic acid (Tonie Wright and Urry 1991, Catak et al. 2009, Manning et al. 2010, Pace et al. 2013). There is a 3:1 selectivity for the formation of isoaspartic:isoaspartic acid (Tonie Wright and Urry 1991, Cournoyer et al. 2007), and both D- and L- aspartic and isoaspartic acid are formed (Li et al. 2003, Manning et al. 2010).

Glutamine can also undergo deamidation, however, typically at a much slower rate compared to asparagine. This is because the six-member ring of the glutamyl-succinimide is energetically less favorable than the five-member ring of asparagyl-succinimide (Manning et al. 2010). Nevertheless, glutamine deamidation can sometimes occur to a significant extent on proteins, as has been observed on an IgG1 mAb at basic pH (Liu et al. 2008).

16.5.5 Effect of Sequence and Structure on Deamidation

The most common mechanism of deamidation (at neutral pH) is via the succinimide intermediate. As this mechanism requires reaction with the amide with the C-terminal adjacent amino acid, reactivity is highly sequence dependent. The two general trends are that C-terminal amino acids with smaller side chains and with proton-donating side chains are more reactive with the asparagine side chain (Manning et al. 2010). As such, the most reactive sequences for deamidation are Asn-Gly, Asn-Ser, and Asn-His (Manning et al. 2010). It should be noted that acid-catalyzed deamidation does not require reaction with this amide; therefore, the sequence has minimal effect on deamidation rates (Manning et al. 2010).

While primary sequence is an important factor, the rate of deamidation at a certain amide of a protein is highly dependent on the exposure of the amide to solution. If an asparagine is buried within a protein, then it is less likely to be deamidated than a surficial residue (Manning et al. 2010). Sydow and colleagues recently developed an *in silico* prediction algorithm for identifying deamidation sites and found that structural flexibility and small size of the side chain of the C-terminal amino acid were most indicative of reactivity (Sydow et al. 2014).

16.6 Discoloration

Visual appearance is an important quality attribute of therapeutic proteins including monoclonal antibodies (mAbs), fusion proteins, and other types of proteins and peptides. According to guidance from the United States (US) Food and Drug Administration (FDA), a qualitative statement describing the color is required in DS and DP specification (FDA 1999). There are also requirements on drug product color in European Pharmacopoeia (Europe 2012). Abnormality in appearance could be potentially associated with purity, stability, and other attributes, which may impact safety and efficacy. The current trend toward higher concentration formulations is gaining popularity due to the emerging market need of subcutaneous administration (Shire et al. 2004). Slight variations in product coloration become more noticeable in high-concentration formulations. Therefore, understanding and controlling abnormal color is critical in the development of protein therapeutic products within the pharmaceutical and biotechnology industries.

To sense the color of an object such as a vial of protein solution, which is not incandescent or luminescent, three components are essential: light source, object, and observer. When light emitted from the light source hits the sample, one portion is absorbed, and the rest is reflected or transmitted. Color may be perceived by the observer from the reflected or transmitted portion, which is the difference between light from the light source and light reflected or absorbed by the sample. If background light is white, then the perceived color is mostly dependent on the light reflected or absorbed by the sample. Human eye color perception of light originates mostly from sensing three primary colors, introduced by Thomas Young in 1790s. Primary colors are sensed by cone cells in retina, the discovered by Max Schultze in the 1860s. More specifically, three visual pigments in cones corresponding to primary colors are able to absorb photons within certain wavelength ranges (Wald 1964), which trigger signal cascade and transmit color information to the brain. Absorption spectra of pigments are similarly bell shaped (Wright 1929, Smith and Guild 1931). Overall sensitivity to visible light spectrum by human eyes is described by luminous efficiency function, combined from individual sensitivity functions of every primary color. Human cones are able to sense light with the wavelength ranging from 360 to 830 nm, and most sensitive to light around 500 nm (Institute of Ophthalmology, Stockman et al. 2008). In colorimetry, the perceived color could be quantified by mathematically transforming visible emission/absorption spectrum to tristimulus values defined by various metrics (color spaces) such as CIE XYZ, CIE RGB, and CIE Lab.

In general, the color of a protein solution can correlate with multiple sources, as discussed in the following text:

First, chromophores could form in amino acid degradation products. Common natural amino acids and peptide bonds are not colored, though some amino acids absorb light in UV region close to visible region. In aqueous solution, UV light is absorbed by tryptophan at 280–305 nm, by tyrosine at 260–290 nm, by phenolalanine at 240–270 nm, and by cysteine at 250–300 nm (Kerwin and Remmele 2007). However, modified amino acids could be colored. It was reported that certain tryptophan derivatives are yellow-brown, such as kynurenine (Kyn, wavelength of maximum absorbance at 360 nm (Fukunaga et al. 1982)), *N*-formylkynurenine (NFK, wavelength of maximum absorbance at 320 nm (Fukunaga et al. 1982)), and hydroxytryptophan (HO-Trp). Wavelengths of maximum absorbance of these tryptophan derivatives are close to or beyond the lower bound of visible region. However, peaks in UV–Vis spectrum of samples in aqueous solution are usually broadened, due to excitation/relaxation of rotational and vibrational states superposed to each electronic state transition. If protein concentration is high enough, it is possible that peaks in UV region extend to visible region. Kyn and NFK could be generated from tryptophan by light irradiation (Pirie 1971, Pirie 1972). Yellow discoloration of proteins was correlated to UV light irradiation in various proteins. For example, it was reported that browning of crystallin was associated with long-term sunlight exposure (Pirie 1971, Pirie 1972, Grossweiner 1984, Hood et al. 1999). Yellow discoloration upon UV light irradiation was also reported in wool proteins (Dyer et al. 2006) and therapeutic protein drug substances (Qi et al. 2009, Li et al. 2014, Xu et al. 2014). Upon absorbing UV light, the excited indole ring of a tryptophan residue could relax to a neutral radical by ejection of one electron and then one proton. The neutral radical could react with oxygen to form peroxy radical, which will further form NFK, Kyn, and other products after ring-opening (Bent and Hayon 1975, Creed 1984, Kerwin and Remmele 2007). NFK and Kyn could act as photosensitizers (Walrant and Santus 1974), generating reactive oxidation species (ROS) and further oxidize other parts of the protein. Antibodies have been reported to be able to generate oxidative singlet oxygen through surface tryptophan residues (Wentworth et al. 2000, Wentworth et al. 2001, Nieva and Wentworth, 2004, Sreedhara et al. 2012). Upon UV light activation, tryptophan could also be excited to triplet and could induce activation of disulfide bonds (Hoffman and Hayon 1972, Davies and Truscott 2001). Disulfide-bond scrambling could lead to protein aggregation and high-molecular-weight (HMW) species has been associated with yellow cataracts (Masters et al. 1978). Tryptophan photolysis was reported to be related to the location of tryptophan residues, with regard to exposure to aqueous solvent, neighboring residue and three-dimensional structure of the protein (Pigault and Gerard 1984). Besides UV irradiation, other oxidizing conditions could induce tryptophan oxidation such as peroxides (Simat and Steinhart 1998, Gracanin et al. 2009, Ji et al. 2009). Tryptophan oxidation could also be catalyzed in Fenton reaction involving transitional metals such as Fe (Uchida et al. 1990, Finley et al. 1998, Gracanin et al. 2009, Ji et al. 2009) and Cu (Cheng et al. 1992). In the presence of Fe, hydroxyl radical could form from peroxides that could originate from singlet oxygen (Stadtman 1993, Itakura et al. 1994, Gracanin et al. 2009), and oxidative hydroxyl radicals could further attack tryptophan residues. It should be noted that iron in the form of EDTA complex, often added to cell culture as nutrient, could still be active as catalyst for oxidation depending on the iron-to-EDTA ratio (Uchida et al. 1990, Stadtman and Berlett 1991). Products from histidine oxidation may contribute to yellow discoloration (Le Brun et al. 2010) too. Yellow discoloration was observed in histidine buffer after storage and discoloration was decelerated by adding EDTA (Le Brun et al. 2010). Similarly to tryptophan oxidation, histidine oxidation could be induced by UV irradiation (Tomita et al. 1969) or metal-catalyzed oxidations (Amici et al. 1989). Mass spectrometry has been widely applied for identification and localization of tryptophan and histidine oxidation in proteins, including therapeutic proteins (Kurahashi et al. 2001, Dyer et al. 2006, Wei et al. 2007, Yang et al. 2007, Ji et al. 2009, Qi et al. 2009, Boyd et al. 2011, Dreaden et al. 2011, Sreedhara et al. 2012, Vijayasankaran et al. 2013, Amano et al. 2014, Li et al. 2014, Xu et al. 2014).

Secondly, chromophores may originate from species other than natural amino acids and are covalently attached to protein primary structure. Some proteins in nature are colored. For example, rhodopsin, a light-sensitive protein in rod cells in retina, appears red-pink. Rhodopsin is constructed with the chromophore retinal, derived from vitamin A, attached to opsin, the protein scaffold (Bownds and Wald 1965). Retinal is covalently linked to a lysine on opsin through a Schiff base. Advanced glycation end products (AGEs), as a result of Maillard reaction between reducing sugars and amino acids, could cause discoloration (Maillard 1912, Sell and Monnier 1989, Yan et al. 1994, Butko et al. 2014). Maillard reaction contains a series of nonenzymatic reactions between amines and reducing carbohydrates (Hodge 1953, Ames 1990). At the start, an amine group such as in lysine or arginine is glycated by a reducing sugar such as glucose to form a Schiff base, also named as Amadori products. Amadori products could rearrange into various structures and further react with more carbohydrate units and amine groups through aldol condensation and aldehyde–amine polymerization. Melanoidins ("brown nitrogenous polymers and copolymers" (Hodge 1953)) are formed in the final stage. Mass spectrometry has been applied to identify glycation products during the initiation stage of Maillard reaction, which was associated with yellow discoloration of therapeutic protein after storage (Chumsae et al. 2013, Butko et al. 2014).

Thirdly, chromophores could be excipient components in the solution, noncovalently attached to protein. For example, riboflavin was found to bind strongly to certain IgGs rendering yellow color (Zhu et al. 2006). As another example, vitamin B12 (cobalamin) was found to add pink color to purified drug substance through noncovalent binding to the protein (Prentice et al. 2013, Derfus et al. 2014). The binding was strong enough that cobalamin was copurified with drug substance during downstream purification process. A side note is cyanocobalamin is commonly added to cell culture as a nutrient. Cyanocobalamin could convert into hydroxocobalamin in the presence of light (Prentice et al. 2013). Mass spectrometry could be used to detect cobalamin molecule separated from protein solution (Derfus et al. 2014).

Lastly, biophysical factors such as Rayleigh scattering could affect appearance of protein solution. In protein concentration measurement by UV spectroscopy, baseline correction is routinely performed to subtract the estimated contribution from Rayleigh scattering (Winder and Gent 1971, Mach and Middaugh 2011). As reported in some cases, an elevated level of aggregation was observed in yellow or brunescent cataracts, triggered by protein denaturing associated with aspartic acid racemization (Masters et al. 1977, Masters et al. 1978).

16.7 Sequence Variants

Central dogma of molecular biology describes the basic process of transferring genetic information (Crick 1970). Three general steps within the multistep process are essential. (i) Replication: from DNA to DNA; (ii) Transcription: from DNA to RNA; (iii) Translation: from RNA to polypeptide. Deviations in any of these steps might result in sequence variants in synthesized polypeptides. As a general term, sequence variance could refer to occurrence of protein molecular variants with amino acid sequences differed from the designed sequence in the final drug product. Protein sequence variance discussed in this section focuses on variants generated before or during polypeptide biosynthesis in the host cells.

When sequence variance is detected at the protein level, it could originate from two major sources: inconsistency between mRNA and target sequence to be expressed (mutation on DNA/RNA level), and mistranslation of mRNA (amino acid misincorporation at the protein level).

On DNA/RNA level, during monoclonal therapeutic protein manufacturing, certain stages could introduce variants. Firstly, designed DNA sequence needs to be chemically synthesized and amplified before fusion to the expression vector. Impure DNA sequences during chemical synthesis could be carried over to vector. Secondly, errors could occur during replication of DNA and transcription in the host cells during or after transfection. However, spontaneous mutation rates are low, on the order of magnitude of 10^{-7} to 10^{-11} for DNA replication in microbes with DNA chromosomes (Drake 1991), 10^{-10} to 10^{-11} for DNA replication in higher eukaryotes (Drake et al. 1998), 10^{-5} for RNA polymerase in *E. coli* (Blank et al. 1986, Ninio 1991). Commonly observed DNA/RNA sequence variants are single nucleotide polymorphisms (SNPs), while substitution of two base pairs in proximity was reported (Harris et al. 1993). Error rates could increase when cells are exposed to stressing conditions such as transfection process (Lebkowski et al. 1984), mutagenic regents (Blank et al. 1986), growth medium that affects growth rates (Smith 1992, Bridges 2001), near UV light (Webb and Malina 1970, Webb and Lorenz 1972, Webb 1977) and alkylating agents (Rebeck and Samson 1991). Other events such as splicing or gene crossover could also cause variance in DNA (Wan et al. 1999). It is possible one DNA variance remains after clone screening and eventually introduces a certain percentage of protein variant in the manufactured protein batch. During clone selection stage, if the selected clone is polyclonal instead of monoclonal, there is a higher probability that the variant DNA is not screened out. For example, during development of rhuMAb HER2, a Tyr-to-Gln protein variant was discovered to originate from DNA variance in polyclonal cell line (Harris et al. 1993). Also, multiple copies of plasmid DNA could be integrated to the host cell chromosome (Ringold et al. 1980, Kaufman and Sharp 1982, Wurm

1990), which indicates even within monoclonal cells there is a probability that the variant DNA is retained. In order to eliminate DNA/RNA sequence variance during cell line selection, analytical tools such as cDNA sequencing and PCR screening (Harris et al. 1993, Guo et al. 2010) can facilitate the detection of abnormalities at the DNA/RNA level.

At protein level, even if DNA/RNA in the cell line is variant-free, sequence variance could occur in produced protein due to mistranslation of mRNA. Mistranslation rate in *E. coli* was estimated to be on the order of magnitude of 10^{-4}, higher than DNA/RNA level error rate (Edelmann and Gallant 1977, Ellis and Gallant 1982). Mistranslation could result from tRNA misaminoacylation (Parker 1989, Beebe et al. 2008, Guo et al. 2009) or mRNA misreading (Seetharam et al. 1988, Brinkmann et al. 1989, Calderone et al. 1996). Misaminoacylation occurs when aminoacyl-tRNA synthetases cannot completely differentiate their corresponding substrates from other amino acids, and consequentially charge tRNAs with incorrect amino acids (Lin et al. 1984, Schön et al. 1988, Swanson et al. 1988). Natural error rate was estimated to be less than 1 in 3000 in chicken albumin biosynthesis (Loftfield 1963). Misaminoacylation in CHO cell expression system has also been reported (Wen et al. 2009). Misaminoacylation results more from structural similarities between certain amino acids, rather than from similarities in codon sequences for these amino acids (Brick et al. 1989, Perona et al. 1989). As opposed to misaminoacylation, misreading is more linked to similarities in codon sequences, such as codons with single-nucleotide differences (Davies et al. 1965). Misreading involves either mismatch between codon and tRNA, or frameshift. Natural misreading rates vary for different amino acids. In *E. coli*, error rates of different amino acids were estimated to be on the order of magnitude of 10^{-4} to 10^{-3} (Parker 1989). The frequency of misreading is largely dependent on the competition between aminoacylated tRNAs (Kramer and Farabaugh 2007) and can increase when correct aminoacylated tRNAs are insufficient in several scenarios. In many cases, amino acid starvation induced more mistranslation in *E. coli* and CHO expression system (Khetan et al. 2010, Feeney et al. 2013). If the medium is deprived of certain amino acids and the corresponding aminoacylated tRNAs, then other tRNAs with similar anticodons could bind to the codon through dominance in abundance. In another case, during recombinant protein production, use of high-yield expression systems and stressing conditions such as oxidative stress (Ling and Söll 2010) also caused more mistranslation. Also, choice of codon could increase mistranslation. After a codon that is naturally rare was engineered in *E. coli*, partial mistranslation was observed, while no mistranslation was detected when an *E. coli* preferred codon was used (Seetharam et al. 1988). A less observed type of misreading is frameshift, which has been observed in *E. coli* and was proposed to be caused by codon recognition by two bases instead of three (Dayhuff et al. 1986). To reduce mistranslation in monoclonal therapeutic protein manufacturing, several aspects should be considered, such as choice of expression system, choice of codons, medium composition, and feeding strategy (Khetan et al. 2010, Feeney et al. 2013), as well as stressing factors for the host cells.

Detecting sequence variants in the manufactured protein could be challenging, since the variant may be low in abundance (on the order of 10^{-4} magnitude has been observed (Wen et al. 2009, Zhang et al. 2013b)) and different by as few as one amino acid residue. Analytical tools need to differentiate low-level variants from the abundant wild type, based on the property of the variant amino acid residues. Charge-based separation such as ion-exchange chromatography (Dorai et al. 2007) and isoelectric focusing (Ling and Söll 2010), size-based separation such as gel-electrophoresis (Guo et al. 2010), hydrophobicity-based such as peptide mapping (Harris et al. 1993) and Edman sequencing (Yu et al. 2009) have been used for detecting protein sequence variance. Mass spectrometry could detect the molecular mass difference induced by sequence variance, except for the case of Leu-Ile substitution. Since molecular mass is an intrinsic property, mass spectrometry could provide complementary information besides chromatography when the variant is not separated by such methods. Mass spectrometry could be applied on protein level (Fu et al. 2012) or peptide level after protease treatment of the protein sample (Que et al. 2010, Yang et al. 2010, Huang et al. 2012b). To improve the performance of MS-based sequence variance analysis, many technical aspects could be considered. For example, mass difference induced by amino acid substitution ranges from 0.0364 Da (between Gln and Lys) to 129.0578 Da (between Gly and Trp), which may require MS instrument with different resolutions to detect sequence variants (Tanaka et al. 2006). Also, since the variant may be low in abundance, improvement of MS sensitivity, both on the experimental and software part (Zhang 2012), could enhance the capability to detect variant protein or peptide. Improved data interpretation could accelerate the identification process too. Several computer algorithms and software packages such as SEQUEST (Eng et al. 1994) and Mascot with error tolerant search (ETS) (Creasy and Cottrell 2002) were designed to match MS/MS spectra to customized or general databases to facilitate identification of sequence variants (Gatlin et al. 2000, Que et al. 2010, Yang et al. 2010).

16.8 Glycosylation

Posttranslational modifications play a critical role within many biological systems and are known to impact both structure and function of proteins. Within eukaryotes, glycosylation represents one of the most complex posttranslational modifications observed. The majority of proteins destined for the cell surface or secretion are glycosylated, and over two thirds of the proteins contained within the SWISS-PROT database contain glycosylation sites. Given the pervasiveness and potential impact on structure and function, the understanding and characterization of glycosylation has become a major focus in the development of protein-based therapeutics (Brooks et al. 2002). Modification to protein glycosylation contributes to protein solubility, stability, folding, and assembly into fully active complexes as well as protection against degradation (Gagneux and Varki 1999, Barinka et al. 2004, Martinek et al. 2010, Mrazek et al. 2013).

N- and O-glycosylation are the two major forms of glycosylation. N-glycosylation is defined as the transfer of an oligosaccharide from the dolichol diphosphate-activated precursor onto the amide of Asn occurring within the consensus sequence of Asn-Xxx-Ser/Thr (where Xxx is any amino acid except Pro). This transfer is catalyzed by the oligosaccharyltransferase enzyme complex (EC 2.4.1.119). This process occurs cotranslationally as soon as the unfolded enzyme polypeptide passes through the secretory protein channel into the oxidative environment of the lumen of ER. After the attachment to enzyme polypeptide, the above oligosaccharide precursor is further modified by many enzymes in ER and Golgi making N-glycosylation one of the most variable posttranslational modifications. Within the Golgi, O-glycosylation may also be added through the attachment of shorter oligosaccharides to Ser or Thr residues, with higher frequency in enzymes rich in these amino acids or containing them in clusters (Mrazek et al. 2013). Thus, many proteins acquire N- and O-glycosylation at several glycosylation sites, which may contribute to their complex molecular architecture and help to maintain their long-term stability (Barinka et al. 2004, Ettrich et al. 2007, Plihal et al. 2007, Ryslava et al. 2011, Vanek et al. 2011).

16.8.1 Glycoprotein Structure

The high degree of oligosaccharide complexity, resulting from the variable composition, linkage, branching points, and configuration of monosaccharides, coupled with the presence of various degrees of glycosylation at different glycosylation sites on glycoproteins, necessitates a diversity of analytical approaches employed in the study of this post-translational modification. Relative to other PTMs, the analysis of protein glycosylation can be particularly challenging. This increased difficulty is the result of several factors. To fully characterize a glycoprotein, information on the peptide sequence, glycosylation site, and glycan structure must be gathered. Glycosylation typically represents a distribution of structures rather than a single structure. The glycans attached to proteins in humans are composed of seven different monosaccharides: mannose (Man), glucose (Glc), galactose (Gal), *N*-acetylglucosamine (GlcNAc), *N*-acetylgalactosamine (GalNAc), fucose (Fuc), and sialic acids (SA) or neuraminic acids (NeuNAc) (Brooks 2009). They can be linked in a linear manner or by branching chains and can be of various sequences and lengths. While glycans such as Glc, Gal, and Man have identical mass and charge, they represent different stereoisomers of the same underlying chemical structure, and their permutation in a glycan structure results in a diverse range of possible glycoforms.

Four types of protein-linked glycans are known, including (i) N-linked, (ii) O-linked, (iii) C-glycans, and (iv) glycosylphosphatidylinositol anchors.

N-linked glycosylation in which the oligosaccharide is linked via a GlcNAc molecule in a β-*N*-glycosidic type bond to the nitrogen of the amide group of an asparagine (Asn) as illustrated in Figure 16.4.

N-linked glycosylation of proteins is a cotranslational event occurring during protein synthesis and is initiated as the newly synthesized polypeptide chain enters the lumen of the ER. N-linked oligosaccharide synthesis continues as the protein is transported from the ER to the Golgi apparatus and is completed by the time the glycoprotein leaves the *trans*-Golgi network (Ryslava et al. 2013). The presence of a consensus sequence within the protein amino acid sequence is a prerequisite for N-linked oligosaccharide incorporation. The consensus sequence for N-linked glycosylation is Asn-Xaa-Ser/Thr, in which Xaa may be any amino acid with the exception of proline. The consensus sequence allows recognition of the glycosylation site by the first enzyme involved in N-linked oligosaccharide production (oligosaccharyltransferase or

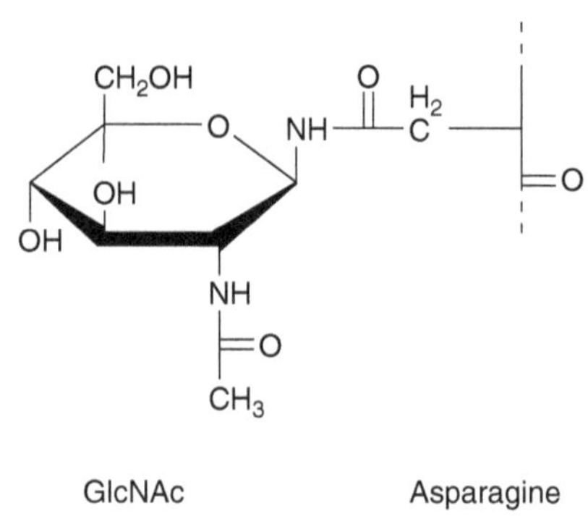

Figure 16.4 N-linked oligosaccharide structure – linkage of GlcNAc to asparagine.

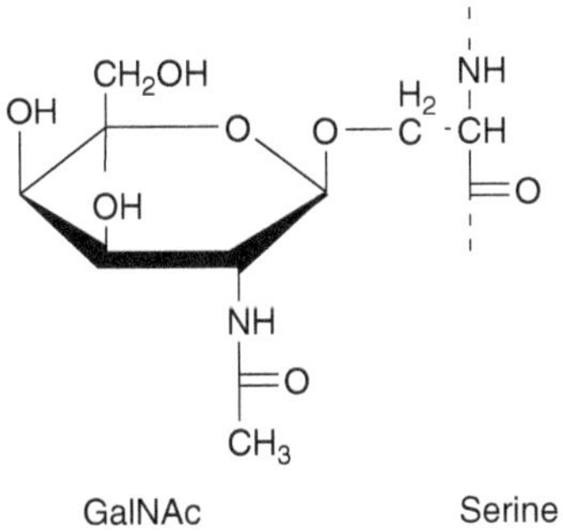

Figure 16.5 O-linked oligosaccharide structure – linkage of GalNAc to serine.

OST) by providing a protein conformation, which enables the enzyme to gain access to the glycosylation site. However, the presence of the N-linked consensus sequence Asn-Xaa-Ser/Thr does not ensure oligosaccharide attachment. The sequence may occur many times in a polypeptide chain with only a small number of the potential sites being glycosylated. Numerous factors influence whether a putative glycosylation site is coupled to an oligosaccharide. Occupied N-linked consensus sequences are frequently those in which the consensus sequence is in a "loop" within the polypeptide chain, which enables access to the OST.

O-linked glycoproteins, in which the first monosaccharide of the oligosaccharide chain, typically GalNAc, is attached through an α-*O*-glycosidic linkage to an oxygen molecule of an amino acid residue, typically serine or threonine and, to a lesser extent, hydroxylysine or hydroxyproline on the polypeptide chain of a protein as illustrated in Figure 16.5.

Both N- and O-linked glycoprotein share common features and many proteins contain both types of glycan attachments within the same protein molecule. C-glycans (Vliegenthart and Casset 1998) with the glycan (mannose) attached to the Trp residues by a C—C bond in a consensus sequence of Trp-Xxx-Xxx-Trp or Trp-Ser/Thr-Xxx-Cys and glycosylphosphatidylinositol anchors with the glycan attached to the carboxyl terminus of certain membrane associated proteins by a phosphoethanolamine bridge with mannose (Willy 2009) are less common.

Within therapeutic proteins N- and O-linked glycosylation are by far the most common forms of glycosylation. Both N- and O-linked glycans are characterized by complex branched structures that vary greatly in form and size. Common core structures for both N- and O-linked glycosylation are provided in Figures 16.3 and 16.4. N-linked glycans contain a common trimannosyl-chitobiose core ($Man_3GlcNAc_2$) with one or more antennae attached to each of the terminal mannose (Morelle et al. 2006). Based on the location and nature of the additional monosaccharides added to the core, N-linked glycans are further classified into (i) the "high mannose" or "oligomannose" type N-glycans that have only Man residues added to the core; (ii) N-glycans of the complex type that contain *N*-acetyllactosamine (Galb1-3/4GlcNAc) within their antennal region; and (iii) the "hybrid type" N-glycans that contain both Man residues and *N*-acetyllactosamine attached to the trimannosyl-chitobiose core residues (Figure 16.6).

O-linked glycans, on the other hand, are characterized by the stepwise addition of sugar residues directly to a protein through a hydroxyl group. In mammals, the

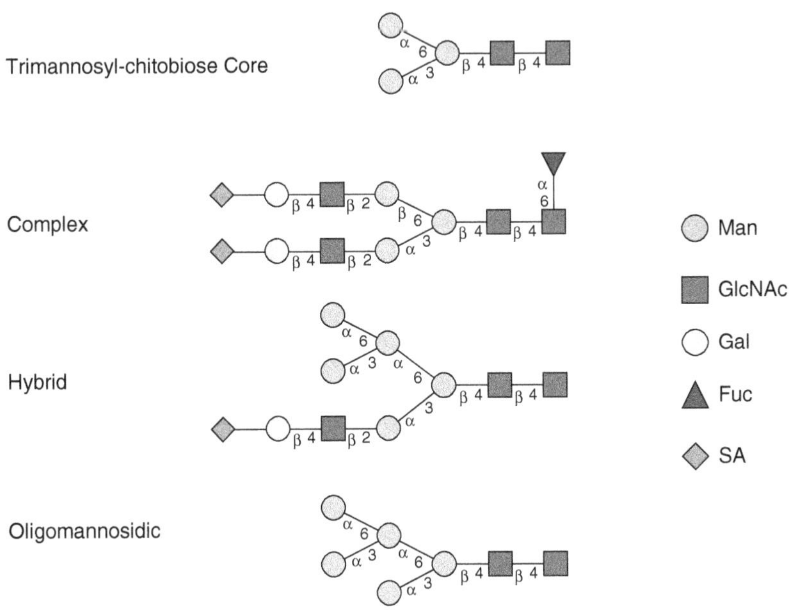

Figure 16.6 N-linked oligosaccharide structural classes.

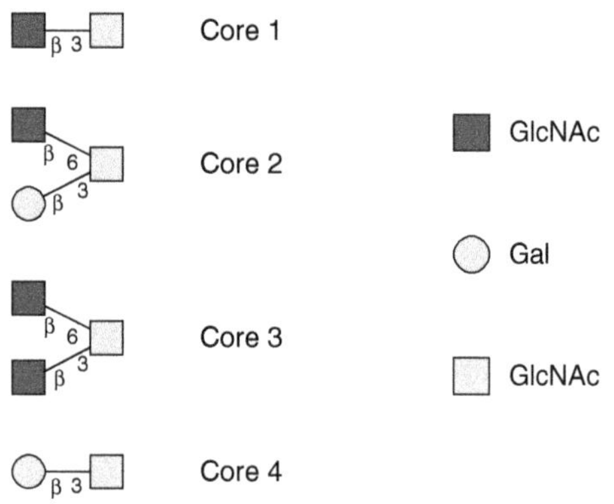

Figure 16.7 O-linked oligosaccharide core structures.

initiating step is typically the addition of *N*-acetylgalactosamine to Ser/Thr residues, although other monosaccharide units, such as GlcNAc, or Man-linked oligosaccharides have been reported to be involved in *O*-glycosidic linkages to hydroxyl amino acids (Morelle et al. 2006). Subsequent addition of Gal and/or GlcNAc leads to the formation of the common O-glycan core structures (Figure 16.7). Biosynthesis of complex N- and O-linked glycans is completed by a variety of capping reactions, the two most important in mammals being sialylation and fucosylation (Gesslbauer et al. 2007). Because of the acidic nature of sialic acid (SA) residues a net negative charge on otherwise neutral glycans is imparted by its presence. Glycans can be further modified by acetylation, methylation, phosphorylation, and sulfation, which can occur at internal or terminal positions in the glycan structure (Brooks 2009).

Because of the complexity and heterogeneity of glycans, no single analytical technology provides a comprehensive set of analytical data to ensure complete glycoprotein characterization. Therefore, a range of analytical methods and strategies have been employed in tandem to provide thorough glycoprotein characterization.

The analysis of glycoproteins is typically divided into three general approaches, each of which provides information on different aspects of glycoprotein structure, as shown in Figure 16.8. First, analysis of intact glycoproteins can provide a global view of the glycan population through a view of the overall profile of the glycan content present on the glycoprotein. It does not, however, provide information on the location of glycosylation sites within the protein, and in many cases information on the oligosaccharide structures are confounded due to isobaric masses of glycoforms as well as the presence of multiple glycosylation sites. Secondly, glycopeptide analysis, performed on proteolytic digest mixtures, allows the identification of site-specific glycosylation and in many cases provides strong data for the determination of oligosaccharide structures when MS^n data is evaluated. Thirdly, analysis of glycans after release from a glycoprotein allows for the use of a wide range of methods and analytical

Figure 16.8 Glycoprotein characterization through mass spectrometry.

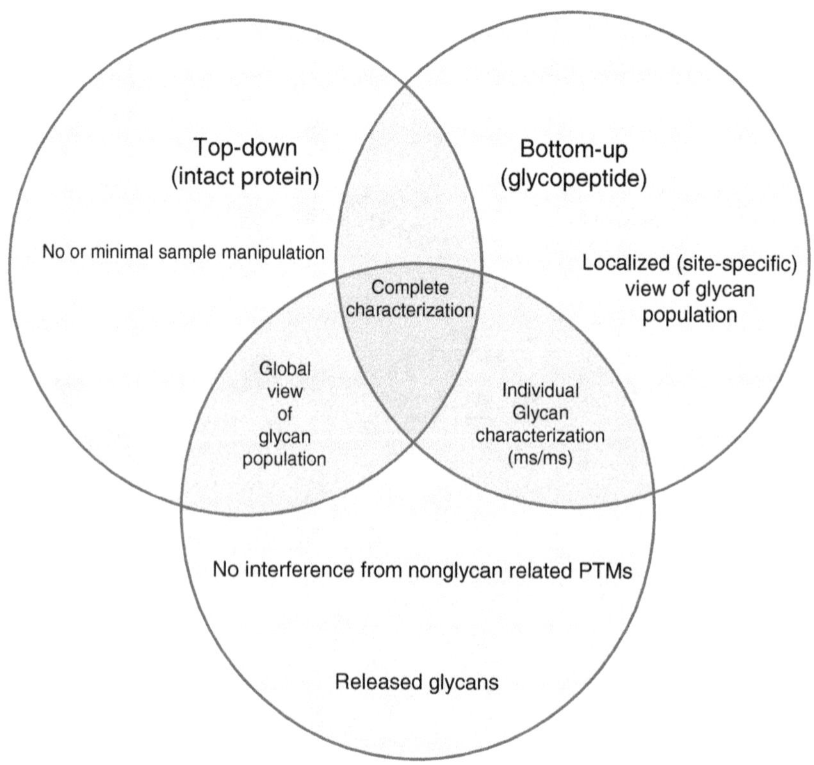

strategies, which enable a detailed structural investigation of individual glycan structures as well as the overall glycan profile without complication from the presence or interference of additional protein modifications. Each of these approaches provides valuable information and when taken together can produce a comprehensive view of the glycan structure of a given glycoprotein.

16.8.2 Intact Glycoprotein Analysis

Electrospray ionization with time-of-flight mass spectrometry (ESI-ToF MS) is a powerful tool for the analysis of glycoproteins. The characterization of glycoproteins by mass spectrometry is inherently more difficult than the mass spectrometric analysis of nonglycosylated proteins, because of the increased structural heterogeneity of glycoproteins as well as the typical reduction in ionization efficiency of glycosylated protein compared to the nonglycosylated proteins. Nonetheless, to some degree, intact glycoproteins can be resolved to their individual glycoforms using mass spectroscopic techniques.

Electrospray ionization coupled to a quadrupole time-of-flight mass analyzer (ESI-QToF MS) has emerged as the predominant technique for intact mass evaluation of glycoproteins. Given the extended mass ranges detectable and the computing power available to rapidly deconvolute the data generated, the mass of an intact mAb can be determined quickly and reproducibly on an ESI-quadrupole instrument with high mass accuracy and resolution.

Modern ToF analyzers achieve a mass accuracy on the order of 2–10 ppm, a resolution of 5000–30,000, and a maximum *m/z* of up to 10,000. These capabilities have made ESI-ToF or ESI-Q-ToF configurations the method of choice for mass determination of intact mAbs.

Using this methodology, ions from a given molecule produce an ion envelope in which a given ion differs by plus or minus one charge from adjacent ions in the series. In measuring the molecular weight, the charge on any one of the ions is first established by solving a series of simultaneous equations for any two consecutive ions in the series. From the combined data the charge for all ions in the series can be deduced and the molecular weight calculated. As with any spectral data, overlapping of peaks may occur within electrospray data, particularly when several species are present, each giving rise to its own series of multiply charged ions. Under these conditions, accurate values for mass/charge ratios of the components in an unresolved multiplet may not be obtainable without some form of data deconvolution.

Maximum entropy (MaxEnt) techniques are particularly well suited for the effective deconvolution of mass spectral data. The MaxEnt deconvolution solution represents the minimum amount of spectral structure consistent with the data and is often capable of providing a level of deconvolution, which enables overlapping peaks to be resolved and accurately centroided (Ferrige et al. 1991).

When appropriately calibrated, ESI-ToF instruments can routinely determine the mass of an intact mAb with a mass accuracy of less than 100 ppm. With experimental conditions optimized to minimize adduct formation, a calibrant analyzed immediately before sample analysis, and carefully controlled deconvolution parameters, mass accuracy that approached 10 ppm has been achieved. This level of accuracy in mass measurement of intact mAbs by a ToF analyzer approaches the natural variation of the protein average mass. The mass range and mass accuracy of an ESI-ToF instrument, combined with an appropriate charge-deconvolution algorithm, make it one of the best instruments available for mass determination of intact mAbs (Zhang et al. 2009).

The high level of mass resolution afforded by modern systems results in enhanced resolution in deconvolution of the ion envelope data. The enhanced resolution allows for resolved detection of individual glycoforms on glycoproteins of relatively large size. A typical example of an ESI-qToF mass spectra of an mAb is shown in Figure 16.9. For mAbs, typically in the 150 kDa mass range, the mass resolution after deconvolution easily allows for separation of different glycoforms.

Typical raw and deconvoluted spectra of an mAb: (a) Raw ESI-qToF spectrum acquired on a Thermo Q-Star Elite. (b) Region of raw spectrum showing charge states 49–53. (c) Deconvoluted mass spectrum, in which clear resolution of glycoforms is identifiable.

It should be noted that isobaric glycoforms cannot be distinguished from one another from the intact analysis data. In the example presented (Figure 16.9), an mAb with only two N-linked glycosylation sites and only three predominant oligosaccharide structures, the two glycoforms G0G2 and G1G1 are indistinguishable from one another. In glycoproteins with only moderate glycosylation, such as typical mAbs, the confounding of structural identification is minor. However, as a protein's glycosylation becomes more extensive, the number of isobaric glycoforms increases dramatically. In such cases, the intact mass data may be useful as a measure of glycan content and a means of monitoring profile consistency, but complementary methods are needed in order to provide a detailed glycan compositional characterization.

Generally, intact mass analysis gives a global view of the protein glycosylation, that is to say that site-specific information is not extractable from the intact view directly. However, if it were possible to fragment the intact protein within the spectrometer, it would be possible to gain some higher level of region-specific information regarding the attached glycan location and local

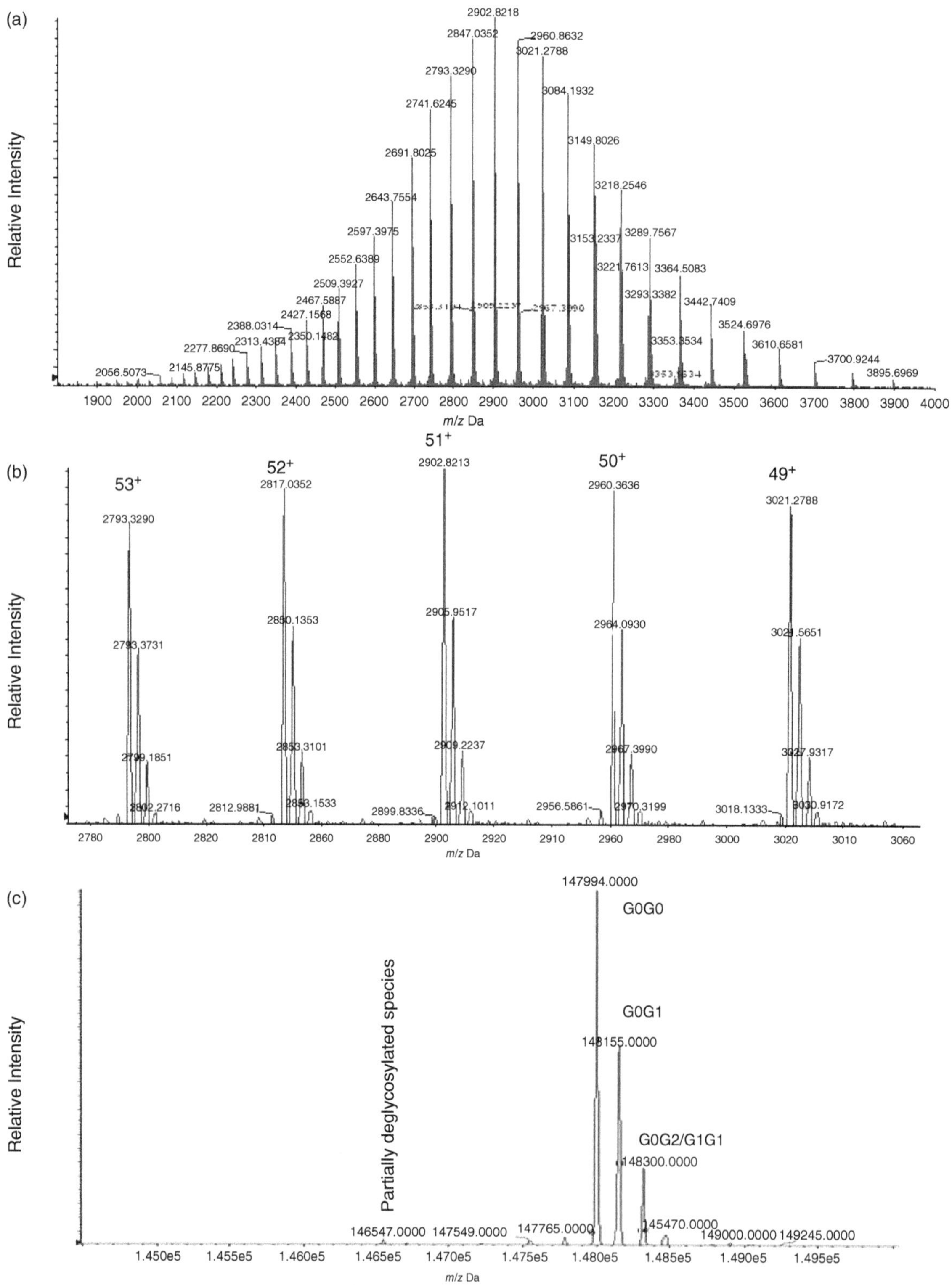

Figure 16.9 Typical intact mass spectra of monoclonal antibody proteins.

glycan populations. Top-down mass spectrometry refers to mass spectrometric evaluation of the instrument-induced fragmentation of biomolecular ions of any size (Reid and McLuckey 2002, Kelleher 2004, Garcia 2010). This approach has the potential to be able to identify and characterize all types of posttranslational modifications, including glycosylation, on the native protein of interest.

To date, the fragmentation of intact proteins the size of intact monoclonal antibodies in the gas phase has been attempted with collision-induced dissociation (CID), both through the use of the dedicated CID cell as well as directly within the ion source itself (Karabacak et al. 2009, Ryan et al. 2010). So far, CID-generated sequence coverage, particularly from IgGs, has been limited. Electron capture dissociation (ECD)- (Zubarev et al. 2000, Zubarev 2003, Zubarev 2004) and electron transfer dissociation (ETD) (Syka et al. 2004)-induced fragmentation of large proteins provides an alternative, radical-based mechanism of fragmentation and creates additional opportunities in top-down MS analysis. The radical-induced fragmentation generally provides more extensive sequence coverage on large proteins and efficient rupture of disulfide bonds (Zubarev et al. 1999, Zubarev et al. 2000) compared to CID and infrared multiphoton dissociation (IRMPD). However, possibly the most significant advantage of ECD/ETD over CID/IRMPD is in the characterization of labile PTMs on peptides (Mirgorodskaya et al. 1999); this is generally less pronounced on proteins, where the distribution of vibrational energy is more easily dispersed in the large protein structure, which increases the chance for labile PTMs to remain intact during the rupture of the protein backbone bonds in CID/IRMPD (Siuti and Kelleher 2007, Mikhailov et al. 2010). Although this approach presents great promise, it is still in its infancy. As the availability of ETD/ECD instruments becomes more accessible, the maturity of top-down analysis with respect to characterization of PTMs including glycosylation will continue to evolve. The approach most commonly used in obtaining detailed localized characterization information regarding site connectivity and localized glycan population distributions remains the analysis of glycopeptides.

16.8.3 Glycopeptide Analysis

Determining the glycosylation site specificity can be problematic in the top-down approach since the approach inherently gives a global, rather than local, view of the glycoprotein being investigated. Advances in fragmentation techniques and data analysis may ultimately allow the top-down approach to be used more routinely for this purpose, but currently the analysis of glycopeptide in a bottom-up approach is more common. This approach typically employs a combination of specific enzymatic proteolysis (usually with trypsin) followed by fractionation of glycopeptides by liquid chromatography or affinity chromatography and ultimately glycopeptide analysis by MS and MS/MS (Huddleston et al. 1993, Geyer and Geyer 2006, Ito et al. 2006, Zhao et al. 2006, Dalpathado and Desaire 2008). Due to its specificity and robust performance, the most common proteolytic enzyme employed in this type of analysis is trypsin. As the analysis of glycopeptides is typically part of a concurrent protein structural characterization, the same protocols for digestion are generally used and no additional optimization of the digestion protocols is required to accommodate glycoproteins. Trypsin readily produces highly predictable peptide masses because of its high activity and specificity. In addition, tryptic glycopeptides guarantee a basic residue in every peptide, which increases ionization efficiency during MS analysis. A significant drawback with this approach is that glycoproteins may be resistant to trypsin. In addition, the resulting glycopeptides may often be too large for effective MS/MS analysis. This problem is often complicated by the presence of missed cleavages particularly near the sites of glycosylation. Despite these limitations, in the characterization of therapeutic proteins, where glycoprotein samples are typically relatively pure and available in significant quantity, this approach can give relatively comprehensive data on the location and micro-heterogeneity of glycosylation.

If needed, enrichment of the glycopeptides can be performed, but this introduces its own set of challenges. One enrichment strategy has been developed in which a cleavage of the carbon–carbon group between the diols of saccharide units produces aldehyde groups that can be captured by reaction with hydrazine functional groups that are immobilized on a solid support (Zhang et al. 2003, Zhang and Aebersold 2006). Capture with immobilized boronic acid has also been employed (Sparbier et al. 2005). Hydrophilic interaction liquid chromatography (HILIC) has also been reported as well as lectin affinity chromatography using ConA, WGA, or a combination of lectins (Zhao et al. 2006). However, despite the wide variety of the methods, there is still no generally effective method for glycopeptide enrichment, as no single method is both comprehensive and highly specific (Calvano et al. 2008).

16.8.4 Tandem MS of Glycopeptides

In theory, tandem MS can provide peptide and glycan sequence as well as the site of glycosylation. In practice, however, tandem MS analysis of glycopeptides can be problematic and far from routine. Studies of glycopeptide fragmentation reactions have focused almost exclusively on protonated tryptic glycopeptides. The typical fragments correspond to the loss of the fragments from the glycan moiety, while information on the peptide sequence and glycan attachment sites is often harder to obtain. Tandem MS is complicated by the sizes of tryptic peptides, which tend to be larger than the mass range that would allow for comprehensive sequence characterization using CID. This coupled with the labile nature of the glycan-peptide bond makes complete characterization using CID alone difficult. A common strategy, therefore, is to determine the overall mass of the glycopeptide and perform tandem MS to yield the peptide mass. To sequence the peptide, the glycopeptide is first deglycosylated and subjected separately to tandem MS.

Multiple reviews on glycopeptide analysis by mass spectrometry (Wuhrer et al. 2007, Dalpathado and Desaire 2008) conclude that the fragmentation behavior of glycopeptide ions under CID tend to vary with the instrument, instrumental parameters, specific peptide composition, charge carrier, and charge state (Hakansson et al. 2003, Ito et al. 2006, Deguchi et al. 2007, Wuhrer et al. 2007, Seipert et al. 2008).

Upon instrumental and experimental optimization, detailed information on oligosaccharide structure can be obtained from the CID MS/MS data. Extensive fragmentation of the glycan can be obtained with the absence of peptide backbone fracture. MS/MS data on glycan stoichiometry and connectivity in combination with knowledge about the protein expression system used can then be useful in assignment of glycan structures. Ultimately, detailed linkage information may require additional experimental data such as enzymatic glycan sequencing data.

Glycopeptides when subjected to CID also yield low-molecular-weight ions such as *m/z* 163 (Hex + H), 204 (HexNAc + H), 292 (NeuAc + H), 366 (Hex-HexNAc) that are diagnostic for the presence of glycosylation (Huddleston et al. 1993, Medzihradszky et al. 1997). In this way, glycopeptides can be readily identified by selective ion monitoring with ion trap MS (Wuhrer et al. 2005) or qToF mass analyzers (Ritchie et al. 2002, Jebanathirajah et al. 2003). In addition, neutral losses of saccharides such as hexose, *N*-acetylhexosamine, fucose, and *N*-acetylneuraminic acid can also be used to identify the presence of glycopeptides in mass spectra.

ECD and ETD have been applied to glycopeptides and show great promise as a complement to CID (Hakansson et al. 2001, Catalina et al. 2007). These methods tend to cleave peptide bonds while leaving the attached glycan unaltered. The usefulness of the fragmentation technique has been demonstrated by a fair number of ECD and ETD studies on N-linked glycans where the glycosylation sites are predictable from the consensus sequence. The true power of ETD and ECD will become evident as application in characterization O-linked glycans becomes more prevalent.

Thus, ETD, ECD, and CID provide complementary information in a thorough characterization of glycopeptides, CID providing data for glycan characterization and ETD/ECD providing peptide structural information including glycan site connectivity.

16.8.5 Free Glycan Analysis

When detailed glycan characterization of the global glycan population is required, it is often most practical to release the glycans from the glycoprotein and analyze the resulting free glycans directly or after derivatization. As underivatized glycans typically do not contain strong chromophores, derivatization is an important consideration when high sensitivity is required. When characterizing a therapeutic protein, it is not uncommon to have significant enough quantities to alleviate the need for derivatization. However, derivatization is often required to detect and characterize glycan structures, which may represent only a small fraction of the overall population or if quantification is needed. Many options are available for derivatization with fluorescent tags being the most common (Anumula 2006). Prior to derivatization, however, glycans must be effectively released from the glycoprotein.

16.8.6 Release of Glycans from Glycoproteins

Different chemistries are used for the release of N- and O-linked glycans. This is due to the fundamental difference in the linkage connectivity. Intact N-linked glycans can be released effectively using relatively mild conditions because the existence of PNGase F cleaves the amide linkage between the protein and the glycan. PNGase F cleaves the linkage between the core GlcNAc and the asparagine residue of all classes of N-linked glycans, with the exception of N-glycans that contain α(1,3) linked fucose on the core GlcNAc directly attached to the protein. This type of structure is more likely found in plant- and insect-derived glycoproteins than those expressed in mammalian systems (Royle et al. 2006). Also available are endoglycosidase D (endo D) that releases all classes of N-linked glycans through cleavage between the two GlcNAc residues within the chitobiose core, and endo H that cleaves at the same location and is selective for oligomannose and hybrid type structures. The ease of enzymatic release for N-linked glycans has made the approach the most commonly used; however, release of N-linked glycans can also be performed by chemical methods, which are typically used for O-linked glycan release.

Unfortunately, an enzyme of comparable activity and general effectiveness as PNGase F is for N-linked glycan release is not known for O-linked glycan release. An enzyme, *endo-α-N*-acetylgalactosaminidase (O-glycanase), which is specific for cleavage of core 1 O-glycan structures, has been reported (Bhavanandan et al. 1976). Given there are eight known O-linked cores and that in cases where the core 1 structure is present the cores are often extended beyond the O-glycanase specificity, O-glycanase cannot be used as a general solution for O-linked glycan cleavage. Therefore, chemical methods of O-linked glycan release are generally used. Two commonly used chemical methods are hydrazinolysis (Merry et al. 2002) and base-induced β-elimination (Carlson 1966). Care must be taken to ensure that the chemical means of removal does not alter the glycans being released. The condition employed in alkaline β-elimination can cause glycan degradation. An ammonia-based nonreductive β-elimination has been reported, which minimizes the potential for glycan degradation and is compatible with subsequent mass spectroscopic analysis (Huang et al. 2001). Reductive β-elimination with $NaBH_4$ results in the release of glycans and reduction of the resulting free reducing terminus to an alditol. This prevents degradation of the glycan but at the same time limits the number of suitable reagents for derivatization if subsequent tagging and quantitation is desired.

16.8.7 Detailed Sequence and Linkage Analysis of Glycans

Complete structural analysis of glycans involves not only monosaccharide and sequence information but also the stereochemistry of each linkage and the level of branching. Regulatory agencies often require such a detailed structural analysis of therapeutic glycoproteins (such as monoclonal antibodies) because the glycan structure can affect a protein therapeutic's immunogenicity, stability, or pharmacokinetics (Raju 2008). Glycan analysis is important during glycoprotein production as a means of mapping process parameters and downstream processing impact on glycan structure (Hossler et al. 2009). Analysis of the primary structure of oligosaccharide is complicated by the number of parameters that must be determined. These include (i) the nature of the individual monosaccharides as well as their ring conformation; (ii) the absolute stereochemistry of individual residues (D- or L-); (iii) the anomericity of the glycosidic bonds (a or b linkages); (iv) substitution patterns and branch points; and (v) the nature and location of any additional chemical modifications (i.e., acetylation, methylation) on a given monosaccharide. Although mass spectrometry can provide information on some of these parameters, by itself it cannot provide a complete characterization of all of the structural detail. Although there are reports of using CID product ion mass spectra to provide information on stereochemistry of individual sugar residues (Mueller et al. 1988), the linkage position (Laine et al. 1988), and branching structure (Carr et al. 1985, Domon and Costello 1988) by evaluating the distinct product-ion patterns specific to oligosaccharide that contain the same monosaccharides linked with a different branching structure, use of this data is unlikely to provide strong enough evidence to make structural and stereochemical assignments with great confidence. Typically, additional information is used to support the comprehensive assignment of oligosaccharide structure. Mass spectroscopic data coupled to enzymatic sequencing experiments is one method in which connectivity and stereochemistry is probed further. The principle of oligosaccharide sequencing is to take advantage of the ability of enzymes (*endo*- and *exo*-glycosidases) to remove terminal monosaccharides from the nonreducing end of oligosaccharides. The *exo*- and *endo*-glycosidases that are used in the structural analysis of oligosaccharides are very specific for the monosaccharide anomericity (α/β) of the glycosidic linkage, and the absolute stereoisomer (D/L) of the glycon. The individual specificities for some of the better defined exoglycosidases are summarized in Table 16.2.

Table 16.2 Commonly employed glycosidases used in oligosaccharide sequencing.

Enzyme	Source	EC number	Specificity
α-D-Sialidase	*Arthrobacter ureafaciens* sialidase	EC 3.2.1.18	Releases α(2-6/3/8)-linked nonreducing terminal *N*-acetylneuraminic acid (NANA, Neu5Ac) and *N*-glycolylneuraminic acids (NGNA, Neu5Gc)
	Streptococcus pneumoniae sialidase	EC 3.2.1.18	Releases α(2-3)-linked nonreducing terminal sialic acids (NANA and NGNA)
β-D-Galactosidase	Bovine testes β-galactosidase	EC 3.2.1.23	Hydrolyzes nonreducing terminal galactose with β(1-3/4) linkages
	S. pneumoniae β-galactosidase	EC 3.2.1.23	Hydrolyzes nonreducing terminal galactose with β(1-4) linkages
β-D-Mannosidase	*Helix pomatia*	EC 3.2.1.25	Hydrolysis of terminal, nonreducing β-D-mannosidase with β(1-4) residues in β-D-mannosidase
β-*N*-Acetyl-D-hexosaminidase	β-*N*-acetyl-D-hexosaminidase cloned from *S. pneumoniae* expressed in *Escherichia coli*	EC 3.2.1.30	Will digest β(1-4)-linked GlcNAc to mannose but not a bisecting GlcNAc β(1-4) linked to mannose
α-L-Fucosidase	Almond meal α-fucosidase	EC 3.2.1.51	Releases α(1-3/4)-linked nonreducing terminal fucose residues except core α(1-6) fucose
	Bovine kidney α-fucosidase	EC 3.2.1.51	Releases α(1-2/6) fucose-linked nonreducing terminal fucose residues more efficiently than core α(1-3/4) linked fucose

16.9 Conclusion

It is challenging to characterize biologics due to the large size and considerable complexity of biomolecules. The analytical characterization of biologics is intended to provide an understanding of the structures and properties of molecular variants, which can be generated during cell culture, protein purification, storage, and *in vivo* after administration. Mass spectrometry has long been a major tool supporting development of each of these process areas with comprehensive information that few other techniques can match. Recent advances in MS instrumentation that provide improved resolution and sensitivity have allowed for the analysis of biologics in greater detail than ever before by allowing for the characterization of variants present at extremely low levels. Continued advancements in instrumentation and software capabilities will allow more detailed and rapid characterization of biologics through improved sensitivity and efficiency, which ultimately leads to faster delivery of therapeutics to patients with improved safety and efficacy of drug product.

References

Adrover M, Marino L, Sanchis P, et al. (2014). "Mechanistic insights in glycation-induced protein aggregation." *Biomacromolecules* **15**(9): 3449–3462.

Agarkhed M, O'Dell C, Hsieh MC, et al. (2013). "Effect of polysorbate 80 concentration on thermal and photostability of a monoclonal antibody." *AAPS PharmSciTech* **14**(1): 1–9.

Amano M, Kobayashi N, Yabuta M, et al. (2014). "Detection of histidine oxidation in a monoclonal immunoglobulin gamma (IgG) 1 antibody." *Anal Chem* **86**(15): 7536–7543.

Ambrogelly A, Liu YH, Li H, et al. (2012). "Characterization of antibody variants during process development: the tale of incomplete processing of N-terminal secretion peptide." *mAbs* **4**(6): 701–709.

Ames JM (1990). "Control of the Maillard reaction in food systems." *Trends Food Sci Technol* **1**: 150–154.

Amici A, Levine R, Tsai L, et al. (1989). "Conversion of amino acid residues in proteins and amino acid homopolymers to carbonyl derivatives by metal-catalyzed oxidation reactions." *J Biol Chem* **264**(6): 3341–3346.

Antes B, Amon S, Rizzi A, et al. (2007). "Analysis of lysine clipping of a humanized Lewis-Y specific IgG antibody and its relation to Fc-mediated effector function." *J Chromatogr B Analyt Technol Biomed Life Sci* **852**(1–2): 250–256.

Anumula KR (2006). "Advances in fluorescence derivatization methods for high-performance liquid chromatographic analysis of glycoprotein carbohydrates." *Anal Biochem* **350**(1): 1–23.

Aono H, Wen D, Zang L, et al. (2010). "Efficient on-column conversion of IgG1 trisulfide linkages to native disulfides in tandem with protein A affinity chromatography." *J Chromatogr A* **1217**(32): 5225–5232.

Arai K, Kumakura F and Iwaoka M (2010). "Characterization of kinetic and thermodynamic phases in the prefolding process of bovine pancreatic ribonuclease A coupled with fast SS formation and SS reshuffling." *Biochemistry* **49**(49): 10535–10542.

Bagal D, Valliere-Douglass JF, Balland A, et al. (2010). "Resolving disulfide structural isoforms of IgG2 monoclonal antibodies by ion mobility mass spectrometry." *Anal Chem* **82**(16): 6751–6755.

Baldwin AD and Kiick KL (2011). "Tunable degradation of maleimide-thiol adducts in reducing environments." *Bioconjug Chem* **22**(10): 1946–1953.

Banks DD, Hambly DM, Scavezze JL, et al. (2009). "The effect of sucrose hydrolysis on the stability of protein therapeutics during accelerated formulation studies." *J Pharm Sci* **98**(12): 4501–4510.

Barinka C, Sacha P, Sklenar J, et al. (2004). "Identification of the N-glycosylation sites on glutamate carboxypeptidase II necessary for proteolytic activity." *Protein Sci* **13**(6): 1627–1635.

Battersby JE, Hancock WS, Canova-Davis E, et al. (1994). "Diketopiperazine formation and N-terminal degradation in recombinant human growth hormone." *Int J Pept Protein Res* **44**(3): 215–222.

Beebe K, Mock M, Merriman E, et al. (2008). "Distinct domains of tRNA synthetase recognize the same base pair." *Nature* **451**(7174): 90–93.

Bent D and Hayon E (1975). "Excited state chemistry of aromatic amino acids and related peptides. III. Tryptophan." *J Am Chem Soc* **97**(10): 2612–2619.

Betz SF (1993). "Disulfide bonds and the stability of globular proteins." *Protein Sci* **2**: 1551.

Bhavanandan VP, Umemoto J and Davidson EA (1976). "Characterization of an endo-alpha-*N*-acetyl galactosaminidase from *Diplococcus pneumoniae*." *Biochem Biophys Res Commun* **70**(3): 738–745.

Blank A, Gallant JA, Burgess RR, et al. (1986). "An RNA polymerase mutant with reduced accuracy of chain elongation." *Biochemistry* **25**(20): 5920–5928.

Bownds D and Wald G (1965). "Reaction of the rhodopsin chromophore with sodium borohydride." *Nature* **205**: 254–257.

Boyd D, Kaschak T and Yan B (2011). "HIC resolution of an IgG1 with an oxidized Trp in a complementarity determining region." *J Chromatogr B* **879**(13–14): 955–960.

Bradshaw RA, Brickey WW and Walker KW (1998). "N-terminal processing: the methionine aminopeptidase and N alpha-acetyl transferase families." *Trends Biochem Sci* **23**(7): 263–267.

Brick P, Bhat T and Blow D (1989). "Structure of tyrosyl-tRNA synthetase refined at 2.3 Å resolution: Interaction of the enzyme with the tyrosyl adenylate intermediate." *J Mol Biol* **208**(1): 83–98.

Bridges BA (2001). "Hypermutation in bacteria and other cellular systems." *Philos Trans R Soc Lond B Biol Sci* **356**(1405): 29–39.

Brinkmann U, Mattes RE and Buckel P (1989). "High-level expression of recombinant genes in *Escherichia coli* is dependent on the availability of the dnaY gene product." *Gene* **85**(1): 109–114.

Brooks SA (2009). "Strategies for analysis of the glycosylation of proteins: current status and future perspectives." *Mol Biotechnol* **43**(1): 76–88.

Brooks SA, Dwek MV and Schumacher U (2002). *Functional and Molecular Glycobiology*, BIOS Scientific.

Brorson K and Jia AY (2014). "Therapeutic monoclonal antibodies and consistent ends: terminal heterogeneity, detection, and impact on quality." *Curr Opin Biotechnol* **30c**: 140–146.

Butko M, Pallat H, Cordoba A, et al. (2014). "Recombinant antibody color resulting from advanced glycation end product modifications." *Anal Chem*, **86**(19), 9816–9823.

Cal PM, Bernardes GJ and Gois PM (2014). "Cysteine-Selective Reactions for Antibody Conjugation." *Angew Chem Int Ed Engl*, **53**(40): 10585–10587.

Calderone TL, Stevens RD and Oas TG (1996). "High-level Misincorporation of Lysine for Arginine at AGA Codons in a Fusion Protein Expressed in *Escherichia coli*." *J Mol Biol* **262**(4): 407–412.

Calvano CD, Zambonin CG and Jensen ON (2008). "Assessment of lectin and HILIC based enrichment protocols for characterization of serum glycoproteins by mass spectrometry." *J Proteomics* **71**(3): 304–317.

Carlson DM (1966). "Oligosaccharides isolated from pig submaxillary mucin." *J Biol Chem* **241**(12): 2984–2986.

Carr SA, Reinhold VN, Green BN, et al. (1985). "Enhancement of structural information in FAB ionized carbohydrate samples by neutral gas collision." *Biomed Mass Spectrom* **12**(6): 288–295.

Carty RP, Pincus MR and Scheraga HA (2002). "Interactions that favor the native over the non-native disulfide bond among residues 58–72 in the oxidative folding of bovine pancreatic ribonuclease A." *Biochemistry* **41**(50): 14815–14819.

Catak S, Monard G, Aviyente V, et al. (2009). "Deamidation of asparagine residues: direct hydrolysis versus succinimide-mediated deamidation mechanisms." *J Phys Chem A* **113**(6): 1111–1120.

Catalina MI, Koeleman CA, Deelder AM, et al. (2007). "Electron transfer dissociation of N-glycopeptides: loss of the entire N-glycosylated asparagine side chain." *Rapid Commun Mass Spectrom* **21**(6): 1053–1061.

Chaderjian WB, Chin ET, Harris RJ, et al. (2005). "Effect of Copper Sulfate on Performance of a Serum-Free CHO Cell Culture Process and the Level of Free Thiol in the Recombinant Antibody Expressed." *Biotechnol Prog* **21**(2): 550–553.

Chen J, Shiyanov P, Zhang L, et al. (2010). "Top-down characterization of a native highly intralinked protein: concurrent cleavages of disulfide and protein backbone bonds." *Anal Chem* **82**(14): 6079–6089.

Cheng RZ, Uchida K and Kawakishi S (1992). "Selective oxidation of histidine residues in proteins or peptides through the copper(II)-catalysed autoxidation of glucosone." *Biochem J* **285**(2): 667–671.

Chumsae C, Gifford K, Lian W, et al. (2013). "Arginine modifications by methylglyoxal: discovery in a recombinant monoclonal antibody and contribution to acidic species." *Anal Chem* **85**(23): 11401–11409.

Chumsae C, Zhou LL, Shen Y, et al. (2014). "Discovery of a chemical modification by citric acid in a recombinant monoclonal antibody." *Anal Chem* **86**(18): 8932–8936.

Cournoyer JJ, Lin C, Bowman MJ, et al. (2007). "Quantitating the relative abundance of isoaspartyl residues in deamidated proteins by electron capture dissociation." *J Am Soc Mass Spectrom* **18**(1): 48–56.

Craig DB and Dombkowski AA (2013). "Disulfide by Design 2.0: a web-based tool for disulfide engineering in proteins." *BMC Bioinf* **14**: 346.

Creasy DM and Cottrell JS (2002). "Error tolerant searching of uninterpreted tandem mass spectrometry data." *Proteomics* **2**(10): 1426–1434.

Creed D (1984). "The photophysics and photochemistry of the near-UV absorbing amino acids–I. Tryptophan and its simple derivatives." *Photochem Photobiol* **39**(4): 537–562.

Crick F (1970). "Central dogma of molecular biology." *Nature* **227**(5258): 561–563.

Dalpathado DS and Desaire H (2008). "Glycopeptide analysis by mass spectrometry." *Analyst* **133**(6): 731–738.

Davies J, Gorini L and Davis BD (1965). "Misreading of RNA codewords induced by aminoglycoside antibiotics." *Mol Pharmacol* **1**(1): 93–106.

Davies MJ and Truscott RJW (2001). "Photo-oxidation of proteins and its role in cataractogenesis." *J Photochem Photobiol B* **63**(1–3): 114–125.

Dayhuff TJ, Atkins JF and Gesteland RF (1986). "Characterization of ribosomal frameshift events by protein sequence analysis." *J Biol Chem* **261**(16): 7491–7500.

Deguchi K, Ito H, Baba T, et al. (2007). "Structural analysis of O-glycopeptides employing negative- and positive-ion multi-stage mass spectra obtained by collision-induced and electron-capture dissociations in linear ion trap time-of-flight mass spectrometry." *Rapid Commun Mass Spectrom* **21**(5): 691–698.

Derfus GE, Dizon-Maspat J, Broddrick JT, et al. (2014). Red-colored IgG4 caused by vitamin B12 from cell culture media combined with disulfide reduction at harvest. *mAbs* **6**(3): 679–688.

Dick LW, Jr., Kim C, Qiu D, et al. (2007). "Determination of the origin of the N-terminal pyro-glutamate variation in monoclonal antibodies using model peptides." *Biotechnol Bioeng* **97**(3): 544–553.

Dillon TM, Ricci MS, Vezina C, et al. (2008). "Structural and functional characterization of disulfide isoforms of the human IgG2 subclass." *J Biol Chem* **283**(23): 16206–16215.

Domon B and Costello CE (1988). "Structure elucidation of glycosphingolipids and gangliosides using high-performance tandem mass spectrometry." *Biochemistry* **27**(5): 1534–1543.

Dorai H, Sauerwald T, Campbell A, et al. (2007). "Investigation of product microheterogeneity: a case study in rapid detection of mutation in mammalian production cell lines." *BioProcess Int* **5**(8): 66.

Drake JW (1991). "A constant rate of spontaneous mutation in DNA-based microbes." *Proc Natl Acad Sci U S A* **88**(16): 7160–7164.

Drake JW, Charlesworth B, Charlesworth D, et al. (1998). "Rates of spontaneous mutation." *Genetics* **148**(4): 1667–1686.

Dreaden TM, Chen J, Rexroth S, et al. (2011). "N-Formylkynurenine as a Marker of High Light Stress in Photosynthesis." *J Biol Chem* **286**(25): 22632–22641.

Dyer JM, Bringans SD and Bryson WG (2006). "Characterisation of photo-oxidation products within photoyellowed wool proteins: tryptophan and tyrosine derived chromophores." *Photochem Photobiol Sci* **5**(7): 698–706.

Edelmann P and Gallant J (1977). "Mistranslation in *E. coli*." *Cell* **10**(1): 131–137.

Ellis N and Gallant J (1982). "An estimate of the global error frequency in translation." *Mol Gen Genet MGG* **188**(2): 169–172.

Eng JK, McCormack AL and Yates JR (1994). "An approach to correlate tandem mass spectral data of peptides with amino acid sequences in a protein database." *J Am Soc Mass Spectrom* **5**(11): 976–989.

Ettrich R, Kopecky V, Jr., Hofbauerova K, et al. (2007). "Structure of the dimeric N-glycosylated form of fungal beta-N-acetylhexosaminidase revealed by computer modeling, vibrational spectroscopy, and biochemical studies." *BMC Struct Biol* **7**: 32.

Europe Co (2012). *European Pharmacopoeia*. C. o. Europe.

FDA (1999). *Guidance for Industry: Q6B Specifications: Test Procedures and Acceptance Criteria for Biotechnological/Biological Products*. U. S. F. a. D. Administration, U.S. Food and Drug Administration.

Feeney L, Carvalhal V, Yu XC, et al. (2013). "Eliminating tyrosine sequence variants in CHO cell lines producing recombinant monoclonal antibodies." *Biotechnol Bioeng* **110**(4): 1087–1097.

Ferrige AG, Seddon MJ, Jarvis S, et al. (1991). "Maximum entropy deconvolution in electrospray mass spectrometry." *Rapid Commun Mass Spectrom* **5**(8): 374–377.

Finley EL, Dillon J, Crouch RK, et al. (1998). "Identification of tryptophan oxidation products in bovine α-crystallin." *Protein Sci* **7**(11): 2391–2397.

Fu J, Bongers J, Tao L, et al. (2012). "Characterization and identification of alanine to serine sequence variants in an IgG4 monoclonal antibody produced in mammalian cell lines." *J Chromatogr B* **908**: 1–8.

Fukunaga Y, Katsuragi Y, Izumi T, et al. (1982). "Fluorescence characteristics of kynurenine and *N*′-formylkynurenine, their use as reporters of the environment of tryptophan 62 in hen egg-white lysozyme." *J Biochem* **92**(1): 129–141.

Gagneux P and Varki A (1999). "Evolutionary considerations in relating oligosaccharide diversity to biological function." *Glycobiology* **9**(8): 747–755.

Gao X, Ji JA, Veeravalli K, et al. (2015). "Effect of individual Fc methionine oxidation on FcRn binding: Met252 oxidation impairs FcRn binding more profoundly than Met428 oxidation." *J Pharm Sci* **104**(2): 368–377.

Garcia BA (2010). "What does the future hold for Top Down mass spectrometry?" *J Am Soc Mass Spectrom* **21**(2): 193–202.

Gatlin CL, Eng JK, Cross ST, et al. (2000). "Automated identification of amino acid sequence variations in proteins by HPLC/microspray tandem mass spectrometry." *Anal Chem* **72**(4): 757–763.

Gesslbauer B, Rek A, Falsone F, et al. (2007). "Proteoglycanomics: tools to unravel the biological function of glycosaminoglycans." *Proteomics* 7(16): 2870–2880.

Gevaert K, Goethals M, Martens L, et al. (2003). "Exploring proteomes and analyzing protein processing by mass spectrometric identification of sorted N-terminal peptides." *Nat Biotechnol* **21**(5): 566–569.

Geyer H and Geyer R (2006). "Strategies for analysis of glycoprotein glycosylation." *Biochim Biophys Acta* **1764**(12): 1853–1869.

Goetze AM, Liu YD, Zhang Z, et al. (2011). "High-mannose glycans on the Fc region of therapeutic IgG antibodies increase serum clearance in humans." *Glycobiology* **21**(7): 949–959.

Gorman JJ, Wallis TP and Pitt JJ (2002). "Protein disulfide bond determination by mass spectrometry." *Mass Spectrom Rev* **21**(3): 183–216.

Gracanin M, Hawkins CL, Pattison DI, et al. (2009). "Singlet-oxygen-mediated amino acid and protein oxidation: Formation of tryptophan peroxides and decomposition products." *Free Radic Biol Med* **47**(1): 92–102.

Gray WR (1993). "Disulfide structures of highly bridged peptides: a new strategy for analysis." *Protein Sci* **2**(10): 1732–1748.

Grossweiner LI (1984). "Photochemistry of proteins: a Review." *Curr Eye Res* **3**(1): 137–144.

Gu S, Wen D, Weinreb PH, et al. (2010). "Characterization of trisulfide modification in antibodies." *Anal Biochem* **400**(1): 89–98.

Guo M, Chong YE, Shapiro R, et al. (2009). "Paradox of mistranslation of serine for alanine caused by AlaRS recognition dilemma." *Nature* **462**(7274): 808–812.

Guo D, Gao A, Michels DA, et al. (2010). "Mechanisms of unintended amino acid sequence changes in recombinant monoclonal antibodies expressed in Chinese hamster ovary (CHO) cells." *Biotechnol Bioeng* **107**(1): 163–171.

Guo A, Han M, Martinez T, et al. (2008). "Electrophoretic evidence for the presence of structural isoforms specific for the IgG2 isotype." *Electrophoresis* **29**(12): 2550–2556.

Hakansson K, Chalmers MJ, Quinn JP, et al. (2003). "Combined electron capture and infrared multiphoton dissociation for multistage MS/MS in a Fourier transform ion cyclotron resonance mass spectrometer." *Anal Chem* **75**(13): 3256–3262.

Hakansson K, Cooper HJ, Emmett MR, et al. (2001). "Electron capture dissociation and infrared multiphoton dissociation MS/MS of an N-glycosylated tryptic peptic to yield complementary sequence information." *Anal Chem* **73**(18): 4530–4536.

Hamblett KJ, Senter PD, Chace DF, et al. (2004). "Effects of drug loading on the antitumor activity of a monoclonal antibody drug conjugate." *Clin Cancer Res* **10**(20): 7063–7070.

Han X, Jin M, Breuker K, et al. (2006). "Extending top-down mass spectrometry to proteins with masses greater than 200 kilodaltons." *Science* **314**(5796): 109–112.

Harris RJ (1995). "Processing of C-terminal lysine and arginine residues of proteins isolated from mammalian cell culture." *J Chromatogr A* **705**(1): 129–134.

Harris RJ, Murnane AA, Utter SL, et al. (1993). "Assessing genetic heterogeneity in production cell lines: detection by peptide mapping of a low level Tyr to Gln sequence variant in a recombinant antibody." *Nat Biotechnol* **11**(11): 1293–1297.

Hatahet F and Ruddock LW (2007). "Substrate recognition by the protein disulfide isomerases." *FEBS J* **274**(20): 5223–5234.

Heads JT, Adams R, D'Hooghe LE, et al. (2012). "Relative stabilities of IgG1 and IgG4 Fab domains: influence of the light-heavy interchain disulfide bond architecture." *Protein Sci* **21**(9): 1315–1322.

Hodge JE (1953). "Dehydrated foods, chemistry of browning reactions in model systems." *J Agric Food Chem* **1**(15): 928–943.

Hoffman MZ and Hayon E (1972). "One-electron reduction of the disulfide linkage in aqueous solution. Formation, protonation, and decay kinetics of the RSSR-radical." *J Am Chem Soc* **94**(23): 7950–7957.

Hood BD, Garner B and Truscott RJW (1999). "Human lens coloration and aging: evidence for crystallin modification by the major ultraviolet filter, 3-hydroxy-kynurenineo-β-d-glucoside." *J Biol Chem* **274**(46): 32547–32550.

Hossler P, Khattak SF and Li ZJ (2009). "Optimal and consistent protein glycosylation in mammalian cell culture." *Glycobiology* **19**(9): 936–949.

Huang SY, Hsieh YT, Chen CH, et al. (2012a). "Automatic disulfide bond assignment using a1 ion screening by mass spectrometry for structural characterization of protein pharmaceuticals." *Anal Chem* **84**(11): 4900–4906.

Huang Y, Mechref Y and Novotny MV (2001). "Microscale nonreductive release of O-linked glycans for subsequent analysis through MALDI mass spectrometry and capillary electrophoresis." *Anal Chem* **73**(24): 6063–6069.

Huang Y, O'Mara B, Conover M, et al. (2012b). "Glycine to glutamic acid misincorporation observed in a recombinant protein expressed by *Escherichia coli* cells." *Protein Sci* **21**(5): 625–632.

Huddleston MJ, Bean MF and Carr SA (1993). "Collisional fragmentation of glycopeptides by electrospray ionization LC/MS and LC/MS/MS: methods for selective detection of glycopeptides in protein digests." *Anal Chem* **65**(7): 877–884.

Huh JH, White AJ, Brych SR, et al. (2013). "The identification of free cysteine residues within antibodies and a potential role for free cysteine residues in covalent aggregation because of agitation stress." *J Pharm Sci* **102**(6): 1701–1711.

Humbard MA, Zhou G and Maupin-Furlow JA (2009). "The N-terminal penultimate residue of 20S proteasome alpha1 influences its N(alpha) acetylation and protein levels as well as growth rate and stress responses of *Haloferax volcanii*." *J Bacteriol* **191**(12): 3794–3803.

Hutterer KM, Hong RW, Lull J, et al. (2013). "Monoclonal antibody disulfide reduction during manufacturing: Untangling process effects from product effects." *mAbs* **5**(4): 608–613.

Hwa J, Reeves PJ, Klein-Seetharaman J, et al. (1999). "Structure and function in rhodopsin: further elucidation of the role of the intradiscal cysteines, Cys-110, -185, and -187, in rhodopsin folding and function." *Proc Natl Acad Sci U S A* **96**: 1932.

Institute of Ophthalmology UCL. "CIE (2008) physiologically-relevant 2-deg V(λ) luminous efficiency functions." Retrieved 06/20/2014, 2014, from http://www.cvrl.org/.

Itakura K, Uchida K and Kawakishi S (1994). "Selective formation of oxindole- and formylkynurenine-type products from tryptophan and its peptides treated with a superoxide-generating system in the presence of iron(III)-EDTA: a possible involvement with iron-oxygen complex." *Chem Res Toxicol* **7**(2): 185–190.

Ito H, Takegawa Y, Deguchi K, et al. (2006). "Direct structural assignment of neutral and sialylated N-glycans of glycopeptides using collision-induced dissociation MS^n spectral matching." *Rapid Commun Mass Spectrom* **20**(23): 3557–3565.

Jebanathirajah J, Steen H and Roepstorff P (2003). "Using optimized collision energies and high resolution, high accuracy fragment ion selection to improve glycopeptide detection by precursor ion scanning." *J Am Soc Mass Spectrom* **14**(7): 777–784.

Ji JA, Zhang B, Cheng W, et al. (2009). "Methionine, tryptophan, and histidine oxidation in a model protein, PTH: Mechanisms and stabilization." *J Pharm Sci* **98**(12): 4485–4500.

Jones LM, Zhang H, Cui W, et al. (2013). "Complementary MS methods assist conformational characterization of antibodies with altered S–S bonding networks." *J Am Soc Mass Spectrom* **24**(6): 835–845.

Junutula JR, Bhakta S, Raab H, et al. (2008). "Rapid identification of reactive cysteine residues for site-specific labeling of antibody-Fabs." *J Immunol Methods* **332**(1–2): 41–52.

Karabacak NM, Li L, Tiwari A, et al. (2009). "Sensitive and specific identification of wild type and variant proteins from 8 to 669 kDa using top-down mass spectrometry." *Mol Cell Proteomics* **8**(4): 846–856.

Kaufman RJ and Sharp PA (1982). "Amplification and expression of sequences cotransfected with a modular dihydrofolate reductase complementary DNA gene." *J Mol Biol* **159**(4): 601–621.

Kelleher NL (2004). "Top-down proteomics." *Anal Chem* **76**(11): 197a-203a.

Kerwin BA and Remmele RL (2007). "Protect from light: Photodegradation and protein biologics." *J Pharm Sci* **96**(6): 1468–1479.

Khetan A, Huang Ym, Dolnikova J, et al. (2010). "Control of misincorporation of serine for asparagine during antibody production using CHO cells." *Biotechnol Bioeng* **107**(1): 116–123.

Kim KM, Yi EC, Baker D, et al. (2001). "Post-translational modification of the N-terminal His tag interferes with the crystallization of the wild-type and mutant SH3 domains from chicken src tyrosine kinase." *Acta Crystallogr D Biol Crystallogr* **57**(Pt 5): 759–762.

Kotia RB and Raghani AR (2010). "Analysis of monoclonal antibody product heterogeneity resulting from alternate cleavage sites of signal peptide." *Anal Biochem* **399**(2): 190–195.

Kramer EB and Farabaugh PJ (2007). "The frequency of translational misreading errors in *E. coli* is largely determined by tRNA competition." *RNA* **13**(1): 87–96.

Kshirsagar R, McElearney K, Gilbert A, et al. (2012). "Controlling trisulfide modification in recombinant monoclonal antibody produced in fed-batch cell culture." *Biotechnol Bioeng* **109**(10): 2523–2532.

Kurahashi T, Miyazaki A, Suwan S, et al. (2001). "Extensive investigations on oxidized amino acid residues in H_2O_2-treated Cu,Zn-SOD protein with LC-ESI-Q-TOF-MS, MS/MS for the determination of the copper-binding site." *J Am Chem Soc* **123**(38): 9268–9278.

Laine RA, Pamidimukkala KM, French AD, et al. (1988). "Linkage position in oligosaccharides by fast atom bombardment ionization, collision-activated dissociation, tandem mass spectrometry and molecular modeling. l-Fucosylp-(.alpha.1.fwdarw.X)-d-*N*-acetyl-d-glucosaminylp-(.beta.1.fwdarw.3)-d-galactosylp-(.beta.1-*O*-methyl) where X = 3, 4, or 6." *J Am Chem Soc* **110**(21): 6931–6939.

Le Brun V, Friess W, Bassarab S, et al. (2010). "A critical evaluation of self-interaction chromatography as a predictive tool for the assessment of protein–protein interactions in protein formulation development: A case study of a therapeutic monoclonal antibody." *Eur J Pharm Biopharm* **75**(1): 16–25.

Lebkowski J, DuBridge R, Antell E, et al. (1984). "Transfected DNA is mutated in monkey, mouse, and human cells." *Mol Cell Biol* **4**(10): 1951–1960.

Li B, Borchardt RT, Topp EM, et al. (2003). "Racemization of an asparagine residue during peptide deamidation." *J Am Chem Soc* **125**(38): 11486–11487.

Li Y, Polozova A, Gruia F, et al. (2014). "Characterization of the degradation products of a color-changed monoclonal antibody: tryptophan-derived chromophores." *Anal Chem* **86**(14): 6850–6857.

Lin S, Baltzinger M and Remy P (1984). "Fast kinetic study of yeast phenylalanyl-tRNA synthetase: role of tRNAPhe in the discrimination between tyrosine and phenylalanine." *Biochemistry* **23**(18): 4109–4116.

Ling J and Söll D (2010). "Severe oxidative stress induces protein mistranslation through impairment of an aminoacyl-tRNA synthetase editing site." *Proc Natl Acad Sci U S A* **107**(9): 4028–4033.

Liu H, Gaza-Bulseco G and Chumsae C (2008). "Glutamine deamidation of a recombinant monoclonal antibody." *Rapid Commun Mass Spectrom* **22**(24): 4081–4088.

Liu YD, Goetze AM, Bass RB, et al. (2011). "N-terminal glutamate to pyroglutamate conversion *in vivo* for human IgG2 antibodies." *J Biol Chem* **286**(13): 11211–11217.

Liu P, Tarnowski MA, O'Mara BW, et al. (2009). "Characterization of S-thiolation on secreted proteins from *E. coli* by mass spectrometry." *Rapid Commun Mass Spectrom* **23**(20): 3343–3349.

Loftfield RB (1963). "The frequency of errors in protein biosynthesis." *Biochem J* **89**(1): 82.

Lu C, Liu D, Liu H, et al. (2013). "Characterization of monoclonal antibody size variants containing extra light chains." *mAbs* **5**(1): 102–113.

Luo J, Zhang J, Ren D, et al. (2012). "Probing of C-terminal lysine variation in a recombinant monoclonal antibody production using Chinese hamster ovary cells with chemically defined media." *Biotechnol Bioeng* **109**(9): 2306–2315.

Mach H and Middaugh CR (2011). "Ultraviolet spectroscopy as a tool in therapeutic protein development." *J Pharm Sci* **100**(4): 1214–1227.

Maillard L (1912). "Action of amino acids on sugars. Formation of melanoidins in a methodical way." *Compt Rendus Geosci* **154**: 66–68.

Mallick P, Boutz DR, Eisenberg D, et al. (2002). "Genomic evidence that the intracellular proteins of archaeal microbes contain disulfide bonds." *Proc Natl Acad Sci U S A* **99**(15): 9679–9684.

Mamathambika BS and Bardwell JC (2008). "Disulfide-linked protein folding pathways." *Annu Rev Cell Dev Biol* **24**: 211–235.

Manning M, Chou D, Murphy B, et al. (2010). "Stability of Protein Pharmaceuticals: An Update." *Pharm Res* **27**(4): 544–575.

Mao Y, Valeja SG, Rouse JC, et al. (2013). "Top-down structural analysis of an intact monoclonal antibody by electron capture dissociation-Fourier transform ion cyclotron resonance-mass spectrometry." *Anal Chem* **85**(9): 4239–4246.

Martinek V, Sklenar J, Dracinsky M, et al. (2010). "Glycosylation protects proteins against free radicals generated from toxic xenobiotics." *Toxicol Sci* **117**(2): 359–374.

Martinez T, Guo A, Allen MJ, et al. (2008). "Disulfide connectivity of human immunoglobulin G2 structural isoforms." *Biochemistry* **47**(28): 7496–7508.

Masters PM, Bada JL and Zigler JS (1977). "Aspartic acid racemisation in the human lens during ageing and in cataract formation." *Nature* **268**: 71–73

Masters PM, Bada JL and Zigler JS (1978). "Aspartic acid racemization in heavy molecular weight crystallins and water insoluble protein from normal human lenses and cataracts." *Proc Natl Acad Sci U S A* **75**(3): 1204–1208.

Medzihradszky KF, Besman MJ and Burlingame AL (1997). "Structural characterization of site-specific N-glycosylation of recombinant human factor VIII by reversed-phase high-performance liquid chromatography-electrospray ionization mass spectrometry." *Anal Chem* **69**(19): 3986–3994.

Melchers J, Dirdjaja N, Ruppert T, et al. (2007). "Glutathionylation of trypanosomal thiol redox proteins." *J Biol Chem* **282**(12): 8678–8694.

Merry AH, Neville DC, Royle L, et al. (2002). "Recovery of intact 2-aminobenzamide-labeled O-glycans released from glycoproteins by hydrazinolysis." *Anal Biochem* **304**(1): 91–99.

Mikhailov VA, Iniesta J and Cooper HJ (2010). "Top-down mass analysis of protein tyrosine nitration: comparison of electron capture dissociation with "slow-heating" tandem mass spectrometry methods." *Anal Chem* **82**(17): 7283–7292.

Min H, Han D, Kim Y, et al. (2014). "Label-free quantitative proteomics and N-terminal analysis of human metastatic lung cancer cells." *Mol Cells* **37**(6): 457–466.

Mirgorodskaya E, Roepstorff P and Zubarev RA (1999). "Localization of O-glycosylation sites in peptides by electron capture dissociation in a Fourier transform mass spectrometer." *Anal Chem* **71**(20): 4431–4436.

Morelle W, Canis K, Chirat F, et al. (2006). "The use of mass spectrometry for the proteomic analysis of glycosylation." *Proteomics* **6**(14): 3993–4015.

Mrazek H, Weignerova L, Bojarova P, et al. (2013). "Carbohydrate synthesis and biosynthesis technologies for cracking of the glycan code: recent advances." *Biotechnol Adv* **31**(1): 17–37.

Mueller DR, Domon BM, Blum W, et al. (1988). "Direct stereochemical assignment of sugar subunits in naturally occurring glycosides by low energy collision induced dissociation. Application to papulacandin antibiotics." *Biomed Environ Mass Spectrom* **15**(8): 441–446.

Murad W and Singh R (2013). "MS2DB+: a software for determination of disulfide bonds using multi-ion analysis." *IEEE Trans Nanobiosci* **12**(2): 69–71.

Nieva J and Wentworth Jr P (2004). "The antibody-catalyzed water oxidation pathway–a new chemical arm to immune defense?" *Trends Biochem Sci* **29**(5): 274–278.

Ninio J (1991). "Transient mutators: a semiquantitative analysis of the influence of translation and transcription errors on mutation rates." *Genetics* **129**(3): 957–962.

O'Connor BD and Yeates TO (2004). "GDAP: a web tool for genome-wide protein disulfide bond prediction." *Nucleic Acids Res* **32**(Web Server issue): W360-W364.

Oien DB and Moskovitz J (2008). "Substrates of the methionine sulfoxide reductase system and their physiological relevance." *Curr Top Dev Biol* **80**: 93–133.

Opanasopit P, Shirashi K, Nishikawa M, et al. (2001). "*In vivo* recognition of mannosylated proteins by hepatic mannose receptors and mannan-binding protein." *Am J Physiol Gastrointest Liver Physiol* **280**(5): G879-G889.
Pace AL, Wong RL, Zhang YT, et al. (2013). "Asparagine deamidation dependence on buffer type, pH, and temperature." *J Pharm Sci* **102**(6): 1712–1723.
Parker J (1989). "Errors and alternatives in reading the universal genetic code." *Microbiol Rev* **53**(3): 273.
Patterson JT, Asano S, Li X, et al. (2014). "Improving the serum stability of site-specific antibody conjugates with sulfone linkers." *Bioconjug Chem* **25**: 1402–1407.
Patterson SD and Katta V (1994). "Prompt fragmentation of disulfide-linked peptides during matrix-assisted laser desorption ionization mass spectrometry." *Anal Chem* **66**(21): 3727–3732.
Perona JJ, Swanson RN, Rould MA, et al. (1989). "Structural basis for misaminoacylation by mutant *E. coli* glutaminyl-tRNA synthetase enzymes." *Science* **246**(4934): 1152–1154.
Pigault C and Gerard D (1984). "Influence of the location of tryptophanyl residues in proteins on their photosensitivity." *Photochem Photobiol* **40**(3): 291–296.
Pirie A (1971). "Formation of *N'*-formylkynurenine in proteins from lens and other sources by exposure to sunlight." *Biochem J* **125**(1): 203–208.
Pirie A (1972). "Fluorescence of *N'*-formylkynurenine and of protein exposed to sunlight." *Biochem J* **128**(5): 1365–1367.
Plihal O, Sklenar J, Hofbauerova K, et al. (2007). "Large propeptides of fungal beta-*N*-acetylhexosaminidases are novel enzyme regulators that must be intracellularly processed to control activity, dimerization, and secretion into the extracellular environment." *Biochemistry* **46**(10): 2719–2734.
Prentice KM, Gillespie R, Lewis N, et al. (2013). "Hydroxocobalamin association during cell culture results in pink therapeutic proteins." *mAbs* **5**(6): 974–981.
Qi P, Volkin DB, Zhao H, et al. (2009). "Characterization of the photodegradation of a human IgG1 monoclonal antibody formulated as a high-concentration liquid dosage form." *J Pharm Sci* **98**(9): 3117–3130.
Qi J, Wu J, Somkuti GA, et al. (2001). *Biochemistry* **40**: 4531.
Que AH, Zhang B, Yang Y, et al. (2010). "Sequence variant analysis using peptide mapping by LC-MS/MS." *BioProcess Int* **8**: 52–60.
Raju TS (2008). "Terminal sugars of Fc glycans influence antibody effector functions of IgGs." *Curr Opin Immunol* **20**(4): 471–478.
Rebeck GW and Samson L (1991). "Increased spontaneous mutation and alkylation sensitivity of *Escherichia coli* strains lacking the ogt O6-methylguanine DNA repair methyltransferase." *J Bacteriol* **173**(6): 2068–2076.
Reid GE and McLuckey SA (2002). "'Top down' protein characterization via tandem mass spectrometry." *J Mass Spectrom* **37**(7): 663–675.
Ringold G, Dieckmann B and Lee F (1980). "Co-expression and amplification of dihydrofolate reductase cDNA and the *Escherichia coli* XGPRT gene in Chinese hamster ovary cells." *J Mol Appl Genet* **1**(3): 165–175.
Ritchie MA, Gill AC, Deery MJ, et al. (2002). "Precursor ion scanning for detection and structural characterization of heterogeneous glycopeptide mixtures." *J Am Soc Mass Spectrom* **13**(9): 1065–1077.
Royle L, Radcliffe CM, Dwek RA, et al. (2006). "Detailed structural analysis of N-glycans released from glycoproteins in SDS-PAGE gel bands using HPLC combined with exoglycosidase array digestions." *Methods Mol Biol* **347**: 125–143.
Ryan CM, Souda P, Bassilian S, et al. (2010). "Post-translational modifications of integral membrane proteins resolved by top-down Fourier transform mass spectrometry with collisionally activated dissociation." *Mol Cell Proteomics* **9**(5): 791–803.
Ryslava H, Doubnerova V, Kavan D, et al. (2013). "Effect of posttranslational modifications on enzyme function and assembly." *J Proteomics* **92**: 80–109.
Ryslava H, Kalendova A, Doubnerova V, et al. (2011). "Enzymatic characterization and molecular modeling of an evolutionarily interesting fungal beta-N-acetylhexosaminidase." *FEBS J* **278**(14): 2469–2484.
Saleem A, Azam SS and Zarina S (2012). "Docking and molecular dynamics simulation studies on glycation-induced conformational changes of human paraoxonase 1." *Eur Biophys J* **41**(2): 241–248.
Santora LC, Stanley K, Krull IS, et al. (2006). "Characterization of maleuric acid derivatives on transgenic human monoclonal antibody due to post-secretional modifications in goat milk." *Biomed Chromatogr* **20**(9): 843–856.
Schön A, Kannangara CG, Cough S, et al. (1988). "Protein biosynthesis in organelles requires misaminoacylation of tRNA." *Nature* **331**(6152): 187–190.
Schumacher FF, Nunes JP, Maruani A, et al. (2014). "Next generation maleimides enable the controlled assembly of antibody-drug conjugates via native disulfide bond bridging." *Org Biomol Chem* **12**(37): 7261–7269.
Seetharam R, Heeren RA, Wong EY, et al. (1988). "Mistranslation in IGF-1 during over-expression of the protein in *Escherichia coli* using a synthetic gene containing low frequency codons." *Biochem Biophys Res Commun* **155**(1): 518–523.
Seipert RR, Dodds ED, Clowers BH, et al. (2008). "Factors that influence fragmentation behavior of N-linked glycopeptide ions." *Anal Chem* **80**(10): 3684–3692.

Sell DR and Monnier VM (1989). “Structure elucidation of a senescence cross-link from human extracellular matrix. Implication of pentoses in the aging process.” *J Biol Chem* **264**(36): 21597–21602.

Shin HC, Narayan M, Song MC, et al. (2003). “Role of the [65–72] disulfide bond in oxidative folding of bovine pancreatic ribonuclease A.” *Biochemistry* **42**(39): 11514–11519.

Shire SJ, Shahrokh Z and Liu J (2004). “Challenges in the development of high protein concentration formulations.” *J Pharm Sci* **93**(6): 1390–1402.

Simat TJ and Steinhart H (1998). “Oxidation of free tryptophan and tryptophan residues in peptides and proteins.” *J Agric Food Chem* **46**(2): 490–498.

Siuti N and Kelleher NL (2007). “Decoding protein modifications using top-down mass spectrometry.” *Nat Methods* **4**(10): 817–821.

Smith KC (1992). “Spontaneous mutagenesis: experimental, genetic and other factors.” *Mutat Res Rev Genet Toxicol* **277**(2): 139–162.

Smith T and Guild J (1931). “The CIE colorimetric standards and their use.” *Trans Opt Soc* **33**(3): 73.

Sparbier K, Koch S, Kessler I, et al. (2005). “Selective isolation of glycoproteins and glycopeptides for MALDI-TOF MS detection supported by magnetic particles.” *J Biomol Tech* **16**(4): 407–413.

Sreedhara A, Lau K, Li C, et al. (2012). “Role of surface exposed tryptophan as substrate generators for the antibody catalyzed water oxidation pathway.” *Mol Pharm* **10**(1): 278–288.

Stadtman ER (1993). “Oxidation of free amino acids and amino acid residues in proteins by radiolysis and by metal-catalyzed reactions.” *Annu Rev Biochem* **62**: 797–821.

Stadtman E and Berlett B (1991). “Fenton chemistry. Amino acid oxidation.” *J Biol Chem* **266**(26): 17201–17211.

Stockman A, Jägle H, Pirzer M, et al. (2008). “The dependence of luminous efficiency on chromatic adaptation.” *J Vis* **8**(16): 1–26.

Sun S, Zhou JY, Yang W, et al. (2014). “Inhibition of protein carbamylation in urea solution using ammonium-containing buffers.” *Anal Biochem* **446**: 76–81.

Swanson R, Hoben P, Sumner-Smith M, et al. (1988). “Accuracy of *in vivo* aminoacylation requires proper balance of tRNA and aminoacyl-tRNA synthetase.” *Science* **242**(4885): 1548–1551.

Sydow JF, Lipsmeier F, Larraillet V, et al. (2014). “Structure-based prediction of asparagine and aspartate degradation sites in antibody variable regions.” *PLoS ONE* **9**(6): e100736.

Syka JE, Coon JJ, Schroeder MJ, et al. (2004). “Peptide and protein sequence analysis by electron transfer dissociation mass spectrometry.” *Proc Natl Acad Sci U S A* **101**(26): 9528–9533.

Tanaka K, Takenaka S, Tsuyama S, et al. (2006). “Determination of unique amino acid substitutions in protein variants by peptide mass mapping with FT-ICR MS.” *J Am Soc Mass Spectrom* **17**(4): 508–513.

Tang L, Sundaram S, Zhang J, et al. (2013). “Conformational characterization of the charge variants of a human IgG1 monoclonal antibody using H/D exchange mass spectrometry.” *mAbs* **5**(1): 114–125.

Tao L and English AM (2004). “Protein S-glutathiolation triggered by decomposed S-nitrosoglutathione.” *Biochemistry* **43**(13): 4028–4038.

Tomita M, Irie M and Ukita T (1969). “Sensitized photooxidation of histidine and its derivatives. Products and mechanism of the reaction.” *Biochemistry* **8**(12): 5149–5160.

Tonie Wright H and Urry DW (1991). “Nonenzymatic deamidation of asparaginyl and glutaminyl residues in protein.” *Crit Rev Biochem Mol Biol* **26**(1): 1–52.

Torosantucci R, Schöneich C and Jiskoot W (2014). “Oxidation of therapeutic proteins and peptides: structural and biological consequences.” *Pharm Res* **31**(3): 541–553.

Tous GI, Wei Z, Feng J, et al. (2005). “Characterization of a novel modification to monoclonal antibodies: thioether cross-link of heavy and light chains.” *Anal Chem* **77**(9): 2675–2682.

Uchida K, Enomoto N, Itakura K, et al. (1990). “Formation of diastereoisomeric 3a-hydroxypyrroloindoles from a tryptophan residue analog mediated by iron(II)-EDTA and l-ascorbate.” *Arch Biochem Biophys* **279**(1): 14–20.

Vanek O, Brynda J, Hofbauerova K, et al. (2011). “Crystallization and diffraction analysis of beta-*N*-acetylhexosaminidase from *Aspergillus oryzae*.” *Acta Crystallogr Sect F: Struct Biol Cryst Commun* **67**(Pt 4): 498–503.

Vanhooren A, Devreese B, Vanhee K, et al. (2002). “Photoexcitation of tryptophan groups induces reduction of two disulfide bonds in goat α-lactalbumin.” *Biochemistry* **41**(36): 11035–11043.

Vijayasankaran N, Varma S, Yang Y, et al. (2013). “Effect of cell culture medium components on color of formulated monoclonal antibody drug substance.” *Biotechnol Prog* **29**(5): 1270–1277.

Vlasak J and Ionescu R (2008). “Heterogeneity of monoclonal antibodies revealed by charge-sensitive methods.” *Curr Pharm Biotechnol* **9**(6): 468–481.

Vliegenthart JF and Casset F (1998). “Novel forms of protein glycosylation.” *Curr Opin Struct Biol* **8**(5): 565–571.

Wald G (1964). “The receptors of human color vision: action spectra of three visual pigments in human cones account for normal color vision and color-blindness." *Science* **145**(3636): 1007–1016.

Walrant P and Santus R (1974). "*N*-formyl-kynurenine, a tryptophan photooxidation product, as a photodynamic sensitizer." *Photochem Photobiol* **19**(6): 411–417.

Wan M, Shiau FY, Gordon W, et al. (1999). "Variant antibody identification by peptide mapping." *Biotechnol Bioeng* **62**(4): 485–488.

Wang W, Vlasak J, Li Y, et al. (2011). "Impact of methionine oxidation in human IgG1 Fc on serum half-life of monoclonal antibodies." *Mol Immunol* **48**(6–7): 860–866.

Ward NE, Stewart JR, Ioannides CG, et al. (2000). "Oxidant-induced S-glutathiolation inactivates protein kinase C-alpha (PKC-alpha): a potential mechanism of PKC isozyme regulation." *Biochemistry* **39**(33): 10319–10329.

Wasylaschuk WR, Harmon PA, Wagner G, et al. (2007). "Evaluation of hydroperoxides in common pharmaceutical excipients." *J Pharm Sci* **96**(1): 106–116.

Waterman KC, Adami RC, Alsante KM, et al. (2002). "Stabilization of pharmaceuticals to oxidative degradation." *Pharm Dev Technol* **7**(1): 1–32.

Webb RB (1977). Lethal and mutagenic effects of near-ultraviolet radiation, in *Photochemical and Photobiological Reviews*, Smith KC, Springer: 169–261.

Webb RB and Lorenz JR (1972). "Toxicity of irradiated medium for repair-deficient strains of *Escherichia coli*." *J Bacteriol* **112**(1): 649–652.

Webb RB and Malina MM (1970). "Mutagenic effects of near ultraviolet and visible radiant energy on continuous cultures of *Escherichia coli*." *Photochem Photobiol* **12**(6): 457–468.

Wei Z, Feng J, Lin H-Y, et al. (2007). "Identification of a single tryptophan residue as critical for binding activity in a humanized monoclonal antibody against respiratory syncytial virus." *Anal Chem* **79**(7): 2797–2805.

Wen D, Vecchi MM, Gu S, et al. (2009). "Discovery and investigation of misincorporation of serine at asparagine positions in recombinant proteins expressed in Chinese hamster ovary cells." *J Biol Chem* **284**(47): 32686–32694.

Wentworth AD, Jones LH, Wentworth P, et al. (2000). "Antibodies have the intrinsic capacity to destroy antigens." *Proc Natl Acad Sci U S A* **97**(20): 10930–10935.

Wentworth P, Jones LH, Wentworth AD, et al. (2001). "Antibody catalysis of the oxidation of water." *Science* **293**(5536): 1806–1811.

Willy M (2009). "Analysis of glycosylation and other post-translational modifications by mass spectrometry." *Curr Anal Chem* **5**: 144–165.

Winder AF and Gent WLG (1971). "Correction of light-scattering errors in spectrophotometric protein determinations." *Biopolymers* **10**(7): 1243–1251.

Woo HJ, Lotz MM, Jung JU, et al. (1991). "Carbohydrate-binding protein 35 (Mac-2), a laminin-binding lectin, forms functional dimers using cysteine 186." *J Biol Chem* **266**(28): 18419–18422.

Wootton SK and Yoo D (2003). "Homo-oligomerization of the porcine reproductive and respiratory syndrome virus nucleocapsid protein and the role of disulfide linkages." *J Virol* **77**(8): 4546–4557.

Wright WD (1929). "A re-determination of the trichromatic coefficients of the spectral colours." *Trans Opt Soc* **30**(4): 141.

Wu J and Watson JT (1998). *Anal Biochem* **258**: 268.

Wu J, Yang Y and Watson JT (1998). "Trapping of intermediates during the refolding of recombinant human epidermal growth factor (hEGF) by cyanylation, and subsequent structural elucidation by mass spectrometry." *Protein Sci* **7**: 1017.

Wuhrer M, Balog CI, Koeleman CA, et al. (2005). "New features of site-specific horseradish peroxidase (HRP) glycosylation uncovered by nano-LC-MS with repeated ion-isolation/fragmentation cycles." *Biochim Biophys Acta* **1723**(1–3): 229–239.

Wuhrer M, Catalina MI, Deelder AM, et al. (2007). "Glycoproteomics based on tandem mass spectrometry of glycopeptides." *J Chromatogr B Analyt Technol Biomed Life Sci* **849**(1–2): 115–128.

Wurm FM (1990). "Integration, amplification and stability of plasmid sequences in CHO cell cultures." *Biologicals* **18**(3): 159–164.

Xu J, Jin M, Song H, et al. (2014). "Brown drug substance color investigation in cell culture manufacturing using chemically defined media: A case study." *Process Biochem* **49**(1): 130–139.

Xu W, Peng Y, Wang F, et al. (2013). "Method to convert N-terminal glutamine to pyroglutamate for characterization of recombinant monoclonal antibodies." *Anal Biochem* **436**(1): 10–12.

Yan SD, Chen X, Schmidt AM, et al. (1994). "Glycated tau protein in Alzheimer disease: a mechanism for induction of oxidant stress." *Proc Natl Acad Sci U S A* **91**(16): 7787–7791.

Yang J, Primack R, Frohn M, et al. (2015). "Impact of glycation on antibody clearance." *AAPS J* **17**(1): 237–244.

Yang Y, Strahan A, Li C, et al. (2010). "Detecting low level sequence variants in recombinant monoclonal antibodies." *mAbs* **2**(3): 285–298.

Yang J, Wang S, Liu J, et al. (2007). "Determination of tryptophan oxidation of monoclonal antibody by reversed phase high performance liquid chromatography." *J Chromatogr A* **1156**(1–2): 174–182.

Yu XC, Borisov OV, Alvarez M, et al. (2009). "Identification of codon-specific serine to asparagine mistranslation in recombinant monoclonal antibodies by high-resolution mass spectrometry." *Anal Chem* **81**(22): 9282–9290.

Yuk IH, Zhang B, Yang Y, et al. (2011). "Controlling glycation of recombinant antibody in fed-batch cell cultures." *Biotechnol Bioeng* **108**(11): 2600–2610.

Zhang Z (2012). "Automated precursor ion exclusion during LC-MS/MS data acquisition for optimal ion identification." *J Am Soc Mass Spectrom* **23**(8): 1400–1407.

Zhang H and Aebersold R (2006). "Isolation of glycoproteins and identification of their N-linked glycosylation sites." *Methods Mol Biol* **328**: 177–185.

Zhang Q and Flynn GC (2013). "Cysteine racemization on IgG heavy and light chains." *J Biol Chem* **288**(48): 34325–34335.

Zhang B, Harder AG, Connelly HM, et al. (2010). "Determination of Fab-hinge disulfide connectivity in structural isoforms of a recombinant human immunoglobulin G2 antibody." *Anal Chem* **82**(3): 1090–1099.

Zhang M and Kaltashov IA (2006). "Mapping of protein disulfide bonds using negative ion fragmentation with a broadband precursor selection." *Anal Chem* **78**(14): 4820–4829.

Zhang H, Li XJ, Martin DB, et al. (2003). "Identification and quantification of N-linked glycoproteins using hydrazide chemistry, stable isotope labeling and mass spectrometry." *Nat Biotechnol* **21**(6): 660–666.

Zhang Z, Pan H and Chen X (2009). "Mass spectrometry for structural characterization of therapeutic antibodies." *Mass Spectrom Rev* **28**(1): 147–176.

Zhang Q, Schenauer MR, McCarter JD, et al. (2013a). "IgG1 thioether bond formation *in vivo*." *J Biol Chem* **288**(23): 16371–16382.

Zhang Z, Shah B and Bondarenko P (2013b). The Main Cause of Amino Acid Misincorporations in Recombinant Proteins. The 61st ASMS Conference on Mass Spectrometry & Allied Topics, Minneapolis, MN.

Zhang B, Yang Y, Yuk I, et al. (2008). "Unveiling a glycation hot spot in a recombinant humanized monoclonal antibody." *Anal Chem* **80**(7): 2379–2390.

Zhao J, Simeone DM, Heidt D, et al. (2006). "Comparative serum glycoproteomics using lectin selected sialic acid glycoproteins with mass spectrometric analysis: application to pancreatic cancer serum." *J Proteome Res* **5**(7): 1792–1802.

Zhou S, Zhang B, Sturm E, et al. (2010). "Comparative evaluation of disodium edetate and diethylenetriaminepentaacetic acid as iron chelators to prevent metal-catalyzed destabilization of a therapeutic monoclonal antibody." *J Pharm Sci* **99**(10): 4239–4250.

Zhu ZC, Chen Y, Ackerman MS, et al. (2013). "Investigation of monoclonal antibody fragmentation artifacts in non-reducing SDS-PAGE." *J Pharm Biomed Anal* **83**: 89–95.

Zhu X, Wentworth P, Kyle RA, et al. (2006). "Cofactor-containing antibodies: crystal structure of the original yellow antibody." *Proc Natl Acad Sci U S A* **103**(10): 3581–3585.

Zubarev RA (2003). "Reactions of polypeptide ions with electrons in the gas phase." *Mass Spectrom Rev* **22**(1): 57–77.

Zubarev RA (2004). "Electron-capture dissociation tandem mass spectrometry." *Curr Opin Biotechnol* **15**(1): 12–16.

Zubarev RA, Horn DM, Fridriksson EK, et al. (2000). "Electron capture dissociation for structural characterization of multiply charged protein cations." *Anal Chem* **72**(3): 563–573.

Zubarev RA, Kruger NA, Fridriksson EK, et al. (1999). "Electron capture dissociation of gaseous multiply-charged proteins is favored at disulfide bonds and other sites of high hydrogen atom affinity." *J Am Chem Soc* **121**(12): 2857–2862.

INDEX

Note: Page numbers followed by "f" and "t" refer to figures and tables.

a

acetylation 130, 225, 234, 239
acid-labile cleavable linker 185, 194, 198, 215
Adcetris® 183, 185, 198
adenosine triphosphate (ATP) 58, 59f, 207
ADME of antibody–drug conjugates 185
ado-trastuzumab emtansine (T-DM1) 183
adsorption 136, 211
advanced glycation end products (AGEs) 226, 230
affinity capture 62, 84, 176, 186
  elution 91
  immunoaffinity capture 58, 64, 68, 148, 150, 151f, 155, 208
  peptide 17
  protein 17, 137
affinity enrichment 58, 129, 131, 135
  immunoaffinity enrichment 129, 157, 210
affinity purification 5, 16, 17, 117, 119, 189, 193
  immunoaffinity purification 148, 208
albumin 16, 84, 100, 132, 136, 148, 152, 176, 191, 205–207, 226
AlphaLISA™ 83
alternative splicing, and human diseases 55–56
Amadori products 226, 230
ammonium bicarbonate 112
AMP-activated protein kinase (AMPK) 58–62, 59f, 60–61t
animal studies
  universal peptide LC-MS/MS assay, for human mAbs bioanalysis 161–170
antibodies
  antidrug, *see* antidrug antibodies (ADAs)
  anti-idiotype 108
  monoclonal, *see* monoclonal antibody(ies)
  polyclonal 108, 110
  therapeutic 83–84
antibody-dependent cellular cytotoxicity (ADCC) 189
antibody–drug conjugates (ADCs) 11, 16–18, 21, 67, 70, 84, 132
  ADME 185
  catabolite quantitation by LC–MS/MS 196–197
  circulating, DAR distribution of 188–189
  deglycosylation of 187–188
  disulfide bond in 224–225, 225f
  drug-to-antibody ratio 84
  future perspectives of 198
  linker chemistry 185
  mass-spectrometry-based methodologies 183–186
  mechanism of action 183–186, 184f
  pharmacokinetic characterization of, during drug development 183–198
  preclinical and clinical pharmacokinetic support 197–198, 197f
  structure and characteristics of 184t
  total antibody quantitation by ligand binding or LC–MS/MS 189–193, 190f, 192t
  total conjugated drug quantitation by ligand binding or LC–MS/MS 193–196, 196t
  toxins 185
  unique bioanalytical challenges to 185–186
antidrug antibodies (ADAs) 11, 136, 146, 161, 173
  bridging assay 107, 109f
  multiplexing direct measurement of 93–95, 94f, 95t
  by quantifying binding measurements, indirect measurement of 95–96, 96f
  semiquantification and isotyping of 91–97
antigen-binding fragments 108
antihuman Fc antibody 210
anti-idiotype antibodies 108
anti-mAb antibodies
  impact on IgG4 clinical LC-MS/MS assay performance 173, 173f
anti-peptide capture 80–81
anti-protein capture 78–80, 79f
Apidra® 82
ApoC3 protein 137–138, 138f
arrival time distribution (ATD) 2
asparagine (Asp)
  deamidation of 24–25
  N-linked glycosylation 29
aspiration ion mobility spectrometry (AIMS) 1
Asp-N 15, 78
automatic gain control (AGC) 13, 22–24, 27–29

b

base-induced β-elimination 239
basic local alignment search tool (BLAST) 16, 146
bead extraction and acid dissociation (BEAD) 96–97
β-amyloid (Aβ) 75, 76f
Biocore 136

Protein Analysis using Mass Spectrometry: Accelerating Protein Biotherapeutics from Lab to Patient, First Edition.
Edited by Mike S. Lee and Qin C. Ji.

biologics license applications (BLAs) 107
biotherapeutics, IA–MS applications in 81–84
bispecific antibodies 16, 21, 156
BMP1, alternative splicing of 58
bone morphogenetic protein (BMP) 58
bovine serum albumin (BSA) 136, 226
brain natriuretic peptide (BNP) 76, 77f
brentuximab vedotin 183
bystander effect 185

c
capillary electrophoresis (CE) 102
capillary electrophoresis coupled to mass spectrometry (CE–MS) 22, 102
capillary electrophoresis-sodium dodecyl sulfate (CE-SDS) 21
capillary isotachophoresis (CITP) 102
capillary zone electrophoresis (CZE) 102
carcinoembryonic antigen (CEA) 80
cardiac troponin I (cTnI) 81
CD2-Fc protein 28
cell-based neutralizing antibody assays, method development of 96–97
cetuximab 28, 209
Chinese hamster ovary (CHO) cells 122, 227, 231
chromophores 229–230, 238
chymotrypsin 14–15
cibacron blue dye 207
citric acid 225
Clinical Proteomic Technology Assessment for Cancer (CPTAC) 81
collision cross section (CCS) 1, 2, 3f, 7, 8, 223
collision-induced dissociation (CID) 4, 5, 22, 23, 25–29, 237–238
colorimetry 229
combined fractional diagonal chromatography (COFRADIC) 225
complementarity determining regions (CDRs) 108, 147–148, 215
complement-dependent cytotoxicity (CDC) 189
complex mixtures, IM–MS and LC–IM–MS analysis of 2–7, 3–7f
cone cells, in retina 228
critical quality attributes (CQAs) 18, 226
critical reagents, monitoring quality of 107–126
  conjugated critical reagent, incorporation ratio of 122, 123f
  integrity of reagent cell line production system 116
  LBA specificity loss during clinical development 116–122, 119–121f
  LC–MS characterization, importance of 123–124
  protein modifications, impact of 111t
  reagent construction prior to LBA method development, confirmation of 114–116, 115f, 117–118f
  structural analysis 113f
C-terminal lysine processing 109, 113
CTLA4-Fc protein 28
cyanocobalamin 230
cyanogen bromide (CNBr) 95, 96, 215
cyclosporin A (CsA) 137
cyclosporin G (CsG) 137
cysteine oxidation 24
cysteine-labeled trastuzumab antibodies 187

d
data-dependent analysis (DDA) 5, 6f
data-independent analysis (DIA) 4, 5, 13, 14
deamidation 24–26, 26f, 27f, 94, 109, 112, 228
  peptide 1 asparagine 164–165, 164f, 165f
  structure and sequence effect on 228
Delta14–15 56
diethylenetriamine pentaacetic acid (DTPA) 227
differential mobility spectrometry (DMS) 1, 39, 40f, 42
differential scanning calorimetry (DSC) 226
dihydroxyimidazolidine 226
discoloration 228–230
DiscoveryQuant™ software 40
disulfide bond, in antibody drug conjugates 224–225, 225f
disulfide isoforms 222–224, 223f, 223t
5,5’-dithiobis-(2-nitrobenzoic acid) (DTNB) 224
dithiothreitol (DTT) 14, 113, 113f, 115f, 116, 122, 123f, 136, 191
drift-time ion mobility spectrometry (DTIMS) 1
drug-to-antibody ratios (DARs) 84, 183, 224
  distribution of circulating ADCs 188–189
dual universal peptide LC-MS/MS assay, for human mAbs bioanalysis 165–170, 166f
  calibration curve/QC replicate acceptance criteria 168
  deployment of 168, 168f
  identification and evaluation of 165–166
  interpreting and reporting study 168–169
  light and heavy chain, quantitative evaluation and comparison of 167
  peptide 1 and peptide 2, quantitative agreement between 167
  related studies 169–170

e
Edman sequencing 45, 231
electron capture dissociation (ECD) 112, 237, 238
electron transfer dissociation (ETD) 22, 112, 237, 238
  posttranslational modification using 26–31, 30f, 31t
electrospray ionization (ESI) 1, 67, 72, 73f, 112, 186
electrospray ionization coupled to a quadrupole time-of-flight mass analyzer (ESI-QToF MS) 235

electrospray ionization–mass spectrometry (ESI–MS) 45, 99, 102–105
  interfaces 104f
  nanoelectrospray 103f
  nanospray, integrating with miniaturized liquid chromatography 46–47, 47f
electrospray ionization with time-of-flight mass spectrometry (ESI-ToF MS) 235
Endo-Leptin 82–83
endoplasmic reticulum (ER) 136
enzyme immunoassay (EIA) 68
enzyme-linked immunosorbent assay (ELISA) 35, 68, 71, 74, 75, 78, 80–83, 99, 129, 136, 137, 140, 145, 150, 155, 156, 189, 192t
Erbitux® 83
erythropoietin (EPO) 28
ethylenediamine tetraacetic acid (EDTA) 227, 229
European Medicine Agency (EMA) 183, 203
European Pharmacopoeia 228
extracellular domain (ECD) 188
extracted ion chromatography (XIC) 16, 29, 139

f

Fab-arm exchange 170, 171f
Fab fragments 21, 108, 113
Fc fragments 113
Fc fusion protein drug candidates bioanalysis
  peptide 1, applications of 162–164, 163f
  surrogate peptide identification 161–162
  universal peptide LC-MS/MS assay for 161–162
Fc-fusion proteins/peptides 17, 21, 28, 82
Fenton reaction 229
fibroblast growth factor 21 (FGF21) 83
field-asymmetric waveform ion mobility spectrometry (FAIMS), *see* differential mobility spectrometry (DMS)
flavin adenine dinucleotide (FAD) 207
Food and Drug Administration (FDA) 11, 18, 107, 203, 228
formic acid 112
Fourier transform ion cyclotron resonance (FT-ICR) 12, 13, 22
Fourier transform mass spectrometers (FTMS) 112, 137
free glycan analysis 238
free sulfhydryl 224
full-scan LC–HRMS, intact protein quantification using 131–133
  data acquisition and analysis 133, 134f
  sample preparation 131–132

g

gel-electrophoresis 231
gemtuzumab ozogamicin 183
generic peptide LC–MS/MS assays
  human mAb-based therapeutic agents 161
  internal standard options for 173–175
  ligand binding assay 161
  limitations of 177
  sample preparation strategies 175–177
  signature peptides 161
Glu-C 15
glucagon-like peptide 1 (GLP-1) 75, 77, 82
glutamine 136, 146, 225, 228
glycans
  release from glycoproteins 238–239
  sequence and linkage analysis 239, 239t
glycation 226
glycopeptide(s)
  analysis 237
  structure elucidation of, using multiple fragmentation mechanisms 28–31, 30f, 31t
glycoproteins
  characterization through mass spectrometry 234f
  intact analysis 235–237, 236f
  release of glycans from 238–239
  structure of 232–235, 232–234f
  tandem MS of 237–238
glycosylation 28–31, 30f, 82, 94
  free glycan analysis 238
  glycopeptide analysis 237
  glycoprotein structure 232–235, 232–234f
  intact glycoprotein analysis 235–237, 236f
  N-linked 109, 113, 137, 232–233, 232f
  O-linked 233–234, 233f, 234f
  tandem MS of 237–238
Golgi apparatus 136
growth hormone (GH) 136
growth hormone-binding protein (GHBP) 136
G-Tenecteplase 29
Gyros™ 83

h

HCD product-dependent ETD (HCDpdETD) method 28, 29, 30f
heated electrospray ionization (HESI) 24
high-abundant proteins, depletion of 131–132
high energy conversion dynode (HED) detection system 39
higher-energy collision dissociation (HCD) 13, 22
  posttranslational modification using 23–26, 23f, 25–27f
high-molecularweight (HMW) 229
high-performance liquid chromatography (HPLC) 131
  intact protein quantification using 132–133
high-pressure, high-resolution separations coupled with intelligent selection and multiplexing SRM MS (PRISM-SRM MS) 101–102, 101f
high-pressure liquid chromatography (HPLC) 35
high-resolution/accurate mass (HR/AM) detection 74
high-resolution accurate mass spectrometry (HRAM) 12, 18, 39–41, 41f
  for immunogenicity assessment 94

high-resolution mass spectrometers (HRMS) 12–13
  data processing 16
  protein quantitation approaches 14t
high-resolution multiple reaction monitoring (HR-MRM) 5, 7
histidine (His) 135
  oxidation 24, 227
Hu-14 100
Humalin® 107
Humalog® 82
human growth hormone analog (hGHA) 136
human mAb kappa light chain candidates 163t, 166
human mAbs bioanalysis, universal peptide LC–MS/MS assay for
  animal studies 161–170
  dual universal peptide LC–MS/MS assay candidates 165–170, 166f
  generic peptide LC–MS/MS assays 173–177
  human studies 170–175
  IgG4 clinical LC–MS/MS assay, development and evaluation of 171–172, 172t
  IgG4 clinical LC–MS/MS assay performance, anti-mAb antibodies impact on 173, 173f
  peptide 1, applications of 162–164, 163f
  peptide 1 asparagine deamidation on quantification, impact of 164–165, 164f, 165f
  peptide 1 variants, potential applicability of 163–164
  potential expansion of 170–171, 171f
  quantification of human mAb and Fc fusion protein drug candidates 161–162
  surrogate peptide identification 161–162
hybrid IA–MS methods 68
hydrazinolysis 239
hydrogen/deuterium exchange (HDX) mass spectrometry 226
hydroimidazolone 226
hydrophilic interaction liquid chromatography (HILIC) 211, 237
hydrophilic small proteins 16
hydrophobic interaction chromatography (HIC) 21, 188
3-hydroxy-3-methylglutaryl coenzyme A reductase (HMGCR) 55
hydroxytryptophan (HO-Trp) 229

i

IgG4 clinical LC–MS/MS assay
  development and evaluation of 171–172, 172t
  performance, anti-mAb antibodies impact on 173, 173f
2-imino-2-methoxyethyl-1-thiomannoside (IMT-mannose) 226
immunoaffinity-based mass spectrometry (IA–MS) 67–85
  applications of 74–84
  future direction of 74–75
  hybrid methods 68
  IA capture formats 71–72, 72f
  liquid chromatography 72–74, 73f, 74t
  mass spectrometry detection 74
  methods, classification of 69–71, 69f, 70t
  sample preparation 148, 150–151, 150f, 151f
  stable-isotope-labeled internal standards 71
immunoaffinity capture (IC) 58, 64, 68, 148, 150, 151f, 155, 208
immunoaffinity enrichment 129, 157, 210
immunoaffinity purification 148, 208–211, 209f
immunocapture, *see also* immunocapture LC–MS, for immunogenicity assessment
  of ADCs from plasma or serum 186–187, 187f
  genetic 210–211
  of specific peptide 208, 209f
  of specific protein 208–210
immunocapture LC–MS, for immunogenicity assessment
  future perspectives of 97
  multiplexing direct measurement of 93–95, 94f, 95t
  semiquantification and isotyping of antidrug antibodies by 91–97
immunocytokines 21
immunogenicity
  assessment, immunocapture LC/MS for 91–97
  defined 91
immunoglobulin A (IgA)
  structure of 93f
immunoglobulin D (IgD)
  structure of 93f
immunoglobulin-degrading enzyme of *Streptococcus pyogenes* (IdeS) 113, 113f, 215
immunoglobulin E (IgE)
  structure of 93f
immunoglobulin G (IgG) 147, 147f, 148
  structure of 93f
immunoglobulin G1 (IgG1) 28
immunoglobulin M (IgM)
  structure of 93f
Immunoglobulin Y14 (IgY14) 100
immunoprecipitation (IP) 70, 82
incurred samples reanalysis (ISR) 167
infrared multiphoton dissociation (IRMPD) 237
in silico prediction algorithm 228
insulin-like growth factor 1 (IGF-1) 78
intact protein quantification, using LC–HRMS 129–140
  adsorption 136
  bottom-up approach 129–130
  calibration and quality control sample strategy 135
  examples of 137–138, 138–140f
  full-scan LC–HRMS 131–133
  future perspectives of 138–140
  internal standard strategy 133–135
  posttranslational modifications 136–137
  specific protein binding 136
  stability 135–136
  top-down approach 130
intact protein quantitation 15–16, 15f
interleukin-21 (IL-21) 81
interleukin-33 (IL-33) 208
iodoacetamide (IAA) 14, 113

ion cyclotron resonance (ICR) mass analyzers 22
IonDrive™ QJet Ion Guide 39
IonDrive™ Turbo V Source 39
ion exchange chromatography (IEX) 21, 231
ion mobility–mass spectrometry (IM–MS) 22
  protein analysis by 1–8
isoelectric focusing chromatography (IEF) 21, 231

k

Kadcyla® 183, 190, 195, 198
Kcnq2 56
kynurenine (Kyn) 229

l

lanthionine 224
Lantus® 82
LC–MS/MS-based peptide quantitation, *see also* liquid chromatography tandem mass spectrometry (LC–MS/MS)
  high-resolution accurate mass 39–41, 41f
  methodologies 37t
  selectivity 39
  sensitivity 39
  workflow, benefits of 38
  workflow strategies 37f
LC-SRM-based triple quadrupole MS 12
Leu-Ile substitution 231
Leu-Leu-Glu-Leu-Ile-Arg (LLELIR) 76, 77f
Levemir® 82
Life Technologies 140
ligand binding assays (LBAs) 11, 35, 36, 38, 68, 79–83, 145, 155, 183, 189–190, 203
  critical reagents, monitoring quality of 107–126
  for immunogenicity assessment 91–93, 92f, 95
  protein reagent quality characteristics for 124t
  total antibody quantization by 189–190, 190f, 192–193, 192t
  total conjugated drug quantitation by 193–196, 196t
LINAC® collision cell 41
Linear Accelerator™ Trap Electrodes 39
linker chemistry 185
liquid chromatography–high-resolution mass spectrometry (LC–HRMS) 11
  intact protein quantification using 129–140
  regulated bioanalysis using 18
liquid chromatography–mass spectrometry (LC–MS) 11, 45–46
  cell-based neutralizing antibody assays, method development of 96–97
  immunoicapture, for immunogenicity assessment 91–97
  internal standard, for protein quantitation 17–18
  protein bioanalysis, sample preparation strategies for 203–216
  sample extraction 16–17
liquid chromatography tandem mass spectrometry (LC–MS/MS) 183, *see also* LC–MS/MS-based peptide quantitation
  catabolite quantitation by 196–197
  of monoclonal antibodies 145–158
  total antibody quantization by 190–193, 192t
  total conjugated drug quantitation by 193–196, 196t
liquid–liquid extraction (LLE) 195
low-density lipoprotein receptor (LDLR) 55
lower limit of quantitation (LLOQ) 67, 75–77, 76f, 79, 79f, 81, 100–102, 119, 148, 152–153, 155, 167, 172, 172f, 204
Lys-C 15
lysine variants 226

m

Maillard reaction 230
maleimide-based drug conjugation 225
malondialdehyde (MDA) 207
MASCOT database 16
  error tolerant search 231
Mason–Schamp equation 1
mass spectrometric immunoassay (MSIA) 17, 71, 74
mass spectrometry-based assay, for high-throughput and high-sensitivity biomarker verification 99–105, 100f
  advanced ESI–MS instrumentation 102–105, 103f, 104f
  sample proceeding strategies 100–102, 101f
mass spectrometry systems
  diversity of, advantages of 41, 42f
  future perspectives of 41–42
matrix-assisted laser desorption ionization mass spectrometry (MALDI-MS) 74, 80, 81, 112, 188
maximum entropy (MaxEnt) techniques 235
MCF-7 cells 56
2-mercaptoethanol 136
Meso Scale Discovery™ (MSD) 83, 91
metal-catalyzed oxidation (MCO) 227
methionine (Met)
  N-terminal 225
  oxidation 24, 25f, 227
methylation 234, 239
micro liquid chromatography, quantitative bioanalysis using 47–49, 48f, 48t
miniaturized liquid chromatography and nanospray ESI-MS, integration of 46–47, 47f
monoamine oxidase B (MAO-B)
  extraction from human plasma 50
  surrogate peptide 50, 51t, 52f
monoclonal antibodies, LC–MS/MS bioanalysis of 145–158
  future perspective of 156–158
  internal standard selection 153–154
  preclinical applications of 154–156, 156t, 157f
  sample preparation 147–152, 150f, 151f
  surrogate peptide selection and optimization 146–148, 147f, 149t

monoclonal antibody(ies) 17, 18, 70, 108–110, 116–120, 183, 203
  heavy chain generic peptides 162t
  LBA critical reagent, structural analysis of 113–114, 113f
  LC–MS/MS bioanalysis of, *see* monoclonal antibodies, LC–MS/MS bioanalysis of
  proteins, intact analysis of 235, 236f
  recombinant 21
  structure of 162f
monomethyl auristatin E (MMAE) 185
monomethyl auristatin F (MMAF) 185
Moore's law 7
MRMPilot 146
mucosal addressin cell adhesion molecule (MadCAM) 83–84
multiple reaction monitoring cubed (MRM^3) scan 36, 39, 40f, 42
multiple reaction monitoring mass spectrometry (MRM–MS) 58, 68, 75, 99, 130
  splice variants quantitation by 62–63, 62t, 63f
MultiQuant™ Software 40
Mylotarg® 183, 185, 198
myosin light chain 1 (Myl3) 78–79, 79f, 208–209

n
nanobore liquid chromatography (nLC) 45
nano liquid chromatography, quantitative bioanalysis using 47–49, 48f, 48t
National Institute of Standards and Technology (NIST) 22
N–C-terminal variants 225–226
Nek2 56
neoepitope of type II collagen (NET2C) 72, 72f
nerve growth factor (NGF) 81, 210
neuron-specific enolase (NSE) 80
  isovariants 210
N-formylkynurenine (NFK) 229
N448 glycoforms 29–31, 31t
nicotinamide adenine dinucleotide (NAD) 207
NISTmAb 22, 24, 25–27f
N-linked glycosylation 109, 113, 137, 232–233
N-maleimidomethylcyclo-hexane-1-carboxylate 185
nonreduced capillary electrophoresis with sodium dodecyl sulfate (NR CE-SDS) 224
NovoLog® 82
NTproBNP 75–76, 77f, 78f

o
oligosaccharide
  core structures, O-linked 234f
  sequencing 239, 239t
  structural classes, N-linked 233f
  structure, N-linked 232f
  structure, O-linked 233f
oligosaccharyltransferase enzyme complex 232
O-linked glycoproteins 233–234, 233f, 234f
online sample preparation 211–213, 212f
Orbitrap Fusion™ Tribrid™ instrument 22, 28
Ortbitrap mass spectrometry 12–18, 133
  posttranslational modifications in therapeutic proteins using 21–31
orthogonal acceleration time-of-flight (oa-TOF) based IM-MS 1, 2
osteopontin (OPN) 56
  domain structure of 59f
  isoform-specific peptides 62, 62t
  isoforms, plasma concentration of 62–63, 63f
oxidation 24, 25f, 94, 109, 112
  metal-catalyzed oxidation 227
  methionine 227
  photooxidation 227–228
  in protein therapeutics 226–227

p
packed-emitter columns 49–51, 49–52f, 51t
para-aminobenzyl (PAB) group 185
parallel reaction monitoring (PRM) 5, 13, 15
PEGylated proteins 16
  protein precipitation for 132
pellet digestion 50, 165, 215
PepFinder 2.0 software 29
peptide 1 asparagine deamidation on human mAbs quantification, impact of 164–165, 164f, 165f
peptide biomarkers
  IA–MS applications in 74–78, 76–78f
  sensitivity 75
peptide 1 LC–MS/MS assay 162–164
peptide(s)
  bioanalysis, challenges of 36–38
  derivatization of 206–207
  mapping 231
  therapeutics 81–83
Phage ELISA-based selection of reactive thiols (PHESELECTOR) 224
phosphorylation 130
photooxidation 227–228
PicoChip™ column 50, 50f
PicoFrit™ column 49, 49f
PicoFuze™ column 49, 50, 50f
PNGase F (Prozyme®) 188
polyclonal antibodies (PAbs) 108, 110
polyethylene glycol (PEG) 70
  conjugation 82
  hydrophilic moiety 132
postacquisition data mining (PADM) 129, 133, 138
posttranslational modifications (PTMs) 21–31, 109, 110, 113, 121, 129, 130, 133, 136–137
  using electron transfer dissociation 26–31, 30f, 31t
  using higher-energy collision dissociation 23–26, 23f, 25–27f
precursor-messenger ribonucleic acids (pre-mRNAs) 55
progastrin releasing peptide (ProGRP) 80, 210
ProntoSil C18 50
proprotein convertase subtilisin/kexin type 9 (PCSK9) 55
prostate-specific antigen (PSA) 81, 102, 208
protease inhibitors 135
protein analog 135
protein C inhibitor (PCI) 81
protein digestion strategies 214–215

protein measurement, importance of 67
protein precipitation (PPT) 16–17, 82, 152, 196
  sample preparation strategies 205
protein therapeutics, characterization of
  discoloration 228–230
  disulfide bond, in antibody drug conjugates 224–225, 225f
  disulfide isoforms 222–224, 223f, 223t
  free glycan analysis 238
  free sulfhydryl 224
  glycation 226
  glycopeptide analysis 237
  glycoprotein structure 232–235
  glycosylation 232–239, 232–234f, 236f, 239t
  intact glycoprotein analysis 235–237
  N–C-terminal variants 225–226
  oxidation 226–228
  release of glycans 238–239
  sequence and linkage analysis 239
  sequence variants 230–231
  tandem MS of glycopeptides 237–238
  thioether/trisulfide bond 224
  thiolation isoforms 222
ProteoExtract albumin depletion kit 207
pyroglutamic acid (pE) formation 225–226

q

Q Exactive hybrid quadrupole Orbitrap
  ion path on 13f
  protein quantitation approaches 14–16, 14t, 15f
  quantitation modes on 13–14
quadrupole/linear ion trap (QTRAP) system 39, 41
quadrupole time-of-flight (Q-TOF) system 41, 42f, 74, 133, 137, 138, 146
qualitative LC–MS systems 45
quantitative bioanalytical assays 45

r

radioimmunoassay (RIA) 68
radioimmunoconjugates 21
Rayleigh scattering 230
reactive oxygen species (ROS) 227
reagent management 122–123
  analytical toolbox and fit-for-purpose approach for 125–126
recombinant human methionyl-leptin (rMethLeptin) 82–83
reflectron TOF 12
reverse-phase high-performance liquid chromatography (RP-HPLC) 21, 22
rhodopsin 229–230
RNA-Seq 56–57

s

sample preparation strategies, for LC–MS protein bioanalysis 203–216, 204f
  assay sensitivity 205–213
  comparison of 204t
  derivatization 206–207, 207f
  free, total, and ADA-bound proteins 213–214
  high-abundance proteins, depletion of 207–208
  immunoaffinity purification 208–211, 209f
  immunocapture sample preparation strategies 213f
  interference from antidrug antibodies or soluble target, overcoming 214
  issues of 204–205
  ligand binding assays 203
  lower limit of quantitation 204
  major challenge of 203–204
  major strategies for 203
  monoclonal antibody-based therapeutics 203
  online sample preparation 211–213, 212f
  protein digestion strategies 214–215
  protein precipitation 205
  solid-phase extraction 205–206, 206f
Schiff base 226, 230
selected ion monitoring (SIM) 13–15
selected reaction monitoring mass spectrometry (SRM MS) 12, 16, 74, 93, 94, 99–100, 104, 145–147, 153, 164
selective peptide derivatization (SPD) 206
SelexION™ Differential Ion Separation Technology 36, 39, 40f
Seppro IgY14 LC10 column 208
sequence variants 230–231
SEQUEST database 16, 231
Serine (Ser)
  O-GlcNAc glycosylation 29
  O-linked glycosylation 29
signal-to-noise ratio (S/N) 207
signature peptides 161, 177
simple mixtures, IM–MS and LC–IM–MS analysis of 2–7, 3–7f
simultaneous Quan/Qual analysis 12, 14, 18
single nucleotide polymorphisms (SNPs) 230
SISCAPA Assay Technologies Inc. 100
size exclusion chromatography mass spectrometry (SEC-MS) 21, 22
  LBA critical reagent, structural analysis of 112, 116, 117, 119
Skyline 146
sodium dodecyl sulfate (SDS) 214
sodium dodecyl sulfate–polyacrylamide gel electrophoresis (SDS-PAGE) 21, 224
  LBA critical reagent, structural analysis of 112, 115–117, 119, 120f, 121f, 125
solid-phase extraction (SPE) 17, 131, 139, 175, 191, 205–206, 206f
  with acid dissociation 91
soluble epithermal growth factor receptor (sEGFR) 83
solution fractionation 132
specific protein binding 136
splice variant proteins
  global profiling of 56–57, 57f
  identification of 56–63
  immunocapture and mass spectrometric identification/quantitation of 58f
  quantitation of, by MRM–MS 62–63, 62t, 63f
  relative expression, characterization of 57–62, 59f, 60–61t

SQI Diagnostics Inc. 92
stable isotope-labeled internal standards (SIL-ISs) 71, 155, 161
  cleavable flanking 154
  monoclonal antibodies 154
  peptide 173–174, 175
  protein 174–175
stable isotope-labeled (SIL) protein/peptides 17–18, 134–135
stable isotope labeling with amino acids in cell culture (SILAC) 17, 71, 135
Stable Isotope Standards and Capture by Anti-Peptide Antibodies (SISCAPA) 69, 80, 81, 100, 176, 208
strong-cation exchange mass spectrometry (SCX–MS) 4, 4f, 206
structure–activity relationship (SAR) 82
subambient pressure ionization with nanoelectrospray (SPIN) 104f, 105
Sulfo-NHS acetate 226
sulfosuccinimidyl acetate 226
SuperMix columns 100
surface plasmon resonance (SPR) 112, 136
Surfactant Protein D (SP-D) 136
surrogate peptide quantitation 50, 51t, 52f
SWISS-PROT database 232

t

target engagement (TE) assay 108, 109f
therapeutic antibodies 83–84
therapeutic peptides 81–83
Thermo Fisher Scientific 140
Thermo Scientific™ benchtop Q Exactive™ series mass spectrometers 13
Thermo Scientific KingFisher Instruments 17
Thermo Scientific™ LTQ™ Orbitrap Mass Spectrometer 13
thioether/trisulfide bond 224
thiolation isoforms 222
threonine (Thr)
  O-GlcNAc glycosylation 29
  O-linked glycosylation 29
time-of-flight (TOF) 112
  analyzers 22
  -based IM–MS 1
  quadrupole 41, 42f, 74, 133, 137, 138, 146
  reflectron 12
tissue inhibitor of metalloproteinase 1 (TIMP1) 80
tissue plasminogen activator (TPA) 22, 29
  recombinant 22
TNFRII-Fc protein 28
total ion chromatogram (TIC) scan 206
toxins 185
traveling-wave ion mobility spectrometry (TWIMS) 1–2, 2f, 4, 5
trichloroacetic acid (TCA) 152
trifluoroacetic acid (TFA) 112, 133
2,4,6-trinitrobenzenesulfonic acid (TNBS) 225–226
Triple Quad MS systems 39, 41, 42f
triple quadrupole mass spectrometer (TQMS) 11–12, 74
tris(2-carboxyethyl)phosphine (TCEP) 113, 113f, 152
trisulfide bond 224
troponin I biomarkers 208
Trp RNA-binding attenuation protein (TRAP) 2, 3f
trypsin 14
  digestion 152–153
  MAO-B extraction from human plasma using 50
tryptophan (Trp) 135
  oxidation 24, 227, 229
  photolysis 229
tyrosine (Tyr) oxidation 24

u

universal peptide LC–MS/MS assay, for human mAbs bioanalysis
  animal studies 161–170
  dual universal peptide LC–MS/MS assay candidates 165–170, 166f
  generic peptide LC-MS/MS assays 173–177
  human studies 170–175
  IgG4 clinical LC–MS/MS assay, development and evaluation of 171–172, 172t
  IgG4 clinical LC–MS/MS assay performance, anti-mAb antibodies impact on 173, 173f
  peptide 1, applications of 162–164, 163f
  peptide 1 asparagine deamidation on quantification, impact of 164–165, 164f, 165f
  peptide 1 variants, potential applicability of 163–164
  potential expansion of 170–171, 171f
  quantification of human mAb and Fc fusion protein drug candidates 161–162
  surrogate peptide identification 161–162, 170f
ustekinumab 210

x

XTENylation 82

www.ingramcontent.com/pod-product-compliance
Lightning Source LLC
LaVergne TN
LVHW062147200926
848484LV00007B/1872